D1165708

Primate Encounters

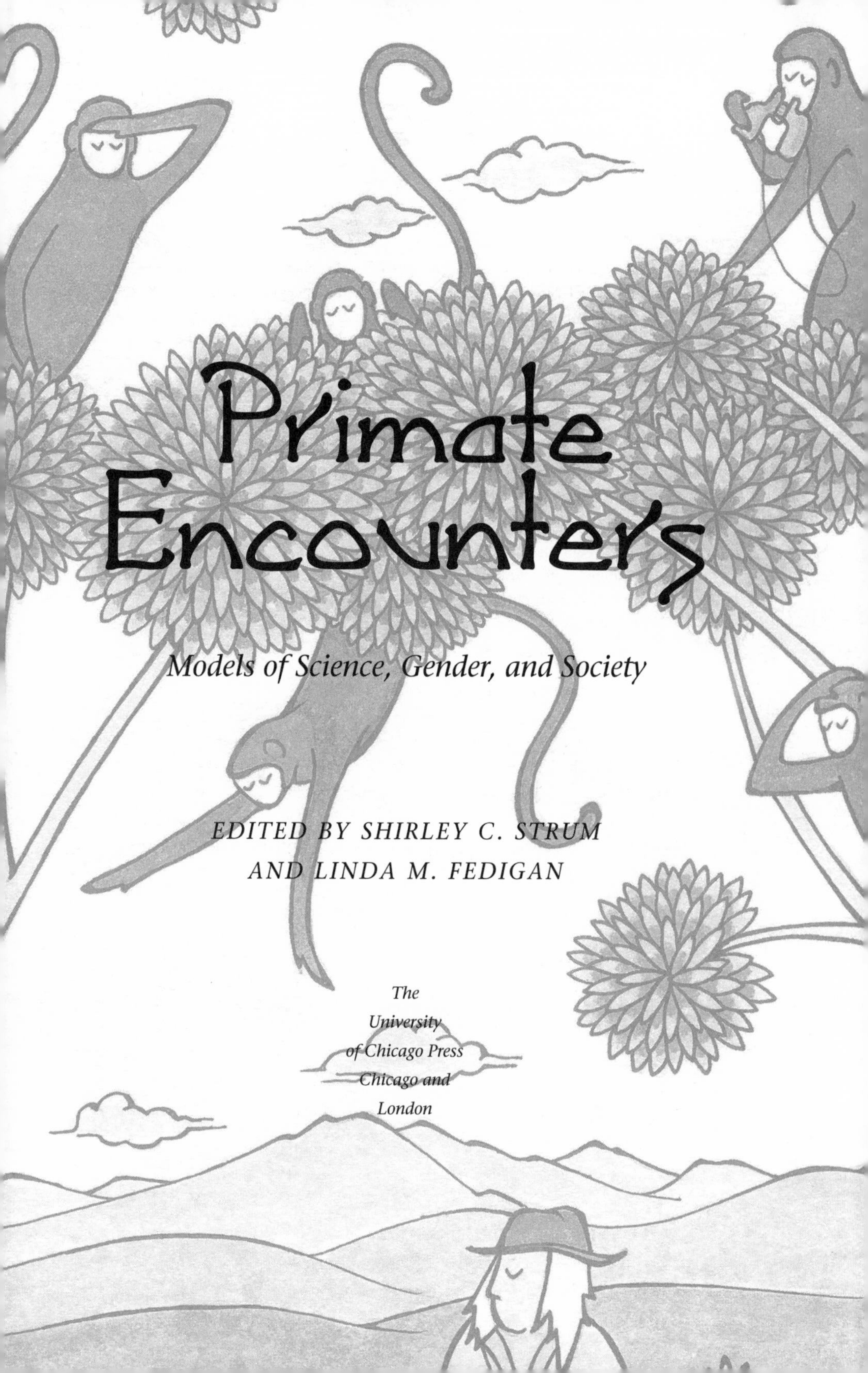

Primate Encounters

Models of Science, Gender, and Society

EDITED BY SHIRLEY C. STRUM
AND LINDA M. FEDIGAN

The
University
of Chicago Press
Chicago and
London

SHIRLEY C. STRUM is professor of anthropology at the University of California, San Diego. She is author of *Almost Human: A Journey into the World of Baboons* and coeditor of *The New Physical Anthropology* and *Natural Connections: Perspectives in Community-Based Conservation.* Strum has studied olive baboons in Kenya since 1972 and is director of the Uaso Ngiro Baboon Project. **LINDA MARIE FEDIGAN** is professor of anthropology at the University of Alberta. She is author of *Primate Paradigms: Sex Roles and Social Bonds,* published by the University of Chicago Press, and coeditor of *The Monkeys of Arashiyama: Thirty-Five Years of Research in Japan and the West.* Fedigan has studied the Arashiyama West group of Japanese macaques since 1972 and the white-faced capuchins of Santa Rosa in Costa Rica since 1983.

The University of Chicago Press, Chicago 60637
The University of Chicago Press, Ltd., London

Printed in the United States of America
09 08 07 06 05 04 03 02 01 00 1 2 3 4 5
ISBN 0-226-77754-5 (cloth)
ISBN 0-226-77755-3 (paper)

Library of Congress Cataloging-in-Publication Data

Primate encounters : models of science, gender, and society / Shirley C. Strum and Linda M. Fedigan, editors.

p. cm.

Includes bibliographical references (p.).

ISBN 0-226-77754-5 (cloth : alk. paper)—ISBN 0-226-77755-3 (pbk. : alk. paper)

1. Primates. 2. Primatologists. 3. Women primatologists. I. Strum, Shirley C. (Shirley Carol), 1947– II. Fedigan, Linda Marie.

QL737.P9 P67245 2000
599.8—dc21 99-087519

∞ The paper used in this publication meets the minimum requirements of the American National Standard for Information Sciences—Permanence of Paper for Printed Library Materials, ANSI Z39.48-1992.

For

JONAH and *JOHN,*

and for

SYDEL,

who helped

to make this book

a reality.

Contents

Preface

In June of 1996, twenty-three people met at Teresopolis, Brazil, to consider an unusual topic: how and why ideas about primate society have changed during the relatively brief history of the field of primate studies. This was a Wenner-Gren Foundation "international" symposium. Following their format, we spent an entire week "isolated together" at a small resort in the mountains above Rio. Both the questions we considered and the process that evolved transformed the symposium into a real "workshop" (we refer to it as such throughout the book)—exploring new domains, building bridges across chasms, and generally poking and prodding each other to find ways to move forward.

A few words on the history and goals of the workshop are warranted as it helps to explain how we came to be at Teresopolis and what was so unique about what happened. Shirley Strum had often encountered two vexing questions in her dealings with the media and the public: Why is it that only women study primates? and, Wasn't it true that our ideas about primates have changed dramatically because of the influence of the women who study them? Her answers to both questions were not what the media or the public expected: Many men, in fact more men than women study primates and many if not most of the ideas she could think of changed because of the influence of specific theories, or new methods, or the impact of long-term studies. It seemed to her that both men and women studying primates contributed to these changes. She began to wonder why her sense of the field was different from the popular images. This led her to questions about how "science" works, about the interaction of "science and the media" and "science and society," about the special position of primate studies. In 1990 she decided to enlarge the discussion by inviting a few others to join in. For one weekend, Linda Fedigan, Thelma Rowell, Donna Haraway, and Elisabeth Lloyd, each with a special interest in either primate studies or women in science or both, met at the University of California, San Diego.

The meeting was the beginning of a collaboration and a journey that has produced *Primate Encounters*. Meanwhile, in a different country on the other side of the world, Linda Fedigan was also being asked repeatedly by colleagues and the media: Why are there so many women primatologists? Like Shirley, Linda doubted that there were more women than men in primatology, and one of her responses was to investigate the actual proportions of women in primate studies as compared to related and parental disciplines. Yet for Linda the question about women in primatology resonated with her long-standing interests in sex and gender differences. She had wondered for some time what role gender might play in science, particularly whether men and women scientists might differ in their approaches to the study of sociality in male and female animals. It seemed to her that it was not a coincidence that scientists began to bring female primates out of the shadows and onto center stage at the same time as the proportions of women in primatology increased.

Shirley and Linda realized after the 1990 weekend meeting in San Diego that the issues were complex; a great deal of groundwork was needed to be able to answer what initially had seemed like straightforward questions. The first step involved finding out which ideas have changed. This meant reviewing ideas about primates and writing a history of them. The history itself suggested that to understand how ideas actually change would require a fairly sophisticated and comprehensive framework. Once the framework was developed, it became obvious to us that the time and expertise involved went well beyond what we, as primatologists, possessed.

At this point, Sydel Silverman and the Wenner-Gren Foundation came to the rescue. The foundation agreed to sponsor an international symposium on "Changing Images of Primate Societies: The Role of Theory, Method, Gender (and Culture)." Over eighteen months, we worked closely together, mapping out the structure of the workshop, choosing participants and topics for the papers which would be circulated beforehand. Throughout, Shirley and Linda's goals for the workshop were modest. We wanted to be able to ask better questions about the issues and to take the various controversies, particularly about the role of women in changing ideas about primate society, onto firmer empirical ground.

Primate Encounters presents the workshop's innovative framework, tries to give a flavor of the meeting of scientists of different disciplines, different generations, and different national traditions. It also illustrates what happens when scientists and those who study them try to talk to each other.

The organization and content of the book departs from the workshop in several ways that reflect the lessons we learned. Chapters have been added representing "missing" perspectives, once we realized that they were missing. The chapters of this book have been substantially revised from the papers circulated before the meeting. The book also focuses on the most productive aspects of our framework: comparison and analysis. For a start, in contrast to the workshop, which began with a consideration of "methods, theory, and gender," the book opens with a historical look at ideas about primate society, first a North American "potted history" (section 1), and then the views of "elders" in primatology who have the advantage of hindsight (section 2). The "factors" highlighted in the framework are now embedded in the chapters. The next two sections take a comparative perspective on the questions we have posed. Section 3 discusses the diversity of primatologies as evidenced in other "national" traditions. Section 4 enlarges the lens beyond primatology for insights from other closely related disciplines. Section 5 assembles the analytic resources of science studies, feminist science studies, and studies of popular culture to consider various "models of science and society" that are specifically relevant to our questions. The rest of the book is new material, either commentary and analysis on the papers and the workshop discussions (section 6: Reformulating the Questions), or new discussions carried out afterwards, during 1996–1998 (section 7: Conclusions and Implications: Future Encounters of the Primate Kind).

Although the workshop was about primate studies and the role of specific factors in changing our scientific ideas about primate societies, it soon became clear that there were fundamental issues about science which had to be considered. What is science? How does it work? And how important is who does it? As a result, the book highlights both the new perspectives of science studies and the tensions generated by the divergent ideas about science that were held by scientists and science analysts. The two groups had difficulty talking to each other during our intimate, week-long association. There were problems created by different backgrounds and lexicons. The most serious obstacle was mutual suspicion. Despite these, we managed to connect, to discuss, to build bonds with each other. The book includes much of this exchange since it seems to us particularly appropriate in understanding the current controversy over science called "the science wars."

Finally, the book has a somewhat unusual format. Sections 2, 3, 4, and 5 end with e-mail exchanges between the participants that took place after the workshop. Section 7 is entirely composed of e-mail exchanges. This is our way of recreating and carrying forward some of the most interesting discussions of the workshop and situating the sections and specific chapters. We hope these exchanges bring alive the debates and give the reader a flavor of what it means to assemble such diverse participants. The e-mail exchange also offered participants an opportunity to rethink and restate their positions in light of their experience at the workshop. E-mails have been edited to highlight key points in the discussion. Individuals voluntarily participated so that, just as in the workshop discussions, some spoke more than others and a few just listened but never commented.

Two workshop participants (Sarah Hrdy and Craig Stanford) did not contribute chapters to *Primate Encounters,* and two additional contributors were invited to present their own unique perspectives (J.A.R.A.M. van Hooff and Charis Thompson Cussins). Hans Kummer circulated a fascinating paper for discussion but was unable to attend the workshop at the last minute. He decided not to submit his paper as he did not benefit from the workshop synergy.

The book is part of a process, one that began more than eight years ago and will certainly continue. It is not an end but a stopping off point where we can take stock of where we have been and where we might go from here. There are many people who have contributed. First, we would like to thank Wenner-Gren for their support of both the workshop and of the book. Special appreciation goes to Sydel Silverman, who shared her knowledge and her skills to keep our intellectual journey on track and productive. The week in Brazil would never have happened without the hard work of Laurie Obbink and Mara Drogan. Laurie and Mara attended to the smallest details with care, precision, and enthusiasm. The staff at Hotel Rosa dos Ventos in Teresopolis, Brazil, made certain we had everything we needed for our workshop to succeed. We also thank Sandra Zohar, Rebecca Feasby, and Ann O'Neill for assisting with the numerous technical and editorial details that are part of polishing a book manuscript. Sandra's editorial skills were especially helpful in the final manuscript. Peter Else produced the wonderful illustrations and Noeline Bridge constructed an effective index; we thank them both. And our accomplishments would have been impossible without the incredible commitment and valuable insights of the participants. We asked them to dedicate themselves to our agenda even though most of them disagreed, to subvert their own questions in order to try to answer ours, and to find ways of overcoming their anxieties about each other and to take the other's point of view. They then rejoined the discussion from the far corners of the globe during the eighteen months of e-mail exchanges. Because of the participants' immense goodwill and creativity, we avoided the potential chaos that could have resulted from combining such diverse people and opinions.

Finally, we would like to acknowledge the importance of our collaboration. At so many critical junctures, it was the divergence and complementarity of our standpoints, and the willingness to be open-minded and explore the how and why of the "other," that proved so fruitful. The years of discussions were always stimulating and enjoyable (and sometimes difficult, as we uncovered the potholes in the information superhighway between Canada and Kenya). What is clear is that we thought thoughts together that neither of us could have thought alone. Our collaboration and the efforts at Teresopolis and afterwards made possible an evolution of thinking about science, gender, and society. *Primate Encounters* documents that process.

SECTION 1

Introduction and History

1 Changing Views of Primate Society: A Situated North American View

Shirley C. Strum and Linda M. Fedigan

Introduction

This chapter has its own history—one that is relevant to its current form. Primatology is a young science that has come into its own since the end of World War II. During this short period, scholars' interpretations of primate behavior and society have changed considerably. These changes have generated a great deal of interest among feminists, historians of science, and the popular media because they have been linked to a provocative claim: that women scientists played a major role in the revisions. This book derives from a workshop that was motivated by the interest in this controversy.

Before it was possible to explore who or what might be responsible for changing our ideas about primates, we needed to ask a more basic question: Have our ideas changed and, if so, exactly which ideas? At the beginning of our collaboration, we spent many hours talking about this. Fortunately, both of us had taught classes on the history of primate studies. To our surprise (and delight), we agreed on most things. Ideas had changed significantly, we believed, from the period of the first scientific field studies to the present; we even agreed on which ideas these were.

Both of our courses segmented the period into the same historical stages demarcating shifts in interpretations and identified a set of enduring issues that persisted despite various transformations. Armed with these tools, we began to compile the history of changing ideas about primate society that is presented below. We intended it to be a springboard to the more important question concerning the agents responsible for change.

Presenting an accurate history of ideas about primate society is a daunting task, particularly for nonhistorians. Our solution was to select and present a more manageable limited history that concentrates on primate field studies (ignoring the majority of research that has been done in captivity), on evolutionary interpretations of behavior and society (ignoring the immense biomedical literature), and on work done primarily by American scientists (who of course are only part of the vast primatological enterprise). Satisfied with our efforts, we wondered whether others would agree. The answer came quickly as our condensed history became part of the position paper for the Wenner-Gren workshop on "Changing Images of Primate Society: The Role of Theory, Method, and Gender (and Culture)." Among the participants were primatologists, ethologists, and behavioral ecologists of different generations and from a variety of national traditions. What we had imagined as a logical set of restrictions for our history aroused strong reactions in others. The reasons became clear only at the end of the week-long workshop. Rather than having touched on some universal truth about the history of ideas, our consensus was more likely a product of cultural, institutional, and disciplinary factors—in short, a similarity in our graduate training. As we discovered, and as this book will demonstrate, there have been and will continue to be many "primatologies" (see chapters 2–10), each the product of specific institutions, disciplinary training, mentors, and historical and sociocultural settings.

The history presented below is therefore comprehensible only if we "situate" ourselves in time and space. We are both part of the same age cohort from a lineage that can be called "North American anthropological primatology descended from Sherwood Washburn." This tradition grew out of the New Physical Anthropology (Strum, Lindburg, and Hamburg 1999) initiated by Sherwood Washburn in the 1950s. Washburn rekindled an interest in primate field studies as part of making better reconstructions of primate and human evolution. Locating ourselves this way changes the question from whether we are right or wrong in our portrayal of ideas to a more interesting set of inquiries: How, when, and why does this lineage and its view of knowledge interact, intersect, displace, and join the other primatologies, both historically and in the present? We no longer naively expect others to agree with our version of history; instead, we are interested in the disagreements. This expansion of perspectives also affects preexisting notions about who might have a different history of primatology. In fact, the deeper we delved during the workshop, the more differences we found, until it became problematic to use general categories like "American" or "European" or "Japanese." Rather than feel defeated by this realization, we see the move into the local situated sites of primatology as a first step in rebuilding a more robust understanding of what happened and is happening. This makes our potted history itself a historical document and a contribution to the newly initiated discussion about the diversity of primatologies. As such, it becomes a useful point on the increasingly complex map of our ideas about primate society.

The Making of a Potted American Anthropological History of Ideas about Primate Society

The naturalistic study of primate behavior has grown dramatically in the last seventy-five years. The primary impetus for modern field studies in the American anthropological tradition initiated by Washburn[1] was the belief that knowledge of our closest living relatives could help us understand the origins and evolution of human behavior. The result has been widespread interest in the conclusions of these studies among both American scientists and the American general public.

Images of primate society have changed extensively during this short history. Simply put, we have moved from a general vision that primate society revolves around males and is based on aggression, domination, and hierarchy to a more complex array of options based on phylogeny, ecology, demography, social history, and chance events. The current image of primate society, if we ventured to make generalizations at all, would be a strong counterpoint to the earlier view. It would highlight the importance of females within society, emphasize tactics other than aggression (particularly those that rely on social finesse and the management of relationships), and argue that hierarchy may or may not have a place in primate society, but that males and females are equally capable of competition and rank ordering.

Curiosity about what caused these changes in images of primate society is widespread. The reasons must relate to the origins of primatology, the abundant interest in its conclusions, and the nature of the modifications. More recently, primate studies also has become embroiled in the controversy about whether knowledge is socially constructed. Feminists, historians, and those who study the interaction of science and society, as well as the popular media, have focused on women scientists and their role in building new images of primates and primate society.

Despite all of this interest, very few comprehensive reviews of the history of the field (Gilmore 1981; Ribnick 1982; Southwick and Smith 1986) and only a few discussions of the impact of specific factors on our ideas about primate society exist. Investigators identified as important—at least at some point—theory (Gilmore 1981; Richard 1981; Sperling 1991), methodology (Burton 1994; Mason 1990; Strier 1994a), gender of the scientist (Adams and Burnett 1991; Haraway 1989; Hrdy 1986), and the larger sociocultural context (Asquith 1986a, 1994; Haraway 1989; Sperling 1991).

Our potted history is part of a larger project, one that resulted in the workshop that is the basis of this book. We sought a way to construct a framework that would help us ask questions about who and what changed our ideas about primate society in a more appropriate manner and guide us in the search for empirical data that might help resolve current controversies. That framework is discussed later; first, in this chapter, we try to define which ideas have changed by sketching their historical progression through major stages of research from 1920 to the present. As already mentioned, we focus on primate field studies and evolutionary interpretations of behavior and society, primarily by American scientists. This is our point of reference, the what and the how (stages and issues) of our quest. The historical review is not exhaustive, but we have tried to make it comprehensive. We have chosen to discuss two representative studies to illustrate each period.

Stages and Issues

We have divided the period from the 1920s to the present into stages to facilitate comparisons. Using primate field studies as our marker, we have delineated

Pre-Stage 1:	Early studies between World Wars I and II;
Stage 1:	1950–1965;
Stage 2:	1965–1975;
Stage 3:	1975–1985; and
Stage 4:	1985 to the present.

To some extent, these divisions are arbitrary, and some stages are characterized by the results of work done in previous years but published during a later time period. The stages do, however, have some integrity in terms of conceptual, theoretical, or methodological Rubicons. For example, the collection of natural-history data characterizes Stage 1. Stage 2 encapsulates the discovery of variability and the dilemma it posed for existing theoretical frameworks. The impact and immediate reaction to sociobiology marks Stage 3, while Stage 4 witnesses a shift to behavioral ecology, the reemergence of animal "mind," and concerns about conservation.

We also propose that throughout this century's primate studies, scientists have been interested in specific questions and issues about primate societies. Over time, the answers to these questions have changed. The ten key questions below appear to be enduring issues:

1. Why do primates live in social groups?
2. What is the social structure of the group, and what holds society together?
3. What is the relationship of the group to its environment?
4. What is the role of aggression, dominance, sex, and affiliation in primate societies?
5. What is the basic nature of males, females, and the relationship between the sexes?
6. What is the pattern of ontogeny, development, and socialization?
7. What is the role of instinct, learning, and cognition in behavior?
8. What is the pattern of intra- and interspecies variability?
9. What is the evolutionary relationship between different social grouping patterns?
10. What is unique about humans, and what do we share with our primate relatives?

These ten questions are stated here in the simplistic terms of their original formulation, although they were rephrased in a variety of ways during later stages (for example, see Strier 1994a). Despite the later modifications, however, we suggest that the essential issue at the heart of each question remains the same. These enduring questions help chart a path through the increasingly complex enterprise of understanding primate society.

Even with the restrictions we placed on our history, space still is insufficient to conduct an extensive review of primate field studies. Instead, we outline only some of the important influences and trends of each stage. Although this risks oversimplification of the many views that existed in each stage (see Dunbar 1990), it does allow us to begin our analysis, itself situated within the specific tradition to which we belong.

PRE-STAGE 1: PRIMATE STUDIES BETWEEN THE WORLD WARS

The variety of studies on the social behavior of primates initiated in the period between World Wars I and II included the establishment of primate colonies and laboratories, experiments in raising apes as part of human families, field expeditions, and field studies (for several versions of the history of primate studies, see Asquith 1986a, 1994; Bramblett 1994; Burton 1994; Gilmore 1981; Haraway 1989; Mason 1990; Ribnick 1982; Richard 1981, 1985; Southwick and Smith 1986; Sperling 1991). Although they were not part of a unified scheme, these diverse efforts shared a common scientific ethos: that the close biological relationship of nonhuman and human primates would allow us to use nonhuman subjects (particularly monkeys and apes) as surrogates or as windows into understanding human behavior and its evolution. This approach had particular relevance and appeal to investigators from the disciplines of psychology and anthropology, and, increasingly, zoologists also became involved.

Zuckerman and Carpenter, baboons and howlers Robert Yerkes was the central figure in primate studies during the 1920s and 1930s. He initiated field expeditions, sponsored behavioral studies, founded the first primate research facility in the United States, and set the explicit and implicit scientific agendas for laboratory and field research. His book, *The Great Apes* (Yerkes and Yerkes 1929), was a landmark, both because it established how little scientists of the time knew about primates, and because it greatly stimulated further research. The scientific issues of Yerkes' time have their own history, which we will not consider here. Instead, we begin with two influential studies of monkey behavior that provided the first clear images of primate society: Solly Zuckerman's study of hamadryas baboons at the London Zoological Garden (1932) and the field study of howlers by Yerkes' student Clarence Ray Carpenter (1934).

When Zuckerman conducted his study of the hamadryas baboons at the London Zoo in the late 1920s and early 1930s, scientists knew nothing of the natural behavior patterns and social system of this species. The zoo colony had been created in 1925 with a hundred males. By 1927, when thirty females joined the group, only fifty-six males remained; the rest had died of injuries incurred during fights. Colony composition continued to change as animals died from stress-related disease or injury and others replaced them. Today we realize that the colony contained too many animals—strangers to each other—in too small a space and combined in inappropriate sex ratios. Field studies since have documented that hamadryas baboons live in polygynous societies with several females to every male, and that neighboring males establish relationships with one another that inhibit them from fighting over females (see Kummer 1971a, 1971b). Zuckerman, however, believed he was watching "normal" behavior and extrapolated from the zoo's baboon social catastrophe a general thesis of sexual competition as the basis of hamadryas social life (1932). He concluded that since female primates were continually receptive (an assumption later refuted) and since males fought vigorously over access to females, sex must be the driving force behind primate society. His model inextricably linked sex, aggression, competition, and male dominance over both females and each other (expressed through a rigid male hierarchy).

Carpenter's study of howler monkeys on Barro Colorado Island in Panama (1934) stands in marked contrast to Zuckerman's work, both in approach and interpretation. Carpenter insisted that data from captive animals could not be interpreted properly without knowledge of that species' natural way of life. His naturalistic study of howlers suggested a well-organized and coordinated society that resulted from a network of mutualistic social relationships. The dyadic analysis of relationships he employed helped describe the process of social integration through affiliation. Sexual relations were basically communal, with females more assertive and sexually motivated during mating than males. According to Carpenter, this communality was adaptive. Through multiple matings, females and males became conditioned (i.e., bonded) to many partners, and thus established important social and emotional ties essential to group cohesion.

Relations within age-sex classes also seemed relaxed in the howlers' communal atmosphere. Competition or aggression between clan (group) males was absent; neither did sharp gradients of dominance and submission exist among either males or females. Carpenter's howlers reserved antagonism for intergroup contact, but even aggression between groups appeared highly ritualized, consisting mostly of mutual avoidance.

Theory and methods The theoretical framework of primate studies was not as explicit at this stage as in later periods. Both Zuckerman and Carpenter considered social behavior adaptive. This functionalism they painted with broad biological and evolutionary strokes yet also linked to social structuralism (Gilmore 1981) and systems theory (Haraway 1983). They saw the selective advantages of social behavior as couched in terms of benefits accruing to the group and to the species.

While they shared the same interpretative framework, the two investigators diverged dramatically in methodology. Zuckerman's methods and conclusions evinced little appreciation for the difficulties of collecting behavioral data, the complexities of social life, or the limitations of his chosen zoo setting. Still, his approach was representative of the undeveloped state of scientific knowledge and methodology in his era. By contrast, Carpenter set a new and lasting standard for data collection. He clarified and pioneered methods of habituation for wild primates, made explicit the standards for the acceptance of naturalistic observations as facts, and developed a new approach to the analysis of complex social interactions.

The themes, issues, questions, and currents embedded in these two models of primate society lay dormant during the period surrounding World War II but reemerged when field studies revived in the 1950s.

STAGE 1: 1950–1965—THE REVIVAL OF PRIMATE FIELD STUDIES

Sherwood Washburn's insistence that proper interpretation of primate functional anatomy and human behavioral evolution required comparative information from living primates in their naturalistic settings (1951, 1962) was the greatest stimulus to the first wave of post–World War II field studies by Americans. Students of Washburn with the fundamental objective of describing as much social behavior from the wild as possible—particularly that relevant to evolutionary issues—carried out many of the influential field studies of this period (but see also Collias and Southwick 1952; S. Altmann 1962).

Washburn and his students used an evolutionary framework with a straightforward underlying argument: All primates, whether human or nonhuman, shared certain adaptive features, a "primate pattern" believed to vary little across contexts and across species. Therefore, they reasoned, scientific generalizations and evolutionary reconstructions could be based on extrapolations from one species to another. These post–World War II field studies searched for the characteristics of the primate pattern in the primate group and its constituent parts: the primate male, the primate female, and the primate immature. Investigators tried to delineate the basic dimensions of the primate pattern including sexual behavior, socialization, infant development, play, social organization, intergroup relations, and daily routine.

The list of primate species studied in the wild grew significantly during Stage 1;[2] some of the more notable were baboons in Kenya, Rhodesia, South Africa, and Uganda; patas monkeys in Uganda; vervets in Kenya; rhesus monkeys in North India and on Cayo Santiago; bonnet macaques in India; and langurs in North India, with some data on populations in South India and Ceylon. British scientists studied chimpanzees and gorillas in Tanzania and Uganda, the French studied lemurs in Madagascar, Swiss biologists initiated fieldwork on hamadryas baboons in Ethiopia, and the first reports of Japanese studies of macaques appeared in English.

Baboons and langurs To illustrate the nature of Stage 1 field studies, and as a heuristic device for discussing images of primate society, we will continue to follow research on baboons. However, we will contrast baboons with research on Hanuman langurs to illustrate the divergence of views about primate society. Baboons probably were the most studied species of the period. Washburn and his student DeVore worked on baboons in

Rhodesia and two places in Kenya: Nairobi Park and Amboseli Reserve (Washburn and DeVore 1961). Later, K. R. L. Hall, an English psychologist, added his data on South African baboons to theirs (DeVore and Hall 1965; Hall and DeVore 1965). Baboon group sizes varied widely, but for ease of observation and data collection, investigators focused on smaller groups. Baboon society appeared to be remarkably consistent across locations, and investigators argued that this was because baboons everywhere had to solve the same basic problems of survival on the savanna. The main selective pressures were the many large predators, the lack of trees, and the limitations on resources such as food, water, shade, and sleeping sites.

Baboon anatomy and society seemed to reveal the evolutionary solution to these basic problems. Male anatomy (big size, greater muscle mass, large canines, and an impressive mantle of shoulder hair) was specialized for aggression, while male behavior was crafted for competition and defense. Aggression between males created and then was modulated by a dominance hierarchy that the three interpreted as the organizing principle of baboon society. The male hierarchy provided structure, stability, and leadership, ensuring that the group would be protected from threats and that peace would prevail. As a result, males were the leaders, policers, defenders, and protectors of the group.

Female relationships seemed more difficult to identify. Female status was thought to be more variable and more subtle than that of the males. A female's position improved, no matter what her status, when she was sexually receptive, particularly when she was consorting with a dominant male. The female role was based on reproduction and the care of infants.

The baboon studies also described details of socialization, communication, mating behavior, ranging, and feeding. These too seemed remarkably similar, despite variations in ecology and in species. The conclusion was obvious: Baboon behavior was specially suited to survival on the savanna; society was cohesive, well structured, and male centered. The selective advantage of such a system accrued to the group and, through the group, to the species.

Hanuman langurs first became subjects of study in Stage 1. Phyllis Jay, also a Washburn student, observed these langurs in North India from 1958–1960 and surveyed them throughout India in 1964–1965 (Jay 1965). Jay's main study site was a woodland habitat, but langurs occupied a wide range of habitats, from forests to more open areas. North Indian langurs lived in smaller groups than baboons, but, like those of baboons, these groups were multimale and multifemale. In addition, male langurs lived in all-male groups whose ranges overlapped those of bisexual groups.

Although the baboon and langur studies used similar methods of observation and the same dyadic age-sex class analysis, the results were very different. Jay described langur groups as peaceful and relaxed; the presence of infants was a major cohesive factor. The style of langur mothering was very different from that of baboons: A new mother was willing to pass her infant to others almost immediately, and babysitting by "aunts" was common. Langur males had a dominance hierarchy, but in most cases the males were quite peripheral to the group, both spatially and socially. Female langurs policed interactions within the group, defended infants against threats, and acted as socializing agents. The males helped coordinate group movements and were leaders in that sense. However, dominance was not particularly visible or important in langur life. Being a member of the group seemed protection enough against predators; the many alert animals gave ample warning for the group to escape to the trees.

The obvious differences between langur society and baboon society were in the context of similarities, particularly in the processes of socialization, communication, play, ranging, foraging, and daily routine. These commonalities lent weight to the idea of an evolutionarily significant primate pattern that transcended species boundaries. But the clear differences reported for langurs and baboons also held a hidden challenge (whose full impact would not be felt for some time) to the belief in an overarching primate pattern.

The contrasting images of primate society offered by the baboon and the langur studies in Stage 1 show striking parallels with the divergent depictions offered prior to World War II based on hamadryas and howlers. Stage 1 observers described savanna baboon society, like Zuckerman's hamadryas society, as male-centered, competitive, aggressive, rigidly organized, and hierarchical. Like Carpenter's howlers, langur society (at least in North India) was low key, with males and females in complementary roles and sex, aggression, and dominance relegated to the background of everyday life.

Theory and methods Stage 1 field studies can be called "improved natural history." Like the earlier Boasian sociocultural anthropology, the mandate of these early field studies was to collect descriptions of as many primate societies as possible. Yet, unlike the cultural ethnographies, researchers couched the animal studies in a strictly functionalist evolutionary framework.

The evolutionary formulations of Stage 1 were more precise than those described prior to World War II. The Modern Synthesis, developed in the 1940s (e.g., Dobzhansky 1944; Huxley 1942; Mayr 1942; Simpson 1949), had a powerful impact on physical anthropology by the 1950s (see Washburn 1951, 1962). This modern version of evolutionary theory explicitly acknowledged the importance of behavior in the process of natural selection, and Washburn argued (see also Hooton 1955) that naturalistic studies of primate social behavior would provide the context for understanding human evolution as part of the primate pattern.

During Stage 1, primatologists trained in anthropology also looked to the social sciences for theoretical orientation. According to Gilmore (1981), Richard (1985), and Sperling (1991), many of these primatologists applied the structural-functional model used by British social anthropologists, particularly Radcliffe-Brown, to the interpretation of societies of nonhuman primates. In brief, social behavior is modeled as part of an ordered, integrated system in which individuals play patterned roles that function to fulfill the needs of the group. According to Gilmore (1981), the structural-functional model was congruent with both the prevailing group selectionist view among evolutionary biologists and Washburn's emphasis on understanding the structural features of societies as adaptive (i.e., functional) mechanisms. Examples of structural-functional approaches from this stage (and the next) in primatology include studies of social roles, play as practice for adulthood, socialization as a mechanism to adjust the behavior of the individual to the needs of the group, and dominance hierarchies as a device to regulate aggression and bring order to social life.

Primatologists who contributed to Stage 1 also set new standards for data collection. Many studies were still of short duration compared to those of later stages because practitioners looking for the primate pattern assumed they could find it in a few months. Yet, for the most part, the organization of these field studies was more scientifically rigorous, extensive, and, ultimately, more comprehensive than previous efforts. Researchers selected "typical" groups of each species/population and based comparisons on age-sex classes, simple food lists, ranging patterns, behavioral ethograms, and social relationships. They largely limited individual identification of animals to groups of smaller sizes. Carpenter's method of dyadic analysis became the standard entry point into the group's organization. One notable piece of research in terms of methods was Stuart Altmann's (1962) two-year study on Cayo Santiago. In this study, he individually marked and systematically tracked animals in two very large groups of rhesus macaques—presaging the later widespread use of checklists and other quantitative techniques.

Stage 1 inherited a controversy from the previous period about the basis of primate society and the nature of social relationships. In some ways, the ascendancy of the idea of the primate pattern, and specifically the baboon model for primate society, reduced some ambiguities. Why live in a group? The simple answer was that it was adaptive. In fact, the major primate adaptation seemed to be group living both as a means of predator defense and as the basis of learning and reproduction. What holds a group together? Stage 1 field studies dealt a blow to the theory that sex is the basis of primate society. Researchers discovered that some primates had breeding seasons, and yet their societies remained intact even when no sexual behavior occurred (Lancaster and Lee 1965). In this new picture of primate behavior that began to coalesce, emotional bonds replaced sex as the societal glue, an acknowledgment of the primacy of the bonds socialization creates between various group members, especially between mothers and offspring.

Other issues remained unresolved. Were social relationships about power and domination (baboons), or were they based on attraction, affiliation, and cooperation (howlers, langurs)? Was the group structured by a male dominance hierarchy or by a network of attraction and affiliation between all individuals? How flexible was behavior, and what was the relative role of instinct (genetics) and learning? These unanswered questions left a controversy about where behavior could best be investigated: in the field or in captivity?

Despite the sophistication our understanding of primates gained during Stage 1, and despite what appeared to be unresolved issues about the nature of society and social relationships, a monotypic view of primate society emerged. This view was baboon based—a toned-down reincarnation of Zuckerman's model. The male dominance hierarchy remained central, but now, in addition to providing the structure of the group, it also provided peace, stability, and coordination. Social roles were still sexually distributed; males were leaders, defenders, and policers, while females cared for infants. These roles, however, now were more elaborate. Competition was no longer the war of all against all witnessed at the London Zoo because group living necessitated a rudimentary mutualism through which animals sometimes subverted their individual interests to group interests in exchange for the benefits of being a member of the group.

A description of this period would be incomplete without mention of the field studies of apes (Goodall 1965b; Reynolds and Reynolds 1965; Schaller 1963, 1964, 1965a, 1965b). New findings about the tool-using and hunting behavior of chimpanzees gave these apes a special position, for the purposes of understanding human evolution, in the study of nonhuman primates. Yet the primate pattern and the baboon model so captured scientific imagination that the chimp data did not really alter the scientific zeitgeist. The tendency during Stage 1 was to assume (often unconsciously) that baboon society could represent all monkey society, and that monkey society could represent all primate society (including that of humans), despite the variability observed between species. Even Carpenter presented a new vision of howler society after a brief restudy (1964b, 1965). Howlers now seemed much more like the competitive, male-dominated baboons.

STAGE 2: 1965–1975—THE DISCOVERY AND ENIGMA OF VARIABILITY

Stage 2 is best characterized as a period that revealed the extent of variability in primate behavior and society (see S. Altmann 1967; Jay 1968). Field studies of new species, and of the same species in different locations, challenged the notion of a unified primate pattern. At the same time, existing theory seemed unable to provide a compelling explanation for what this variability meant or what generated it. Consequently, the rationale for studying nonhuman primates, as it previously had been articulated, faltered. How could we make evolutionary generalizations across species, particularly from nonhuman to human primates, if we could not accurately predict the behavior of the same species in different locations? Some primate researchers turned to other ways of solving the evolutionary puzzle such as studies of hunter-gatherers (e.g., Lee and DeVore 1968) and ethological studies of children (e.g., Blurton Jones 1967; McGrew 1972).

The number and variety of field studies in Stage 2 are too large to enumerate here. Publishers brought out upward of twenty new books on primate social behavior. Some of these focused on individual species (Altmann and Altmann 1970; Goodall 1967b; Jolly 1966a), but many volumes were edited collections of field reports (S. Altmann 1967; Dolhinow 1972; Holloway 1974; Jay 1968; Morris 1967; Tuttle 1975).

Baboons and langurs Studies of baboons and langurs continue to be good illustrations of the data, issues, and dilemmas of this era. Researchers found that Hanuman langurs in South India differed from those of the North Indian populations. They lived at higher densities in smaller troops and had smaller, better defended home ranges than North Indian langurs (Jay 1968; Sugiyama 1965a, 1967). In the south, groups contained only one male. According to Sugiyama (1965a), the extra males lived in large, roving, all-male bands. The description of South Indian langurs' behavior made them seem like a different species. All-male bands aggressively attacked bisexual groups until they defeated the male leader. Forced out, he took the group's immature males with him when he left. The new leader was thought to kill the group's infants. Mothers who had lost their infants seemed to become sexually receptive soon thereafter and mated with the new male.

Explaining the striking differences in behavior between Jay's langurs in North India and those in South India was difficult. Certainly population density was extremely high at the southern sites (220 to 349 langurs per square mile in South India versus 7 to 16 individuals per square mile in North India). Higher densities, smaller groups, smaller home ranges, and daily intergroup encounters all seemed interrelated in some way. But the causal links and adaptive explanations were not yet well formulated.

The South Indian data generated a host of questions. Do langurs form the peaceful, relaxed, multimale society of Jay's description, or do they live in polygynous groups subject to frequent aggressive attacks and subsequent episodes of infanticide? Which pattern was normal? How could the variation be explained? And how could such a range of behaviors occur within one species?

Stage 2 field studies of baboons also documented variability. Females had a stable hierarchy of their own (Ransom 1979; Hausfater 1975; Strum 1975a, 1975b), the baboon social network was complex including "special relationships" between males and females and between infants and males (Ransom 1979; Ransom and Ransom 1971; Ransom and Rowell 1972; Strum 1975a, 1975b), and the male hierarchy was not so obvious (Rowell 1966; Strum 1975a, 1975b) or so adaptively related to predator protection as had been assumed (Rowell 1966, 1972; S. Altmann 1979a, 1979b; Rhine 1975). Unlike the langur case, the newly discovered variability among baboons did not yet seriously call into question the prevailing male-centric model of baboon society for at least two reasons. Some studies did support the model (Hausfater 1975); furthermore, the new divergent data did not yet offer a compelling alternative model. The discovery of variability among baboons, however, did contribute to the growing number of irregularities that eventually would topple the idea of a single primate pattern.

More variability The prevailing image of baboon society and of a baboonlike primate pattern were under threat from another source. New studies of monkey species in a variety of habitats offered a range of alternative images. For example, the success of patas monkeys on the savanna was not based on aggressive males but on other specializations in anatomy, group organization, and sexual division of roles (Gartlan 1974; Hall 1968); vervets exhibited varying social relations, ranging patterns, and territorial behavior in different habitats (Gartlan and Brain 1968); and rhesus and Japanese macaque societies were found to be based on female genealogical relationships (Koyama 1967; Sade 1967). Gartlan (1968) and Rowell (1972, 1974) questioned the assumption that dominance hierarchies are a widely and naturally occurring adaptive structure in African monkeys in particular and in primate societies in general.

The plethora of social systems reported for our closest relatives, the apes, revealed the troubling aspects of variability (Reynolds 1967). During this stage, researchers documented ape social organization in a range from solitary individuals (orangutans: Galdikas 1979) to monogamous pairs (siamangs: Chivers 1974; gibbons: Ellefson 1974) to age-graded male groups (gorillas: Fossey 1979; Harcourt 1979b; Schaller 1972) to fission-fusion communities (chimpanzees: Goodall 1968; Nishida 1968). The existence of such a diversity of social systems among our closest genetic kin seriously compromised the assumption that we could easily make evolutionary models of human society based on our knowledge of primate relatives.

Theory and methods The diversity of primate behavior and social organization exposed by Stage 2 field studies raised the question of whether recently exposed anomalies were real or artifactual. By now, scientists recognized the vulnerability of natural history to observer bias. Researchers made concerted efforts to find methods that would minimize all forms of bias. J. Altmann's (1974) review of sampling options proved extremely useful. Her discussion of the various methods that were in use in the study of behavior and their relative strengths and weaknesses influenced subsequent generations of primate fieldworkers.

The selection of species for study was another possible bias. Although more than 90 percent of primate species are arboreal forest dwellers, the primary criteria for the selection of species or study sites often had been convenience and ease of observation. This skewed studies toward terrestrial or semiterrestrial species, with a preponderance of baboon and macaque species in the corpus of our knowledge about primates. There was reason to be concerned about what this skewed sample might be doing to our ideas about the primate pattern and our understanding of primate society.

Skepticism extended as well to the evolutionary framework. Even if observed differences were real, were they all adaptive? Could some behavior be random and not necessarily interpretable in evolutionary terms (S. Altmann 1965)? Could some of the variant behavior be abnormal rather than normal? Should abnormal behavior be excluded from evolutionary interpretations (e.g., Gartlan 1968)? Or should we think in terms of "species potential" (Kummer 1971a, 1971b), seeing all behavior as part of the range of possible responses under different circumstances?

Although each of these arguments (observer bias, species bias, random variation) could explain some of the reported variability, none was comprehensive. Given the same raw material (the group as the main primate adaptation and primates born social), what are the causes of different forms of social organization? Scientists applied two lines of argument to explain variation during Stage 2: the ecological and the phylogenetic. One of the first ecological approaches created grades of social organization related to differences in environment (Crook 1970a, 1970b; Crook and Gartlan 1966; Eisenberg, Muckenhirn, and Rudran 1972). This ecological model and its subsequent revisions, however, accounted for only a portion of the documented variability. More important, variability within grades was as great as that between them (Clutton-Brock 1974). Advocates of ecological models admitted that some aspects of social organization probably were phylogenetic (see Struhsaker 1969). Yet phylogenetic analyses were not much more successful, and scientists had not yet fully articulated promising new ideas about the interaction between ecology, mating strategies, and social organization (Goss-Custard et al. 1972). No one theory or model seemed fully able to resolve the growing problem of intra- and interspecific variability, in part because there were so many variables, and they were still loosely defined.

The primate pattern that had been handed down from Stage 1 decayed rather slowly, perhaps because of the time lag between the studies themselves and their impact. Many old ideas persisted, including stereotypes of male and female primates. In Stage 2, however, the social nexus in our image of primates enlarged beyond the males to include families, special relationships, and individuals in tripartite (Kummer 1967; Kummer, Goetz, and Angst 1974), as well as dyadic perspective. Confusion over the meaning of variability dominated the view of primates at the end of Stage 2. The lack of a good theoretical framework that could explain variation motivated new forays into theory, methods, and interpretations that were to bear fruit in subsequent stages.

STAGE 3: 1975–1985—THE SOCIOBIOLOGICAL ERA

The ten years between 1975 and 1985 can be termed the era of sociobiology because of the crucial impact this scientific approach had on primate studies and on images of primate society. But sociobiology was not the only important contributor to ideas during Stage 3. At least four other factors played a significant role: long-term studies of primates, new studies of a variety of previously unexamined primate species, studies of female primates by female scientists, and the emergence of theories of animal mind.

A change in theory The impact of sociobiology on primate behavioral research was radical not in its insistence on an evolutionary framework, but because it shifted the unit of selection from the group to the individual and, ultimately, to the gene. The theoretical framework of sociobiology attempts to explain social behavior entirely in terms of self-maximizing biological processes. The most important of these are self-replication and reproductive success. The tenets of sociobiology had two consequences for primate studies. First, the addition of modern genetics to behavioral explanations breathed new life into the evolutionary approach. Sociobiology promised to resolve the anomalies of Stage 2 through the application of new theoretical models that would explain intra- and interspecific variability. Second, it created a new way to talk about social behavior and generated testable hypotheses about individuals and groups. Space does not permit a complete discussion of the application of sociobiological principles to primate behavior (see Gray 1985). We will simply highlight some of the most important consequences of this new, powerful theory for interpretations of primate society.

Sociobiology tackled many pressing issues about primate social behavior, often by reframing the questions and shifting explanations away from social-science interpretations toward more biological ones. Sociobiologists demonstrated that conditions for group selection were limited and focused questions about adaptation on the individual rather than on the group or species. Darwin's struggle for survival became much more precisely defined: selfish competition between individuals or between genes for reproductive success under conditions of limited resources. "Fitness" now had both individual and inclusive components. Altruism and cooperation became selfish strategies that at times could be used to improve reproductive success. This meant that kin should help each other, but only to the degree that they are related. Kin-selection theory (Hamilton 1964) seemed a robust explanation of many social behaviors in primates; reciprocal altruism and parental investment theories (Trivers 1971, 1972) appeared to explain most of the others. Primate sociability emerged in a fresh guise.

The several theories developed under the rubric of sociobiology generated many new hypotheses, some of which were hotly contested. Reproductive strategies, including mating strategies and rearing strategies, assumed center stage. Differences in the costs of reproduction for males and females produced a "battle of the sexes" (Dawkins 1978). Parents and offspring also had their "battle of the generations" (Dawkins 1978; Trivers 1972). Researchers often interpreted examples of cooperation as competition in disguise. Behaviors that had seemed strange and abnormal acquired novel and acceptable evolutionary explanations. For example, some scientists interpreted both infanticide and cannibalism as evolutionary tactics—ways that individuals could improve their reproductive success at the expense of others. Aggression was also an adaptive tactic, but whether it was advantageous to be aggressive depended on a range of factors including age, relative rank, and your opponent's behavior. Evolutionary Stable Strategies, or ESS (Maynard Smith 1978), provided a new way to describe the accurate prediction of options. Through these avenues, sociobiology addressed and reassessed almost every dimension of primate behavior including the nature of competition and cooperation, the importance of family and kinship, and the relationships between individuals, particularly between parents and offspring, between males, between females, and between males and females.

The language of explanation also changed. Sociobiologists combined economic analysis with game theory to generate metaphors about trade-offs of costs and benefits and a warfare of tactics and strategies. Some of the new terminology carried emotional baggage when applied to animals (e.g., *rape, prostitution, infanticide, exploitation, deception, infidelity, coyness*), and some was simply more precise than anything previously part of the scientific lexicon (e.g., *parent-offspring conflict, reciprocal altruism, evolutionary stable strategies*).

Studies of primates flourished in this new intellectual environment; scientists initiated more investigations in Stage 3 than during any time before or after (Southwick and Smith 1986). Rather than try to review them all, we will focus on what happened to ideas about baboons and langurs during the sociobiological era.

Langurs and baboons Sarah Hrdy intermittently studied langurs at Mount Abu in South India from 1971 to 1975 (Hrdy 1977). The langurs of Mount Abu lived at exceedingly high densities in all-male groups or one-male heterosexual groups. The turnover of leader males resulted in a high incidence of infanticide. One of Hrdy's innovations was to use sociobiological theory to suggest an adaptive explanation for langur infanticide. In her view, the killing of unweaned infants was not abnormal behavior brought on by the stress of high-density living but a male reproductive strategy honed by selective forces. Killing infants improved the new male's reproductive success by making it possible for him to sire infants earlier (and hence more often) than if he waited for females to come into estrus naturally. In addition, by eliminating the offspring of his male rivals, the new male improved the competitive chances of his own future progeny.

Infanticidal males obviously gained at the expense of the females who lost their infants. This made sense because of the asymmetry in parental investment between male and female mammals. The sexes almost seemed like "two different species" (Hrdy 1977). Female langurs were said to have counterstrategies, although these did not appear very effective. Older females risked their lives to protect infants. Reproductive females, whether or not they were already pregnant, copulated with the invading male either to confuse him about paternity and thus protect their infants from future attack or in the hope that their sons, sired this way, would inherit the winning male reproductive strategy. Hrdy also framed these counterstrategies in sociobiological terms. For example, the old females who provided aid were thought to be relatives who were already beyond their reproductive prime. By saving the infants of younger female kin, they increased their own inclusive fitness.

Other aspects of langur behavior also underwent reevaluation. Hrdy and Hrdy (1976) reported that female dominance rank declined with age and suggested that rank was related to a female's reproductive potential. Such a system could only occur in groups of closely related females in which both the dominant and subordinate animals would benefit from inequality. Langur infant sharing was interpreted, at least in part, as a tactic for increasing the caretaker's fitness by disrupting the reproductive success of the mother. This was mothering "to death" rather than the earlier mothering "to learn" or "helping" hypotheses (Hrdy 1976).

The sociobiological view of langur society explained many anomalous behaviors. In the process it told an evolutionary story as well-integrated and compelling (although controversial) as the baboon scenario proposed during Stage 1. Langur society was a battleground of individuals whose reproductive interests often conflicted. Females were active strategists, as ready to compete, exploit, and make choices as the males. But male interests dominated nonetheless because the inequalities of mammalian reproduction limited female options.

Sociobiology did not create such a radical revision of baboon society. Perhaps the already well organized adaptationist evolutionary scenario resisted. Or perhaps the large number of baboon studies presented too much data, too much variation, and too much noise to fit neatly into the sociobiological framework. A few studies did try to reassess specific aspects of baboon behavior using sociobiological principles. For example, Popp, a student of DeVore who studied baboons in the Maasai Mara in Kenya, reinterpreted several social patterns including aggression and agonistic buffering (1978). He suggested that males should use aggression selectively because it carried a high risk of injury. Young males who, if injured, had the most to lose in terms of future reproductive potential should avoid aggressive tactics. By contrast, males nearing the end of their reproductive lives should be aggressive since the risk was worth the relative gain. In all populations, baboon males sometimes pick up infants when they are in the midst of an agonistic encounter. A variety of interpretations had been proposed for this behavior. Popp added a new argument to past interpretations by suggesting that "agonistic buffering" was effective because the infant was actually the challenger's offspring. Once the infant was in place, aggression would be against the challenger's reproductive interests.

Baboon males sometimes form alliances in which two lower-ranking males, acting together, can defeat a more dominant male. Packer's (1977) study of baboons at Gombe Stream Reserve in Tanzania suggested that male coalitions were based on reciprocal altruism both because the assistance was reciprocated at a later date and because coalition partners were more successful together than if they acted alone.

Wasser's study of baboons at Mikumi National Park in Tanzania (1983b) interpreted female aggression toward pregnant females as a reproductive strategy that often resulted in miscarriages among competitors.

Strum's long-term study of Pumphouse baboons near Gilgil, Kenya, suggested that male investment in special relationships with females had greater reproductive payoffs than did a male's rank in an agonistic dominance hierarchy (1982) and that the success of agonistic buffering depended on a system of strategically reciprocating social relationships between males and females and between males and infants (1983a, 1983b).

Smuts's study of the Eburru Cliffs troop in the Gilgil baboon population (1983a, 1983b, 1985) applied sociobiological principles to argue that female choice had an impact on male reproductive success through the influence of special relationships on consorting.

Although sociobiological interpretations of baboon behavior tinkered with certain aspects of baboon society, they primarily reinforced rather than changed the previous emphasis on competition and aggression. Yet the picture of baboon society did begin to change. Perhaps the most important perturbations were the consequence of long-term studies.

Long-term studies Most long-term studies of primates were not preconceived but began as short-term projects that simply continued. Studies of chimpanzees (Goodall 1965b; Nishida 1968) were among the first. Stage 3 witnessed more long-term studies than ever, and the longitudinal baboon projects (e.g., Amboseli, Gilgil, Gombe, Mikumi, Okavango) were outnumbered only by Japanese studies of Japanese macaques (see Asquith 1994). All long-term projects share some crucial characteristics. Foremost is the identification of individuals. Individuals become more than just members of age-sex classes, and following them through their lives produces rich detail and essential information on a variety of topics. Long-term studies demonstrated the value of life history data, of demographic data, and of socioecological data to interpretations of behavior. It was only in this diachronic perspective that emergent properties of social organization such as dominance, kinship, and friendship could be assessed. The result was the discovery of a new dimension of primate society: social complexity.

Long-term studies of baboons all suggested a similar *structure* to society, one that confirmed the findings of long-term studies of macaques (e.g., Koyama 1967; Sade 1972; papers in Fedigan and Asquith 1991). Since males migrated between groups, females—not males—were the stable core of the group. Consequently, baboon society was matrilineal and kinship was the key to understanding both the outcome of interactions and the structure of the group. Within their extended families, females performed many of the roles that observers earlier had attributed to males: policing, protecting, and leading. Females also had a clear and stable dominance hierarchy rather than subtle relationships dependent on association with dominant males as suggested by DeVore, Washburn, and Hall. Jeanne Altmann's (1980) work at Amboseli in Kenya helped to flesh out the picture of female baboon daily life. Females spent some time and energy in direct reproductive activities but expended the most effort by far acquiring enough food to sustain themselves and their unweaned young. At the same time, they had to maintain social relationships in order to survive successfully in the group. These "dual career" mothers had a knowledge of home range, of group history, of social relationships that ensured female social power. Males were not irrelevant to social life, but the data from a number of studies enabled Altmann to argue convincingly that baboon society was not as male-focused as it had seemed in the earlier studies.

A long-term study at Gilgil (Strum 1975a, 1982, 1983a, 1983b, 1987) reinforced the view that females were central to baboon society and challenged the earlier baboon model in other ways, most notably on the role of aggression, the evolutionary significance of the male dominance hierarchy, and the nature of the relationship between males and females. Males in this population did not have a stable dominance hierarchy, and male rank did not provide access to limited resources (1975b), particularly when these were receptive females (1975a, 1982). Pumphouse baboons had nonaggressive alternatives, social strategies that were less risky than aggression but just as effective in competition and defense (1982, 1983a, 1983b, 1987).

The existence of baboon social strategies gave the relationship between males and females a new complexion (Strum 1975a, 1975b, 1983a, 1989). Not only was baboon society female-based and female-centered, but females could both exert influence over males and successfully compete with them for some limited resources, despite their smaller size. Males needed female cooperation in reproduction and in defense. Females needed male protection and the improved access to resources that they provided. Males and females created, monitored, and carefully managed an unwritten social contract. A similar social contract existed between males and infants (Strum 1983b). These were the special relationships that Ransom (1979) had earlier described for Gombe baboons but whose significance was not well understood. Now it seemed that "friendships" were evolutionarily important for both males and females and that mutual need generated complementary and more balanced power relationships (see also J. Altmann 1980; Seyfarth 1978; Smuts 1983a, 1983b, 1985).

The complexity of baboon social strategies required skillful actors who could dexterously manipulate relationships and situations (Western and Strum 1983; Strum 1987). The existence of such social sophistication and skills, later termed "Machiavellian intelligence" (Byrne and Whiten 1988), suggested (but did not yet cause) a reinterpretation of baboon society almost as radical as the sociobiological revision of langur society. Ironically, the new concept of baboon society seemed more egalitarian and more socially flexible, whereas the new version of langur society seemed more asymmetrical, with only a few biologically determined social options.

Diversity of primate species and other theory During Stage 3, studies on a great variety of species proliferated, while numerous studies, particularly of apes, continued. We consider only a few of the new findings and interpretations here.

The long-term study of chimpanzees at Gombe during this decade produced abundant and provocative data. In 1977, Goodall announced that Gombe chimpanzees had a high rate of beatings, killings, and infanticides. Nishida later confirmed such violent behavior among the Mahale chimpanzees (1979). Wrangham (1979) and Pusey (1979) now characterized the social organization of chimpanzees as communities of related males with dispersing females. Mating was not entirely promiscuous, as previously reported. Gombe chimps formed consort pairs, sometimes initiated by a dominant male (McGinnis 1979) and sometimes by the female (Tutin 1979).

A new emphasis on arboreal species in both Old World and New World monkeys helped to correct the earlier bias toward terrestrial cercopithecoids. Among most arboreal species, however, details of social behavior were simply hard to see. Investigators therefore turned their attention to demography and ecology. Until now, ecological studies of primates had lagged behind studies of social behavior. Early primate ecology was rudimentary, often consisting of simple food lists and descriptions of habitats. As investigators tried to grapple with the variability of primate behavior at the end of Stage 2, they turned to a consideration of socioecology.

Sociobiological theory predicted that social organizations were the outcome of the interaction of reproductive strategies and ecological factors such as predation, diet, food distribution, and the energetic costs of traveling in groups of different sizes (Clutton-Brock and Harvey 1977). Optimal foraging theory (Pyke, Pulliam, and Charnov 1977) suggested that animals employ feeding strategies to optimize or maximize caloric and nutrient intake. Researchers examined these predictions in primates (S. Altmann 1974; Sussman 1977). The results had implications for many old issues such as territoriality, sexual dimorphism, and home range.

On the social side of the socioecology equation, Altmann and Altmann (1979) argued for the role of demographic constraints on socioecological patterns, and Wrangham (1980) explored the wide-ranging implications for primate society of the pattern of female philopatry and male dispersal found in many Old World cercopithecines (female-bonded primate groups).

Despite the difficulty of getting details of social behavior from arboreal species, the new studies on prosimians and New World monkeys produced provocative results. For example, prosimian data raised the possibility that females can be dominant over males (Jolly 1984), and that even among these supposedly primitive primates, a variety of social organizations can exist. Research on New World monkeys demonstrated that some primate societies, such as those of howlers, can function without a reliance on kinship (Clarke and Glander 1984). This decade's studies provided even more evidence of species diversity and, in so doing, further softened the focus on baboons as *the* primate society.

The series of review chapters commissioned for the edited volume *Primate Societies* (1987) well exemplifies the growth of primate studies on many fronts in this time period. Twenty years after the first primate study group convened in 1962–63 at the Center for Advanced Study in the Behavioral Sciences at Stanford (leading to DeVore's 1965 *Primate Behavior*), a second study group assembled at Stanford to review the status of primate field research. *Primate Societies,* with forty-six authors and five editors, surveys much of the state of the art up to that time. The book's chapters illustrate a range of perspectives: sociobiology, socioecology, and cognitive ethology, as well as those derived from long-term field studies. *Primate Societies* emphasizes the diversity of social patterns found among both familiar and newly studied species.

Methods Sociobiological research broke with the traditional natural history approach and its vague group selectionism during Stage 3. Instead, the new body of theory offered tightly constructed predictions amenable to quantitative testing using behavioral data. Long-term studies became particularly valuable since individuals were known and at least matrilineal kinship was certain. Biological data were also critical, stimulating biosocial projects (Melnick and Kidd 1983; Turner 1981) that took biological samples from wild animals. Blood samples, in particular, were essential to determining relatedness within a group and the genetics of the population.

Ecological research methods changed as well. Optimal foraging theory provided more precise methodology for assessing the physical environment. The emphasis shifted to making predictions about how individuals should interact with their habitat, requiring finer levels of ecological analysis. Studies of primates now also measured aspects of foods, such as nutritional content and secondary compounds, that might influence consumption (Glander 1978; Oates, Swain, and Zantovska 1977). Cost-benefit analyses generated time and activity budgets that potentially could be linked to survival and reproductive success (see studies in Clutton-Brock 1988; Sussman 1979).

Other issues: females studying females and animal mind Two other trends helped to make Stage 3 images of primate society different from those of previous periods and set the stage for what followed. The first was the impact of female scientists studying female primates and the second, the emergence of animal mind as a topic of investigation. Although females always had been subjects of primate field studies, the disproportionate influence of the prevailing baboon model with its focus on males had made it seem as though females were unimportant in primate societies. By the early 1980s, female scientists began to change this view (J. Altmann 1980; Fedigan 1982; Hrdy 1981; Small 1984).

A variety of studies documented the significance of females and the diversity of their roles and behaviors. Other elements reinforced and interacted with the new view during Stage 3. For example, sociobiological theory recognized the evolutionary importance of female strategies. Long-term studies suggested that matrilineal networks were at the core of many primate groups and that females, not just males, were social manipulators. The changing view of animal mind (see below) implied that both males and females were capable of strategic decision making.

Stage 3 females, moreover, were different from their earlier counterparts. These females struggled with their own problems of survival amid myriad conflicting demands; they were both more cooperative and more competitive and exercised more power and control—even over males, since female choice and sexual selection could mold male behavior.

Changing views of the cognitive abilities of animals also contributed to shifts in the image of primate society during Stage 3. Earlier, Jolly (1966b) had argued for the central importance of social behavior in the evolution of primate intelligence, and Humphrey (1976) had proposed that the large brains of primates were an adaptation to the complexities of social life. But it was really Griffin (1976) who spearheaded the new movement with his claim that behaviorism had robbed animals of the cognitive abilities that were clearly essential to their survival. His goal was to reverse this bias and make the investigation of animal mind part of the study of animal behavior. By the second edition of his book, scientists already had widely accepted Griffin's previously heretical position (1984). Cognitive ethology subsequently emerged as a new discipline (Griffin 1992).

The shift in attitudes toward animal mind seemed to take place overnight. But the groundwork had been laid by the "cognitive revolution," which started more than a decade earlier among the human behavioral sciences (see Gardner 1985), as well as by provocative primate data. Studies of captive apes, particularly language experiments, presented compelling evidence for the humanlike abilities of some pongids (see reviews in Parker and Gibson 1990; Ristau and Robbins 1982). Long-term studies of wild chimpanzees and baboons showed them to be naturally sophisticated tacticians whose negotiation of social complexity seemed to require mind.

The emerging cognitivist position, which posited active actors, stood in opposition to the prevalent sociobiological view of animals as "gene machines" in which smartness resided in the genetic material, not in the mind. This fundamental clash of perspectives was weaker than might have been expected since cognitivists did not strongly articulate their view until the end of Stage 3, by which time the sociobiological gene-strategist position was beginning to wane.

By the end of Stage 3, the idea that there was one primate society had disintegrated. Variability, unmasked over the last twenty years, became enshrined in theory. Now, any answer to the question, "What is the nature of primate society?" required investigators to consider which species, which group, which habitat, and at what point in the group's history. Although the baboon model lingered (especially in the minds of the public and those in other scientific disciplines) within primatology, it had been both transformed and overwhelmed.

During the sociobiological era, those studying primates abandoned many issues as irrelevant while restating others using the new lexicon. For example, "family" became "kin" with a focus on inclusive fitness, "sex" got resurrected as "mating strategies," and "weaning" was reconfigured as "parent-offspring conflict." The theory also generated new topics for primate investigations, including female choice, infanticide, foraging strategies, and life history, to name just a few.

Theory provided some important answers, but it also renewed and generated controversies. Is a group an accidental selfish herd, a cooperating kin group, or an assembly of individuals for whom the benefits of living together outweigh the costs, despite their conflicting interests? Do groups stay together because of genetic calculations, evolutionary stable strategies, evolutionarily based emotions, facultative interactions, or all of these? Theory-generated primate nature seemed to become ever more selfish, competitive, manipulative, and exploitative. Male and female reproductive interests appeared irreconcilable—yet the sexes sometimes made peace. New data and new theory left investigators to wonder whether it was really only the female who had to pay the price of divergent mammalian reproductive physiologies.

Sociobiology brought humans back into the picture after the primate-pattern rationale collapsed. The diversity among primates meant that no other species could model humans. The only bridge left to help understand human behavior would have to be evolutionary principles. Toward that end, Wilson (1975) argued that sociobiology, as the unified theory, would absorb all the human social sciences including anthropology, sociology, and psychology.

STAGE 4: 1985 TO PRESENT—ECOLOGY, COMPLEXITY, COGNITION, CONSERVATION, ANIMAL RIGHTS, AND BEYOND

The sociobiological era ended with a multitude of new answers to old questions. Yet despite the unifying evolutionary vision, anomalies persisted, presaging shifting ideas in Stage 4. For example, the existence of social complexity in primates suggested that individuals used cognitive abilities regardless of the efficacy of gene strategies. Social strategies included exploitation and competition, but they also involved cooperation and assistance. Despite the evolutionary reasons for adversarial relations between males and females, the sexes at times complemented each other, even cooperated. Each shift created another shift, so that as the theory, methods, and data of Stage 3 redefined female options, they also transformed male options.

Stage 4 may be the hardest period to characterize for a number of reasons. The most obvious is that it is our current history and thus still unfinished. Without knowing the future, the present is harder to define. But the difficulties of depicting the present stage result also from the growing fragmentation and specialization within the discipline, as well as from the increasingly dense and entangled interactions within the science of primatology, and between primatology and its larger context.

Stage 4 is perhaps best categorized as an era that has moved away from a strongly reductionist application of sociobiological theory. Crook (1989) has argued that in the history of studies of the evolution of behavior, explanations have oscillated between the genetic and the environmental. Classical ethology's focus on innate factors was a reaction to behaviorist environmentalism; then the emergence of socioecology led to a renewed interest in social relations as flexible biotic systems. Next, sociobiology once more emphasized genetic determinism. Most recently, behavioral ecology has proposed a more holistic model of adaptation that relates environmental and societal processes to those of genetic selection. Behavioral ecology as well emphasizes multicausal analyses. Changes in how scientists see individual animals reinforce this multifactored approach. No longer applying the parsimonious principles of ethological behaviorism or the reductionist models of sociobiology, researchers see individual primates as sentient beings who play a much more determinative role in their own behavior.

An increasingly dense and complicated entanglement of science and society also characterizes this era. For example, both primate conservation and animal welfare movements have now achieved some scientific legitimacy as a result of shifts in societal concerns. The political climates of source countries also have influenced which species are studied. Foreign researchers are now studying New World monkeys and lemurs in Madagascar because these areas are open to them and because of growing political stability in Latin America. By contrast, many parts of Africa and Asia have become closed to primate research because of political unrest. The shift in geography has entrained a shift in scientific emphasis. The predominance of studies of arboreal species in the last decade has brought ecological and demographic research to the fore, at least in part because the details of social behavior are often difficult to observe in tropical forests. During Stage 4, the composition and complexion of primate studies has changed markedly from prior research. Investigators have not abandoned Old World monkeys; in particular, the work on terrestrial species such as baboons and macaques continues. But the real excitement is about a deluge of other species: arboreal monkeys of both the neotropics and the Old World, pygmy chimpanzees (bonobos), lowland gorillas, and endangered primates such as the lemurs of Madagascar and the muriqui of Brazil.

Theory: From sociobiology to behavioral ecology The scientific pursuit of evolutionary explanations in primatology during our last stage has diverged along several related but distinct lines. Sociobiologists have continued to concentrate on mating and rearing strategies, on sexual selection, and on the variety of social behaviors that determine differential reproductive success. In particular, during Stage 4 the field of human sociobiology has expanded, reframing human cultural behavior using evolutionary principles (Alexander 1986a, 1986b; Barash 1986; Betzig et al. 1988; Mealey, Young, and Betzig 1985; Turke and Betzig 1985). Other investigators (Dunbar 1988, 1989; Garber 1987; Isbell 1991; Janson 1992; Richard 1985; Terborgh 1983; van Schaik 1989) have turned away from the strongly genetic approach of the 1980s toward the perspectives of behavioral ecology.

The primary concerns of behavioral ecology are survival and reproductive strategies: how animals find enough food, avoid predators, and balance the conflicting demands that their environment places upon them. Behavioral ecology is an outgrowth of both socioecology and sociobiology (Crook 1989; Foley 1986). Behavioral ecologists apply evolutionary principles to behavior, shifting away from both the environmental determinism of early socioecology and the genetic determinism of early sociobiology. Optimal-foraging theory and life-history theory have been crucial in explaining how animals interact with their living and nonliving environments (e.g., Krebs and Davies 1993). The fact that primates are slowly reproducing species should have major consequences for primate patterns of fecundity, mortality, and survivorship, and, through these, for demographic processes and population dynamics (Clutton-Brock 1988; DeRousseau 1990; Dunbar 1988; Fedigan et al. 1986).

The broader perspective of behavioral ecology has placed primates in communities, widening the ecological context to include consideration of community structure and dynamics (Gautier-Hion 1988; Richard 1985; Standen and Foley 1989; Terborgh 1983). This contrasts with previous approaches, which treated each species (and sometimes each behavior) separately and viewed primates as unique, interpreting their adaptations in isolation from the nonprimate ecological context.

Behavioral ecology also has reconsidered the question of why primates live in groups. During Stage 4, primatologists have pursued a lively debate about the relative importance of the determinants of social groupings. Although most theorists agree that group living involves a trade-off between increased predator protection and decreased foraging opportunities, some researchers (e.g., van Schaik 1989; Terborgh and Janson 1986) have stressed the importance of predator defense strategies as the basis of the size and composition of primate groups; others have emphasized resource exploitation and defense (Isbell 1991; Rodman 1988; Wrangham 1987). Obviously, the answer to the question of why primates live in groups is multifaceted and at a minimum involves predator defense, resource defense, foraging efficiency, and rearing strategies. For each species and in each situation, the balance of costs and benefits changes with the specifics of evolutionary history, habitat, group size, and group composition.

Phylogenetic analyses of social systems are less common but still important. Phylogenetic hypotheses sometimes represent a last resort when attempts to explain social patterns in terms of local ecological circumstances have failed (e.g., Rodman 1988). Phylogenetic constraints may best explain the distribution of certain behavioral patterns across species. At the very least, researchers see phylogenetic analyses as an integral part of behavioral ecology (Chan 1992, 1993; Di Fiore and Rendall 1994; Garber 1994; see also discussion in Gautier-Hion et al. 1988). Di Fiore and Rendall (1994), for example, performed a cladistic analysis identifying a highly uniform and conservative suite of social organization traits in Old World monkeys that all hinge on female philopatry and male dispersal. The similarity and persistence of these traits in the face of considerable ecological diversity among cercopithecoids suggests that a strong and conservative phylogenetic influence has molded the social organization of these related species. Ecological adaptations are still important: The cercopithecoid ancestor itself presumably evolved social behaviors in partial response to past environments. But primate societies are not infinitely free to vary in response to contemporary ecological conditions. Social structure is a composite result of evolutionary, historical, epigenetic, and situational factors.

Social complexity within groups New perspectives on the internal social dynamics of groups also have impacted significantly on Stage 4 ideas. Research in this period has explored social complexity, particularly its implications for competition and cooperation. Earlier investigations had transformed the diversity and complexity of social relationships from social "noise" into an important evolutionary resource for individuals. Social strategies quickly became primate "politics," changing the context for both competition and cooperation. Now not only size or strength gave an individual a competitive edge, but also the ability to assess and manipulate the social situation. This meant that other factors could play a role, including age, temperament, tenure in the group, the history of previous interactions, and the current social context. Equally significant, social strategies require cognitive as well as social skills. Actors have to perceive the multiple dimensions of social relationships to predict the combined effects of sequential or simultaneous polyadic interactions, in order to plan and manipulate such interactions advantageously. The work of de

Waal (1989), Goodall (1990), Kummer (1967), Strum (1987), and Whiten and Byrne (1988) presents strong evidence that primates are capable of such skills. The rather mechanistic model of brute force in dominance and competition is modified in a way that finds many parallels in the study of human politics with its intricate motivations, objectives, and manipulations (see discussions in Byrne and Whiten 1988; Mason and Mendoza 1993; Silverberg and Gray 1992; Schubert and Masters 1991).

The politics of social strategies has become a subject for study, mainly focused on alliances and coalitions, on reconciliation, and on how aggressive and nonaggressive social strategies interact. Studies of coalitions and alliances primarily have emphasized Old World monkeys and apes (Datta 1986; de Waal 1984; Dunbar 1984b; Harcourt 1988, 1989; Hunte and Horrocks 1987; Moore 1982; Seyfarth and Cheney 1984; and Silk 1982; see especially papers in Harcourt and de Waal 1992). In addition to these observations largely from the field, Chapais (see summary in 1991) conducted an elegant series of experiments on alliances in Japanese macaques. He documented the key role of these associations in competitive success, particularly in the acquisition and maintenance of dominance rank. Investigators also have given more attention to primate cooperation and how social relationships and social systems are maintained (see Hinde 1983). De Waal (1986a, 1986b, 1987a, 1987b, 1989), for one, has argued that the strong prior focus on conflict has been at the expense of understanding how conflicts are socially mediated in order for primates to successfully live in social groups. This stage has witnessed the blossoming of the study of reconciliation, or peacemaking, as part of this new focus (Aureli 1992; Cords 1988; de Waal 1993; Judge 1991; Kappeler and van Schaik 1992; York and Rowell 1988). We now recognize that aggression is not the only option for individuals within groups. Current work has begun to address how aggressive and nonaggressive options emerge and are integrated during an individual's lifetime (Strum 1994a), and how they are implemented according to constraints of space and time during interactions (Forster and Strum 1994).

Another aspect of the study of social dynamics during Stage 4 is research on mate choice. The priority-of-access model (reviewed in Bercovitch 1991; Cowlishaw and Dunbar 1991; Fedigan 1983; Shively 1985; Smuts 1987b) has been at the center of a long debate. The model assumes, among other things, that dominant males will have greater reproductive success by monopolizing mating. Controversy over the model has not been resolved by extensive research or comparative analysis (see discussion in *Animal Behavior,* vol. 44, 1992). Instead, some investigators have turned their attention to other factors that may influence mate choice for both males and females. Huffman (1991b), Janson (1984), Keddy (1986),

Manson (1992), and Small (1990), among others, have conducted studies of female mate choice, and Small (1989, 1993) has reviewed the central question for female choice: What do female primates want from males? Resources, parental care, protection from aggression, familiarity, novelty, status, superior genes? Although hardly any experimental work is yet available, and observational research on female choice in primates is in its infancy, it seems that primates have preferences for certain sexual partners. Huffman, for example, found that female Japanese macaques, unlike baboons, do not prefer to mate with their friends (1991b). While they may repeat consorts with a given male from one year to the next, they begin to avoid mating with these same individuals after a few years. Case histories of individual males suggest that as their consort frequencies decline, even top-ranking males are prone to emigrate. Such findings provide us with a more dynamic view of what determines individual preferences than does the traditional dominance model, which has tended to assume that dominance determines mating success, irrespective of the history of relations or preferences between individuals.

Animal cognition Studies of behavioral ecology and social complexity have set the image(s) of primate society(ies) on a new course, and the recognition of animal mind has created a new kind of actor. This happened in several steps. First came the idea that mind was an important determinant of behavior. Next was an interest in how and why that mind evolved. Jolly (1966b) and Humphrey (1976) earlier had presented the case for the social function of intellect. Then Milton (1981, 1988) revived the discussion by suggesting the importance of an ecological hypothesis focused on the importance of mapping and tracking seasonally variable food resources essential to survival. A variety of investigators (reviewed in Byrne 1995) subsequently have explored the evolution of cognition among primates. In general, the social-origins hypothesis has had more advocates (cf. Byrne 1995; Parker and Gibson 1990), and the social manipulation of strategic partners has been transformed into the Machiavellian intelligence hypothesis (Byrne and Whiten 1988). Experimental work, particularly field experiments, have filled in some of the details of social intelligence. For example, Seyfarth and Cheney (1988; see also Cheney and Seyfarth 1990) demonstrated that vervets not only recognize other individuals from their calls, but can also recognize relationships between other members of the group. When they played back an infant's lost call to the group, the mother oriented toward the hidden speaker, while the other group members oriented toward the mother, suggesting that they recognized the relationship between the mother and the distressed infant. Research on rhesus macaques by Gouzoules, Gouzoules, and Marler

(1984) demonstrated that the screams of monkeys during conflicts convey information about the participants' relative ranks and levels of agonism, and that listeners act upon this information. These and other studies have shown that primate calls are not just expressions of arousal—a long-standing assumption—but contain and convey specific information about the environment and about the social contingencies of an interaction. By using experimental protocols and sophisticated technology for recording, analyzing, and playing back vocalizations, this research has demonstrated that primates remember past interactions, recognize relationships along kinship and dominance lines (see Dasser 1988), communicate contextual information, and predict the responses of others on the basis of this knowledge.

Animals' knowledge of social relationships has also led to discussion about whether primates possess a "theory of mind" (see Byrne 1995; Premack 1988; Whiten 1993); that is, whether they respond to what they think another individual might be believing and desiring. Tactical deception, which occurs when individuals deliberately mislead conspecifics, may provide some of the best evidence for the existence of theory of mind among primates (see Byrne 1995; Byrne and Whiten 1988, 1990; Cheney and Seyfarth 1990; Mitchell and Thompson 1986; Snowdon 1990a). While primate tactical deception is still controversial, evidence exists from all families of monkeys and apes.

Cognitive issues are a major focus of primate research in Stage 4. The controversy about the evolutionary origins of primate intelligence continues, but any cognitivist stance implies a major shift in images of primate society; individual action has become more intricate, more variable, less deterministic.

Methods In the period since 1985, technological developments have continued in both data collection and analysis. Sociobiology, behavioral ecology, and cognitive ethology all generate testable predictions that require quantitative data. New technology, particularly computer equipment and software, has made it possible to record, analyze, and model both old and new kinds of data.

Biological data are increasingly important in resolving many debates in the study of primate society, including controversies about the sources of individual reproductive success that can be settled only through data on genetic relatedness. For the more easily observed baboons and macaques, information on paternity is needed. The newer research projects on more difficult-to-study species also have need of biological data, for they often are not able to obtain basic observational data on kinship, either maternal or paternal. To solve this problem, study animals are captured for samples that can provide essential genetic analyses of population and group structure (Glander et al. 1991; Hildebolt, Phillips-Conroy, and Jolly 1993; Melnick, Kidd, and Pearl 1987; Richard, Rakotomanga, and Schwartz 1991; Sussman 1991, 1992). Biological samples from free-ranging primates also can be used to study growth, development, physical condition, and the physiological correlates of behavior such as the relationship between the endocrine stress response and rank in baboons (Sapolsky 1989, 1990, 1993; Steklis 1993).

The development of noninvasive techniques that do not require the capture of animals is also an exciting new source of biological data. These include DNA analyses from hair follicles (e.g., Morin et al. 1994) and hormonal assays from fecal analyses (Strier and Ziegler 1994; Wasser, Rister, and Steiner 1988; Wasser, Monfort, and Wildt 1991).

Field experimentation, although limited, is an important aspect of Stage 4 methods. This approach combines the controlled environment of a laboratory with the evolutionarily appropriate context of the natural setting. As early as the 1950s Washburn (1951) argued for experimentation as a way to tease out causal factors, but the approach was not popular. Kortlandt's (1967) early imaginative manipulations with chimpanzees (e.g., dressing up as a chimp, automated leopard) were sometimes derided. Later, Kummer's sophisticated series of transplantation experiments with anubis and hamadryas baboons (1973; Kummer, Goetz, and Angst 1970, 1974) clearly demonstrated the value of field experimentation, but still not many followed his lead. Stage 4 field experiments (see references in Cheney and Seyfarth 1990) have tended to focus on cognitive issues, translating captive experimental design into a natural setting and in the process addressing the criticism leveled at both captive studies of cognition and the anecdotal nature of naturalistic studies of primate cognition.

Ecological methods have become more sophisticated during Stage 4, particularly techniques for monitoring habitats and measuring substrates. Improved technology in remote sensing, in precise location and mapping using GPS (Global Positioning System) equipment, and in integration of data using GIS (Geographical Information System) has made geographical and topological data a more important aspect of primate research (e.g., Provost et al. 1993; Sprague 1993; Sussman, Green, and Sussman 1994). Ecological monitoring programs have become routine parts of most field projects, providing essential background data for studies of ranging, foraging, and social behavior. Monitoring also has been crucial to conservation work since deciding on the long-term viability of primate populations requires, at a minimum, an assessment of habitat quality.

As standardized methodology increasingly has become part of social, ecological, and socioecological studies of primates, interdisciplinary boundaries have faded. Zoologists, psychologists, and anthropologists—even cognitive scientists and philosophers—now seem to be interested in the same issues, not just the same subjects. More than ever, it is the topic, not the disciplinary background of the investigator, that dictates the methodology.

Baboons We return to baboons, the species that began our discussion of primate society. In Stage 4, it is impossible to propose one simple model of baboon society, and within the diversity of baboon research no single study can be taken as representative. Yet baboon research is typical of research on other species in some respects. There is a wide range of theoretical interests focused on increasingly specialized topics. In other ways, studies of baboons remain distinct by virtue of their large number, their long duration, and the quality of data generated by excellent conditions of observation.

Long-term field sites continue to be important, including sites in Kenya (Amboseli: Altmann et al. 1988; Gilgil/Chololo: Strum 1987, Barton et al. 1992; Maasai Mara: Sapolsky 1990), in Tanzania (Gombe: Packer et al. 1995; Mikumi: Norton et al. 1987, Wasser and Wasser 1995), in Botswana (Okavango: Bulgar and Hamilton 1988), and in Namibia (Brain 1992). Even recently initiated sites like those at Tana River in Kenya, Kibale Forest in Uganda, and numerous locations in South Africa are organized as "long-term" projects by piggybacking shorter cross-sectional studies to create continuous project records that benefit subsequent short studies. Stage 4 baboon research has a strong theoretical basis and is constructed around hypotheses from sociobiology, behavioral ecology, and cognitive ethology. The issues include life-history strategies, reproductive strategies (mating and rearing), social complexity, ontogeny of social skills, animal cognition, communication, reconciliation, foraging strategies, ontogeny of foraging skills, demographic processes, and socioecology. Studies concentrate on both males and females, with special interest in sex-related constraints and patterns such as those imposed by motherhood on females and by life-history patterns on male sexual and social options.

Despite the lack of consensus about baboon society, researchers have widely observed some characteristics. The group is primarily a multimale unit organized around a core of female matrilines arrayed in a relatively stable hierarchy. Group size, age structure, and socionomic sex ratio are affected by ecological conditions. Males are dominant over females by virtue of their larger size, but females are not powerless. They create their own options through social tactics, special relationships, kinship, and the female hierarchy. The male dominance hierarchy is not stable for very long. The result is that a male holds a variety of ranks during his lifetime. Aggression is a basic part of the male repertoire, but so are nonaggressive strategies. Male competitive tactics depend on life history factors such as age and residency status, as well as on size and physical condition. Sexual competition is part of baboon life for both males and females. Males must compete over access to a limited number of reproductive females, and females must compete over access to preferred males. Female choice can influence male consort success, while at the same time, males can play an active role in molding female preferences. Sexual competition, like any type of baboon competition, is embedded in a matrix of social relationships that must be created, managed, serviced, and repaired. This means that competition rests on a foundation of assistance, coordination, and cooperation.

Baboons may be among the most socially complex nonhuman primates. The amount of complexity within a group depends on its size and composition, factors sensitively tuned to key resources like food, water, and sleeping sites. The result is a dynamic set of socioecological relationships in which size, dimorphism, survivorship, age structure, patterns of migration, foraging strategies, and social strategies (to name just a few factors) are inextricably linked in determining individual evolutionary success. Today baboons everywhere have more options in our images of their societies than ever before.

Howlers What has become of our perceptions of howler society in the fifty to sixty years since Carpenter's original study? Scientists long considered Carpenter's 1934 monograph the type report not just for the mantled howler species, but for the entire genus of howlers, and, according to Neville et al.'s review (1987), many of Carpenter's original conclusions still stand. Today we do, however, know a good bit more about dispersal patterns, intragroup aggression, and foraging patterns in howlers. Studies of the behavioral ecology and demographic patterns of howlers are fairly common and indicate that the neotropical forests do not provide the rich smorgasbord that was once assumed. Howlers are selective leaf eaters who carefully choose new leaves on individual trees of species that contain lower levels of secondary compounds or toxins (Glander 1982).

Data from longitudinal studies are now available for mantled howlers in Costa Rica (Clarke and Glander 1984; Glander 1992; Jones 1980) and red howlers in Venezuela (Crockett 1985; Crockett and Eisenberg 1987). From these we know that both sexes disperse from natal groups and that the adults of a group are largely unrelated. Although red and mantled howlers differ in some ways, in general, females disperse a little earlier and more often than males. In mantled howlers, Glander (1992) reports that 96 percent of females and 79 percent of males disperse. Some males remain in their natal groups and eventually take over breeding positions from their presumed fathers. Most individuals of both sexes spend from one to four years of their lives as solitaries because it is not easy to successfully transfer to a new group.

Carpenter found little competition or aggression between group members and no sharp gradient of dominance. Today we know that the picture is rather more complicated. Observers can rank both males and females in a hierarchy, primarily on the basis of supplantations rather than overt fights. Why Carpenter should have concluded that howlers are peaceful animals is easy to see. Social interactions of any type are rare in howlers compared to other primates, and howlers spend most of their waking hours quietly feeding and resting while digesting their largely folivorous diet. Neville et al. (1987) concluded that competitive interactions between howlers are subtle, often consisting of one animal supplanting another from food or a resting place. Even more common is for one howler to avoid an interaction or move out of the way of another. On rare but notable occasions, however, howlers of both sexes do fight, and then the interactions are very intense, resulting in wounds and falls from high in the canopy. In both Venezuela and Costa Rica, researchers have reported occasional infanticide of young infants by newly resident males (Clarke and Glander 1984; Crockett and Sekulic 1984).

One of the several ways in which howlers are different from Old World monkeys is that their dominance hierarchies are constituted in reverse order of age: Younger females rank over older females, and younger males rank over older males. This is primarily related to the pattern of bisexual dispersal, in which young individuals of both sexes must fight their way into established groups. According to Glander (1992), a newly immigrated female either will receive some support from one or more of the resident males and fight her way to the top of the female hierarchy or else leave to try another group. For red howlers, Crockett (1985) has argued that each group has a limited number of female breeding positions, which the resident females defend. A young female either must fight her way in or form an entirely new group with other young immigrants. In both mantled and red howlers, the unrelated resident females of the group may form an alliance to keep the newcomer at a distance, but the immigrant female tries to take on each female one at a time, rising step by step in rank.

Thus howler monkeys, as noted by several of the researchers who study them, raise many challenges to the general assumptions about primate societies—assumptions based largely on studies of ground-dwelling cercopithecines. Strier (1994a) has reviewed how studies of New World monkey species should lead us to rethink the many generalizations in the myth of the "typical primate." Some taxonomists (e.g., Rosenberger 1979) place howlers (Alouattini) and their close relatives, the Atelini (wooly monkeys, spider monkeys, and muriquis), in the subfamily Atelinae because these species share many morphological and behavioral features (Rosenberger and Strier 1989). Research on atelines, especially in the last decade, challenges at least the following assumptions about the typical primate social pattern:

1. *The common primate pattern is one of male dispersal and female philopatry.* In contrast to many Old World monkeys, howlers of both sexes disperse, and in muriquis, spiders, and wooly monkeys, females disperse (Moore 1984, 1992).
2. *Most primate societies are female-bonded, and matrilines/nepotism form a central structural feature of the social system.* In howlers, unrelated resident females form alliances against immigrant females. In spider monkeys, females disperse to new ranges and different communities from their mothers, whereas males occupy large ranges overlapping those of their mothers (Chapman et al. 1989; Fedigan and Baxter 1984; Fedigan et al. 1988; Symington 1988). In muriquis, patrilineal associations and male affiliation are essential to social structure (Strier 1994a; Strier et al. 1993).
3. *Dominance hierarchies are based on inherited rank for females and aggressive interactions for males.* In howlers, the newest immigrants of both sexes usually fight their way to the top rank or leave the group. In muriquis, intragroup relations are largely egalitarian with co-dominance between males and females and no female dominance hierarchies.
4. *Intragroup competition over food is frequent.* In howlers, food competition mainly takes the form of subtle avoidance and supplantation interactions. In the howlers of Santa Rosa, large groups fission in the seasons when food is only available in small patches and fuse when food is more densely distributed (Chapman 1989; Fedigan 1986a). Similarly, muriquis and spider monkeys avoid direct confrontations over food, either by avoiding interactions or through fission-fusion feeding groups (Kinzey and Cunningham 1994; Strier 1990, 1994a).

Conclusions about baboons and howlers The baboon and howler studies, although very different in their scope and focus, point to some similar conclusions. First, we can no longer talk about "primate society." Primate societies are highly varied and affected by both local circumstances and individual species' phylogeny. The longer we continue to study the familiar species such as baboons, macaques, and chimpanzees—and the more we branch out to study the lesser-known primates such as prosimians, neotropical monkeys, arboreal colobines and cercopithecines of the African and Asian forests, lowland gorillas, and bonobos—the more apparent it is that past generalizations about primates were premature and perhaps biased by limited data. During Stage 4 we find ourselves facing old dilemmas about primate society, albeit with new data and new methods: What generates primate diversity, and, given this diversity, how can we generalize about primate patterns?

Second, simplistic models are inadequate to explain the social, the ecological, or the interaction of the two. We need more comprehensive ways of taking into consideration the broad array of factors that complicate and yet determine primate society.

Third, the presence of new actors in primate society confound the task. Individuals, regardless of sex or age, are strategists in an intricate evolutionary game. Their options, choices, and successes depend on a variety of factors, including environment, demography, age, sex, development, personality, biology, and historical accident. What is still unclear is how much flexibility and variability exist in individual actions over a day, a week, and a lifetime—and how much these matter.

The preceding discussion of baboons and atelines illustrates that our understanding of primates and primate societies today is based on more information on more species in more environments. These data are both cross-sectional and longitudinal; they include more ecology and more biology than in any previous period. Sociobiology, behavioral ecology, and cognitive ethology provide heuristic and intelligible ways for beginning to explain the complexity that has been unearthed. Better methods of data collection, analysis, and modeling have allowed us to agree on what constitutes the data, even if we do not always agree on interpretations. Many questions remain, yet no one can doubt that today we understand the diversity of primate behavior and societies better than did Zuckerman, Carpenter, and their contemporaries.

Enduring Issues

We began our historical review wanting to know whether and how ideas about primate society have changed in the last seventy years. Our framework included stages and enduring issues that had some bearing on images of primate society. We are now in a position to draw some conclusions from this history. Because of space limitations, we review only the first five issues, which contain elements crucial to images of primate society.

WHY DO PRIMATES LIVE IN GROUPS?

Zuckerman's conclusion (from Pre-Stage 1) that sexual competition is the basis of primate society is our starting point. Stage 1's position is more reminiscent of Carpenter's view than of Zuckerman's: Primates live in groups for a variety of reasons, the most important of which are reproduction and predator defense. But this stage's researchers also see the group as the repository of traditional knowledge that goes beyond the individual and is critical to individual survival. The vague socioecological models of primate groups promoted during Stage 2 are replaced, in Stage 3, by precise sociobiological arguments based on genes and couched in terms of reproductive success. Next, behavioral ecology contributes an additional set of integrative socioecological principles about how and why the group is a primary primate adaptation. Controversy surrounds the relative importance of predator protection and resource competition as explanations for group living. In retrospect, the original question appears to have changed somewhat. The question is no longer, "Why live in a group?" but, "Why live in a group of a particular size and composition?" The answer(s) still contain(s) elements that both Zuckerman and Carpenter would recognize.

WHAT HOLDS SOCIETY TOGETHER?

Zuckerman and Carpenter did not agree. One claimed that sexual instincts were the cement of the group, while the other argued for social conditioning, social attraction, and affiliation. Stage 1 arguments already contained an implicit distinction between explanations about the motivation of individuals and about evolutionary advantages. Generally, investigators agreed with Carpenter that strong emotions and affiliative interactions created social cohesion. But evolutionary payoffs to group living existed, for the group and for the species. Natural selection insures that animals are motivated to do what was essential to their survival in the past—that they would *want* to be social. Stage 3's sociobiological framework shifted the focus from the group to the individual and from proximate to ultimate causation. Society then became the result of gene-based individual selfish strategies, which would change when conditions changed. Stage 4 contributed more complexity and more active agency to this strategic view of society.

WHAT IS THE NATURE OF SOCIETY: AGGRESSION AND DOMINANCE, SEX AND AFFILIATION?

Society exists and holds. These realities, however, don't make arguments about the basic nature of this society any less problematic. Seventy years of primate research begins and ends with controversy. Is society aggressive and competitive or peaceful and cooperative? Zuckerman and Carpenter each championed a different view, but for decades Zuckerman's claims and the baboon model captured the collective imagination. In its various incarnations, this view emphasized aggression, competition, and dominance (primarily over things sexual and primarily by males). Peaceful chimpanzees briefly replaced aggressive baboons as the standard primate society in Stage 2. But sociobiology in Stage 3 revived the competitive view, making individuals and their societies seem even more selfish and exploitative than earlier. At the same time, researchers reported that chimpanzees turned to warfare, infanticide, and cannibalism, exploding our image of their peaceful society.

After 1985, in Stage 4, our trail becomes more contorted. More and more good reasons emerge for primates to cooperate, assist, affiliate, and reconcile, as well as compete and exploit. Meanwhile, at least in some places, the old villains, baboons, turn into social managers who shun aggression and follow the golden rule. Chimpanzees and other species now make peace after they make war. Although the controversy continues, scientists cannot reasonably assume that primate society has a basic nature because there is no longer one basic primate society.

WHAT IS THE BASIC NATURE OF MALES, FEMALES, AND THE RELATIONSHIP BETWEEN THE SEXES?

The Zuckerman/Carpenter contrast is, again, a useful starting point. In one view, males are central, controlling, and domineering, while females are peripheral, submissive, and dominated. In the other, both males and females are communal and cooperative and their relationship and societal roles complementary. For quite a while, Zuckerman's position—or a variant—prevailed.

Then, in Stage 3, sociobiology retooled the role of females, making them competitive and aggressive actors, central to the evolutionary story, only to reinsert them as pawns in male games by using mammalian physiology to redefine male and female nature. At the same time, other voices made themselves heard. Social complexity, primate social strategies, and politics gave both males and females new opportunities and, for some species, created new constraints on males. The availability of cognitive tactics helped to level the playing field since smarter, weaker animals could win against stronger but dumber opponents.

Perhaps most important, by Stage 4 the diversity of species, social organizations, and social relationships suggested that regardless of the basic nature of mammalian males and females, relationships between them could be exploitative, cooperative, complementary, or some combination of all three.

WHAT IS THE RELATIONSHIP OF THE GROUP TO THE ENVIRONMENT?

During our sweep of time, primatologists have moved from largely ignoring and oversimplifying the effects of environmental factors to developing an integrated socioecology in which the ecological molds the social and the social modifies the ecological. Whereas Zuckerman gave no consideration to the impact of the abnormal social and physical environment in which he observed hamadryas baboons, Carpenter felt it was important to study primates in a naturalistic setting. According to Richard (1981), Carpenter laid the foundations of primate ecology: a meticulous and almost completely atheoretical description of feeding, ranging, and habitat. This approach virtually isolated primate ecology from relevant advances in general ecology. In Stage 1, primate ecology moved toward a simplistic form of environmental determinism. The assumption of the early field studies of baboons, for example, was that because both baboons and hominids evolved on the East African savannas, baboons would make a good model for humans. This position—that there is one social organization best suited to each habitat, and that the environment can unilaterally determine social systems—carried over into the next stage's attempts to explain the variability of social systems between species. In Stage 2, the enigma of intra- and interspecific variation brought environmental issues to the fore. Variations in local environmental factors such as food supplies, predation levels, and population densities might be the cause of intraspecific differences since baboons, vervets, and langurs behaved differently in different locations. Variations in social groupings *between* species also might be attributed to major ecological features such as open-country or tropical-forest living. Various schemes attempted to categorize all known primate species into a few ecological grades or levels. Thus the solution to the puzzle of variation in primate societies was a typological perspective with little theoretical underpinning.

In the 1970s and 1980s, primate socioecology became much more sophisticated, particularly as it drew on formal ideas from the discipline of ecology. Primatologists recognized that social systems are interrelated with many factors: demographic and life-history processes, phylogenetic constraints, and the presence of other species, as well as abiotic aspects of the environment. Researchers now saw primates as part of ecological communities—actors that both affected and were affected by their environments. Hypotheses from models developed by theoretical ecologists replaced simplified adaptive stories.

In the most recent decade, behavioral ecologists initiated the first integrated studies of social systems and their environments. Washburn originally had envisioned primatology as a holistic discipline—one that would integrate knowledge of the biology, evolutionary history, social behavior, and ecological context of its study subjects (Steklis 1993). Until recently, however, studies with proximate perspectives remained separate from those focused on ultimate causes of behavior. Stage 4 research is beginning to realize the potential of a combined perspective and the need for multiple standpoints: biological, behavioral, ecological, and phylogenetic.

Conclusions

Having traced the changes in ideas about primates and the impact of these changes on images of primate society through the several stages of primate field research, we ran into unexpected difficulties. Although we shared a vision of which ideas were important, we disagreed about the factors that had been critical. This disparity motivated us to develop a more comprehensive framework for understanding the process of change (see below; and chapter 22, *Science Encounters,* and chapter 23, *Gender Encounters,* this volume). The contrast in interpretations was baffling. Only now—after writing the history, participating in the workshop, and engaging in the postworkshop discussions—do we have a better sense of why we could not agree. The reasons fit nicely with the discussion that developed in Brazil. First, we needed to understand ourselves as individuals with a particular situated history. This unique involvement with issues and processes of science creates a standpoint that, in turn, fashions opinions about what matters. Although we both descended from the same scientific lineage (probably the reason we agreed on ideas), our research lives were nonetheless dramatically different. The result was distinct standpoints and contrasting interpretations.

Shirley began her field research in 1972 keenly interested in how information about nonhuman primates could be used to create better reconstructions of the life of early hominids. Until then, baboons were the preferred model because they were one of the few primates (besides humans) to adapt successfully to life on the savanna. Their social organization and sexually dimorphic anatomy seemed to make perfect evolutionary sense. Survival on the savanna depended on the protection and political leadership offered by adult males through their aggressive abilities and the male dominance hierarchy. Females had another role: They were the socializers of the young. This powerful model already had pervaded both scientific and popular accounts of human evolution. Yet there was reason to be skeptical. By the early 1970s, data from other primate studies, particularly the long-term work on Japanese macaques in Japan and on rhesus monkeys on Cayo Santiago Island, suggested that females, not males, were the stable core of primate groups. New evidence for a diversity of social organizations and sex roles also had emerged. Was the baboon model correct about male and female roles, and were baboons a legitimate template for interpretations of humans? This was the basic question of Shirley's first field study. Male baboons had been well studied; not so females. What would happen to the picture of baboon society if improved methods were applied to the investigation of the behavior of both males *and* females? Shirley even intended to trap out all the males at the end of the study, but that wasn't necessary since, by then, the behavioral data had provided compelling answers.

The first ten years of baboon research placed Shirley in the middle of a controversy about both male and female roles and about the existence and importance of a male dominance hierarchy. Ironically, although initially interested in both males and females, her research increasingly focused on the enigmatic males. Yet each new shift in the interpretation of male behavior also reconfigured females. Shirley's framework (derived from the data and perspective of continuous long-term observation) was embedded in newly discovered "social complexity." From there, her research moves through an array of topics: from social complexity to matters of primate mind to the impact of history and process on socioecology and the individual, with detours through conservation (baboon crop raiding, translocation of primates, and community-based conservation). The intense controversy over the male dominance hierarchy stimulated Shirley to think about "science" (What is science? How do we choose questions and accept answers?) and get involved in the new field of science studies. Science issues emerged in each subsequent research topic (e.g., cognition, historicity, complexity, conservation).

Given Shirley's experience, it is not hard to see why long-term studies seemed to her one of the most important influences on changing ideas about primate society. They fundamentally changed the rules of the game for interpretations of primate behavior and society by radically shifting the frame of reference. Discoveries about social complexity reopened old questions about animal mind; the acceptance of the existence of mind refashioned interpretations of primate tactics and strategies and, through them, evolutionary explanations. What is more, long-term studies and important methodological and theoretical moves in primatology appeared to be intimately connected.

Shirley's research topics suggested as well that the process of interpreting primate society had also to refer to the larger societal context as this seemed to influence what science thought about, for example, male and female roles, aggression, hierarchy, animal abilities, and conservation. In the end, it was the way research was transformed, for example the progression from a focus on gene strategies to social complexity to cognition to conservation, that suggested to Shirley the complex interaction of an array of factors in changing our images of primate society. This included the impact of methods, theory, gender, and culture. From her experience, gender seemed the junior partner.

Analyzing Linda's history demonstrated how different interests yield divergent perspectives. Linda's primate research began at almost the same time as Shirley's, nearly thirty years ago, and developed from a similar anthropological interest in sociality and the roles that males and females play in society. There, however, the similarities end. Whereas Shirley began and continued to work with baboons in Kenya, Linda studied a variety of primate species, all living in multimale, multifemale social systems. She was particularly influenced by the longitudinal, collaborative approach Japanese primatologists had taken to the study of monkeys and has mainly worked with Japanese macaques and neotropical capuchin monkeys. One of the most striking features of Japanese monkeys and capuchins is how prepossessing the males appear as they strut about the social landscape while the smaller females and juveniles forage and rest and socialize. But, returning to the same groups over many years, Linda discovered that few if any of the same males are still in the group, whereas many of the adult females are still present, foraging and resting and socializing. Japanese scientists long ago recognized this phenomenon and were instrumental in our growing awareness of the matrilineal structure of macaque society. As it turns out, not all multimale, multifemale societies exhibit this pattern of male dispersal and female philopatry, but it is quite common in Old World monkeys and New World capuchins. These unassuming female monkeys who remained to carry the plot (while the males entered briefly to play cameo roles) immediately intrigued Linda. Early on, she decided to focus on sociality from the female perspective and address the question of how and why female primates successfully carry out their lives in the social company of males. This orientation led her to studies of sex differences, sexual selection, sexual dimorphism, life histories, and reproductive success, always pursuing the theme of how females grow up, choose mates, produce and rear infants, and grow old and die in the continual social presence of males.

In the same way that Shirley's long history of research with baboons led her to recognize easily the importance of longitudinal studies to changing ideas in science, Linda's many years of focusing on sex differences in primates caused her to be very sensitive to gender issues in science. Scientists, after all, are a social community of primates in which individual males and females compete for success. Linda's participation in a local chapter of an international group called WISE or WISEST

(Women in Scholarship, Engineering, Science and Technology), which encourages more women to become and remain scientists, led her to reflect on whether or not primatology has been a more welcoming discipline for women than other sciences. This connection also led her to consider the role women and the women's movement may have played in changing our ideas about female primates and primate sociality. Indeed, many feminist historians and philosophers of science have singled out primatology as a discipline in which the presence of women has made a major difference. Linda considers the question of whether primatology indeed can be considered a "feminist science" open and researchable. It seems to her quite possible that the atmosphere of intellectual goodwill toward women and toward feminist issues that increasingly prevails in primatology means that shifting sex ratios of scientists, as well as shifting attitudes about gender, have influenced changing ideas of primate society.

Our situated case histories, when put together, suggest complementary viewpoints that have different core concerns. In light of what we know now, rather than what we assumed at the beginning of this project, neither interpretation is wrong, yet each is insufficient on its own.

The historical review we have just presented bolsters the view that multiple factors may have played an important role in our understanding of primates during each stage of research, and in the shifts in orientation between stages. Single factors may have been influential at times but always existed in complex interaction with other factors—minimally, theory, methods, gender of the scientists, and the cultural/historical/social context.

These conclusions provide a useful first step toward our goal, but we wanted something more comprehensive and complete. To meet this challenge, we developed a framework that integrates two approaches. The first is comparative. Comparison with closely related disciplines and other national traditions of primatology could help highlight developments unique to this history of primatological ideas and those that might be explained more broadly. Comparison also should simplify the analysis of factors. Many fields closely related to primatology share some—but not always the same—characteristics. We selected cultural anthropology, comparative psychology, animal behavior, and archeology based on their characteristics—and because, in some cases, they were the parent disciplines. Variation in configurations of theories, methods, and practitioners allow us, as investigators, to hold some factors constant while segregating and highlighting the effects of another. For example, we should learn something about the impact of theory, as opposed to that of women scientists, when we compare two fields that have a large proportion of women investigators but do not share the same theoretical orientation, something about the impact of methods when we compare two fields that use the same theory but employ different methods, something about the influence of practitioners when we compare fields that share theory and methods but diverge in terms of the demographics of their scientists, and so forth. Aligning disciplines according to certain common historical signposts also should be instructive. The composite picture has the potential to offer better clues about the role of specific factors, and possibly even about the interaction between them, than the study of any one field in isolation.

Contrasting national traditions of primatology comprises another useful comparison, one that also illustrates the framework's second approach: setting the history of primatological ideas in its larger cultural, historical, and social context. Japanese primatology already has been compared to the North American tradition (Asquith 1986a, 1991). The provocative Japanese case study seemed an excellent starting point from which to extend the analysis more broadly, including other cultural/national traditions such as those of British, Swiss, Brazilian, Dutch, French, Spanish, and Mexican scientists.

Just as producing a good history requires special training, we felt that understanding the larger context and its effect on ideas about primate society required special expertise and specialized analytic tools. In addition, primate studies' distinctive place in the border zones of many disciplines and many cultural categories raises essential questions about what science is and about the relationship between science and society. To deal with this, we turned to science studies and feminist studies as resources in our examination of the intersection of society and science in primatology. We also wanted to consider the relationship between scientific and popular ideas about primates, recognizing that popular culture and the media, in particular, were an important part of the larger context for primatology.

Throughout our collaboration, our basic assumption has been that the controversies surrounding interpretations about who or what changed our views of primate society can be resolved with facts. First, however, we had to ask the right questions and be guided to the right places to look for the answers—our ultimate goal in creating this history and its companion framework.

Acknowledgments

We thank Pam Asquith, Charis Thompson Cussins, Mary Pavelka, David Western, and Sandra Zohar for many helpful suggestions that improved the manuscript. Sandra Zohar also provided excellent editorial assistance. Linda Fedigan's research is funded by an ongoing operating grant (#A7723) from the Natural Sciences and Engineering Research Council of Canada (NSERCC).

NOTES TO CHAPTER ONE

An earlier (and more extensive) version of this paper titled "Theory, Method, Gender and Culture: What Changed Our View of Primate Society" can be found in Strum, Lindburg, and Hamburg (1999).

1. From here on, we will cease to continually situate our statements, assuming that the reader is fully aware of their very specific ontogeny.

2. Our understanding of primate studies in this stage is much influenced by the summaries of field studies in *Primate Social Behavior,* edited by Southwick (1963), and *Primate Behavior: Field Studies of Monkeys and Apes,* edited by DeVore (1965). The latter was the result of a nine-month-long conference held at the Institute for the Advanced Study of the Behavioral Sciences at Stanford in 1962–63. These volumes describe a number of different primate societies and present synopses of important concepts and a vision of what future research would entail.

SECTION 2

What Do the Pioneers Say? The Advantages of Hindsight

What are the views of those that lived this history? We ask the pioneers.

Primate studies is such a young discipline that very little of its history is yet codified in the literature. Having written our own selected history of ideas about primate society and about what might have caused these ideas to change, we wanted to know whether other scientists would agree. In particular, we were interested in the views of those scientists who have experienced much of this history firsthand—people who were part of primate field studies from its beginnings in the 1950s and 1960s. We hoped they would tell us what we had missed and what we might have misrepresented in the history. And since a number of these scholars were nearing or just past retirement age, we wanted to capture on paper some of their local oral histories of primatology.

We thought of North American and European primate studies as a fairly homogeneous historical tradition. Thus, we invited senior scholars from the United States, Britain, and Continental Europe to act as both informants and analysts. We felt that they would have the expertise to expand and critique the framework of stages/issues/contributory factors that we had developed. They would also be informants on the history of ideas about primate society from the standpoint of their own experience and from the situated perspective of the "schools of thought" in which they were trained or which they themselves established. Jan van Hooff, for example, is the founding father of the "Utrecht school" of primate studies in the Netherlands. He was influenced by classical European and British ethologists such as Niko Tinbergen and Desmond Morris, and he in turn trained many Dutch primatologists who are now influential researchers in both Europe and North America. Across the English Channel and around the same time period, Robert Hinde supervised a long line of students at Cambridge University who are now a prominent part of the current generation of fieldworkers. Although the Madingley Laboratory at Cambridge was originally founded by W. H. Thorpe as an ornithological field station, Hinde's work on primates and that of his many students transformed "Madingley" into an influential school of primate studies. Thelma Rowell worked on rhesus macaques with Robert Hinde at Madingley, went on to carry out many years of field research on baboons and guenons in Uganda and Kenya, and taught at the University of California at Berkeley until her recent retirement. With her experience of both British and American primate studies she could provide a "multiply situated" view of the history of ideas. Similarly, Alison Jolly is an important founding figure in primatology who was trained and influenced by both British (Evelyn Hutchinson, Richard Andrew) and American (John Buettner-Janusch, S. Dillon Ripley) scientists. Many years of experience with lemurs in Madagascar and long-standing interests in primate cognition give her a valuable historical perspective. Robert Sussman also began his career with an interest in lemurs under the influence of John Buettner-Janusch, and has gone on to train a large number of primate ecology students. Unlike the other contributors to this section who were educated primarily as ethologists and zoologists, Sussman's entry point into primate studies was through physical anthropology. We looked to him to provide a view that was different from our own and from that of the others in this section.

Thus, the chapters in section 2 represent the views of five "pioneers" about what they see to be significant factors in the history of ideas about primate society. The chapters were written partly in reaction to our own history paper (presented in section 1), but are also about issues that these senior scientists feel are important to understanding the history and nature of this discipline. The chapter by Thelma Rowell, "A Few Peculiar Primates," begins this section. Rowell argues that our ideas about primates have been overly influenced by the work on a few species: baboons, macaques, and the great apes. She also suggests that from a very early stage, there have been two parallel structures in primate studies. The first is composed of serious studies of primates, by people who actually watch the details of behavior. The second is an extravagant media production that produces a parallel world whose vivid images have probably had the most influence on ideas about primate behavior. Rowell explores the right and wrong uses of the comparative method and the problems posed by the fuzzy edges of primate behavior. She uses her own work on baboons, macaques, and guenons, (and Europe-based primate studies as her reference point), to draw the conclusion that the most studied primates, representing only 5 percent of the order, are "peculiar." Thus, in her view, we risk making serious errors if we generalize from them. Instead, we need to consider whether we should segregate these few peculiar primates from all other primates, and whether there are really consistent differences between primates and other social mammals if we eliminate the peculiar ones. Finally, she asks what would happen to our interpretations if we widened the boundary to include all gregarious long-lived vertebrates capable of mutual recognition?

In her chapter, "The Bad Old Days Of Primatology?" Alison Jolly reacts to some common assumptions about the way primate studies have changed over time. She argues that primatology as written in the early

1960s is often taken as a baseline, and retrospectively characterized as male-biased, nonquantitative, and guilty of underestimating the complexity of primate societies and primate individuality. Upon reflection, Jolly suggests that this is an exaggeration. In her view, there was not male bias but gender unconsciousness. The change in perspective that occurred in primate studies is better described as recognizing the different agendas of the two sexes rather than as a shift toward female viewpoints. Equally important, in her opinion, and perhaps even more clear-cut, is the shift toward respecting animal consciousness and individuality. This has helped both scientists and the public identify with other primates as individuals and as species. Empathy with individuals and with nature as a whole has done much for primatology and for primate conservation. Jolly concludes that rational empathy is ever more needed to bridge nature and human nature.

Robert Sussman, in "Piltdown Man: The Father of American Field Primatology," draws a startling connection between two seemingly unrelated events—the Piltdown forgery in hominid paleontology that occurred in the early part of this century and the lack of field research on primates until after World War II. The fundamental premise of his chapter is that science goes awry when we allow our preconceptions about human nature to determine our theories of behavior and when we fail to adequately test these theories. Sussman holds a strong belief that science should be data-driven. He provides several examples from the field of biological anthropology. In his opinion, here preconceptions concerning human nature have often influenced views of primate behavior, then these behavioral patterns have been misused to reinforce concepts about human nature. He begins with a cautionary tale about the widespread acceptance of the Piltdown forgery. He argues that scientists had a preconception that the earliest humans were distinguished from other primates by their large brains. This led them to reject the early australopithecine finds and accept the fraudulent, large-brained Piltdown fossil. Because of this, they also failed to show much interest in the study of our primate relatives in their natural habitat until the 1950s, when early humans were accepted as morphologically very similar to other primates. Sussman provides two further examples of influential theories based on inadequately tested preconceptions about human nature: "man the hunter" as a model of human evolution, and sociobiological explanations of aggression, territoriality, and male dominance.

"Some Reflections on Primatology at Cambridge and the Science Studies Debate," represents Robert Hinde's unique perspective. He suggests that there are psychological forces that generate distinct scientific schools of thought (e.g., we are attracted to those who see the world as we do) and counteracting forces that work to unify scientists (e.g., a common literature). To reconcile such schools of thought, we need to understand their origins. To that end, Hinde presents a short history of one school that he is considered to have founded—the Madingley Laboratory at Cambridge University. He describes how primatology at Madingley began with a rhesus macaque colony that was established to gain insight into human issues. Later, the recording techniques developed for use with these captive animals were applied to the field projects on African monkeys and apes in which Hinde's many students were involved. He outlines the ethological, ornithological, and ecological influences on the Madingley school and highlights some of the differences between Madingley and other schools (such as Washburn and his students in the United States). Having supervised a large number of the women primatologists who are now prominent in the discipline, Hinde also gives us his view of sex, gender, and female scientists. He concludes that the differences he has outlined need to be seen against the backdrop of commonalties. The challenge is to understand how divergent research programs and agendas are in fact interrelated.

Van Hooff's "Primate Ethology and Socioecology in the Netherlands" gives us a view of primatology in one of the nations of Continental Europe. We see some of the interplay between Dutch, German, Swiss, and British ethologists, psychologists, and zoologists that form the early context of primate studies in Europe. Van Hooff begins by describing an important controversy in Dutch psychology during the 1930s and '40s between the subjective vitalists and objective mechanists and the early influence of Niko Tinbergen (who advocated a strictly objective approach). He describes his training in Holland and Britain and his subsequent return to Utrecht University where he began to supervise a series of students that were to become important theoreticians and fieldworkers in primatology. Tracing the history of primate research at Utrecht through some of the models of social behavior that were tested (e.g., reconciliation, game theory, "social homeostasis"), he introduces us to the key players in Dutch primate studies. Van Hooff concludes that in the Netherlands, the history of ideas about primate society have advanced not in linear progression, but by "meandering forward" with many oscillations between mentalistic and mechanistic approaches. He also touches on the countervailing forces of global science and national traditions for scientists in small countries, a subject that will be pursued in more detail in the next section.

These papers and our discussions demonstrated the extent to which our initial assumption that North American, British, and Continental European primate studies have a fairly homogeneous history is incorrect.

Euroamerican primatology has had multiple origins with diverse configurations of influences and routes. Certainly, there are some intriguing connections in the patchwork, for example the influence of Tinbergen and other founding ethologists on both Hinde and van Hooff, the role of Buettner-Janusch in drawing both Jolly and Sussman to study lemurs, and the many research interests held in common across schools and generations. Also important is the continuing if intermittent dispersal of primatologists to countries and institutions beyond their natal groups. However, what stands out most are the quite different entry points into primate studies that affect the multiple trajectories of ideas about primate society. We were surprised that we would perceive the history(ies) of our field so differently. The other participants were startled as well, reflected in oft-repeated versions of the phrase, "but surely we all think that. . . ." The e-mail exchanges presented after the chapters in this section demonstrate our continuing attempts to come to a common understanding of what primate studies are about and why we study primate society in the first place.

2

A Few Peculiar Primates

Thelma Rowell

Introduction

It is notoriously difficult to define the order Primates. The formal definition requires a cluster of characters, none of which are unique to the order, and there is a straggle of species that have at various times been counted in or out of it.

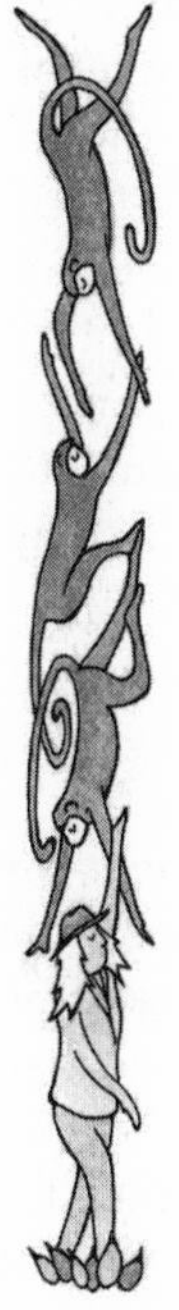

Primates in the narrowest definition is still an extraordinarily diverse order by any measure, including size, locomotion, diet, reproductive physiology, and, of course, behavior. This diversity is expressed by a rather small number of species, which makes it a convenient miniature model to represent zoological theories and methods to new students.

To a zoologist, then (and this essay is written to give a zoological perspective), the obvious thrust of "primatology" would be a study of diversification. There are, however, few people who might be described as primatologists in that they study the range of primates, and most students of primate behavior are particularly limited, confining their investigation to a single genus.

Even in the face of such specialization, I find it difficult to believe, as was suggested at the start of this conference, that anyone who was seriously interested in primate behavior ever believed that there was a single order-wide pattern of social organization. Extreme research specialization makes fertile ground for invalid generalizations, however, to which primatology seems especially prone.

This is an essay about boundaries. We need to consider whether the taxonomic unit, the order, provides the most valid, or even convenient, unit of study of social behavior. Are there more interesting or useful boundaries to be cast instead, either wider or more narrowly? The question is important if we are to use the comparative method that has been so successful in zoology. With inappropriate boundaries, we shall waste time looking for special reasons for what are in fact general characters. If, for example, we were to ponder why primates, in particular, have tails, without attending to the fact that nearly all vertebrates have post-anal extensions of the vertebral column, we would be using too narrow a boundary for our deliberations. On the other hand, too wide a boundary can lead to over-generalizing and missing important differences within that boundary: some primates, like a few other vertebrates, do not have tails as adults, and this might possibly make an important cluster to study. Note that that cluster "tail-less vertebrates" is not taxonomically defined.

Ecologists have long recognized that boundaries are the most interesting and productive parts of ecosystems, and the same is true of academic disciplines. It is important that useful boundaries are recognized so that productive comparisons can be made across them, and valid generalizations made about what they enclose: recognizing boundaries is not limiting.

Primatology has resulted from crossing disciplinary boundaries, but confusion can arise when implicit assumptions are not identified in the process. As people more and more distant from the sweaty and often boring coalface of direct observation and experiment have joined in the primatological discussion, a parallel, fantasy "primatology" has developed. It is important to maintain a boundary between what we have seen and counted and the vivid but edited images of popular "primatology."

Finally, our primate subjects also recognize boundaries, both physical and social. These are not necessarily the ones that seem obvious to us, and their variety and significance is a promising, though difficult, area for study.

Generalizations and Comparisons

Because of the diversity of the order, it is difficult to generalize about any aspect of primate biology, especially when the object is comparison, as in "all primates show this, unlike all other mammals." Nonetheless, generalization has been popular. One of the first to attempt it was Zuckerman (1932, 1933). He was writing before there was any serious study of primates in the wild, and he was an anatomist and physiologist interested in the menstrual cycle. Like all of us he was happy to generalize outside his own specialty. He wrongly concluded that the menstrual cycle is a defining primate characteristic. Having seen baboons in South Africa (as well as people—about both of which he was right) and the London Zoo Collection, he also concluded that no primates breed seasonally. His interest in behavior was, at most, an expression of his interest in physiology, but of course his generalizations had important implications for the study of social behavior.

Thirty years after his books on primate behavior, in 1962, Zuckerman chaired a symposium at the London Zoo on primate behavior in the wild and in captivity. His chairman's concluding remarks were breathtaking in their arrogance when I heard them, and make amusing reading still: in summary, he said that no one had found out anything interesting in the last thirty years and fieldwork was all anecdotes anyway, not science. This last opinion allowed him to discount the mounting evidence even then available from tropical field naturalists, of seasonal breeding by a variety of primates.

Zuckerman was, perhaps, an early example of people on the fringe of a subject picking up fragments that suit ideas they have developed from rigorous studies elsewhere. This phenomenon was penetratingly discussed by Lehrman (1974), in an essay that is still highly relevant today. He was talking about ethology, then developing as a hybrid discipline from parental zoology and psychology. He addressed the results of psychologists taking isolated observations of animal behavior that seemed analogous to their questions about human behavior, without appreciating or understanding the whole biology of the animal. There is an analogy, I think, with the multidisciplinary origins of primatology. Lehrman showed that the problem lies with the right and wrong ways to use the comparative method. As an example of the wrong way, he used the comparison of captive rhesus macaques and human mothers. Rhesus monkeys kept in small cages are typically extremely protective of their infants, restraining them and guarding them from contact with others. This observation was used as a political tool in the 1950s by those who thought that it proved that it was natural for women to remain in continuous contact with their infants. But suppose, asked Lehrman, that the data on primate maternal behavior available in the 1950s had been from langurs, or even bonnet macaques, both of which share their infants and leave them with other group members very soon after birth?

The comparative method is the standard approach of all nonexperimental science, and that includes nearly all evolutionary studies. To be effective it requires the widest possible range of observation, of as many aspects of as many subjects or species as possible. Thus to compare one aspect of social behavior (maternal) in two not very closely related species (rhesus and human) in quite dissimilar habitats, is unlikely to prove at all fruitful and may well be downright misleading.

In that 1962 symposium chaired by Zuckerman, the first paper was presented by K. R. L. Hall. His subject was variation in the ecology of the chacma baboon. At the end of it he declared his continuing interest in "group change and stability, and in the effects of social learning and recognition involved in the development of individuals within the group" (Hall 1963, 26). Thus, he intended to use the comparative method, both by looking at different groups of the same species, and by looking at the same individual or group as it changed over time. He and his student Gartlan began further work on this theme on vervets and patas monkeys in Uganda shortly afterwards, and Gartlan and Brain (1968) attempted to compare two types of guenons (vervets and samangas), trying to cover every aspect of their natural history. The ultimate goal, I think, was to find correlations between changes and differences among groups and their circumstances, and so to suggest causal mechanisms. These are surely entirely modern concerns, already voiced in the early 1960s.

It was clear, in the early 1960s, that Hall's picture of baboon social organization did not correspond to that of DeVore; it was in fact much more consistent with recent findings (Hall 1962, 1963, 1966). Tragically, he died suddenly, leaving unfinished a joint manuscript with DeVore which DeVore finished and published. Hall and DeVore (1965) was much quoted and anthologized, especially in the United States. It gave a misleading impression that the results from East and South Africa were essentially the same, the account given being based primarily on DeVore's study in Kenya. I do not believe Hall would have put his name on the paper in that form had he lived. His primary interest in analyzing the causes of variation in baboon behavior using the comparative method was lost, and baboons were homogenized for years to come.

An observer can always say "I saw a baboon do this." The statement "baboons do this" requires observations of several baboons, and the more observations and the wider the circumstances in which this was seen, the stronger the generalization. "Baboons do this only in the dry season" is a comparative statement, and requires equivalent observation in dry and wet seasons. "Only baboons do this" requires equivalent amounts of observation of as many other species as possible to be credible.

Similarly, any generalization "primates do this" requires observation of a wide range of primates before it is credible, especially important in this order because of its diversity. Zuckerman's sample was far too small to support his generalization. "Only primates do this" requires equivalent amounts of observation of comparable nonprimates, and this is lacking from nearly all statements of this form or intent that are made.

Popular Primates

DeVore made films of baboon behavior in Nairobi National Park, which have been widely available ever since; there can be few students in the United States who have not seen them in class in the last thirty years. The commentary points out very clearly the central position of the dominant adult male. If you turn off the sound, the students are more likely to spot the peanuts being thrown during filming. The center in this case was defined by the trajectory of the peanuts, which were mostly intercepted by the adult males. There is included a beautifully drawn cartoon of the social organization of the clustered walking or foraging troop, centered round the alpha male surrounded by mothers and babies. This cartoon was visually compelling, and made a deep impression. It was also wrong. I hope it is finally dead, but it took years to kill, as observer after observer reported troop formations with adult males at front and rear of linear progressions (Rhine and Westlund 1981). These vivid visual images stuck at the back of the mind of all the journalists and people in other disciplines who took Anthro 1 as a breadth requirement when they were in college. This is the way baboons, all baboons, were "known" to behave.

Thus was crystallized, in the United States, a picture that was already outdated by the time it was first published. Of course it was attractive if you had an appropriate axe to grind. Washburn and DeVore (1961) also made some interesting suggestions about the parallels between the way of life of modern baboons and early man, which were enthusiastically received and widely disseminated to general audiences. An influential American journalist of the time, Robert Ardrey (1966), wrote a fanciful account of human evolution from ancestors whose male dominance and aggression could be inferred from baboons. Ardrey came to see me in Uganda after I had been working with baboons for a few years and had data that did not support that picture at all (this must have been about 1965). I explained it all to him at length—but of course he did not in any way alter the story that he was selling so successfully. Nor was he at all interested in the variation in baboon behavior that Hall had pointed out. Ardrey had his own, right-wing political agenda, and selected fragments of primate research could be incorporated with effect. This is a good example of the distortion of the comparative method that Lehrman decried.

In 1970, when I first came to the United States, people were using Ardrey's books, and also Desmond Morris's *Naked Ape* (1967), as the basis for seminar discussions in the U.S. I remember being amazed at that, having taken even the latter book as pure entertainment, almost as good value as having lunch with its author. While they might have been good starting points for highly critical discussion based on published research,

I suspect that without very skilled leading, most students would be left with clearer impressions of the wit of Morris and the turgid prose of Ardrey than the presentation of real observations and experiments.

Further input affecting the popularity of primates came from "Leakey's ape girls," via National Geographic–sponsored films. Fossey, Galdikas, and Goodall were/are excellent and patient observers, and they have each made critically important contributions to our understanding of their respective apes. None of them was trained as a scientist (or overtrained, as Leakey would have said) and none, I think, was primarily interested in the rigorous analysis of behavior. Films about their field studies made an enormous impact on the general public and stimulated the emulative fantasies of generations of young women. I have disappointed multitudes over the years by not being Jane Goodall, even though I studied monkeys in Africa, and I have had to turn away droves of students who wanted to become Jane Goodall, since to do it through me would have required calculus and further unromantic effort in laboratory and library.

From a very early stage, then, there have been two parallel structures: serious studies of primates, by people who actually recorded what the primates in front of them did; and an extravagant media production, a sort of parallel world whose vivid images, verbal and photographic, stuck in people's minds. Students in other disciplines picked fragments of both these structures to illustrate and reinforce ideas developed in their own fields—just as we did to theirs. I wonder if any other discipline has had quite such a vociferous band of camp followers? Perhaps nowadays ecology comes close.

Biologists who study primates need to be especially aware of the popular primates, which of course have seeped into our consciousness as well. It is easy to forget that films are cut and edited to illustrate the theories of the producer, so that what you see can be, even if unintentionally, misleading. We need to be constantly monitoring from whence an opinion came to us, and to be rigorous about what we accept or reject or keep an open mind for more evidence. This vigilance is perhaps more customary for social scientists than it is for natural scientists, but a boundary between popular primates and rigorous primatology needs to be maintained.

Starting Points

Monkeys must have been an obvious choice of subject for Japanese scientists. As native wild animals there would be common knowledge to build on, and some habituated groups were already available. Free-ranging monkeys were accessible without great expense at a time when resources were few. In Europe and America, people began by studying captive primates. We forget the expensive inaccessibility of tropical habitats at the end of the Second World War. Of course people were curious as to whether primates behaved the same in the wild and in captivity, but we owe the start of field studies primarily to reliable antimalarial drugs and jet aircraft, and secondarily to a new prosperity, particularly in the U.S., which made it possible for Washburn, (and Leakey) in particular, to prise money out of the system to support long stays in the tropics. It took nineteen hours to fly (in a Britannia turboprop) from London to Uganda in 1962—and that was a dramatic new advance on the three days' trip by flying boat down the Nile, or the two-week journey by ship followed by two days in the train. Expeditions to the tropics were not undertaken lightly.

Studies of the other tropical mammals and birds, fish, insects, and plants, as well as ecology and paleontology were developing just as rapidly and for much the same reasons. Where we went in the tropics was largely determined by old colonial commitments. In Africa there was the stimulus of newly achieved, or soon-to-be, independence, and the development aid that went with it; and the creation of national parks, which also stimulated a lot of natural history research. It was all tremendously exciting because we were breaking new ground.

The book *Primate Ethology,* edited by Morris and published, eventually, in 1967 (but written a couple of years earlier), provides a snapshot of what European primate studies felt like in the early 1960s. The contributors were all zoologists by training and all had studied other animals, mostly birds or fish, and mechanisms of communication and mother-infant interactions were the prominent topics. I was still interested in variability of social organization within species in different habitats and over time—probably influenced by Hall. None of us made a sharp distinction between studies of captive and wild animals and we tried to ask appropriate questions for the circumstances. Captivity provided a sort of experiment in itself, and provided possibilities of social manipulation; we were all brought up as experimentalists. Thus, at this time, there was no "primatology"—no boundary between studies of primates and nonprimates. Rather, what had been happening was an extension of ethological method and theory to primates. Ethology had been developed especially with studies of birds, fish, and insects, although since there was already much work on a variety of mammals, mainly published in German, it was no great step to extend ethology to include primates.

I saw no evidence, in the early 1960s, of the idea of a "primate pattern"—on the contrary, inter- and intraspecific diversity were accepted and studied from the beginning, even though the information available was still limited. Zuckerman's massive overgeneralization was already decisively rejected.

It has often been said that naive early students thought in terms of group selection, or species selection, and therefore missed the point. I do not believe that was generally true, although I concede that it is possible to find some damningly sloppy writing here and there. We, zoologists at least, were taught as students, and should have been remembering, that selection must act on the individual. The advent of sociobiology certainly laid out the full implications of that, and was invaluable in tightening up the logic of all studies of behavior. Before the sociobiological revolution, we did not believe that our studies of behavior could reveal anything about the process of evolution at all. I still doubt it. On the other hand there was, from the turn of the century, interest in evolution of behavior on the taxonomic scale—the result, not the process, of evolution (Whitman, quoted in Lorenz 1950). Indeed, that was one of the main interests in early ethology, and it is exemplified in Morris (1967b) in the essays by van Hooff and by Moynihan. At that time we were asking the humbler question "How does the system work?" rather than the perhaps ultimately unanswerable "Why does it work?"

While there have been distinguished American primatologists whose training was in biology, it is my impression that the interest in primates in the United States has been primarily driven by an interest in human evolution: jobs for primatologists are in anthropology departments. Using monkeys as "little furry people" to throw light on human origins is unlikely to be the best way to find out about nonhuman primates—it is an unsatisfactory distortion of the comparative method (Rowell 1999). The distortion can be well seen in anthropological paleontology, where the search for a linear evolutionary progress to Man (through a series of "missing links") seemed for years to blind students to the expectation, by then obvious to biologists, that there would be adaptive radiation at all stages of the history of the Hominoidea. A result of being housed in anthropology has been use of the comparative method exclusively within the order Primates, without attention to nonprimate outgroups and analogues which could be illuminating.

Biologists, anthropologists, and psychologists can talk about primates together without realizing that ideas they take for granted, about evolution for example, are not necessarily held in common. Cross-disciplinary interchanges are stimulating, but they can also be dangerously misleading if not undertaken with great care.

As a biologist, I believe that the isolation of the study of primates from the study of other social mammals and birds, the invention of "primatology," has done a disservice to our understanding of the social behavior of animals. This is a boundary, founded in academic politics, we should be better off without.

Peculiar Primates

While primate diversity, in social behavior as in every other aspect of biology, was already recognized in the earliest stages of modern studies by zoologists (although it was much longer before medical researchers stopped referring to "the monkey" as their research organism), research was already heavily concentrated on chimpanzees, and the papionines. There were and are a lot of practical reasons for that. Macaques and baboons seem to be tough survivors in often primitive captive conditions, so they were available in zoos and from dealers when studies began. Most baboon populations live in relatively open country, accessible by vehicle and found pleasant by people. So why not start there? The great apes, as our nearest living relatives, merited exceptional effort. To some extent this is a self-perpetuating bias: the first information generates new questions and it makes sense to try and answer them using the same or similar species, and so the subject develops its own restricting boundaries.

Most primates, of course, live in dense vegetation and often high in the forest canopy. Fewer people are attracted to that sort of place, and it is almost impossible to make a continuous record of behavior in these environments because animals are only intermittently in view. Our knowledge of the majority of primate species therefore lags far behind our knowledge of the baboons and macaques in both quantity and quality. This means in practice that generalizations about primate behavior tend to be based on about five percent of an order that is noted for its diversity.

There is another reason to prefer to study baboons, macaques, and chimpanzees: they are more fun. Most animals spend the majority of their time doing nothing, but there is usually something worth noting going on in a baboon or macaque group. This is only partly because the groups are fairly large, and visible in an open place. The same is true of elk herds, where the rate of overt social interaction is very low. For example, I estimate the frequency of grooming bouts in a group of baboons to be an order of magnitude more than that in a blue monkey group, and that is many times higher again than the grooming frequency in a capped langur group—taking into account the number of available partners.

Qualitatively, the difference is enormous. The part of a Kenyan forest where I study blue and red-tail monkeys is occasionally invaded by baboon groups looking for figs. We hear them coming half a mile away, my guenons and I. They are always squabbling among themselves, with screams and threat barks and grunts, and the calls of temporarily lost infants. My monkeys make quiet alarm growls and freeze, since baboons eat monkeys as well as figs, if they can get them. They respond similarly to large groups of schoolchildren that are occasionally brought to the forest, and the noise level produced by the baboons and children is roughly similar. Which brings us to the final reason for concentrating on these peculiar primates: *Homo sapiens* is clearly another one of them.

How can baboons afford to devote so much time, energy, and concentration to interactions with other members of their group? I imagine the sheep I studied allowing themselves to be so preoccupied with intragroup squabbles and their resolution, and the thought experiment produces instant mutton. They would be wiped out by predators taking them by surprise. The same would surely happen to most other mammals.

To be distracted is very dangerous: the sport of hawking is based on the ability of hawks to use ground predators (dogs and people) that distract prey and so allow a successful surprise attack from the air. Each spring, suburban cats make a killing from distracted house sparrows at their courtship meetings. The Kenyan forest guenons are territorial, and females daily defend their boundaries against neighboring groups of females. These encounters are noisy, energetic affairs typically lasting fifteen minutes or so. Sometimes, the eagles which are the main predator of forest monkeys take advantage of the distraction of an encounter to make an attack. Territoriality, therefore, is a distraction that has its costs for the monkeys.

It seems that the guenons' territorial boundary is also the boundary between "us" and "them," and the relationships among "us" are hardly in dispute. It is therefore quiet and peaceful, not to say a little boring, to follow a forest guenon group. Not much different, in fact, from following a group of sheep, or elk.

The difference in style between the baboons, macaques, and chimpanzees, on the one hand, and nearly all other primates, on the other, is far greater, I suggest, than any difference between primates and other social mammals. The first hypothesis of a zoologist is that these animals must be either unusually successful at evading predators, or for some reason be under very little threat from them. That might permit the luxury of making alliances and counteralliances within groups and of putting so much effort and attention into squabbling with other members of the group.

Many of the studies that have been most revealing about intragroup competitive interactions have been of groups in corrals or on islands, places where we might look for relaxation of normal antipredator precautions, but the same characteristic social hyperactivity is also apparent in the wild. Antipredator behavior must be very resistant to decay through lack of use. Coss and Goldthwaite (1995) have demonstrated persistence over geological timescales in ground squirrels, and Goodman (1994) has suggested similar persistence of response to raptors in lemurs after extinction of possible predators. This small subset of primates, then, presents its own questions, both in captivity and in the wild, and is open to particular methods of observation and analysis not appropriate to more cryptic and less demonstrative species. There is a practical boundary, then, between them and other primates.

Organizational Boundaries

At first, when we went out to watch monkeys, we thought of social groups as closed breeding units, that is as "groups" in the sense of population genetics. (Certainly I was slow to recognize evidence of the movement of males between groups of baboons; it came as something of a revelation long after my own observations should have been compelling, because I wasn't expecting it). It was natural, then, as well as practical, to see boundaries around groups in field behavior studies (in captive studies, the cage bounded the group). These were social boundaries, since the best studied baboons and macaques have large, overlapping home ranges and do not defend territorial boundaries. Many other species do defend territories, however, and as we have seen, territorial boundaries as defended by guenon females are highly interactive zones—indeed there may be more overt communication between than within groups for long periods. Most of the agonistic behavior we see occurs during these encounters, and it is directed outside the group. A high proportion of grooming bouts occurs in the same context, but that is all within the two opposed groups.

Recently, I have been suspecting that primates may perceive themselves as part of a much wider population beyond the group in which they live—much as human extended families see themselves as part of communities. Baboons and mangabey groups usually avoid their neighbors, but they listen to and recognize the long-distance calls of their males, and may observe them at a distance. Kummer (1968) was the first to address the question of nesting levels of affiliation, as seen in hamadryas baboons, so this is not a new interest. Among the guenons, the females' territorial interactions with adjacent groups provide them with experience of five or six times as many other females and juveniles as they interact with within their group. In expanding populations, adjacent groups may be sister groups that divided within living memory. Males provide longer distance links with their loud calls, and longer term social and genetic links by moving across many groups. I am curious to know whether there are social boundaries within continuous populations that enclose several territorial groups and describe the limits of the wanderings of adults males, so that the population is composed of a series of breeding units, or demes, each divided into several territorial groups. Or, is the deme defined geographically, by the extent of the habitat patch, with adult males wandering as far as they physically can? The relative importance of the boundaries between different levels of grouping is unlikely to be the same for all species. The practical problems of studying them, especially in forest habitats, may be overwhelming. This is moving towards Imanishi's concept of "the specia" (Asquith, pers. comm.) from the other side, as it were.

In concentrating effort on the entertaining primates, we have disregarded the larger population structure and concentrated on the pattern of interaction within groups, and, since 1975 and the sociobiological revolution, on the supposed enhanced fitness of those who successfully compete aggressively for status within the group.

I say supposed, because we are talking of animals which may be breeding for twenty years, but which are unlikely to produce as many as ten (nonfractionable) offspring each during that time, and it is sometimes a bit difficult to retain faith that success in this or that skirmish is going to make a detectable difference in the long run.

I say supposed in the light of the published genealogies of Japanese macaque groups (Watanabe, Mori, and Kawai 1992), in which whole matrilines that were successful at one time fade and go extinct within a generation or two as conditions change, being replaced in the ascendant by previously insignificant lineages.

I say supposed as paternity data confirm what observation had implied—that monkeys that win fights are often, but not necessarily, those that sire offspring, or even the most offspring (de Ruiter et al. 1993).

The whole attempt to explain intragroup interaction in terms of competitive reproductive advantage may have been a distraction from the original question (about how the system works), which remains unanswered. In the first place, it generates untestable hypotheses because the numbers of animals and the duration of research projects do not generate large enough numbers to override the statistical noise in the system. Secondly, it has set many people off chasing unlikely hypothetical genes for highly complex behavior, in disregard of the interactive processes of development. Thus, we see distraction, by the ultimately unanswerable question of why a system might work, from the original, humbler question of how it works.

Where Are the Boundaries?

The first issue that needs to be addressed, then, is why a handful of primates, the squabbling species we might call them, behave in an apparently dangerous way quite unlike other mammals. The key might not lie with the well-studied papionines and chimpanzees themselves, but with the nonconforming species closely related to them: quiet macaques like *M. silenus,* some mangabeys, and gorillas come to mind. On the other hand, there are also a few exceptionally rumbustious species of other taxa. *Presbytis entellus* stands out among the generally exquisitely boring Asian colobines, for example, and *Cercopithecus aethiops* among the guenons. This is a boundary that is not taxonomically defined, but which I think could profitably be recognized and investigated.

Putting the entertaining squabbling species to one side, we are left with a majority of primate species whose social behavior does not seem at first sight to be outside the range shown by other mammals, either quantitatively or qualitatively. It seems to me that there is a great deal written about why primates are different—generally cleverer, especially socially—from other mammals. Most of it is apparently wishful thinking, since there are no comparable studies of the social organization of most social nonprimates. A few, like elephants, hyenas, and some whales, have been accepted as "honorary primates" as far as social sophistication goes.

The problem is largely one of method, in which I include attitude. Until recently I would have said that the same ethological methods were applied to primates as to other animals. Then I watched sheep in the same way I have been watching monkeys, and tried to publish the results (Rowell and Rowell 1993). And, since the sheep were in the title, the manuscripts apparently went to sheep experts to review. They were generally appalled by what they saw as anthropomorphy, and had difficulty understanding why I might be interested in "our" sorts of questions about social organization. Sheep behavior studies are mostly to do with what they eat, and sheep are not, generally, permitted to organize themselves. So now I have to admit that we have, over time, developed our own approach—perhaps we even have a separate discipline. The origin of this divergence is in our expectations: we expect social sophistication in our relatives, so we ask more sophisticated questions, and get appropriately sophisticated answers.

The second question that needs to be addressed, then, is whether there really are consistent differences between primates and other social mammals, or whether, as I suspect, they intergrade in their social sophistication. If there are differences, what are they and how large are they? To investigate this requires applying the primatological approach to other social mammals. That is what has been done, I believe, in the case of the elephants, the hyenas, and the whales already. To tackle these questions, the anthropocentric attitude that assumes that primates must be the cleverest—because they are after all our relations, aren't they?—must be discarded. It seems that there is a political boundary between behavior studies of primates and other social mammals, and this boundary is detrimental to both sides and needs to be eliminated.

Not that there is any reason to confine ourselves to mammals—crows, parrots, and babblers come to mind as obviously serious contenders in social sophistication. Opening the taxonomic boundary leads us to address the next question: What in fact is social sophistication, and how could we measure it so that comparisons might be made? It allows us to avoid using a "more like us" criterion, whether explicitly or not. In any case, that criterion would primarily reinforce the boundary between the squabbling species and the rest. What if we widen the boundary of our interest to include, say, all gregarious long-lived vertebrates capable of mutual recognition? With a wide basis for comparison, we should be less likely to make spurious correlations.

3

The Bad Old Days of Primatology?

Alison Jolly

Sexual intercourse began
In nineteen sixty-three
(Which was rather late for me)—
Between the end of the *Chatterly* ban
And the Beatles' first LP.

Philip Larkin, "Annus Mirabilis"

Introduction

Primatology has been often depicted as a field where (1) women predominate and have had a major influence, (2) the direction of research in social behavior has shifted from an emphasis on males to an emphasis on females, and (3) there has been progressive increase in respect for and empathy with our sister-creatures, making anthropomorphism (sorry, I mean gynomorphism) a permitted approach to scientific understanding.

Far too often the three have been conflated. It should be hardly necessary to point out the contributions of male primatologists to female-centered studies, or toward respect for other primates' minds. I think of Hans Kummer, Robert Hinde, Richard Wrangham, Donald Griffin, Richard Byrne, and Andrew Whiten, among many others. Therefore I shall not single out the contributions of male or female scientists as such. Neither shall I retell the shift to awareness of female strategies, which owes so much to people represented in our workshop and this volume, including Sarah Hrdy, Thelma Rowell, Linda Fedigan, Pamela Asquith, Karen Strier, and Shirley Strum.

What I do want to discuss is my own viewpoint on each of these two themes from a limited perspective in time: the early sixties, when the first wave of postwar field studies were done. This is a frankly personal account. Others who were also working in those early days will have different impressions that will enlarge and correct my own. Those who have written later, revisionist texts may have even read this period more closely than those of us who rely on memory.

From our present perspective those were the bad old days of male-dominated views. It was also a period when ecology was an abstruse academic discipline, before the environment became an emotional and political commitment, and before the popularization of the tele-primate. In a sense the early 1960s is the period that gives the baseline. It does at least for me, since I was then just beginning professional life.

However, writing out my remembered impressions and checking through the published conferences I find the fanged male primatology of legend in only a few exemplars within a wide spectrum of views. I conclude that the shift toward female-centered primatology is arguably much more a shift in *consciousness* of gender than a shift in actual *content,* concomitant with the rise of female consciousness in many other spheres.

Perhaps a more profound shift has been toward identification with primates through the scientific acceptance of animal minds, and through the popularization of primatology. This fundamental change of viewpoint is also now politically crucial. Our perception of primates is deeply bound to the perception of nature, especially of nature as something to be respected and preserved in part because there are sentient beings out there which are "so much like us." The changing ethos of identification with primates as sentient individuals is bound up with the fortunes of both conservation and the movement for animal welfare. These in turn affect the future of our discipline, as well as the lives and survival of wild primates.

Primatology in the Sixties

PRECURSORS

When I opted to study lemurs in 1960 there were few books on primate social behavior. The four which made most impression on me were *The*

Great Apes (Yerkes and Yerkes 1929), *A Field Study of the Behavior and Social Relations of Howling Monkeys* (Carpenter 1964a), *The Mentality of Apes* (Köhler 1927) , and *The Social Life of Monkeys and Apes* (Zuckerman 1932).

Yerkes is now pilloried as the imposer of the IQ test on the US military, for extrapolating the IQ results to keep out "inferior" immigrants, and for promoting eugenics in the name of science. I remember his ape studies, instead, as full of gusto and empathy for chimpanzees, the first two of which were really his pets. He speculated on the special intelligence of Chim the bonobo—and even more peculiar for a psychologist, that a gorilla was smarter than he could actually prove, being too introverted by its species' nature to respond well to formal testing (Yerkes 1927, 1928).

Köhler had a theoretically based mission to show that primates and humans shared leaps of knowledge to new Gestalts. His description of social and emotional life in the Tenerife colony remains a delight. He prefigured the recent emphasis on mirrors as a test of "sense of self" (Gallup et al. 1995). Köhler's chimpanzees not only recognized themselves, but would manipulate bits of broken glass or stare in a puddle of urine to gain a new view of the world outside the window. Although Yerkes and Köhler both attempted to systematize psychological testing, they had no doubts they were dealing with complex minds.

Carpenter's mantled howler study was then the only model of primate fieldwork based on long-term observation of behavior and demography. On first visiting Barro Colorado in 1962 I was amazed, though, that Carpenter had chosen the most phlegmatic of primates as his subjects. Even then it seemed possible that the calmness and lack of competition which were being vaguely attributed to all New World monkeys, in contrast to Zuckerman's (and soon DeVore's) baboons, was actually something about howlers, or even Barro Colorado howlers, as opposed to the cebus, squirrel monkeys, and tamarins on the same island.

When we turn to the "masculine" origins of primatology, then, we do not really mean the prewar studies, but the postwar ethos of Washburn and DeVore. The one real prewar precursor for that view was Sir Solly Zuckerman. Zuckerman started out from *Totem and Taboo:* Freud's urhuman primal horde held together by females' continuous sexual receptivity. It must be admitted that he tested this theory of primate behavior against all the evidence available at the time. It was bad luck that the zoo primate literature then dealt with monkeys and apes transferred to the northern hemisphere, disrupting their normal seasonality, and so confirming Freud's theory (Zuckerman 1933). Jean-Jacques and Arlette Petter (Petter 1962; Petter-Rousseaux 1962) had not yet chronicled the social behavior of Malagasy lemurs, so Zuckerman could safely claim that the seasonally breeding prosimians all lived solitary lives—no primal horde allowed with seasonal estrus. He did a pioneering field study, but calculated that nine days would suffice, punctuated by shooting females for anatomical specimens. No one but Carpenter suspected then that years of observation could still be too short to give all the answers. The baboons trapped for London Zoo's Monkey Hill just happened to be hamadryas—the only species besides gorillas that form male-centered harems, yet again bolstering Freud's view of social primates. It was, however, inexcusable (but good shocking journalism) that the London hamadryas killed about thirty females in "sexual fights" (the females being in all reproductive stages)—at least once continuing so fiercely that keepers only managed to retrieve the female's body twenty-four hours after she died.

LONDON AND NEW YORK, 1962

Zuckerman could not accept the presentations by K. R. L. Hall and a new generation of primate watchers at John Napier's 1962 Symposium of the Zoological Society of London (Hall 1962a, 1962b, 1963). Zuckerman categorically rejected the new findings on sex, seasonality, and society, and on peacefulness within baboon populations. I remember his reference to "so-called meat eating," following a close-up film (a section of DeVore and Washburn's?) of male baboons with blood-smeared muzzles and the intestines of a baby antelope dangling from their teeth. Unlike Yerkes, Köhler, and Carpenter, Zuckerman's was primatology red in tooth, claw, and ideology—but still vegetarian.

At that stage I was a fledgling zoologist from Yale. My mentor and inspiration was Evelyn Hutchinson, who considered ecology to be "the study of the universe," conducted preferably from an intellectual home in Cambridge University. Richard Andrew, my advisor, came fresh from his own Ph.D. at Cambridge under Robert Hinde. John Buettner-Janusch had brought lemurs and lorisoids from Madagascar and Kenya to Yale for studies of their serum proteins and hemoglobins. My final committee member was S. Dillon Ripley, a Yale ornithologist, not yet leading the Smithsonian. Richard Andrew and I took one look at Buettner-Janusch's fabulous array of prosimian species and began work on their behavior. I wrote a thesis exploring the Wood-Jones idea (Wood-Jones 1941) that the arboreal hand was a preadaptation for tool-use and the evolution of intelligence (Bishop 1962; Jolly 1964a, 1964b). My English leanings were further reinforced: the dark-haired man doing sleight-of-hand tricks at the pre-conference party in the British Museum of Natural History was John Napier, the organizer, whose anatomical study of precision and power grips in primates underlay my own work (Napier 1960, 1961).

I realize now that my background stemmed from a zoological tradition in which humans are assumed to be continuous with other mammals. I missed a major thrust within physical anthropology (Cartmill 1990). After the Nazi horrors of World War II, with the revulsion against eugenics, genocide, and racial prejudice, anthropologists found it politically correct to stress the unity of the human species. This came to mean searching for a Rubicon: a single adaptive shift which transformed our lineage into humans—perhaps tool-making, or hunting, or language. Up until Napier's conference in June of '62, the Rubicon was "Man the Tool-Maker" by Kenneth Oakley, fresh from unmasking Piltdown (Oakley 1961).

But at the side of the room at the pre-conference party stood a poster of blurry photographs of chimps fishing for termites (Goodall 1963a). Jane Goodall was modest about putting forward her tool-making apes as a frontal assault on the current definition of "man": she did not say much unless asked (I think it was her first international meeting, too). She presented a different paper the next week, in New York, at a meeting of the New York Academy of Sciences organized by John Buettner-Janusch, where many of the London crowd turned up again like a traveling house party. There Goodall offered a study of a "fashion" at Gombe for building nests in palm trees. It already exemplified her interest in innovative, learned, cultural behavior (Goodall 1962). Then she went off to the *National Geographic* to beg for research funds and a better camera.

At the New York meeting Richard Andrew described primate vocal communication (Andrew 1962), Phyllis Jay (later Dolhinow) spoke on maternal behavior from her langur study in India, Stuart Altmann on the catalogue of behavior patterns (which he already called "sociobiology") of Cayo Santiago rhesus, Jean-Jacques Petter on lemurs in Madagascar, Cynthia Booth on West African *Cercopithecus,* and François Bourlière gave a ringing plea for primate conservation (Altmann 1962; Andrew 1962; Booth 1962; Bourlière 1962; Jay 1962; Petter 1962). Ray Carpenter showed a tantalizing black-and-white film of howlers in treetops, which was why I rushed off to Barro Colorado to meet a real forest. I look in vain for male bias. Basically people reported as much as they could see—it was all so new.

MADAGASCAR, 1963

For me there was then a break while I also went to the field, to study the social behavior of *Lemur catta* and *Propithecus verreauxi.* "Don't you think you should put in an autobiographical note?" suggested Richard Jolly at the time. "After all, the fact that we are engaged must influence your views about sex and grooming and such in your lemurs."

"Oh, no," I said. "That's not *science*!"

NAIROBI AND MONTREAL, 1964

I came back into primatologists' society at Makerere University in Uganda. There were Thelma Rowell and Steve Gartlan and Neil Chalmers; in Nairobi were Louis Leakey, and sometimes Jane Goodall; in Amboseli the Altmanns and Tom Struhsaker. Leakey organized a primate conference in Nairobi, marked mainly by an acrimonious argument between Cynthia Booth and Thelma Rowell over what, precisely, counts as a forest. The importance was that Thelma focused on understanding social differences between baboon populations, particularly those in differing habitats. One way to avoid this thought was to deny that baboons actually could live in different habitats (Rowell 1966, 1972).

Rudi Schenkel spoke on the one-male groups of black and white colobus monkeys, their leaf-eating and their spacing calls within the forest. How odd to take notice of monkeys that were not models of prehuman ancestors; neither baboons nor chimpanzees nor even open-country langurs, rhesus, or vervets (Schenkel and Schenkel-Hulliger 1967). So, in that meeting there was the first paper (that I knew of) on population differences and the first paper since Carpenter with detailed behavior of African monkeys, which were considered interesting in their own right and not for their reference to humanity.

Louis Leakey showed us an early print of the National Geographic film "Miss Goodall and the Wild Chimpanzees." This may have been before the sound track, even—I remember Orson Welles's plummy baritone more by the reactions of later students than from that occasion. We did not then know we were seeing the future.

Meanwhile Stuart and Jeanne Altmann organized quantitative sampling of behavior, which Tom Struhsaker explained to me at length as the way of things to come. Jeanne Altmann's groundbreaking paper on sampling methods with its call to arms for equal observation time for the shyer and more retiring and overlooked group members—i.e. females—was not published until a decade later (Altmann 1974). In 1963–64, most field primatologists hardly thought of sampling as necessary to understand the behavior of animals so obvious as primates. Amassing data, yes, but sampling to reduce observer bias, no. One exception was Neil Chalmers, whose mangabeys kept so well out of sight that he did correct for visibility—the adult males were the ones who dared observe him (Chalmers 1986).

Of course there was a decided bias toward concentrating on adult male monkeys: large, active, aggressive animals with helpful identifying scars. However, most of us thought we were also looking at females. Some, like Jane Goodall and Phyllis Jay and Jeanne Altmann, actually did so. The bias was subtle, not a proclamation of who was important in primate troops.

In December of 1964 Stuart Altmann organized a conference in Montreal on primate social communication. I was too pregnant to reach Montreal from Uganda, so Thelma Rowell kindly read my paper on breeding synchrony in ring-tailed lemurs, followed immediately by her own on female reproductive cycles in baboons and macaques (Jolly 1967; Rowell 1967). When I asked how it went, she said, fine, except that some participants thought she studied all sorts of primates and was obsessed by sex. We might add that both of us talked of the female role in soliciting and even mate choice. What is interesting is my omission of ring-tailed lemur female dominance over males from this paper, and its presentation in my 1966 book as a simple fact, not a political manifesto (Jolly 1966). It simply did not occur to me then that it was more interesting than male dominance, just different.

Hans Kummer presented his landmark study of tripartite relations in hamadryas in Montreal: protected threat was largely a behavior of females manipulating the social situation by "use" of the male harem leader to their own advantage. Neither male nor female bias, but insight and observation. At last Zuckerman's mistake became clear—hamadryas baboons did indeed live in one-male harems, but their society in the wild was a functional one, not the hideous mêlée of the London Zoo captives (Kummer 1967; Kummer and Kurt 1963).

In his wonderful book *In Quest of the Sacred Baboon* (Kummer 1995), Kummer describes his own shift toward sociobiological reasoning. His student Christian Bachmann's experiments on female choice in hamadryas followed that shift (Bachmann and Kummer 1980). However, the transplantation experiments to analyze relative male and female contributions to hamadryas "marriage" were as early as 1966–67 (Kummer, Götz, and Angst 1970). And Kummer's wondering about the "morality" of male possessiveness, as opposed to raw aggression, started even in 1960–61, his first year in Ethiopia (Kummer 1995).

The Montreal meeting launched other initiatives in understanding the mental complexity of primates. Not only did Kummer lay out tripartite relations, but Struhsaker catalogued vervet calls, including the alarm calls later made famous by the studies of Cheney and Seyfarth. In fact, Struhsaker heard five categories, not just three: snake chutter, "Uh!" to minor mammalian predators, "Rraup" to avian predators, threat-alarm-bark to proximity of major predators such as leopards, and chutter-to-observer. Perhaps observers count as a kind of snake? (Struhsaker 1967).

Stuart Altmann was already aware of the importance of Japanese work on primate society and culture (Imanishi and Altmann 1965). He invited

Japanese primatologists to the West. Tsumori (1967) described protocultural behavior in Japanese macaques, especially transmission within matrilines. At Montreal, Yukimaru Sugiyama revealed the existence of infanticide in Hanuman langurs (Sugiyama 1967). So far as I know, neither Phyllis Jay Dolhinow nor Suzanne Ripley disputed Sugiyama's observations—a far cry from the vituperation heaped on the "Langurs of Abu" thirteen years later (Hrdy 1977).

PRIMATE SOCIETIES, PRIMATES, AND CAMBRIDGE, 1965–67

It seems strange, then, that the DeVore photo of the yawning baboon with its dagger canines, and his ubiquitous film, have come to characterize the whole era.

This was what we now think of as the male heyday of primatology. We keep coming back to that baboon film (DeVore and Washburn 1960). So many students and present-day professionals have been influenced that it must stand as a landmark, whether you like it or not. I remember it discussing the three-male central coalition as a variant on tooth-and-claw competition. It also focused on play—those young males clambering onto the Land Rover, including the youngster with one arm, whom I recognized later as an apparently successful adult male in Nairobi Park. The soundtrack said that play established or facilitated later dominance ranking, but it seemed a great way to go about it! What we may stereotype in retrospect as mere focus on male aggression seemed richer on first viewing.

In *Primate Behavior* (DeVore 1965) which was to stand as *the* reference for a decade to come, Hall and DeVore (1965) concluded:

> The main characteristics of baboon social organization, as revealed in the Kenya and southern Africa studies, are derived from a complex dominance pattern among adult males that usually ensures stability and comparative peacefulness within the group, maximum protection for mothers and infants, and the highest probability that offspring will be fathered by the most dominant males. With all the variations so far apparent in groups of differing constitutions, it still remains to discover accurately the kind of relationship among the many adult males of a large group of 80 or more animals. The nature of the social structure of adult females, and its periodic variations, also remains to be worked out over a longer period than has so far been available.

You can see this as a glass half empty or half full: they clearly don't think the female structure will be so interesting, but they do think it needs work. Of course, what people write is necessarily more cautious than what they say.

A nine-month collaborative project at the Stanford Institute for Advanced Studies during 1962–63 laid the foundations for *Primate Behavior.* Besides the two baboon chapters, there were reports by Southwick, Beg, and Siddiqi (Indian rhesus); Koford (Cayo Santiago rhesus); Simonds (bonnet macaques); Jay (Hanuman langurs); Carpenter (mantled howlers); Petter (lemurs), Schaller (mountain gorillas); Reynolds and Reynolds (Budongo chimpanzees); and Goodall (Gombe chimpanzees); and the survey of seasonal breeding by Lancaster and Lee which finally put paid to Zuckerman and Freud. Washburn and DeVore's summary stressed the importance of social grouping, variability, adaptability, and learning—not a word about male dominance.

I read the book on returning from Africa to Cambridge, along with the wildly exciting English translations of the Japanese journal *Primates.* It was true that Japanese monkeys seemed remarkably oriental and formal with their hierarchies of central and peripheral males, and their young taking dependent rank from the mothers' status. However, much later research on Arashiyama West confirmed that central and peripheral males exist even when Westerners watch Japanese macaques. And of course, dependent rank is now clear in most primates—though not in the same sense as in lemurs (Fedigan and Asquith 1991; Pereira 1995).

At that time Robert Hinde and his group were studying captive rhesus groups at the Madingley Sub-Department of Animal Behavior. They focused on mother-infant behavior, and then on elaborating the ideas of John Bowlby about the devastating effect of short-term separation of a child from its mother (Hinde 1983). Hinde had a practical agenda—to help Bowlby improve the treatment of small children in hospitals. These studies of mother and baby rhesus were, perhaps, somewhat inhumane because their goal was to show that similar treatment of humans was inhumane (much less so, of course, than Harlow's earlier work, since Hinde was aware of what he was doing). They were hardly male-centered in any case! Neither did they treat primates as objects removed from human concerns: empathy was part of their rationale.

Somewhat later Hinde took the extraordinary step of admitting Jane Goodall to do her Ph.D., and then Dian Fossey, because the importance of their work far outweighed the niggardly standards set for academic admission. To skip much further toward the present, Richard Wrangham wrote his thesis at Madingley, and it was from Cambridge that he published the landmark paper on female-bonded primate groups that crystallized the shift to seeing females as central to most primate societies from generation to generation (Wrangham 1980).

The Cambridge Sub-Department of Physical Anthropology in 1966–67 was as amazing as the Madingley group: David Pilbeam, then a young lecturer, its effective head, Bernard Campbell teaching the human fossil story, David Chivers a graduate student about to leave for his first siamang fieldwork, and Alison Richard and Tim Clutton-Brock surprisingly enterprising undergraduates.

If you look at who held power in the sixties, it was obviously men. The enormously influential patrons I have quoted include Hutchinson, Leakey, Washburn, and Hinde. All are famous for their sponsorship of women. The conference organizers were John (not Prudence) Napier, John (not Vina) Buettner-Janusch, Stuart (not Jeanne) Altmann, though all three wives worked with their husbands as professionals. The Nairobi conference was organized by and with Kenyans themselves. It was still quite male: Thelma Rowell recalls Stuart Altmann and Tom Struhsaker as "a huge black beard and a huge red beard rolling along side by side" (pers. comm). Among the 1960s primatologists I count that I have named twenty-seven men and thirteen women.

Women were far from equal institutionally, but already present. And, as I have said, the view of a primate group as a rather integrated whole led us to study both sexes of primates. The relative blends of aggression or quiet authority, masculinity and femininity, objectivity and empathy in these assorted characters does not seem to me to have changed very much over the years—they continue as themselves from decade to decade. It does not seem to me that we have swung dramatically, as sometimes believed, toward a female-centered science.

Three Changes Since the Sixties

RAISING GENDER CONSCIOUSNESS

I have rather surprised myself in writing this retrospective view. I began it expecting to show that we have "feminized" our field in the course of discovering feminism. Looking at my memories and at the conference volumes, I conclude that we have not. We have, instead, become gendered.

We have indeed drastically changed our view of primates and of primatology (e.g., Fedigan 1982; Haraway 1989; Hrdy 1977, 1981; Smuts 1985, 1992; Strier 1994a; Strum 1987). I think, though, that the change has not been in the balance of attention to males and females, nor even about the importance of female behavior in society. I think instead it is a militant stance. Since Sarah Hrdy wrote of the separate agendas of the two sexes (itself, of course, rising from Trivers and the sociobiology so often maligned as masculine), we have thought of females as often opposed to males, of individuals in uneasy tension between competition and cooperation within their family and their group. This is not particularly "feminine." If anything this is a tough-minded view of female, and male, interests.

Why then do we look back at early periods as so masculine? We have perhaps invented a male bugaboo: a chimera of Zuckerman, DeVore, and the kind of aggressive baboon who is new to the troop and throwing his weight around. Richard Wrangham may soon be assimilated into the monster for daring to write *Demonic Males* (Wrangham and Peterson 1996). In part there were practical contemporary reasons to respect or fear the monster: Zuckerman in England, Washburn and DeVore in the U.S. had enormous institutional and intellectual power, not least within the Wenner-Gren Foundation that sponsored this book. However, I think that we have also oversimplified history for two other reasons. First, it is convenient to sum up previous decades in a simple rubric so we can get on with the present one. But second, a good story needs a good villain. As we now emphasize the conflict of interests of the two sexes, we have created a worthy opponent.

It would be fascinating to take as a central theme the history of ideas of conflict and cooperation in primatology. Owen Lovejoy's (1981) picture of an essential step toward humanity being the separation of male and female feeding zones, and then division of labor, with far-ranging males bringing home provisions to the more sedentary females, could from one point of view be considered a rather helpful move on the males' part. It would inaugurate some sense of males' responsibility for a monogamous family. However that scenario predictably infuriated feminists as yet another attempt to relegate females to hearth and home—in this case, even before we had come down from the trees. Barbara Smuts's (Smuts and Smuts 1993) picture of male baboons and chimpanzees coercing female sexual submission by physical violence has won a good deal of female primatologists' agreement, though I noted a certain unease when Craig Stanford recounted just the same thing during the conference leading to this book. The one stable factor in the shifting views, to me , is that whatever scenario you describe now, you need to be clear what the costs and benefits are to each sex separately, without an assumption of the good of the family, and of course not the good of the group.

Sociobiology's paradox is showing how altruism arises from selfishness, and groups from disparate individuals. Might it just be that gendered primatology's next intellectual twist will be to emphasize how the two genders actually can cooperate to build up a family or a group (de Waal 1996)?

THE ACCEPTANCE OF ANIMAL MIND

The rise of a gendered approach to primates has coincided in time with a second trend: admitting the scientific respectability of animal minds.

When I was in graduate school, opting for a thesis in animal behavior, Evelyn Hutchinson suggested it would be wise to study the works of the learning theorists if only to be able to refute them. I confess that I lasted about ten pages before fleeing to Lorenz and Tinbergen. I never opened a learning theory text again. Constraining behavior on that procrustean bed seemed to me then and now a pathological way to repress one's own capacity for emotion: by denying the necessity of mind in an albino rat!

Again, looking back at the sixties, most primatologists quietly assumed their animals were thinking. Many of us simply sidestepped the warring schools of psychology. As Asquith has explained to the West, Japanese scientists instead were free to be explicitly anthropomorphic, starting from Zen rather than Cartesian traditions. Jane Goodall was the lightning rod. She had to battle to keep the names of individual chimpanzees in her first *Animal Behaviour* monograph (Peterson and Goodall 1993). In retrospect, this seems odd—Irven Devore told us names for his six male baboons in *Primate Societies*. But Jane has remained suspect to many academics, and vice versa. I remember her fuming that an entire afternoon at Madingley was wasted with otherwise apparently intelligent people earnestly discussing why cats purr.

In the end, her approach greatly humanized the discipline. We all now admit that individual personalities are important and fascinating—integral to understanding any complex primate behavior. The academic turning point was Donald Griffin's *The Question of Animal Awareness,* followed by *Animal Thinking*. The second great leap to me is Richard Byrne's and Andrew Whiten's *Machiavellian Intelligence*. The long struggle over the teaching of language to chimpanzees has now arrived at the feats of Kanzi and his family, and as much so, the feats of Sue Savage-Rumbaugh (Byrne and Whiten 1988; Griffin 1976, 1984; Savage-Rumbaugh and Lewin 1994).

We have come a long way from the learning theorists. There are, of course, cautions to the uncritical acceptance of similarity—for instance the work of Tomasello and of Povinelli, and spelling out of the stages toward imitation in *The Thinking Ape* (Byrne 1995). However, it would be almost unthinkable now to return to Cartesian dualism for any reason but a mystical faith in some non-neurological essence of soul. (Incidentally, Descartes himself was not so far from the modern view as we now imagine. He credited animals with emotions and even dreams as part of their animal souls. He saw language as one of the few uniquely human attributes, but even then he foresaw that an automaton might be built which would produce many words appropriate to given situations. Descartes proposed a primordial Turing test—humans alone show flexibility and innovation in language [Descartes 1637, repr. 1993]).

The admission of animal consciousness to scientific respectability has opened the way to empathy as a method of understanding. This is dangerous ground. I take violent exception to Sy Montgomery's argument that Jane Goodall, and to a lesser extent Dian Fossey and Birute Galdikas, have pioneered a new, intuitive, female science which can transcend the need to tabulate data and check one's observations (Montgomery 1991). I doubt if Goodall would have written *The Chimpanzees of Gombe* if she herself thought that empathy substitutes for observation (Goodall 1986). She might have written other books, including the missionizing *Visions of Caliban* (Peterson and Goodall 1993). However, as Pamela Asquith argued as early as her thesis work, anthropomorphism of some general form is fundamental to *studying* primates, and anthropomorphism toward particular animals or incidents is fundamental to *understanding* primates.

In short, the gendering of our attitudes toward primate social behavior seems to me a much less complete transition than the sea change in scientific attitude toward animal individuality and consciousness.

THE PUBLIC IMAGE OF PRIMATES AND THE BRIDGE TO NATURE

I hesitate even to start on this subject, after Donna Haraway's dissection and reassemblage of primatologists in our social context (Haraway 1989). Haraway has profoundly changed our awareness of how our background shapes our views. There is one point, however, that must be made about the wider public.

The huge success of primate films and books has drawn many, many people with us into what might have been a rather esoteric little field. The wave of identification with primates and field primatologists is one peak on the groundswell of the environmental movement. It also is certainly Haraway's "Simian Orientalism"—the search for an Other on which to project our own images. But whatever and whoever is responsible, the results are action in the real world.

Some primatologists object that television, in particular, bowdlerizes our field. Television leaves out almost all the richness of intellectual detail, and all our cautious qualifications. Primates mostly sleep and eat high in the forest canopy. Filmmakers demand sex and violence, in slow motion, in the open, in the golden light before sunset. Furthermore, filmmakers chronically get the humans all wrong. If there is a face on camera it is still a European or an American, the surrogate for the viewer's being "in touch" with unspoiled nature. This scientist is frequently gritting her teeth and thinking, "I have to submit to this interview to get my animals' story told." There are very, very few films that show trackers from local communities informing the scientists what to look for, or hauling them out of the swamps. Even fewer show any colleagues or students from the actual countries—probably underestimating the viewers' own understanding and empathy.

But what television does is tell stories visually. The information in a visual image is far richer and more compelling than the pared-down text. This is particularly true for animals. You cannot describe an alien species to a person who has never seen its form or motion. Television lets you *see* white sifaka dance over the ground in an intertroop chase, *see* a ring-tailed lemur mother return to lick the ear of her juvenile's dead body (Warren 1997a, 1997b). You *see* a chimpanzee discover that the image in the mirror is herself, from her first tentative taps on the glass to full-blown posturing, as she peers to gain a first real view of her own genital swelling (Bass 1995). Television infects others with the wonder and joy that are the privilege of primatologists. The same medium that oversimplifies (and thereby inevitably biases) the story, conveys primate emotions with a complexity beyond mere words.

The effort to save savanna and forest, to improve the welfare of captive creatures, to educate the young of Antananarivo or Dar es Salaam or New York City to care for their environment, links up to public excitement over these wonderful animals. Whatever our feelings about the intrusion of society into science, our science is a part of society.

Today, our identification with other primates, male or female, cute or ugly, strangely clever or uncomprehending, is one of the strongest bridges we can hope to build between nature and human nature. Science and scientists cannot be innocent of the need to keep building that bridge, for the sake of people and all the other primates.

4 Piltdown Man: The Father of American Field Primatology

Robert W. Sussman

Large Brains and Early Concepts of Human Evolution

Preconceptions concerning human nature and evolution have influenced views of primate behavioral patterns, and these behavioral patterns have been misused to reinforce concepts of human nature and evolution. The general acceptance of Piltdown man, and the rejection of australopithecines, as our earliest ancestor was a result of the belief that the earliest humans must have been separated from their nonhuman primate ancestors by the possession of large brains. I believe that this perceived gap between nonhuman and human primates led to the delay of anthropological interest in the study of primates in their natural habitats.

Even though Darwin (1874) and Huxley (1959 [1863]) emphasized the continuity between animals and humans in the processes of evolution and in certain behavioral attributes, they both reiterated the importance of the spiritual and intellectual gap existing between humans and other animals. What separated the biology and behavior of all humans from other animals was the presence of a large brain.

No doubt man, in comparison with most of his allies, has undergone an extraordinary amount of modification, chiefly in consequence of the great development of the brain. (Darwin 1874, 150)

The idea that a large brain was the defining characteristic of humans and the feature that separated the earliest humans from their nearest brutish ancestors was repeated by many scientists at the turn of the century. Sir Arthur Keith, Sir Grafton Elliot Smith, and Sir Arthur Smith Woodward were among the major proponents of this theory (Millar 1972; Reader 1988; Spencer 1990; Tattersall 1995). Smith believed that "the brain led the way" in human evolution, beginning as an adaptation for arboreal existence in nonhuman primates. Erect posture came later, and was "not the real cause of man's emergence from the Simian stage, but . . . one of the factors made use of by the expanding brain as a prop still further to extend its growing dominion" (Smith 1912, 575–98).

Keith (1912) believed that the two existing human fossil types, Neanderthal and those from Java, were not direct ancestors of modern humans but contemporaries and cousins of our large-brained ancestors because "one was brutal in aspect, the other certainly low in intellect." But this being so, it must be accepted that "in the early part of the Pleistocene, within a comparatively short space of time, the human brain developed at an astounding and almost incredible rate." Woodward (1913, 733), speaking of the search for "missing links," stated that "we have looked for a creature with an overgrown brain and an ape-like face." Thus, to many early scientists, the earliest human ancestor was expected to be generally apelike, perhaps quadrupedal, but with a large brain, much larger than extant apes.

In this light, it was not difficult to accept *Eanthropus dawsoni,* the Piltdown man, as the missing link when it was described (Dawson and Woodward 1913; Smith 1913, 1917). Although many were skeptical of the juxtaposition of an apelike jaw with such a modern-looking cranium, the most influential and well respected human biologists of the day were convinced of the relevance of this fossil and its intermediate position between modern humans and their apelike ancestors. After all, the brutish mandible combined with the human brain case fit preconceived notions of how human evolution must have proceeded.

It is not surprising that in the first quarter of the century, neither the finds in Java (Dubois 1898) nor the Taung discovery (Dart 1925) were regarded as important to the study of human evolution, since they did not fit the preconceived notion of the "missing link." They had small brains and therefore were regarded by most students of human evolution as early apes (Dart and Keith 1925; Keith 1931; Keith et al. 1925). They could not have possessed the intellectual capacities necessary for any of our human ancestors.

These beliefs profoundly affected the study of human evolution for nearly fifty years. By the 1930s and 1940s, even after much fossil evidence was accumulating which suggested that the brain had not led the way in hominid evolution, the most respected anthropologists remained convinced of the validity of the Piltdown specimen. For example, Weidenreich stated: "All that has been known of early man since the discovery of the Piltdown fossils proves that man cannot have had an ancestor with a lower jaw of a completely simian character" (Weidenreich 1946, 22–23). To this Keith (1949, 229) replied:

> A leading authority on such problems, Dr. Franz Weidenreich, has recently proposed that the solution is to deny the authenticity of the Piltdown fossil remains. . . . That is one way of getting rid of facts that do not fit into a preconceived theory; the usual way pursued by men of science is, not to get rid of facts, but to frame theory to fit them.

Keith still did not consider australopithecines as early humans because "the presence of small brains in early members of the human lineage made it necessary to define a 'cerebral Rubicon' in brain size: a threshold which had to be exceeded by anything with a claim to being human" (Tattersall 1995, 73). Similarly, Hooton (1947, 288) stated that australopithecines "lacked the brain overgrowth that is specifically human and perhaps should be the ultimate criterion of a direct ancestral relationship to man of a Pliocene precursor." In typical Hootonian fashion he added his own poetic summation:

> Cried an angry she-ape from Transvaal,
> Though old Doctor Broom had the gall
> To christen me Plesi-anthropus, it's easy
> To see I'm not human at all.

Thus, Piltdown man contributed to a consensus theory of human evolution that dominated anthropology and paleontology between the 1920s and 1940s. This theory led to "the neglect of many significant discoveries because they did not conform with accepted beliefs, while others, less accurately founded, were welcomed because they conformed only too well" (Reader 1988, 70). Reader referred to this as the "Piltdown effect."

Early Studies of Primates in Their Natural Habitat

As long as this gap between any human ancestors and their primate relatives was perpetuated, the study of nonhuman primates did not greatly interest scientists involved in the study of human behavior, and anthropologists showed little interest in the study of primates in their natural habitat before the 1950s. Until 1950, only six individuals conducted field studies of any note on nonhuman primate behavior. In the 1890s, R. L. Garner, an American amateur zoologist and animal collector, during an ape collecting expedition in Gabon, spent 112 days in a cage waiting for gorillas and chimpanzees to wander by (Garner 1896). A South African journalist, lawyer, and amateur naturalist, Eugene Marais studied baboons in his backyard in the early 1900s, but his work remained relatively unknown until his unfinished monograph appeared after his death (Marais 1968). In 1930, Sir Solly Zuckerman, a medical doctor and anatomist, spent nine days in the field observing chacma baboons to supplement his study of a captive colony of hamadryas baboons (Zuckerman 1932).

The other three students of primate field behavior, H. Bingham, H. Nissen, and R. Carpenter, were stimulated and sponsored by R. M. Yerkes, the American psychologist, and were all psychologists themselves. In 1929, Yerkes established a great ape breeding facility in Florida as an extension of his Yale Primate Laboratory. Cognizant of the paucity of information on the natural behavior of apes, he sent Bingham to study gorillas in the Congo and, in 1930, Nissen to French Guinea to collect and study chimpanzees (Bingham 1932; Nissen 1931). Bingham spent two weeks and Nissen forty-nine days in the field; neither returned for further studies. In fact, not one of these individuals was interested in dedicating his life to the study of primates in remote habitats and none contributed to field primatology in any permanent way. Carpenter, however, was an exception.

After earning his Ph.D. in psychology, Carpenter was granted a fellowship to work with Yerkes (Haraway 1989). Yerkes and F. M. Chapman, an ornithologist, convinced Carpenter to study monkeys at Chapman's field site at Barro Colorado in Panama. This began the first long-term involvement of a scientist in field primatology, or as Carpenter called it, "the naturalistic behavior of nonhuman primates" (Teleki 1981). From 1931–35, Carpenter spent a number of months studying howlers, and made some observations on spider monkeys, on Barro Colorado Island (Carpenter 1934, 1935).

In 1937, Carpenter was invited to collect behavioral data on the multidisciplinary "Asian Expedition" to study gibbons in Thailand. This expedition also included two physical anthropologists who were to become very important to primatology: Adolph Schultz and Sherwood Washburn. Carpenter spent four months in the field studying white-handed gibbons, which led to his classic monograph on this species (Carpenter 1940). In 1938, Carpenter was instrumental in exporting between 450 and 500 rhesus monkeys from India and releasing them on Cayo Santiago, an island off the coast of Puerto Rico, which began the first long-term research on semicaptive primate populations.

Carpenter continued to make contributions to primatology until his death in 1975, but he appears to have been a lone figure in primate field research during the 1930s. Cultural anthropologist Alfred Kroeber, in a 1928 paper entitled "Sub-human Culture Beginnings," had suggested that anthropologists would gain a better understanding of human culture through the study of nonhuman primates, and the physical anthropologist Hooton called for primate field studies in his two classic books, *Apes, Men and Morons* and *Man's Poor Relations* (Hooton 1937, 1942). Theoretically the time was not right, however, for such studies did not fit the extant paradigm, and neither of these early requests made much of an impression within anthropology. It was not until after World War II that the subdiscipline of field primatology became solidly established. Studies of primates in their natural habitats began again in the 1950s, but field primatology did not really take off until the next decade.

The Uncovery of the Hoax and the Beginning of Field Primatology

Two developments in the 1950s were instrumental to the enormous growth of this field. First was the exchange of ideas, especially concerning human evolution, between population biologists and anthropologists. This is exemplified by the Cold Spring Harbor Symposium of Quantitative Biology in June 1950, attended by 129 of the most influential biologists and anthropologists in the world. The proceedings, published in 1951, called for an integrated science of paleontology, systematics, and population genetics. A number of the participants pointed to the need for a detailed look at primate behavior in interpreting nonhuman and human primate fossils.

By 1951, because of the work of Gregory (1949) and Le Gros Clark (1947, 1948), most scientists began to believe that australopithecines were indeed hominids. However, it was difficult to interpret how these small-brained fossil hominids fit into the human evolutionary scheme with the supposed similar-aged, large-brained Piltdown fossils (Boaz 1982), a subject debated at the time of the Cold Spring Harbor meeting (see papers by Schultz, Simpson, Washburn, Howells, and Mayr). Thus, the second development that stimulated primate field research was the confirmation, in 1953, that the Piltdown man was a fake (Weiner et al. 1953). This clarified the fact that our earliest ancestors were indeed small brained and probably more like nonhuman primates than modern humans in much of their behavior. The idea that there was a major evolutionary and behavioral "gap" between ancestral humans and nonhuman primates was no longer tenable, and the continuity between ourselves and our ancestors was emphasized.

These developments stimulated biologists, and especially biological anthropologists, to begin thinking about the behavior and ecology of early humans and to see living primates as potential windows into the study of the evolution of human behavior. In 1951, Washburn wrote an article entitled "The New Physical Anthropology," in which he stated:

> Recently, evolutionary studies have been revitalized and revolutionized by an infusion of genetics into paleontology and systematics. The change is fundamentally one of point of view, which is made possible by an understanding of the way the genetic constitution of populations changes. . . . Physical anthropology is now undergoing the same sort of change. Population genetics presents the anthropologist with a clearly formulated, experimentally verified, conceptual theme. The application of this theory to the primates is the immediate task of physical anthropology. (p. 298)

In 1953, in the *American Anthropologist,* biologist G. Bartholomew and an anthropologist, J. Birdsell (who, with Washburn, had attended the 1950 symposium), coauthored a paper entitled "Ecology and the Protohominids," in which they developed a method for reconstructing the behavior of early hominids by extrapolating from the behavior of extant primates and other mammals. I believe that these two papers were central to the new interest in the study of primates in their natural habitat.

In the 1950s and 1960s, many conferences and books focused on the relationship between primate behavior and human evolution. In 1953, a symposium was presented at the AAAS annual meetings entitled "The Non-human Primates and Human Evolution," the proceedings of which were dedicated to Hooton (Gavan 1955). In his paper entitled "The Importance of Primate Studies in Anthropology," delivered shortly before his death, Hooton stated: "If these Australopithecinae were men, we shall have to enlarge the zoological scope of anthropology" (Hooton 1955, 2). Other volumes to appear were *The Evolution of Man* (Tax 1960), *Social Life of Early Man* (Washburn 1961), *Ideas on Human Evolution* (Howells 1962),

Classification and Human Evolution (Washburn 1963), and *African Ecology and Human Evolution* (Howell and Bourliere 1963).

One of the prime movers for this interchange was Washburn. Papers by him, stressing the need for primate field research, appeared in each volume, including now-classic papers by Washburn and DeVore (1961; Devore and Washburn 1963) on baboon and early human ecology and social behavior. The emphasis of these volumes is exemplified by the introductory material for *Social Life of Early Man,* which resulted from a Wenner-Gren symposium. On the cover jacket Washburn stated:

> The social relationships that characterize man cannot have appeared for the first time in the modern human species. . . . Since man is a primate who developed from among the Old World simian stock, his social behavior must also have evolved from that of this mammalian group. Thus the investigation of man's behavior is dependent upon what we know of the behavior of monkeys and apes.

These works stimulated students to study primates in their natural habitats. By the early 1960s, a number of conferences on free-ranging primates were held, and related books began to appear. The first two volumes were based on international conferences held in 1962 in New York (Buettner-Janusch 1962) and London (Napier and Barnicot 1963). Other early collections were edited by Southwick (1963), DeVore (1965), Jay (1968), and Altmann (1967), with the latter three based on meetings held between 1962 and 1965. The major figures in primatology contributed to these books; Washburn again was a major catalyst for many of these meetings and his influence on primate field biology cannot be overemphasized. In fact, the first eight dissertations in primatology after 1960, and fifteen of the first nineteen, were stimulated by Washburn (Gilmore 1981). By 1981, Washburn and his students had "probably produced more than half of the present number of anthropological primatologists" (Gilmore 1981, 388). This is still true today, especially if one considers third-generation students.

In sum, I believe it was the change in the paradigm of human evolution that led to a renewed and reoriented interest in primate field studies. In the early 1950s, the discovery that the Piltdown fossils were fraudulent, and the recovery of more early Asian, and especially australopithecine, fossils in the 1940s ended any idea of a major evolutionary hiatus between the earliest humans and nonhuman primates. From this time on, the only way to view the evolution of human behavioral and intellectual attributes was as part of a continual process from apelike ancestors to more and more modern human types. The brain, and thus the behavior, of the earliest members of this continuum, the australopithecines, was undoubtedly more apelike than it was like modern, culture-dependent humans—a factor that still seems to be difficult for some students of human evolution to understand.

However, it was, I believe, the acceptance of this fact that led to the first anthropologically oriented studies of primates in the early 1950s, which ultimately resulted in the modern subdiscipline of field primatology in the United States. In light of these events, biological anthropologists began to see living primates as potential windows into the study of the evolution of human behavior. However, just as preconceptions of human biology made the study of primates uninteresting to early students of human evolution, preconceptions, also related to our view of human evolution and the biological basis of human behavior, often cloud our interpretation of the data from this area of research. In fact, as we have seen, data are often ignored because they do not fit well-established and widely accepted theory.

> When preconception is so clearly defined, so easily reproduced, so enthusiastically welcomed and so long accommodated, as in the case of Piltdown Man, science reveals a disturbing predisposition towards belief before investigation (Reader 1988, 78).

In the remainder of this paper, I explore some of the ways in which this trend towards belief before adequate investigation tends to lead us to ignore data within the field of primatology. In attempting to understand human behavior and evolution, the windows that nonhuman primates provide often become mirrors that merely reflect what we believe ourselves to be.

Man the Hunter, or Man the Dancer?

Although Washburn was the father of modern field primatology, he also helped perpetuate one of the early preconceptions of early human behavior. In 1966, Lee and DeVore, students of Washburn, organized a Wenner-Gren symposium, *Man the Hunter.* In the resulting volume, Washburn and Lancaster (1968) wrote the often quoted and reprinted article "The Evolution of Hunting." The hunting theory of early human evolution was supported by presumed evidence of hunting by australopithecines, and early observations of social behavior and "hunting" in baboons and chimpanzees. However, the evidence was meager, and the theory was accepted and became popular largely because it coincided with preconceived ideas about the nature of modern humans and their earliest ancestors.

In fact, the view of human nature promoted by this theory differed little from that of Darwin (1874) and had a distinct nineteenth-century Victorian quality (Fedigan 1986b). Many of the features that defined *men* as hunters again actually separated the earliest humans from their primate relatives. Washburn and Lancaster did not amass a large amount of evidence to support their theory. Rather, they relied upon a nineteenth-century concept of cultural "survivals" developed by E. B. Tylor (1871); behaviors that are no longer useful in society but that persist and are pervasive are survivals from a time when they were adaptive.

> Men enjoy hunting and killing, and these activities are continued in sports even when they are no longer economically necessary. If a behavior is important to the survival of a species (as hunting was for man throughout most of human history), then it must be both easily learned and pleasurable. . . . (Washburn and Lancaster 1968, 299)

Using a similar logic, I have developed an alternative, but no less feasible theory—"Man the Dancer." After all, men *and women* love to dance: it is a behavior found in all cultures and has less obvious function in most cultures than does hunting. In the theory of "Man the Hunter," those behaviors and aspects of human nature assumed to be inherited from our ancient ancestors reflected the ideals and values of modern Western society—such as dominant males bringing home the bacon to subordinate females within pair-bonded family groups. "Man the Dancer" would require no such culturally bound restrictions.

Although it takes two to tango, a variety of forms of social systems could have developed from various forms of dance: square dancing, line dancing, etc. It is likely that the footsteps at Laetoli represent not two individuals going out to hunt, but the *afarensis* shuffle, one of the earliest dances. In the movie *2001,* it was wrong to depict the first tool as a weapon when it could as easily have been a drumstick, and the first battle may not have involved killing at all but merely a battle of the bands. Other things such as face-to-face sex, cooperation, language and singing, bipedalism (it's difficult to dance on all fours), and even moving out of the trees and onto the ground might all be better explained by our propensity to dance than by our desire to hunt. In fact, at least two earlier publications, *The Dancing Chimpanzee* (Williams 1967) and "Did Australopithecines Sing?" (Livingstone 1973), tried to enlighten us to this possibility, but alas to no avail.

On a more serious note, although the hunting hypothesis has often been modified, repeated, and resurrected, data gathered by archeologists, primatologists, and ethnographers have not supported early theories of "Man the Hunter." Fedigan (1986b) gives an excellent review of some of the problems with these models. Newer hunting models based on baboon or chimpanzee analogies often "stack the deck"; that is, they choose to include only those behaviors that fit the model (see Kinzey 1987). There is no particular reason why early hominids should behave like any specific living primate. It is only by looking at evolutionary, ecological, and behavioral patterns in specific environmental situations among nonhuman primates, and other relevant animals, that we can understand patterns of behavior and adaptation, and from these possibly develop meaningful models of early hominid behavior. Tooby and DeVore (1987) discuss the value of "conceptual models" over "referential models" in developing theories of human behavior and evolution.

Sociobiology or So-so Biology?

The next widely acclaimed attempt to explain early hominid behavior and human nature using nonhuman primate models was that of E. O. Wilson in *Sociobiology: The New Synthesis* (1975). Wilson, an entomologist, cites over 500 papers in his chapters on insects. His citations on nonhuman primates and on humans are far less numerous, often secondary sources, and often biased in their selection. Why would a biologist, exceptional in his own field, consider it unnecessary to do adequate research on nonhuman primate and human behavior before he writes about humans, and think himself qualified to make broad generalizations about human nature? I believe it is because Western "civilized" humans often think of themselves as experts on the subject. After all, they are human as are their friends and neighbors—personal experience will suffice. As stated earlier, the window of primate behavior often becomes a mirror reflecting one's own image.

Wilson (1975, 551), in his final chapter entitled "Man: From Sociobiology to Sociology," theorizes that characteristics that are biologically conservative among primates are likely to be a part of human nature.

> Characters are considered conservative if they remain relatively constant at the level of the taxonomic family or throughout the order Primates, and they are the ones most likely to have persisted in relatively unaltered form into the evolution of *Homo*.

Wilson comes up with a list of human traits that are assumed by many sociobiologists to be constant throughout the order Primates and/or species-typical human behaviors. These traits include: territoriality, aggressive dominance hierarchies, permanent male-female bonds, male dominance over females, and extended maternal care leading to matrilineality. Just how common are these traits among humans and nonhuman primates? Again preconceptions and stacking the data deck appear to be the order of the day.

TERRITORIALITY

The concept of "territory" was first developed in studies of birds. The essence of the concept is that an animal or group of animals "defends" all or part of its range. Thus there are two major components: space and the active defense of that space. Many animals maintain exclusive areas by vocalizing, displaying, or in some way signaling to possible intruders, and very rarely, if ever, by actually fighting at borders (Waser and Wiley 1980). The concept is not in any way simple, and there are real difficulties in relating various spacing methods used by different animals to the strict concept of territoriality (see Lawes and Henzi 1995). In any case, group spacing mechanisms are extremely variable. Groups of gibbons and the titi monkey could be considered territorial in that they actually have ritualized battles at the borders of their almost exclusive ranges. A number of other primates have specialized loud calls that presumably help maintain exclusive areas (e.g., *Colobus, Indri,* howler monkeys, orangutan males). Some species have overlapping ranges in some localities and exclusive ranges in others (e.g., Richard 1978; Harrison 1983). However, most species of primates have overlapping group ranges and often share resources. This is especially true of many savanna forms such as baboons and chimpanzees. In ring-tailed lemurs and gorillas, several groups may have almost coincident home ranges. Thus in primates, territoriality in the strict sense of the word is rare (Fedigan 1992).

In humans, the concept of territory, as used to define defense of space by birds, is not at all useful. Most hunters and gatherers do not have exclusive, defended ranges and agricultural peoples have a multitude of ways of dealing with land use. Lumping these into a simple concept of territory is nonsensical. Finally, modern warfare often has little to do with directly defending borders. How is a political decision to send troops to Haiti or ethnic "cleansing" in Bosnia similar to a bird or a gibbon displaying at the border of its range?

AGGRESSIVE DOMINANCE HIERARCHIES

Again we are dealing with a complex concept. Dominance hierarchies in animals are defined by a number of criteria, including priority of access to food, space, or mates, grooming direction, leader of group progression, or winner in aggressive encounters. These are often not positively correlated (Bernstein 1981)—i.e., the animal who wins fights does not always lead the group. In fact, defining the group hierarchy by any one of these criteria usually does not help us to understand the complexities of group structure (see also Strier 1994a).

Furthermore, there are many primate species in which dominance hierarchies are unclear, ambiguous, or absent altogether (Walters and Seyfarth 1987). They have not been demonstrated in most prosimians, in many New and Old World arboreal monkeys, in patas monkeys, or in gibbons. They do seem to be present in baboons, macaques, and chimpanzees. However, even among these primates, hierarchies are often unstable, and the genetic influence and consequences of hierarchies are unknown. For example, in baboons, rank changes may occur on average every two weeks among males and every two months among females (Hausfater 1975); and in many studies of baboons and macaques in which paternity is known, little correlation between rank and reproductive success has been found. Generally, the relationship between rank and reproductive success remains obscure (Bercovitch 1991; Rowell 1995).

When we consider humans, the presence of dominance hierarchies based on aggression becomes even more problematic. In ethology an aggressive dominance hierarchy usually is determined by winners and losers of head-to-head aggressive encounters and is normally defined within a closed social group. Is status in human society based on fighting ability or aggressiveness? As humans walk down the street or the halls of their work place, do they display aggressively or are they forced to give way to others as they pass? How many face-to-face fights do normal humans have in their lifetime? What is a person's status (based on aggressive encounters) in his or her social group? And what is a social group within human society? Are these questions meaningful in human societies? I think not.

MALE DOMINANCE OVER FEMALES

Male dominance over females is not a conservative trait in primates. Smuts (1987a) has identified five major types of dyadic dominance relationships between adult male and female nonhuman primates. In three of the five, males are not dominant. These include: (1) species in which sexual dimorphism in body size is slight and in which females are clearly dominant to males (e.g., many lemuriforms); (2) species in which sexual dimorphism is slight and the sexes are codominant (many prosimians, callitrichids, many New World monkeys, and gibbons); and (3) species in which males are larger than females but females sometimes dominate, often through female-female coalitions (squirrel monkeys, talapoins, vervets, many macaques, and possibly patas, Sykes's monkeys, and pygmy chimpanzees). In fact, the only species in which females rarely, if ever, dominate males are those in which males are much larger than females. Sexual dimorphism in humans is slight, and female coalitions are common. Furthermore, ethnographic reviews provide evidence that this simplistic view of male dominance in human societies is false (e.g., Leacock 1986; Mukhopadhyay and Higgins 1988; Quinn 1977).

PERMANENT MALE-FEMALE BONDS OR THE NUCLEAR FAMILY

These are extremely rare among primates, since most primates have promiscuous mating systems. Among the 862 human cultures listed in Murdock's (1967) ethnographic atlas, 16 percent have pair bonds, whereas 83 percent are polygynous (Pasternak 1976).

MATRILINEALITY

Sixteen of the 179 hunting and gathering societies listed in Murdock's atlas are matrilineal. Thus, these evolutionarily and biologically conservative, universal traits are neither conservative nor universal. The best criticism of this approach to modeling human nature from the primate literature is one that Wilson (1975, 551) himself uses to criticize earlier social Darwinists: "Their particular handling of the problem tended to be inefficient and misleading. They selected one plausible hypothesis or another based on a review of a small sample of animal species then advocated the explanation to the limit."

Sociobiology by Any Other Name . . .

It is commonly believed that sociobiology is no longer a major component of primatology, but many of the preconceptions now pervasive in the field are derived from sociobiological premises. Although the current proponents of this approach rarely refer to themselves as sociobiologists, the theoretical underpinnings are the same (e.g., Daly and Wilson 1988; Dunbar 1988; Emlen 1995; Scarr 1993; Tooby and Cosmides 1992). Furthermore, the perceptions of human nature are much the same as they were in the 1950s (e.g., Brown 1991; Degler 1991; Russell 1993; Wrangham and Peterson 1996; Wright 1994).

This approach has reached a wide popular audience, especially through the writings of Robert Wright, a senior editor of *The New Republic,* who has had recent articles in *Time, Newsweek, The New Yorker,* and *Atlantic Monthly* on the new sociobiology. In his book *The Moral Animal, Why We*

Are the Way We Are: The New Science of Evolutionary Psychology, Wright (1994, 6–7) writes:

> Wilson's book drew so much fire, provoked so many charges of malign political intent, so much caricature of sociobiology's substance, that the word became tainted. Most practitioners of the field he defined now prefer to avoid his label. Though bound by allegiance to a compact and coherent set of doctrines, they go by different names: behavioral ecologists, Darwinian anthropologists, evolutionary psychologists, evolutionary psychiatrists. People sometimes ask: What ever happened to sociobiology? The answer is that it went underground, where it has been eating away at the foundations of academic orthodoxy.

Although the basic premises of sociobiology have been criticized extensively, I will briefly discuss, for the remainder of this paper, two widely accepted concepts within this new sociobiology that are based more upon preconceptions about nonhuman and human behavior than upon existing data: infanticide as an evolutionary strategy, and the genetic basis of infidelity among humans.

INFANTICIDE—THROW OUT THE BABY

In 1977, Hrdy developed an elegant theory of sexual selection to explain a handful of cases of infanticide among langurs. She suggested that an infanticidal male gains reproductive advantage by selectively killing the unweaned offspring of his rival males. In addition to the relative gain in genetic representation, the infanticidal act terminates lactational amenorrhea, thereby shortening the interbirth interval. This ensures the earliest possible opportunity for the infanticidal male to mate with and inseminate the infant-deprived female. The advantage, then, being that the female can conceive again sooner—the presumed functional cause for infanticide (Hrdy 1979; Hrdy and Hausfater 1984; van Schaik and Dunbar 1990).

The sexual selection hypothesis has become entrenched as an explanatory hypothesis for infant killing in nonhuman primates and as an explanation for the patterns of child abuse in human society (Daly and Wilson 1988; Emlen 1995; Russell 1993). The theory also has been expanded to include seasonally breeding species, such as the ring-tailed lemur, in which the infanticidal male cannot immediately mate with the dead infant's mother. If a male's infant is the subject of infanticide, he is not likely to be chosen as a mate in subsequent years (he becomes an "incompetent father") (Kappeler 1993; Pereira and Weiss 1991). Thus, females select infanticidal males to father their offspring.

The threat of infanticide has increasingly been viewed as a central factor in theories of primate social evolution, as evidenced by the increasing number of papers that identify infanticidal behavior as a major determinant of primate sociality (Dunbar 1984a; Newton 1988; Pereira and Weiss 1991; van Schaik and Dunbar 1990; van Schaik and Kappeler 1993; Watts 1990). Accordingly:

> van Schaik and Dunbar (1990) juxtaposed several possible male services that would be so valuable to the female that they may select for continuous association between individual males and females. . . .Their preliminary tests supported only the infanticide-prevention hypothesis. . . . This idea can be extended to cover all situations, regardless of whether females are gregarious or solitary. (van Schaik and Kappeler 1993, 250)

This has also been extended to explain the human family. "Longterm pairbonding is found in many human societies, and we ask whether in humans, too, females prefer bonding to a particular male in order to reduce the risk of infanticide on their children" (van Schaik and Dunbar 1990, 55).

Two major problems have been identified with the sexual selection explanation for infanticide (Bartlett et al. 1993; Sussman et al. 1995; but see Hrdy et al. 1995). The first involves the data and the second the theory itself. In an examination of the literature up to 1993, we found that there were only forty-eight cases of infant killing in which the death of the infant was actually observed. These cases occurred in thirteen species of primate, and almost half of the killings (twenty-one) were done by Hanuman langurs. More than half of the langur deaths occurred at one Indian site, Jodhpur (Sommer 1994).

A more serious problem with the data is the fact that the context rarely fits the pattern predicted for sexual selection. The infanticidal male was observed mating with the mother in only eight of the forty-eight cases. In two of these, the male was the most likely father of the infant that he killed! Only six cases involved direct attacks on independent infants, and in an additional three cases a mother-infant pair was the subject of direct repeated attacks. The majority of infant deaths occurred during general aggressive episodes.

Theoretically, the fundamental assumption of the sexual selection hypothesis concerns the genetic basis of infant killing. Although the inheritance of the "infanticidal trait" (Hrdy 1979, 1984a; Newton 1988) is crucial for the operation of the model, there is no evidence supporting its genetic inheritance. Are the sons of infant-killing males more likely to be infant killers themselves because of genes they inherit from their fathers? Furthermore, selection for infant killing has never been demonstrated. What is the increase in relative fitness associated with infanticidal behavior? Selection can be measured by quantifying the covariance between the character and relative fitness in a population that includes infanticidal males (Arnold and Wade 1984; Phillips and Arnold 1989; Schluter 1988).

Selection for infant killing, if it exists at all, is likely to be weak. Some of these infants would have died anyway. Shortening of the interbirth interval due to infant killing needs to be discounted by the underlying death rate so that selection is much weaker than indicated by interbirth interval differences reported in the literature. Furthermore, differences of a few months in the timing of offspring born to infant killers compared to nonkilling males will only have a slight effect on relative lifetime intrinsic rate of increase. Even given a slight increase in fitness for infant-killing males, this increase could well be due to selection on other correlated traits (Lande and Arnold 1983). Individual traits normally are not independent evolutionary entities but are parts of integrated character complexes. Differences in fitness associated with infant killing may actually be due to selection on other functionally related characters, such as overall aggression (Lande and Arnold 1983; Moore 1990).

In rebutting this argument, Hrdy et al. (1995, 154) imply that the critics of sexual selection theory are too interested in the data and the "sacredness of the context." Those supporting it seek general patterns and use theory to explain them; they "derive their greatest pleasure from noting that so many findings could have been correctly predicted on the basis of pitifully incomplete data sets merely by relying on logic, comparisons, and extrapolations guided by evolutionary theory." This is methodologically inadequate. As emphasized by Fischer (1970), in pointing out fallacies in methodology, a statement is not true by simply establishing the possibility of its truth. To date, the data do not support the theory. For the sexual selection theory to be demonstrated, it must be shown that infanticidal males are more successful in leaving offspring than those that do not kill infants, and that this trait is inherited by their offspring. The investigator "must not merely provide good relevant evidence but the best relevant evidence," and "the burden of proof . . . always rests upon its author. Not his critics, not his readers, not his graduate students, not the next generation" (Fischer 1970, 62–63). The sexual selection hypothesis may not be incorrect, but it must be formulated so that it is testable and it remains to be tested, not unilaterally accepted as fact.

IS THERE A GENE FOR INFIDELITY?

A recent cover of *Newsweek* (August 15, 1994) reads: "Infidelity: it may be in our genes." This view of "natural" human behavior rests on two simple predictions of modern evolutionary psychology: that, to spread their genes, the most fit men attempt to sire the maximum number of children and that women should attempt to mate with males who will make the maximum investment in children. We could call these two types of males cavorting males and nurturing males. This view of human society is based on a widely accepted sociobiological tenet. A female mammal, once fertilized, gains little by repeated matings, whereas males are less constrained by their reproductive biology and can increase their reproductive output by continuing to mate with as many females as possible (Dunbar 1988). Furthermore, it is believed that most primates show male-biased dispersal and that males provide little direct paternal care (Bradbury and Vehrencamp 1977; Pusey and Packer 1987; Trivers 1972; Wrangham 1980).

Just as with the infanticide hypothesis, there are problems with the data used to support this theory, and with the theory itself. In a recent article, "Myth of the Typical Primate," Strier (1994a, 233) points out that "data on species that were once considered peripheral to questions about human behavioral evolution are now challenging many long-standing perceptions of comparative behavioral ecology." Among these perceptions is that male-biased dispersal and female philopatry is "typical" in nonhuman primates. In fact, Strier points out that, even when just comparing polygynous and polygamous species, over 50 percent exhibit dispersal by females or by both sexes.

The assumption that the costs of reproduction are extremely high for females and low for males has been challenged by Hrdy (1988) and Tang-Martinez (1997). The data reveal that the costs for males have been greatly underestimated: "Such data have also forced us to reinterpret the behavior of males. . . . by shifting our focus from the production of infants to the survival of infants, we are forced to take into account a whole range of male and female activities that have drastic repercussions on the survival of offspring" (Hrdy 1988, 167).

Theoretically, the hypothesis is contradictory and leads to circular reasoning. If the most fit male (the cavorter) spends his time searching for females with which to mate and increases his fitness even further by killing the infants of his rivals, then he has little time to help any one female nurture her offspring, or protect it from infanticidal males. However, the nurturing male is not fit because he doesn't have time to cavort! And, if he decides to cavort, an infanticidal male may kill his offspring. What type of male is a female to choose (if she has any choice)?

The evolutionary psychologists have an answer. Fit males are infidels, and females choose males that can either fool them (fit cavorting males that at least show a semblance of supporting their offspring), or unfit, nurturing males (who don't cavort). In the latter case, the female's best strategy is to fool the unfit male by cavorting with more fit, cavorting males and cuckolding her nurturing partner. "The theoretical upshot of all this is another evolutionary arms race. As men grow more attuned to the threat of cuckoldry, women should get better at convincing a man that their adoration borders on awe, their fidelity on the saintly" (Wright 1994, 72). Thus, according to the evolutionary psychologists, yes, we are by nature infidels, and are evolving to get better at it all the time!

Allen (1994) lists a number of problems with assuming a significant genetic basis for various social behaviors. These include: (1) poorly defined phenotypes—many of the terms being used, such as dispersal and nurturing, include very different sets of behaviors in different species, and other value-laden terms (e.g., infidelity, cuckoldry, infanticide) are defined by the mores of a given society at a given point in history; (2) reduction of complex processes to a single entity—while mating is a single event mating behavior is not and it is made up of a complex of intellectual, emotional, and behavioral components that cannot be reduced to a single trait as would be expected in standard genetic analysis; (3) uncritical and selective use of information—"stacking the deck," as discussed above; and (4) the tendency in all such research to resort to extreme forms of genetic reductionism, including references to "*the gene* for infidelity," "*the gene* for infanticide," "*the gene* for (fill in behavior of your choice)." This level of simplification rarely applies to relatively simple morphological traits, much less to complex, plastic traits such as social behavior. For example, Greenspan (1995, 76, 78), studying the genetic components of male courtship in fruit flies, notes: "Behavior is regulated by a myriad of interacting genes, each of which handles diverse responsibilities in the body. . . . the genetic influences on behavior will be at least as complicated in people as they are in fruit flies. Hence, the notion of many, multipurpose genes making small contributions is likely to apply." Few if any of those investigators propounding theories of a genetic basis for nonhuman and human social behaviors are actually trained geneticists. As Allen (1994, 13) concludes: "Their naiveté about making genetical analyses and corresponding claims of genetic causality is thus all the more blatant because it would not stand up to any standard genetic scrutiny."

The philosopher of science F. S. C. Northrop (1965) suggests that any healthy scientific discipline goes through three stages during its development: the first involves the analysis of the problem, the second is a descriptive natural history phase, and finally is the stage of postulationally prescribed theory. As Northrop (1965, 37–38) states:

> If one proceeds immediately to the deductively formulated type of scientific theory which is appropriate to the third stage of inquiry, before one has passed through the natural history

type of science with its inductive Baconian method appropriate to the second stage, the result is inevitably immature, half-baked, dogmatic and for the most part worthless theory.

It appears that many of our current preconceptions are clearly defined, enthusiastically welcomed, and easily reproduced, such that we still have that "disturbing predisposition towards belief before investigation." Although the Piltdown man is no longer with us, the Piltdown effect remains. It is time to get out the "Windex" of good scientific method and clean off our window into nonhuman and human social behavior so that it is no longer simply a clouded mirror of our preconceptions.

5

Some Reflections on Primatology at Cambridge and the Science Studies Debate

Robert A. Hinde

Introduction

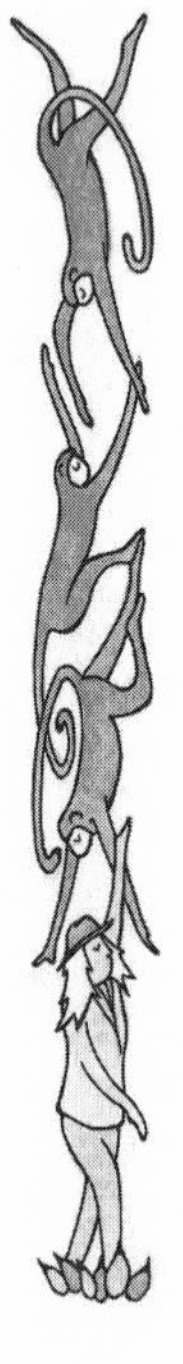

What we perceive and what we record is inevitably influenced by what we are. For instance, our understanding of primate behavior is certainly impoverished by our poor sense of smell. But while our sensory abilities are effectively pancultural, through differences in nature and experience each of us constructs a psychological world that is in some degree idiosyncratic. To say that we "construct a psychological world" is of course a metaphor, implying that the world to which we respond may not be exactly the same as the world "out there," and that each of us has perceptions and associations that may differ from those of others. Fortunately, it is in our nature (for obvious adaptive reasons) to try to achieve a feeling of security and coherence, and to that end we seek a psychological world that is compatible with the views of others. For instance, we attempt to validate our view of the world by comparing it with the views of others. If we find a discrepancy, we are likely to change our own view or to attempt to change others (Backman 1985). A large degree of commonality in the psychological worlds of the members of any given society is therefore to be expected (Hinde 1998a and 1998b).

As scientists, each of us was trained in a particular tradition or traditions—traditions that were themselves nurtured in particular social climates. Close colleagues are therefore likely to share our scientific outlooks. The more our perceptions of the world in general, or of scientific data in particular, coincide with those of our colleagues, the more likely are we to accept them as corresponding with the world "out there." And, because they validate our perspective, we are attracted to those who see the world as we do (Byrne 1971). This is especially the case with views that are not immediately verifiable (Byrne, Nelson, and Reeves 1966), such as basic orienting attitudes to data or scientific hypotheses not yet proven. Thus there are psychological forces that tend to generate distinct scientific schools of thought.

To the extent that such schools grow up around particular individuals in particular cultural climates, they may differ in their orientations and in the ways in which they "explain" (in some sense) the data. Specialists in science studies tend to delight in such differences and, by inducing scientists to examine the differences between their orientations, magnify their importance. But the differences provide no case for postmodern hand-wringing. There are also powerful forces that tend to neutralize their effects. One of these is the scientific literature. The journals and books in which studies are reported can be read by all, and since we all want to validate our perceptions, we either change our own or seek to change those of others when we discover a discrepancy. Another lies in the fact that science, while taking many forms, must at least attempt to build up an organized body of data about repeatable events—and in primatology those repeatable events are out there for all to see.

In no way, therefore, should the existence of different schools be seen as a matter for despair. Rather we should see them in a positive light: their very existence provides us with different perspectives on our goal. We must, however, try to reconcile them, and one route to that end is to understand their sources. To facilitate that process, some information about the backgrounds of primatologists who worked at or from the Sub-Department of Animal Behaviour at Madingley may be of some use. In this contribution, therefore, I am focussing on a microcosm within primatology—though it can be claimed that some influential primatologists came out of the Madingley stable. The causes of that are a separate but interesting issue. It was probably largely due to a historical contingency—the setting up of the Madingley rhesus colony was followed by an opportunity to help in the opening up of primate studies in Africa; and the fieldworkers were able to use recording techniques derived from those developed for the captive groups. I should also add that my own involvement with nonhuman primates took place principally between 1959 and the mid seventies, and that I am well aware of the extent to which the narratives we make about our experiences are subject to retrospective distortion (Murray and Holmes 1996).

The Madingley Laboratory

The Laboratory was originally established by W. H. Thorpe (e.g., 1956/63), an entomologist, as an Ornithological Field Station, and our theoretical background came ultimately from David Lack, the ornithologist, and from Niko Tinbergen, the ethologist. Some of those who studied nonhuman primates at Madingley were primarily zoologists, and saw primates simply as another group of nonhuman species. For instance Thelma Rowell had done her Ph.D. on hamsters, and had had a long-standing interest in sheep; Yvette Spencer-Booth came from bee research; and John Crook did his Ph.D. on weaverbirds at Madingley before turning to primates when he went to work with K. R. L. Hall at Bristol (e.g., Crook 1966), subsequently influencing Robin Dunbar. Others came from first degrees in a variety of other disciplines including psychology, anthropology, and even political science (Dorothy Cheney) and occupational therapy (Dian Fossey).

Primatology started at Madingley for a rather unusual reason. John Bowlby, a London psychoanalyst, had noted that many of the juvenile delinquents that he studied had had disruptions in their early family life (Bowlby 1944), and he was concerned about the restricted nature of the parental visiting hours then prevalent at many hospitals. His evidence was necessarily retrospective and he needed experimental data which, for obvious ethical reasons, could not be obtained from human subjects. Accordingly he helped me to set up a rhesus monkey colony which came to include six captive groups each of one male, three or four breeding females, and their young. Knowing nothing about nonhuman primates, we chose rhesus on grounds of availability. In accordance with our ethological roots, we first carried out a number of descriptive studies (e.g., Hinde, Rowell, and Spencer-Booth 1964). We showed that the behavior of individuals was influenced by their social companions (Rowell and Hinde 1963; with hindsight, it now seems extraordinary that we thought that to be necessary), and we were eventually able to provide evidence that a short period of maternal deprivation could have long-term effects on personality development (Hinde and McGinnis 1977).

The setting up of this colony had unexpected sequelae: Louis Leakey, the paleontologist, believed that clues to human evolution could be obtained from the study of the great apes. Accordingly he arranged for Jane Goodall, and later Dian Fossey, to come to Madingley because, with K. R. L. Hall's untimely death from Virus B, Madingley was the only laboratory studying primate social behavior in the U.K. at the time. Through them came a considerable number of other students and postdocs who worked at Gombe, Karisoke, or elsewhere in East or South Africa, and on Cayo Santiago.

Both the experiments on maternal deprivation, and the field studies of the great apes, involved the use of nonhuman primates to obtain insight into the human condition. But apart from that, anthropological influences were not important. More important were influences from ethology and ornithology.

Ethological Influences

To ethologists it was second nature to distinguish the questions of causation, development, function, and evolution, and at the same time to see them as questions of equal status, all necessary for full understanding, and, although distinct, as interfertile (Huxley 1942; Tinbergen 1963). These distinctions are of fundamental importance for an understanding of animal (and human) behavior, and the neglect of causal and developmental approaches was a major shortcoming of the early sociobiologists. This issue is clearly illustrated with an example: one of the digits on my hand moves differently from the others because (a) the nerves, muscles, etc., are connected differently (causation); (b) the embryonic rudiment grew out differently (development); (c) it makes grasping easier (function); and (d) I am descended from primate ancestors who also had an opposable digit (evolution). Although these questions are distinct, they can also be interfertile —for instance (b) helps one to understand (a); (a) to understand (c); (c) to understand (d); and (d) to understand (a). It would seem that anthropological primatologists did not share this orienting attitude of the ethologists.

It is also a basic principle for ethologists that the description of behavior in the natural environment is a first priority, and most of the early studies by ethological primatologists, like those of the anthropological primatologists, were primarily descriptive in their focus. But beyond that the emphasis was on causation and development. For most, I believe, function came a little later, as the differences between species posed functional questions more clearly: the objective study of function implies differences, differences between species or individuals. Evolutionary questions were inevitably neglected at first in the absence of adequate comparative data. Perhaps because of their focus on causation and development, most ethologists, while recognizing the importance of Wilson's *Sociobiology* as a synthesis, had severe reservations stemming from its neglect of developmental analysis. Perhaps because anthropological primatologists were primarily interested in functional and evolutionary explanations, they accepted the whole of the sociobiological approach more easily. In case I am misunderstood, this is not to underestimate the enormous achievements of Hamilton (1964) or Wilson (1975), nor to deny the importance of functional questions. But one must insist that they are not the only questions, and that answering functional questions is in any case often facilitated by asking also questions about causation, development, and evolution.

With regard to functional questions, biologically trained primatologists distinguish amongst beneficial consequences between those that were and were not likely to have been material for natural selection. Any particular beneficial consequence may be a byproduct of others: one needs evidence that variation in that consequence, in the absence of variation in others, affects or has affected survival or reproduction. That is hard to come by, but only then can one argue that the consequence is one through which natural selection acts or has acted to maintain the behavior in the repertoire of the species.

Another issue with which ethologists (with some German-speaking exceptions) came to terms in the 1950s, and influenced the orientation of ethological primatologists, concerns the sterility of the opposition between instinct and learning (e.g., Lehrman 1953). This view was apparently not shared by anthropologists, who still used (and some still use) a dichotomy between instinct and learning as if it provided a fertile way of conceptionalizing developmental problems. This is a very outmoded approach. First, all behavior depends on both genes and experience. Second, since the early 1970s it has been recognized that animals (and humans) do not learn everything with equal ease—there are constraints on, and predispositions for, learning. This important issue came up early on in ethology (e.g., Baerends 1941), was most clearly demonstrated in the context of avian song-learning (Thorpe 1961), and was widely accepted by psychologists soon after (Hinde and Stevenson-Hinde 1973; Seligman and Hager 1972). It has now been beautifully demonstrated in the work of Mason and Mineka on the responses of rhesus monkeys to snakes (Mineka 1987) and of Cheney and Seyfarth (1990) on vervets.

Ethological influences were apparent in many other aspects of the Madingley work. For instance we felt that, even for captive animals, conditions as near to natural as possible were essential; that so far as possible description should be quantitative; that studies of communication were of special importance; and so on.

Ornithological Influences

Of almost equal importance were inputs from ornithology. By the earlier 1960s, and indeed earlier, ornithology was already much influenced by ethology, and had also made considerable contributions to ecology. It was thus well placed to make a contribution to field studies of nonhuman primates.

Many of the anthropologically trained primatologists attempted to place primate behavior into a linear scheme of human evolution. But no one at all familiar with the ornithological literature could ever have believed in "one pattern" of nonhuman primate social behavior. The socio-ecological approach of Lack (e.g., 1966), developed by Crook in his studies of weaverbirds, was surely an important influence on the theoretical approach to field studies in general and on the problem of social systems in particular, and influenced such workers as Clutton-Brock, Dunbar, and Wrangham. (This, of course, is not to say that any of these workers accepted any of the earlier conclusions, but I am confident that they inherited the problem of the diversity of social systems, and never thought in terms of one basic pattern.)

Indeed ecological studies played an important part in primatology from the mid-1960s onwards. It was during this period that behavioral ecology started to give rise to new perspectives on species differences in social structure, activity patterns, time budgets, fertility, and reproduction. Social behavior is inextricably linked to behavior of other sorts.

In addition, Lack had been pushing the view that natural selection acts on individuals since at least 1954, W. D. Hamilton published in 1964, and in 1966 Lack demolished Wynne Edwards's group selectionist approach. G. C. Williams's critique was published in the same year. Within primatology, the influence of the group selectionist point of view was surely an anthropological phenomenon: no primatologist influenced directly or indirectly by Lack could have subscribed to it after the early 1960s. E. O. Wilson's *Sociobiology* (1975), building primarily on the work of G. C. Williams, may have made anthropological primatologists aware of the importance of individual selection, but others had taken this approach much earlier. (I am aware, of course, that some debate continues, and that there are still some contexts in which group selection may operate, but that is not the issue here.)

Perhaps both of these sources, ornithology and ethology, are encompassed by saying that those with an anthropological bias seem to carve nature so as to separate primates from other animals, while for biologists, nonhuman primates are animal species, to be studied from a biological perspective. This has also meant that those with a biological heritage have not been tempted to interpret primate adaptations in isolation from the nonprimate ecological context: the influence of ecologists such as Charles Elton (1927) and the principle of animal and plant communities were part of their background. (As an anecdote here, I must record how I learned from hearing Richard Wrangham giving a talk on the relations between nonhuman primates and fruits in Gombe in 1968—or was it 1970?)

Social Relationships and Structure

In addition to the issues mentioned so far, there were others that played an important role in the early work at Madingley. As soon as we started to keep rhesus monkeys at Madingley, the importance of the concept of dominance, familiar from ornithological work, became immediately apparent. We accepted that much behavior is ambivalent (Tinbergen 1952), and that aggressive behavior, and its suppression, have been adapted through individual selection to further the interests of the individual, not the group. The bases of the oft-repeated claim that the function of dominance hierarchies is to reduce the amount of aggressive behavior in the group do not bear examination: it involves a confusion of consequence with function (Hinde 1978). We accepted further that, because high rank in a particular context did not guarantee access to particular resources, that does not mean that it may not be influential in obtaining other resources or in other contexts. I cannot believe that anyone ever believed that dominance *determined* mating success, though it is seen as likely to influence it in some cases. In any case, one cannot prove the null hypothesis. Related to that, Madingley primatologists had difficulty with the way in which the concept of "social role" was used by some primatologists (Gartlan and Brain 1968). Role is used in two senses in common speech—to refer to the causes of behavior and to the consequences of behavior. The former usage involves the implication that the behavior is relevant to the functioning of a larger social unit. There is no evidence, as most primatologists came to recognize, either that nonhuman primates in general direct their behavior to furthering the good of the group, or that they have been adapted to do so. Thus role is a descriptive term best not used in the causal analysis of the behavior of nonhuman primates with respect to the group.

As a result of work both in the Madingley colony and in the field, we approached the structure of the group in a different way. It became apparent that it was necessary to distinguish a number of levels of social complexity—short-term interactions, relationships consisting at the behavioral level of a series of interactions between two individuals known to each other, groups, and (in some cases) superordinated categories involving a number of groups. As argued elsewhere (Hinde 1991), each of these levels has properties that are simply not relevant to the level below: for instance a group has a sex ratio but an individual does not. And each level requires explanatory concepts that are not relevant to the level below: thus a group has a two-dimensional spatial structure, and a dyad does not. Furthermore each level has dialectical relations with those on either side so that, for instance, the nature of a dyadic relationship both affects and is affected by its constituent interactions, and affects and is affected by the group in which it is embedded. (I think that, because we were focussing on individuals, we tended to see the importance of individual behavior in determining group structure as much greater than that of group structure in influencing individuals, and that made it difficult at first for us properly to assimilate the early Japanese work: that, however, is a post hoc impression.)

In the human case it is meaningful also to speak of the sociocultural structure—the system of beliefs, values, institutions, and so on shared by more or less all members of the group. But, is that meaningfully applied to nonhuman primates? The members of a monkey group may share some knowledge of the relationships within the group, and of their range, but that is surely rather a different issue.

This model of levels of social complexity enables one to pose some questions a little more precisely. Thus we have evidence that at least some nonhuman primates can respond to an individual in terms of that individual's relationships with others. Do they respond in terms of groups which are more than collections of individuals and their relationships?

In this context, Thelma Rowell (especially) and I became conscious of the importance of social relationships very early on. And in the field the work of both Goodall (1968) and Fossey (1972) was concerned with the idiosyncrasies of individuals, and both laid emphasis on interindividual relationships. Simpson published a detailed study of relationships between chimpanzees, with special reference to grooming, in 1973. Harcourt and Stewart were studying the relationships between gorillas in the mid to late 1970s (e.g., Harcourt 1979b). These are, of course, selected examples, but others could be given.

Members of the Madingley laboratory took part in joint discussions with members of the Psychology Department, where Weiskrantz and others were studying learning in monkeys. And of course we were well aware of Köhler's studies of chimpanzees. Thus we were at least prepared for considerable cognitive complexity—though I think I was personally somewhat surprised by what Jane Goodall revealed to me at Gombe.

Some Other Orientational Issues

Theory (and that term has a wide spectrum of meanings) was I suppose always there in the background, but did not play a major role in the Madingley work. In my own earlier work with Bowlby I had learned a very important lesson. In the 1950s, when Bowlby was still formulating his ideas, he collected together a small group that used to meet regularly in London. It contained a Freudian analyst, a Kleinian analyst, a Hullian, a Skinnerian, a Piagetian, often an antipsychiatrist, myself as an ethologist, and some psychiatric social workers. We had nothing, absolutely nothing, theoretically in common. What we did have was a common problem—the nature of the parent-offspring relationship. (I should say that I had done a very little work on imprinting in birds.) Bowlby was picking what was useful from each of our widely divergent orientations, and constructing his subsequently highly influential attachment theory. What I learned was that it is the *problem* that matters, not the theory. Of course in setting up the rhesus colony the problem was given to us by Bowlby's prior hunch that the parent-child relationship matters, but that was then hardly a theory. And we were guided by our ethological orientations, not by any specific theory.

Furthermore, the problems we tackled seemed to emerge from each other, rather than from any overarching theory. As I have mentioned already, sociobiology, at any rate in its early days, did not greatly influence Madingley primatologists, who were put off by Wilson's early extravagant claims and its neglect of proximate mechanisms and especially of problems of development—though Trivers's (1974) work on parent-infant conflict was based in part on Madingley data.

None of this should be taken to imply that a theory-free science is practicable or possible. Of course we had the overarching theory of evolution through natural selection in the background, but beyond that I think at that time we were guided by hunches rather than theories—and some of us also, perhaps, by a small admixture of anthropomorphism. However we did feel that the close relation between human and nonhuman primates makes a strict discipline in the use of terms even more necessary than in the study of other animals. The use of terms such as "politics" is a case in point. Given that nonhuman primates use deception, manipulation, and so on, does it really help to equate their behavior with that of humans operating within the elaborate structure of, say, a parliamentary democracy? We would hesitate to apply the term "politician" to an individual who manipulates loyalties within a family or small group, so why cloud differences by applying it to primates?

The use of the term "social contract" in reference to nonhuman primates is similarly regrettable: it plays a critical part in some theories of human social behavior (e.g., Holmes and Rempel 1989) which are clearly not applicable to nonhuman primates, and its use in a different context to nonhuman species could easily direct research into a blind alley.

Altruism is another case, but it is probably too late to complain about that.

Female Primatologists and Feminism

An issue important to several of the contributors to this volume is that of the role of women in primatology. What I am sure about is that working with women had a great impact on my own thinking. Carol Berman, Dorothy Cheney, Saroj Datta, Dian Fossey, Jane Goodall, Phyllis Lee, Kathy Rasmussen, Thelma Rowell, Yvette Spencer-Booth, Joan Stevenson-Hinde, Kelly Stewart, and others all taught me to be more sensitive to individual differences and to social relationships than I would otherwise have been. Stevenson-Hinde and Zunz (e.g., 1978) devised tests for individual differences in "personality" in monkeys. That I have learned from female colleagues in this way is not surprising: there is a mass of evidence to show that (on average, of course) women are more expressive and more sensitive to relationships than are men (e.g., Noller 1987), and that femininity (or androgyny) in both partners is conducive to happy marriages, at least in Western cultures (e.g., Acitelli and Young 1996).

But it is this sensitivity that I would emphasize, not a role in pointing out the importance of females in primate troops. That issue has been very much overdone, and seems to have stemmed primarily from the earlier writings of DeVore. At Madingley, by the nature of our initial problem, our work centered on females (as mothers) from the start. Men and women were equally represented among the students, and most theses concerned both sexes. In any case, sex differences were a commonplace in both ethology and ornithology, and I do not believe that the fact that, from the time that Thelma Rowell and I first established the Madingley colony, women were involved in the work influenced our perceptions of the relative contributions made by males and females to social structure.

As an aside, at the meeting on which this volume is based, there was considerable misunderstanding about the relations between the concepts of sex and gender. This suggests that it is worth mentioning the considerable clarification of these issues achieved by personality and social psychologists. In brief, sex is a biological issue, with anatomical correlates. The vast majority of individuals, and all wild-born nonhuman primates, are either male or female. Gender differences, on the other hand, refer to differences in psychosocial orientation that may stem from either biology or culture or both. Reliable instruments are available for assessing gender. These are based on (a) identifying dimensions of behavior, attitudes, etc., in which there are clear *average* differences between biological males and females, and (b) assessing *individuals* on scales of masculinity and femininity derived from these dimensions. A common procedure is to distinguish individuals according to whether they are above or below the median on these two dimensions. This yields four categories: (a) feminine (above the median on femininity and below the median on masculinity); (b) masculine (the reverse); (c) androgynous (above the median on both); and (d) undifferentiated (below on both) (e.g., Bem 1981; Spence and Helmreich 1978). Thus in discussions on this issue it is important to distinguish between the question of whether the sex of the researchers makes any difference, and whether the gender orientation of either male or female researchers does so.

Conclusion

The Madingley enterprise thus had a number of characteristics, not all of which were shared by other workers at the time. Perhaps the most fundamental were that the research workers recognized that Tinbergen's "four whys" were distinct yet interdependent; that "innate" and "learned" could be applied to differences between characteristics, but not to the characteristics themselves; and that primates were just another group of species. With the exception of the chimpanzee work, there was little interest in human evolution. The research workers involved came from a variety of disciplinary backgrounds, but the atmosphere in which they worked had an orientation that stemmed from ethology and ornithology. Although the work initially concerned captive animals, there was soon an opportunity for fruitful interchange between field and the captive colony. Finally, it is important to add that all the individuals involved had their own inclinations and idiosyncrasies. . . .

But the publications of these workers, who are now scattered in a number of research centers in different parts of the world, has become part of primatology. It makes one wonder whether, in view of the forces that draw primatologists nearer to the facts "out there," the differences beloved by science studies specialists really matter for long.

It is relatively easy to document differences in the approaches, backgrounds, theoretical orientations, and so on of those who study nonhuman primates. But it is unbalanced to do that without placing them against a background of the commonalities. If that is done, they may be seen as relatively trivial. Of course the identification of differences in researchers' orientations may help in reconciling theoretical differences or in producing agreement about what is really "out there"—this collection of essays will no doubt help to that end. But many of the differences that do arise come not from differences between backgrounds but simply because the researchers are studying different problems—the nature of social structure and communication, the development of behavior, behavioral ecology, the evolution of species, and so on. What we need to do is to see how those problems, effectively Tinbergen's "four whys," interrelate.

One issue seems to me to be paramount, both from my own experience and from the discussions on which this book is based, and that is that we should avoid phrases like "when primatology achieved independence from its parental disciplines . . ." (Strum and Fedigan 1996, unpubl. ms. circulated before workshop). Our aim must be not to set up a separate empire, but rather to contribute to the unification of knowledge. Primatology's contacts with anthropology, psychology, ethology, behavioral ecology, psychiatry, and paleontology put it into an excellent position to do just that.

6

Primate Ethology and Socioecology in the Netherlands

Jan A.R.A.M. van Hooff

Fascination with primates had its dim beginnings in the Netherlands in the times when Dutch ships began to sail to unknown coasts, and when discoverers brought home reports, if not specimens, live or dead, of the marvels of nature they found (e.g., Tulp 1641; Bontius 1658). Fascination gradually turned into scientific interest, as when the first live orangutan was brought to Holland and accurately portrayed by Vosmaer (1778) in his "Description of the equally rare as peculiar ape-species, named Orang-outang, from the island Borneo" (translated title). As director of the Natural History Collection of the Prince Stadtholder of the Republic of the United Netherlands Provinces, he produced many zoological descriptions. Other primate examples are those of a South American "Quatta" or spider monkey (Vosmaer 1768) and a long-tailed monkey, named "the howler" (Vosmaer 1770). And then there is Petrus Camper, a famous Dutch anatomist, who not only gave the first detailed anatomical descriptions of the orangutan (Camper 1782), but also made a pioneering contribution to the ethology of the human primate. He produced an anatomically based ethography of human facial expression (Camper 1792). It truly was applied ethology *avant la lettre,* for it was meant to help actors in portraying human emotions. However, these early descriptions were general contributions to the *historia naturalis,* and primates were not singled out in a specific way.

We have to turn to more recent times to find studies specifically directed at primates. They are of two kinds. In the Netherlands, as elsewhere, there always have been scientists who primarily work *with* primates as well as those who work *on* primates. The first operate and have operated primarily from a medical-biological perspective. Most of this research has been concentrated for many years in the Biomedical Primate Research Center in Rijswijk in the province of South-Holland. This is one of the major European primate centers and it has a long scientific tradition. The institute harbors well over a thousand primates, the chimpanzees alone numbering more than a hundred individuals. The budget involved in these primate research projects is a multiple of that for work *on* primates. Yet the researchers associated with this primate research center would refer to themselves as immunologists, neurobiologists, virologists rather than as primatologists.

Those who work *on* primates and would prefer to be referred to as primatologists have always formed only a handful of researchers in the Netherlands. They are part of a long, but fragmented, tradition. Their interest was certainly inspired by the similarities of primates to ourselves, and arose first in psychology, and, only later, also in zoology.

Early Reminiscences

To describe characteristic developments in Dutch primatology, I shall follow the development of my own interest and experience, which began in the mid-1950s. I only realized the fragmentation of primatology after I had become immersed in the discipline for some time. This had everything to do with my earliest interest, which was in behavior in general. Having grown up as the son of a zoo director, namely of the Burgers Zoo in Arnhem, I had been in contact with animals all my life, and a fascination for the way different animals behave had come naturally.

In the late fifties, when I studied biology in Utrecht, animal behavior was not taught there as such. One of my professors, Sven Dijkgraaf, was a comparative physiologist and a disciple of Karl von Frisch, the discoverer of the dance language of bees. Dijkgraaf worked in the tradition of von Frisch, exploring the nature and biological significance of sensory functions. He was a pioneer in the study of the labyrinth and the lateral line organs in fish (e.g., Dijkgraaf 1952, 1963), was one of the first to sense the existence of electroperception in fish (Dijkgraaf and Kalmijn 1963), and, perhaps most importantly, he rediscovered the unpublished pioneering work of Spallanzani from the eighteenth century on the mysterious "sense in the dark" of bats (Dijkgraaf 1949, 1960). During the war, when Griffin (1958) in the U.S., with all the technical facilities made available to him, discovered the echolocation of bats, Dijkgraaf, completely independently, in isolated occupied Holland, came to similar, very precise conclusions on the basis of simple but cleverly designed experiments. He conditioned bats in complete darkness to find and discriminate between objects of different form and texture in order to define the domain of their perceptual world (Dijkgraaf 1943, 1946, 1957). Thus Dijkgraaf was interested not only in the mechanisms of sensory perception, but also in the biological significance of these capacities, and, in the "Umwelt" the animal could structure for itself (e.g., Dijkgraaf 1946, when he wrote about the "Sinneswelt," the sensory world, of the bat).

This research stimulated my interest in the animal's view of the world and its behavior, and it brought me in contact with the young and flourishing discipline of ethology, in particular with one its most famous representatives, my compatriot Niko Tinbergen, who had moved to the University of Oxford a few years earlier.

I had become intrigued by Tinbergen's "The Study of Instinct" (1951) and by the objective and conceptual clarity of the ethological models presented therein. Tinbergen emphasized that animal behavior could be studied as a biological phenomenon in an objective way (Tinbergen 1942). Behaviors could be considered as process-organs, of which the causation, the function, the ontogeny, and the phylogeny could be analyzed as for any other biological process or structure. This formed an appealing contrast to other approaches to understanding animal behavior which had their proponents in the Netherlands in those days, e.g., the subjectivistic and phenomenological approaches.

Subjectivistic Vitalists and Objectivistic Mechanicists

In the thirties and forties the Netherlands were the arena of a fundamental controversy which radiated widely outside the national boundaries. A vitalistic and, later, a phenomenological stance were characteristic for the "animal psychologists" of the Amsterdam school, who were prominent in the debate at that time. The pioneer was F. Buytendijk, the first in Europe to found a laboratory for animal psychology, at the Free University in Amsterdam (Roëll 1996). His work ranged from habit formation in unicellular organisms (1919) to the structuring of behavior in a primate (1920) and was strongly theoretically oriented (e.g., Buytendijk 1920, 1928, 1932, 1953). Influential also were his disciples, A. Portielje and, in particular, J. Bierens de Haan. All of them believed the study of animal behavior should have as its ultimate goal the understanding of the animals' subjective experiences and consciousness (Bierens de Haan 1929, 1937a, 1940, 1945; Portielje 1927, 1938). They viewed animal behavior as goal-directed processes steered by purposive instincts. Throughout his career, Bierens de Haan argued that an objectivistic position, such as that taken by von Uexküll in Germany with his "Umweltslehre" and by the behaviorists in the U.S., blocks the way to understanding the essentials of the animal's psyche. Whilst he rejected naive anthropomorphism and acknowledged "Morgan's canon" (1894) as a valuable heuristic principle, he warned at the same time that objectivistic parsimony could easily lead to underestimating the real psychological content and to "interpreting away" ("hinausinterpretieren") the essential psychic factors. Buytendijk and Bierens de Haan also were explicit vitalists; they maintained that the psychic basis of the apparent holistic purposiveness of animal behavior could not be reduced to mechanistic, physiological constituents. The 1929 treatise "Animal Psychology for Biologists" of Bierens de Haan contains three parts: (1) "the psychology of animals as an *independent* branch of zoology" (italics are mine), (2) "the animal as a knowing subject," and (3) "the animal as a feeling and striving subject" (see also Bierens de Haan 1948). This focus is evident in all his work, also in his studies on primates, such as those on tool use and tool manufacturing (1931), on numerical understanding and the structuring of action (1935, 1936), and on the understanding of spatial relationships (1937c). The latter experiments were meant to argue against behavioristic interpretations of habit formation and learning (see Bierens de Haan 1937b; and compare Tolman 1932). Of course he expressed a congeniality with Yerkes (e.g., 1916), Craig (1918), and McDougall (e.g., 1923) in the U.S., and with Köhler (e.g., 1921, 1922) in Germany; they also were scientists who didn't shrink from referring to the purposiveness of animal behavior and the "mentality" of animals.

On the other side there was, in Leiden, an emerging school of "ethologists" with Tinbergen and, later, Baerends and van Iersel as its exponents. Bierens de Haan and Tinbergen figured prominently in the debate about the aims and methods of animal behavior studies. A detailed and monumental account of the debate as it raged in the Netherlands is presented, in Dutch, in the thesis of Roëll (1996) with the translated title, *The World of Instinct: Niko Tinbergen and the Rise of Ethology in the Netherlands*. A concise account in English has recently been given by Burkhardt (1997).

The Tinbergen school advocated a strictly objective approach which sought to explain animal behavior without indulging in what were thought to be unfounded speculations about the subjective world of animals, their consciousness, and their intentions. This view not only regarded the psychological inside of animals as scientifically inaccessible but it also did not attribute any explanatory significance to it. Although the European ethologists and the American behaviorists were scientific opponents in many essential respects, they explicitly agreed on this. Whereas Tinbergen clearly sided with Watson (1913) and his followers by adopting a strictly agnostic position, Lorenz (1963) seemed to take a more moderate position by confirming the existence of subjective experiences and feelings in animals. Yet Lorenz sided with the Tinbergian view when it came to the methodology of explaining behavior and the analysis of its causal fabric. He also explicitly rejected the teleological stance of the animal psychologists (Lorenz 1937, 1942).

This objectivistic perspective dominated the ethological scene for years to come and had an enormous influence. It was also very fruitful in that it generated an enormous number of studies. The Amsterdam school went into oblivion and Bierens de Haan, who in the thirties was *the* exponent of European animal psychology, is now almost forgotten. He did not live to see the writings of Griffin (e.g., 1976) about animal awareness and mental experience, which heralded a new era. In the wake of the first "language" experiments with apes (Gardner and Gardner 1969; Premack 1971, 1975; Fouts 1973; Rumbaugh 1977; Savage-Rumbaugh 1986), primatologists began to develop objective methods to probe "the minds" of monkeys and apes (e.g., Gallup 1970; Gallup et al. 1971; Cheney and Seyfarth 1990; Povinelli 1993). Bierens de Haan would have been astonished to see how the tide has turned and that purposiveness, that is goal-directedness, has been accepted as an essential aspect of animal behavior. He would have been equally astonished to find that this purposiveness is no longer taken as an argument that psychic phenomena must rest on metaphysical principles. The minds of animals have become a subject of objective research, and talking about the mental processes of animals is no longer synonymous with adopting a vitalistic stance. The latter after all was the main reason why the European ethologists, opposing themselves to a "geisteswissenschaftliche" and phenomenological psychology which dominated the continental European scene, were so reluctant to create room for the concept of purposiveness and goal-directedness in their instinct models. And, as Kortlandt (1959) has pointed out, even though the ethologists acknowledged spontaneity of behavior and accepted the notion of appetitive behavior, this didn't mean that they abolished an essentially stimulus-response-chain view of the causation of behavior. In their view "spontaneous" (that is: determined by autonomous internal processes) changes in certain motivations created a specific unrest that led to appetitive behaviors. These brought the animal into situations where the likelihood was increased that it would encounter specific stimuli for which its sensitivity had been increased, and which would thus release the next phases in the behavioral chain and, eventually, the consummatory act. Such a process could be imagined to lead to a certain end state without assuming some internal "consciousness of the goal."

Only after cybernetic principles had become familiar and the vitalistic teleological associations of goal-directedness had faded away did feedback models of goal-steered behavior systems appear (e.g., Wiepkema 1977). Then "purpose" came into perspective as a factor in the regulatory system, namely in the form of an internal set-point or norm value, and was respectable again.

Original, Provocative, and with an Uexküllian Streak

The dichotomy in scientific thinking, in terms of two opposing schools, is, of course, an oversimplified representation of the diversity of opinions. Adriaan Kortlandt, an original and provocative representative of Dutch primatology, has always been eager to point this out (Kortlandt 1959). Kortlandt himself, educated not as a zoologist but as a geographer and a psychologist with roots in the Bierens de Haan tradition, made some fundamental theoretical contributions to ethology. From his inspirer and tutor, the zoologist Portielje (1927), he inherited the typically biological knack for accurate and precise description as well as his research object, the cormorant (Kortlandt 1940a). He studied the "interaction of instincts" (Kortlandt 1940b, 1955) and developed models of the hierarchical organization of behavioral functions which were contemporary with Baerends's 1941 and preceded Tinbergen's 1950 and 1951 exposés on this subject. He also developed the concept of "allochtonous behavior" (Kortlandt 1940c), independently of Tinbergen (1940). And finally he shared with Tinbergen and Baerends, his colleagues and rivals, a vivid enthusiasm for the observational biological approach, with a strong emphasis on field experimentation. He differed from them in accepting a hierarchical system of *goal-directed* instincts as underlying animal behavior. In this respect, he was more in line with McDougall, Craig, and Tolman, and with his compatriots and teachers Bierens de Haan and Portielje. However, as he emphasized in 1959, he did not embrace the vitalistic orientation, adopted by implication by his Dutch colleagues. Instead he situated himself in the cybernetic explanation that emerged in those days.

When Kortlandt gave his inaugural lecture at Amsterdam University, he entitled it "The Cosmology of Animals" (1954). It reflected his congeniality with the approaches taken by von Uexküll and Hediger. These researchers were interested not so much in the causal mechanisms of behavior but rather in exploring the nature of the experiential world of animals, that is in revealing the aspects of the world an animal takes notice of (the "Merkwelt") and which are significant in that the animal deals with them (the "Wirkwelt"). Thus Kortlandt's cosmological psychology or cosmology of animals was a paraphrase of von Uexküll's "Umweltslehre" (von Uexküll 1921). Kortlandt saw Hediger as an exponent of this direction, made explicit in the latter's studies on animals in captivity (1942, 1954). Hediger was interested in the way zoo animals see the structures of the artificial environment they live in and how this compares with the way they see the natural environment to which they have been adapted. A most relevant issue for a zoo researcher is the way animals see the humans they meet in different capacities, e.g., as keepers and visitors, but also, conversely, how humans see the animals they encounter. This interest clearly has a phenomenological streak.

The shift in Kortlandt's interest from "the mechanisms of animal behavior" to the "cosmology of animals" became evident when he moved to chimpanzees in the late fifties. From 1960 onwards he conducted a number of chimpanzee expeditions. The gist of these is rendered nicely in two 1967 papers, one on hand use and the use of instruments and the other on experimentation with chimpanzees in the wild. He explored chimpanzee attitudes towards such diverse environmental features as familiar and strange food items, ornamental and decorative objects, potential prey animals, as well as dead and seemingly-dead animals, dummies and pictures, snakes, spiders, etc. He has become most well-known for the spectacular experiments in which he investigated the use of weapons by chimpanzees in fighting a stuffed exemplar of their major enemy, the leopard.

I had the opportunity to accompany Adriaan Kortlandt on one of his later expeditions in 1974, when he wanted to repeat an earlier experiment and see whether armed defense would indeed be a natural, and therefore, biologically significant reaction towards a live leopard. I admired his stamina and zeal when we transported a tame, almost adult leopard from the Netherlands to the Guinean brush. There an ingenious "leopard launching construction" was built at a site where chimpanzees were expected to come and forage. The leopard was trained to roll-walk in a wire-mesh ball, strong enough to withstand the blows of chimpanzees. The leopard would be launched from a hide and roll to the chimps down a gentle slope. Unfortunately, a heavy brush fire brought a devastating end to the enterprise. The same enterprising imagination was displayed when he addressed the question of how the small-sized early protohominoids might have defended themselves on the savanna against large predators

(Kortlandt 1980). The film footage of the pseudo-protohominoid, constructed out of the window-wiper mechanism of an automobile, waving a thorny acacia branch and thus keeping some lions away from a scared goat, will stick on the retina of whoever has seen it.

These were all elements he used to prove his "dehumanization" hypothesis. It supposed that chimpanzees had progressed on a humanoid evolutionary track, when they moved into the Western African savanna and developed capacities and skills, analogous to those of protohumans when the latter adapted to an existence on the Eastern African savanna. Thus, present-day chimps should still bear the vestiges of this human-like track, after having been pushed back into the forest when early hominids crossed the fluvial barriers to the West (Kortlandt 1972). This position was heralded in a series of short articles in Dutch entitled "Chimpanzees, not humans, but also not animals—but then, what are they?" (Kortlandt 1961). Kortlandt perceived a discrepancy between the capacities revealed in studies of captive, "humanized" chimpanzees (e.g., Hayes and Hayes 1951) and the adaptive use of these same chimpanzee capacities in nature.

The dehumanization hypothesis, although fascinating, has not become convincing. The "almost-human" cultural and cognitive capacities of present-day chimpanzees have been confirmed in recent field and laboratory work (reviews in Wrangham et al. 1994), but at the same time there are more and more indications of their adaptive significance in the wild (van Hooff 1994; Matsuzawa 1996). Still Kortlandt's contributions are of great value. He has emphasized the importance of combining field experimentation with observation, and has criticized his contemporaries for not following this method. In this he is proven right by those few researchers who trod this insight-providing road. Notable examples are Kummer and his collaborators (Kummer et al. 1972; Götz et al. 1978), Cheney and Seyfarth (1990), Matsuzawa (1994), as well as the group of my former collaborator Ronald Noë (1990). Adriaan Kortlandt has always been an individualist with a markedly critical and provocative stance that has sometimes hindered the acceptance of his ideas and their merits, and the formation of a "school."

Primate Ethology in the Tinbergian Tradition: The Utrecht School

In line with its objectivistic and mechanistic orientation and its biological perspective, the emphasis of Dutch ethology was on the precise analysis of the observable input-output relations of behavior processes. As accentuated by Tinbergen (e.g., 1963), this required a precise description of the motor patterns or "organs" of behavior to start with. He strongly voiced the strongly held opinion that such an inductive ethography was often neglected, especially in psychology, because one jumped to problem-oriented analysis too early and easily. Behavior was seen as a structure of effector activities, a system of genetically encoded, fixed action patterns or "Erbkoordinationen" (Lorenz and Tinbergen 1938; Lorenz 1953). This asked for a digitalization of behavior in the form of an ethogram, a repertoire of more or less discrete behavioral elements organized in specific temporal patterns. Descriptions of these elements and their structure had been successfully used in studies on insects (e.g., Baerends 1941), fish (e.g., Baerends and Baerends-van Roon 1951; Wiepkema 1961), and birds (e.g., Tinbergen 1959).

However, there was also the feeling that it might be much more difficult to represent the behavior of mammals in such ethograms. Mammalian behavior was supposed to be organized in a much more flexible manner. The classical study of Schenkel (1947) on the expression movements of wolves, a detailed descriptive analysis, seemed to bring this point home. Lorenz (1951), however, maintained that in spite of this seemingly fluid plasticity in the variation of behavior, an approach analogous to that developed for insects and fish with their more stereotyped patterns would be equally possible and successful, namely by assuming that the variation in behavioral patterns could be understood as a result of the interaction of a few motivational systems. Thus, he reduced the seemingly endless variation in the facial mimics of the dog to varying balances of the underlying motivations of aggression and fear, or, in more empirical terms, of the tendencies to attack and to flee (Lorenz 1953).

For me, the evolutionary aspects of the ethological approach were especially fascinating. If behavior patterns can be studied as morphological structures, then a comparative approach can elucidate the phylogenetic relationships (Baerends 1958). There were the inspiring examples of Lorenz (1941), who reconstructed the evolution of some specialized courtship movements in ducks, and of Schenkel (1956, 1958), who deduced the evolutionary ritualization of courtship displays in phasianid birds. Displays are, of course, the most eminently suited for this, because of the considerable interspecific variation in their fixed action components. Amongst mammalian displays, the facial expressions of primates seemed to offer the best possibility for a similar endeavor.

When I made contact with Niko Tinbergen in 1960, with a study of this kind in mind, he answered that he did not feel very much at home with hairy animals, but more with feathered and scaly ones. But he recommended me to Desmond Morris, one of his first pupils. Morris had just become curator of mammals at the London Zoo, even though his scientific work until that time had been mainly on scaly animals as well (Morris 1970). This gave me the opportunity to do a comparative study of primate facial displays, making use mainly of the enormous collection of monkeys and apes which the London Zoo still had in those days (van Hooff 1962, 1967a). And, of course, Darwin (1890) was my bible, in particular with regard to his conviction that evolutionary continuity encompassed the human species, and that it also encompassed the domain of the mind. It gave me great satisfaction to show that one of the mental characteristics of Man considered to be uniquely human, namely his sense of humor, is associated with one of his most species-characteristic and "instinctive" motor patterns, laughter, and that it is possible to use the comparative ethological method to study homologies of laughter in other primates and trace its evolutionary background. Moreover, the comparative, evolutionary method provided insights on human behavior that could not have been gained otherwise, namely as to the independent origin of the human behaviors "smile" and "laughter" from the primate "grin" and "play face" (van Hooff 1972, 1976, 1989). This typically ethological approach has recently been continued in a methodologically refined elaboration concerning the contextual differentiation of smile and laughter in macaques (Preuschoft 1992; Preuschoft and van Hooff 1997a, 1997b; Preuschoft and Preuschoft 1995).

PCACs and Reconciliations

I had the opportunity to continue my work at Utrecht University and took up the challenge of analyzing the structure of the social behavior of chimpanzees, applying the methods of multidimensional analysis employed so successfully by Piet Wiepkema (1961) in his study on the structure of the reproductive behavior of the bitterling. My intention was not just to develop an ethogram of the social behavior patterns of chimpanzees, but also to describe the relationships between these elements on the assumption that behavior patterns can be ordered in a multilayered hierarchical structure of systems reflecting common underlying causal factors (van Hooff 1970, 1973a) or—to put it differently—reflecting the interaction of different, conflicting social tendencies.

Thus the primatology group that gradually grew in Utrecht had its roots firmly in classical ethology of the Dutch tradition. In accordance with Tinbergian objective rigor, this meant the avoidance of terminology that referred to subjective and emotional categories or that carried functional connotations. In my study of the facial expressions of primates (van Hooff 1967a), the expressions were characterized most puristically in descriptive terms and any allusions to possible emotions were avoided. Unlike Nadia Kohts (1937), when she described the facial expressions of the chimpanzee, I typified the motivation and meaning of the facial displays in terms of the tendencies observable in the concomitant changes in the behavior of both actor and receiver of the displays.

However, studying our closest relatives, the primates, in particular the hominoids, creates an empathic unrest. More than other more remote species, such as fish, birds or rats, primates evoke the subjective appreciation of animals as experiencing, judging, and striving beings. To what extent then does the undeniable purposiveness that we recognize in their behavior justify the interpretation of their behavior in terms of intentions, and "calculated" intentions, for that matter? When we say "calculated" we do not mean, of course, that there necessarily is an explicit reasoning. On the other hand, there is no reason to deny that there is at least an emotional appraisal and valuation underlying behavioral decisions, if only because such denial would be at odds with the view of evolutionary continuity between humans and the other animal species.

This point has been explicitly asserted by my first Ph.D. student, Frans de Waal. His later works, in particular his books, which carried his fascination to a large audience in many countries, testify of his eagerness to demonstrate the common principles underlying both animal and human behavior. After completing a thesis on the coalition behavior of long-tailed macaques in the Utrecht laboratory colonies (e.g., de Waal 1977a, 1977b), he was appointed on a postdoctoral position in the mid-seventies to study coalition behavior in the chimpanzee colony of the Arnhem Zoo (see de Waal 1982). This colony had been established in 1972 and was intended to recreate as much as possible the natural social situation in which there were more adult males in a group. It was designed on the basis of my experiences with the colony of the Holloman Air Force Base in New Mexico, which I had studied in 1966 (van Hooff 1967b). Facilities for scientific research were integrated in the set-up of this zoo colony (van Hooff 1973b). De Waal and van Roosmalen (1979) noted the remarkable phenomenon that the likelihood of an affiliative contact between two individuals was comparatively high after there had been an agonistic conflict between them. I remember the discussions as to whether we could, therefore, call these contacts reconciliations. Because of its anthropomorphic connotations, and even more because we didn't want to preempt the issue of justifying such a functional interpretation, we stuck to simply talking about a "post-conflict-affiliative-contact" or, in short, a "PCAC."

Conceptual Parsimony or Evolutionary Parsimony?

There are two issues here. The one is—and it would please the old phenomenologists to hear that this is being discussed again—to what extent we are able to read the meaning and intent of the behavior of closely related species when we indulge in the phenomenological appreciation of our objects as analogous, if not homologous intentional subjects. In fact, what we do is rely on the intuitive mechanisms of understanding that we have evolved in order to quickly and adaptively deal with our own conspecifics. The justification for this reliance is the conversion of the proven principle of scientific parsimony (e.g., de Waal 1982). Normally it is used in a reductionist, even if heuristic, manner, to only accept more complex interpretations if the more simple, less assuming interpretation is proven to be insufficient. It can be rightly argued that from an evolutionary perspective the most parsimonious assumption is that we share analogous functions and the analogous subjective processes underlying these. The argument of "evolutionary continuity" is now widely taken as a valid justification for studying animal minds as intentional systems (e.g., Allen and Bekoff 1998). However, in any particular case it cannot be merely a matter of such an inference by analogy to determine whether an interpretation is correct. It requires empirical investigation, namely by comparing all accessible process characteristics in the other species to our own to see whether there is a correspondence of form and of demonstrable causes and functions. And then it is interesting for its own sake to find out with what species and in what respects we accord most in our intuitive understanding. For instance, it is interesting to note that we have a good intuitive understanding of the bared-teeth facial expressions of chimpanzees, but that the same expressions by drills and mandrills bewilder us at first. But then, a more detailed comparison of the ethomorphology and the contexts of the expressions in both species shows that this must be due to a confusing resemblance of the "friendly" (i.e., contact facilitating) grin of the mandrill to the human "angry sneer" (van Hooff 1976).

The second issue is one of methodology. Is the phenomenological position correct that our intuitive understanding can be a tool for our scientific understanding? Our intuition and the "Gestalt perception" involved is certainly our best inspirer of hypotheses, as it is in any domain of science (cf. Lorenz 1959). However, a PCAC only comes to be called a reconciliation after, again, the context analysis and comparison has revealed an objective functional correspondence. This is what the researchers involved have, of course, done, namely in developing refined methodologies to detect these behaviors and their effects (e.g., de Waal and Yoshihara 1983; Veenema et al. 1994; Aureli et al. 1989).

Thus, it was found in long-tailed macaques that after a conflict the likelihood of new acts of aggression against the victim was increased. Such aggression did not come only from the former aggressor, but also from the aggressor's kin, and even from other group members (not related to the aggressor). However, if the conflict had soon been followed by an affiliative contact, then the probability of renewed aggression was diminished (Aureli and van Schaik 1991a; Aureli 1992). Similarly the expressions of anxiety and stress were diminished (Aureli and van Schaik 1991b). Similar results have been obtained for chimpanzees (de Waal 1992) and for other species (Kappeler and van Schaik 1992; de Waal 1993; de Waal and Aureli 1996).

Good Relationships and the Roots of Morality

The "Dutch school" has been specifically interested in these matters of what I would call "social homeostasis" in primates (van Hooff and Aureli 1994; de Waal and Aureli 1997), the maintenance of relationships of mutual benefit, as manifested in reconciliation and consolation behaviors, and believed to be resting on the principle of reciprocity. After his pioneer observations on the Arnhem chimpanzees, de Waal continued his investigations in this field in the U.S., whereas in Utrecht in particular Aureli and Das worked on this subject. The interest in these matters has been stimulated by the feeling that they touch on the biological bases of moral behavior (de Waal 1989, 1996a). One indication of this is the following: not only do PCACs reduce the likelihood of a continuation of the conflict, and thus of stress and uncertainty, their occurrence is also more likely after conflicts between animals who have a good relationship (de Waal and Yoshihara 1983; Aureli et al. 1989, 1994, 1997; de Waal and Aureli 1997; Aureli 1997). All this justifies referring to them as "reconciliations." Moreover, the fact that dominant group members can take the initiative of an affiliative contact, involving not only the aggressor towards the victim (Das et al. 1998), but even dominant noninvolvees in the first place towards the aggressor (Das et al. 1997), apparently indicates that the behavior is motivated by a "concern" about the consequences of the conflict for the relationships with and between the opponents. In other words, this concern stretches wider than the immediate context of the conflict; it reveals a sensitivity for the long-term aspects of social relationships. This widens our perspective on the evolution of morality.

The prerogative of humans to a moral conscience has been discussed on many occasions in the past, and explanations have been sought for the development of normative systems as biological phenomena (e.g., Sent 1978; Ruse 1986; Alexander 1987). Morality can be described as the partial overruling of egotistic impulses and interests by complying with a system of norms and rules in order to save and strengthen the social network from which one profits in other contexts. The expectation that the evolution of morality could be understood has been strengthened ever since Trivers (1971) formulated the model of reciprocal altruism. This supposes that individual animals are able to monitor the exchange of favors and to adjust their own altruistic investments accordingly (e.g., de Waal 1989b). Especially de Waal (1996a; 1989c) has suggested that this can shed light on the development of social rule systems and the origin of morality. Again the analogy can be made more convincing by precisely analyzing the conditions in which such concern is manifest.

Game Theory and Biological Markets

Social animals live in groups because the trade-off between benefits and costs is positive. A more direct indication that at least primates, but perhaps many other mammalian species, may be able to appraise the importance of the social network in which they participate is to be found in the investments they are willing to make in maintaining the relationships on which they depend.

The view that biological processes are tuned by natural selection to contribute in the most effective and the most efficient way to achieving fitness has almost the status of an axiom. In various domains animals behave "economically." They appear to make choices and investments according to rules which lead to near-optimal benefit-cost ratios. This has become most evident through studies on foraging behavior.

In the social realm, individuals have to do the same thing, namely when they decide on some form of interaction or relation. Again, this decision better be as good as possible. This means a weighing of benefits and costs of diverse cooperative and competitive options, involving various different partners. And each one of these, in her/his turn, tries to make the best of her/his inclusive fitness interests while interacting with the others. In so doing, individuals must, at the same time, minimize the costs of competition, and maximize the benefits of cooperation. The interaction of the partly conflicting and partly concurring interests are best described in game theoretical models and market models. Another representative of the Utrecht tradition, Ronald Noë, studied coalition behavior of baboons in Africa (Noë 1986; Noë and Sluijter 1990) after having been involved, as a student, in the chimpanzee research in Arnhem. He became aware that market models which allow for multiple players provide better explanations than traditional two-player games, and a review of the available evidence in the literature yielded examples supporting this view (Noë et al. 1991). An example is his "shopping for profitable partners" (Noë 1992), where leverage differences between partners can even lead to situations where the market can take the form of a "veto game" (Noë 1990). Ronald Noë has since then continued working in this field at the Max Planck Institut für Verhaltensphysiologie, "the Institute of Konrad Lorenz," in Seewiesen, Germany. The work continued not only in an empirical socioecological direction, but also in a theoretical direction, exploring the socioeconomic principles of relationships (Noë and Hammerstein 1994).On the one hand there is the model approach. It investigates under which conditions and with what rules a mutually beneficial exchange of services could be stably upheld in the face of egotistic threats, which would yield a greater short-time benefit to a cheater (e.g., Axelrod and Hamilton 1981). These studies show that a very simple rule based on the tit-for-tat principle competes successfully with more complex rules. Animals in which such a simple rule or modifications of it (Nowak and Sigmund 1993) would be installed could, therefore, keep up mutually beneficial reciprocal altruism.

The Mechanisms of Negotiations: Rules of Thumb, or Evaluations?

Modeling studies can tell us which rules are the winning and stable ones. They do not, however, answer the question of which rules animals actually follow in the reality of their lives and what mechanisms of evaluation are involved. This question comes up, because the real situation seems to be more complex. Animals do not always immediately respond. Nor do they seem to reciprocate with qualitatively and quantitatively similar responses; services of a very different nature might be exchanged, such as support in conflicts, grooming, sexual permissiveness, tolerance in feeding competition, etc. The question then is whether all these different currencies are taken into account and translated as a function of some kind of mental bookkeeping. This is an empirical question, which is surrounded by a lot of debate, both with respect to the methodology of measuring the transactional processes and the interpretation of results. In studying our nearest relative, the chimpanzee, Frans de Waal supports the hypothesis that these animals have the cognitive capacities and the associated emotional dispositions to monitor one another's cooperative inclinations and to adjust their own behavior accordingly, either by granting favors or by taking punishing revenge. His research focuses on the precise conditions of such reciprocity, and the results are taken as an indication that the animals have an awareness of the "economics" of their behavior (e.g., de Waal 1996b).

A cautionary stance has been taken by Charlotte Hemelrijk. She studied the same Arnhem chimpanzee colony, where Frans de Waal did his studies on coalition behaviors and where he portrayed these animals as "politicians" involved in complex social maneuvering. Hemelrijk critically investigated the methods used to establish reciprocity in interactions, emphasizing the risk that side-effects of simpler association processes are taken as evidence, and she developed matrix statistical methods (Hemelrijk 1990a, 1990b, 1991). Using such methods she analyzed long-term data from the Arnhem colony, but could not demonstrate the kind of reciprocity in interactions of various kinds that would correspond with "balanced barter" between individuals (Hemelrijk et al. 1991a, 1991b, 1992).

Instead she invoked modern models of self-organization as developed by the biotheoretician Hogeweg (Hogeweg and Hesper 1979; Hogeweg 1988). In these models complex and secondarily adaptive structures can emerge from local interactions of entities without there being a specially selected structuring process. Hemelrijk proposed that much if not all of the relevant structure in such relationships can be explained in this way, which is much more economical in terms of hypothetical assumptions. This would mean that a structure of reciprocal interactions might arise in the absence of any specific motivation to reciprocate (Hemelrijk 1996, 1997).

An extensive discussion of this intriguing problem is beyond the scope of this review. In my opinion the main message emanating from these discussions is that nonparsimonious hypotheses that view animals as beings endowed with complex cognitive and emotional systems of evaluation and appraisal should, of course, be held dear. The reversal of the principle of parsimony, based on evolutionary analogy, if not homology, is a prima facie justification for this. At the same time traditional parsimony, avoiding complexity in concepts and assumptions, should be honored, not only because it fosters the possibility of falsification, but also because the alternatives might turn out to be irrefutable. We should remain very sensitive to the fact that seemingly complex phenomena may arise from entities operating on the basis of simple rules and the self-structuring processes emerging out of their interaction. This should protect us from overenthusiastically adopting mentalist interpretations.

Again the question is: If the interaction resembles a process of negotiation, how are these negotiations executed? Must we see participating individuals as calculating players, who somehow intuitively are aware of the different positive and negative risks? Or are they to be seen as simply and "blindly" following certain "rules of thumb," evolutionarily installed in them? Or perhaps simpler still, is the negotiation-like pattern a side-effect of local interactions on the basis of general rules? It is clear that the answer can only be determined if the details of the temporal organization of the elements of interaction and their mutual feed-backs are taken into account. This is indeed the direction this research is taking (de Waal 1996b).

Comparative studies have suggested (e.g., Byrne and Whiten 1988) that we see such "strategic behaviors" especially, if not only, in species which in other contexts have shown cognitive competencies, such as numerical and attributive competencies, that would become useful in evaluations of the kind supposed (e.g., Boysen and Capaldi 1993; Matsuzawa 1985, 1996; Savage-Rumbaugh et al. 1993; Povinelli and Eddy 1996). It is quite likely that the animals use these capacities in their " socioeconomic" evaluations. On the other hand, an overestimation due to a somewhat impatient and enthusiastic bias lurks, especially when conclusions rest on—it seems—increasingly easily accepted anecdotal evidence (Kummer et al. 1990). However, such anecdotes need not be neglected, as long as they lead to systematic operationalizations (de Waal 1991). A warning note is warranted, since not all capacities claimed have been demonstrated unequivocally (Premack and Woodruff 1978; and compare Povinelli et al. 1990, Povinelli 1994, and Povinelli and Eddy 1996). Irrespective, therefore, of which criterion of parsimony one adheres to as the most valid, thinking of simple models and testing more pretentious explanations against these remains a high priority. I think it is neither a waste of time, nor flogging a dead horse.

The Costs and Benefits of Social Patterns: In Search of Socioecological Principles

In 1971 Herman Rijksen went to Sumatra with the task of improving the management and conservation of the Gunung Leuser reserves, in particular to increase the protection of the highly endangered orangutan, one of the flagship species of the Indonesian fauna. One of his tasks was to establish a rehabilitation station for confiscated animals. This led to the foundation of the Ketambe Research Station. At the same time the project grew into a major ethological and socioecological study. It resulted in an impressive monograph in 1978 and was the beginning of a new flourishing phase of primatological field research in the Netherlands. The initiative was taken by the Department of Nature Conservation of the Wageningen Agricultural University. It led to a growing collaboration with the Utrecht group. The Wageningen department also initiated another pioneer study, namely the one by van Roosmalen (1980) of the socio-ecology of the black spider monkey in Surinam. Both Rijksen and van Roosmalen have since then returned to their original mission, nature conservation.

The Ketambe Research Station has grown into a major center of primate socioecological research, where Indonesian and Dutch researchers, in particular, collaborate. As a Sumatran counterpart of the well-known Borneo study sites, the Ketambe project contributed to further knowledge on the orangutan (e.g., Schürmann 1981, 1982; Schürmann and van Hooff 1986; Sugardjito and van Hooff 1986; Sugardjito et al. 1987; van Hooff 1996; van Schaik and van Hooff 1996; Utami and Mitra Setia 1996; Utami and van Hooff 1997; Utami et al. 1997). It also was the basis for research on other primates, such as that by Carel van Schaik and Maria van Noordwijk, and later, by Jan de Ruiter and Liesbeth Sterck on long-tailed macaques (*Macaca fascicularis*), and more recently, by Liesbeth Sterck, Romy Steenbeek, and Serge Wich on Thomas's langurs (*Presbytis thomasi*).

The emphasis in these studies has been on the evolutionary explanation of variation in social structure amongst primates. Carel van Schaik has been foremost in this field because of his theoretical contributions (e.g, Dunbar 1988). In contrast to the correlational approach, which previously was common, he emphasized deductive modeling in terms of natural selection dynamics. He showed that this can lead to the formulation of testable predictions concerning ecological variables and fitness estimates (e.g., van Schaik 1983, 1989; van Schaik and van Hooff 1983). Impressed by empirical observations on competition between female long-tailed macaques (van Noordwijk and van Schaik 1987; van Schaik and van Noordwijk 1986, 1988), he noted the important role played by the type of resource competition which members of a group experience. The essence of this model is that the distribution of resources will determine whether competition will be mainly of the contest type or of the scramble type, with different effects depending on whether the competition is intra- or intergroup. Van Schaik has reasoned that the type of competition will have far-reaching consequences for the nature of social relationships: whether these are more egalitarian and individualistic or more despotic and nepotistic; and which sex will be exogamous, with consequences for genealogy and population genetics (van Schaik and van Noordwijk 1988). This model is promising because of its elegant coherence and comprehensiveness. Thus it applies to female relationships through competition for resources such as food and its effects (van Schaik 1989), as well as to male relationships through competition for sociosexual partners and its effects (van Hooff and van Schaik 1992, 1994). A further major advance in our understanding of male-female relationships has been generated by acknowledging the important role that infanticide by male primates may play in structuring female-male relationships. The role of infanticide has offered an explanation for the evolution of sexual bonds, in particular monogamy, in some primate species (van Schaik and Dunbar 1990). Thus far monogamy in primates other than the Callitrichids, that is the siamangs and the gibbons, had been a riddle.

The model has sparked a lot of focused empirical research. In the Ketambe project, this has concerned female mating strategies and partner preference, and female migration patterns and their interaction with male mating strategies. Thus the investigations of Jan de Ruiter were concerned with the genetic causes and consequences of sociosexual strategies. In an enormous undertaking, he collected blood samples from more than 230 wild long-tailed macaques, having captured almost complete groups in the Ketambe and surrounding populations. These he subjected to blood protein variation analysis and DNA fingerprinting. One of his interests was the relation between paternity and the proactive promiscuous behavior of female long-tailed macaques. This has been interpreted as an anticipatory response by females to the risk of infanticide by incoming peripheral males. These genetic techniques revealed that in spite of their promiscuousness the females were mainly inseminated by the resident male and did not put into jeopardy their chance of bearing offspring by a proven male (de Ruiter et al. 1991, 1994; Scheffrahn 1996). De Ruiter and Geffen (1998) have further analyzed this large data set from the Ketambe population, a fully wild primate population, which still forms part of an immense population continuum. They found a surprisingly good fit of the population genetic relatedness patterns theoretically expected and those actually found. The analysis even revealed unexpected but explainable features, such as a higher level of paternal relatedness among high-ranking group members. This is due to the observed fact that the α-male prefers to monopolize mating with the dominant females.

Projects on Thomas's langurs (Steenbeek 1996; Sterck 1997, 1998; Sterck and Steenbeek 1997; Steenbeek and Assink 1998) are testing the hypothesis that in this species, where females may migrate to a new male several times during their reproductive career, the moment of migration is tuned to the risk of infanticide and the choice of partner to the protective power that the male has demonstrated by effective harassing. The importance of infanticide is still under dispute. Yet the evidence is growing that it is a factor in structuring female relationships. Sterck et al. (1997) have argued that, in addition to protection against predation, protection against infanticide may be a motive for females to associate with and to join a powerful male. Van Schaik (1996) and van Schaik and Kappeler (1997) see it even as an explanation for the fact that in primates, much more than in other mammals, males and females remain in permanent association.

An issue still under discussion is the role of predation as a factor in primate sociality (e.g., Janson 1998; Hill and Dunbar 1988). Van Schaik (1983) maintained that it must have been the primary factor, even though predation had only rarely been observed. Circumstantial evidence came from comparative studies of subspecies or groups living under conditions of different predation pressure (e.g., van Schaik and van Noordwijk 1985) or experiments under natural conditions (van Schaik 1983; van Schaik and Mitra Setia 1990). Predation may not only influence the size and cohesion of groups but also their composition. Van Schaik and van Noordwijk (1998) and van Schaik and Hörstermann (1994) have argued that group members may allow additional males to join a group because of their vigilance input (see also Stanford 1998).

Ronald Noë addressed this problem in another way. Together with his collaborators in Seewiesen he studies three sympatric species of the colobus monkey in the Ivory Coast's Taï Reserve (e.g., McGraw and Noë 1995; Holenweg et al. 1996; Bshary and Noë 1997a, 1997b; Höner et al. 1997; Noë and Bshary 1997; Zuberbühler et al. 1997). These colobus species differ remarkably in their ecology and in the degree and way in which they engage in polyspecific associations with Diana guenons. It is evident that their behavior is to a large extent a response to predation pressure, not only from leopards and raptors, but especially from other predatory primates, the chimpanzee, and recently man. Comparison of the three species shows that the response of diurnal primate species to predation risk is not always the formation of large associations. Whereas one species, the red colobus, seeks refuge in large numbers and association with Diana monkeys, another species, the green colobus, seeks the solution in crypsis; it "hides" in small groups and gets lost in associations with other species. Their study shows the power of experimentation in the wild, elegantly demonstrating the effect of predation risk by playback experiments with predator vocalizations of predator signals. These had marked effects on the association patterns of colobus monkeys. The effects were as predicted if the associations serve as an antipredator strategy.

Orangutans and "Orangs," Chimpanzees and "Chimps"

A cautionary stance has also been taken in the field of socioecology and the evolutionary explanation of complex social structures on the basis of adapted characteristics. There is a growing appreciation that some complex systems may be explained as necessarily emerging from the interaction of entities operating with simple rules. Te Boekhorst, studying the social dynamics of orangutans (te Boekhorst et al. 1990), wondered about the way their associations could be explained. In subsequent model studies, based on the paradigms of self-structuring developed by Hogeweg and Hesper (1979; see also Hogeweg 1988), te Boekhorst and Hogeweg (1994a) modeled a "mirror world" in which "artificial orangutans" behaved according to a set of simple general rules regarding traveling, foraging, looking for partners. These rules are executed in an environment with specifications corresponding with those characterizing their natural ecology, such as the spatial and temporal (seasonal) variations in fruit trees, their sizes and densities, etc. A social pattern emerged out of the interactions of the "individuals" in this local environment. This model study, as well as an analogous study on "artificial chimps" showed that the emerging pattern corresponded well with that of their natural counterparts, although no assumptions had been made about specially evolved adaptive traits. Hemelrijk (1996) considers this type of modeling extremely useful to test whether inferences about traits adapted in response to specific selective pressures are necessary.

Some might regard such studies as sterile falsification attempts, given many investigations in the field of behavioral ecology where processes of selection have become visible, and where these processes work in directions which are in accordance with a supposed adaptive explanation of specific traits (as in the studies of van Schaik and van Noordwijk and their colleagues). Such model studies should, however, be seen as means to generate alternative explanations, to explore the conditions and constraints of various models and to test their explanatory value. In this way they may in turn lead to more pertinent and decisive questions to be tackled empirically.

Turning in Circles, or Meandering Forward?

This review of the developments and contributions of "Dutch" primatology is not exhaustive. I might have elaborated on a few other studies that have taken place in other university institutions. For instance, the pioneering information-theoretical studies by Cools and van den Bercken at Nijmegen University on the organization of the behavior of long-tailed macaques (e.g., van den Bercken and Cools 1980a, 1980b), or the laboratory studies on the ontogeny of social behavior, in particular the phobia studies by the Timmermans/Vossen group at Nijmegen University (e.g., Timmermans et al. 1986, 1988; Vochteloo et al. 1993, 1997).

I have selected to show those developments I am most familiar with, and which demonstrate certain fundamental issues concerning the changing way scientists have regarded phenomena such as the organization of behavior, of interactional strategies and social patterns. It may seem as though we turn in circles. Mentalist approaches have given way to more objectivistic approaches. These latter have contributed greatly to a sound operationalization of ideas and questions. They were necessary to make behavioral biology into a science. Now, we are reconsidering the questions of animal consciousness and mental processes. We are clearly meandering forward as long as we keep a critical "Tinbergian" outlook alive. This should guard us against overenthusiastic overinterpretation (cf. Povinelli and Eddy 1996; Hobson 1996; Tomasello 1996) and the kind of "new anthropomorphism" which confuses mentalist representations with causal interpretations (Kennedy 1992).

Is There Still a Dutch Primatology?

Science has always crossed frontiers. Yet we could always identify national or regional orientations, sometimes differing in areas of interest, sometimes differing more fundamentally in underlying theoretical visions and paradigms ("schools"). The ease of worldwide communication is working towards a global and universal science. Those scientists working in small countries experience this most clearly. Many of the young primatologists who arose in the Utrecht group and remained active in the field have dispersed widely internationally. For example, in temporal order: Frans de Waal—first Madison, Wisconsin, then Atlanta, Georgia (U.S.A.); Carel van Schaik and Maria van Noordwijk—Durham, North Carolina (U.S.A.); Ronald Noë—first Zürich, Switzerland, then Seewiesen, Germany; René te Boekhorst and Charlotte Hemelrijk—Zürich, Switzerland; Jan de Ruiter—first London, then Winston Salem, North Carolina; Filippo Aureli—first Atlanta, Georgia, then Liverpool, UK; Signe Preuschoft—Atlanta. Still nest odor sticks, if I may use this figure of speech, although it is less appropriate in the context of primates: those who have dispersed, often did so as do macaques, namely towards destinations where they encountered kin. Primatologists, after all, are primates.

E-Mail Exchanges

The diversity of histories apparent in the chapters written by the pioneers suggests that there were different entry points into the study of primates and primate society. Not surprising, the e-mail conversations illustrate how heterogeneous points of view influence answers to the most basic questions: *why study primates, did our ideas about primate society change,* and *how do ideas change?* Generational, disciplinary, institutional, and cultural dimensions are clearly visible in the responses. The existence of multiple interlinked "realities" challenged our attempts to reach a consensus on even these fundamental issues but it also stimulates new and interesting questions about the study of primates and the trajectories of ideas about primates.

Question 1: Why Study Primates?

To: <teresopolis@majordomo.srv.ualberta.ca>
From: "Dick Byrne" <rwb@st-andrews.ac.uk>

I would assert that, with the exception of those biologists who are interested in seed dispersal or guts, everyone who studies specifically primates does so for only one good reason . . . to know more about humans. Of course, we all have different personal motivations for what we do. It's great that primatology is a meeting point for zoology, ecology, anthropology, genetics, psychology, etc. . . . but the common denominator is the relevance of primatology to humans. If we deny that, we should surely find species to study instead that are more convenient, cheap, fast reproducing, nearby, nonendangered, etc.

To: <teresopolis@majordomo.srv.ualberta.ca>
From: "Thelma Rowell" <thelma@ingleton.demon.co.uk>
Well, I do deny it. My interest in people is nonscientific and I first got into primates grudgingly, because there was a job—I would rather have studied rodents at the time. . . . I was interested in menstrual cycles, which are supposedly peculiar to primates and seemed (still seem) irrational and so worth investigation. If you want to study social organization of long-lived mammals, they are a lot more accessible than bats or dugongs and people will give you money to study them. . . . I never believed all that stuff about baboons being just like early man—did anyone?

To: <teresopolis@majordomo.srv.ualberta.ca>
From: "Dick Byrne" <rwb@st-andrews.ac.uk>
Thelma, quite so: as I said, our personal motivations may be all over the place. But WHY WAS THERE A JOB? Because, I'd assert, money was available for a project whose ultimate scientific justification was to learn more about humans by some form of comparative analysis. . . . Assuming baboons have converged so much on early hominids that they are a "model" for them is indeed pretty hopeful! However, using the comparative method to interpret a cladistically derived evolutionary history is not only possible but safer than clutching at models . . .

To: <teresopolis@majordomo.srv.ualberta.ca>
From: "Alison Jolly" <ajolly@arachne.Princeton.edu>
I disagree too. I am fascinated by creatures which are different from people, hard to understand, and come in radiations of beautiful species. OK, little cute animals, but if it had not been lemurs it would have been polychaete worms, who are very beautiful. . . . Does my attitude and Thelma's map onto those who come from biology as opposed to anthropology or psychology?

To: <teresopolis@majordomo.srv.ualberta.ca>
From: "Pamela Asquith" <Pamela.Asquith@ualberta.ca>
. . . regarding Dick's comment—this is true of Japanese primatologists too who are not only studying primates with questions about "hominization" (their term) of behavior and social structure in mind (even if they say nothing about it, still building up a picture), but who are in many cases also initiating studies of humans, especially mother/infant behavior of various sorts in their research.

. . . there is an interesting situation in Japan that no biology department provides for primate studies. People come out of anthropology, psychology, and their subdisciplines . . .

To: <teresopolis@majordomo.srv.ualberta.ca>
From: "Robert Sussman" <rwsussma@wustl.edu>
Thought I would add my own two cents to the discussion. . . . within the U.S. at least, primatology began out of an interest in models for human evolution and that is why most

students of primatology are "related" to Washburn. As we (U.S. primatologists) continued to study primates in the field, we got interested in theories of ecology and behavior in general and many young primatologists may now be . . . only peripherally interested in how this may relate to human behavior and evolution. . . . I think, in anthropology at least, it is important for primatologists to be interested in what their animals can tell us about human behavior and evolution—if not, they shouldn't expect to be in anthropology . . .

To: <teresopolis@majordomo.srv.ualberta.ca>
From: "Shirley Strum" <strum@tt.sasa.unep.no>
. . . despite the diversity of motivations and origins there seems to be a real convergence of perspectives today. I marvel at my colleagues who drift from one discipline to another, from anthropology to zoology. I don't think that would have been possible twenty years ago. I can't do it because, as Bob suggests, I find myself to be an anthropologist who has an agenda, even though this agenda may have changed in the last twenty years. Still, many of us from different disciplines and even from different national traditions are now speaking in the same language and asking some of the same questions. . . . It may be a generational issue so that young primatologists don't now even identify themselves with a discipline, but rather with an approach or set of questions . . .

The bedrock upon which the other Teresopolis issues rested was that our ideas about primate society have indeed changed. At the workshop, we couldn't agree about this; nor could we agree in the e-mails below.

Question 2: Did Our Ideas about Primate Society Change in the Last Fifty Years or Did We Know It All Before?

To: <teresopolis@majordomo.srv.ualberta.ca>
From: "Thelma Rowell" <thelma@ingleton.demon.co.uk>
Well, who is we? Of course we have found out a lot about primates in the last fifty years, we the people who study them. Most ordinary people have probably gone beyond the King Kong image of gorillas, for example. Some ideas seem very resistant to the onslaught of mere facts though. Sociobiology took the myth of the dominant male fighting for females and holding them until ousted straight from Victorian romantic naturalists and has perhaps increased the resistance to alternative scenarios . . .

To: <teresopolis@majordomo.srv.ualberta.ca>
From: "Karen Strier" <skbstrier@facstaff.wisc.edu>
I agree with the consensus that our ideas about primate society have changed, but that answer seems so obvious I began to puzzle further about how much of these changing views are due to us, and how much to the primates themselves. . . . I will argue that one of

the reasons our ideas of primate society have changed is that the primates themselves are different today than they were fifty years ago. Apart from things like habitat disturbance and human interference and in some sites, the presence of ecotourists or sustainable development projects, the elimination of natural predators, local human conflicts, etc. (which could be called EXTERNAL to both us as observers and the primates), there are at least three other factors involved which could be considered INTERNAL (to the observer-primate relationship): (i) what habituated and unhabituated primates do in front of observers; (ii) how adept we've become at following and tracking primate behavior (including our attention to kinship, our experimental and sampling methods, etc.); and (iii) the dynamics of primate societies independent of observers' presence. Focusing on the sociocultural context of our changing views about primates seems to involve an underlying assumption that our primate subjects are static, when in fact we all know that they're dynamic and changing as well.

To: <teresopolis@majordomo.srv.ualberta.ca>
From: "Dick Byrne" <rwb@st-andrews.ac.uk>
Just in case the REALLY obvious needs stating, recall that in 1947 we'd have been relying on Zuckerman's book, itself based on a brief study of the London Zoo hamadryas population in which there was a massively male-biased sex ratio and frequent deaths from the continual violent fights.

It's interesting that today not just the findings are discredited (groups held together by male violence in pursuit of sex . . .), but the results of captive studies would always be interpreted in the light of knowledge of the species' behavioral ecology obtained from fieldwork. If captive groups were constituted differently than occurs normally in the wild, there'd have to be some good reason for doing so. . . .

The root of these changes is presumably that we now see primate society as a result of evolution by natural selection . . . it almost looks as if fifty years ago people thought primate societies were created . . . ! ! ! . .

To:<teresopolis@majordomo.srv.ualberta.ca>
From: "Thelma Rowell" <thelma@ingleton.demon.co.uk>
. . . Karen's point about primates changing is important and shouldn't be confounded by changing monkey-watchers' skills as she went on to do. . . . No, I refuse to make too much of the existentialist view. Our ideas have changed about primates, and most other animals, because we know more about them and they had to change. What people want to project onto animals also changes, but that is a quite separate activity.

To: <terespolis@majordomo.srv.ualberta.ca>
From: "Dick Byrne" <rwb@st-andrews.ac.uk>
I am puzzled by the enthusiasm for the idea that "the primates have changed" (as opposed to the uncontroversial fact that our view of them has changed enormously with improving

habituation, study design, techniques, etc). What is the evidence? I took Karen's original mention to mean something like "muriquis in tiny isolated forest blocks aren't likely to behave like primeval muriquis in the vast Atlantic forest," and very reasonable that is too . . .

To: <teresopolis@majordomo.srv.ualberta.ca>
From: "Thelma Rowell" <thelma@ingleton.demon.co.uk>
I don't want to defend Zuckerman . . . but it isn't fair to judge the young prosector of the London zoo of the thirties with hindsight. He had no basis for any expectations of how monkeys would behave, and it was very daring to put them out on the terraces . . . there was no way to find out it wasn't a good idea except by trying. The absence of information fifty years ago is hard to imagine today . . . people had no ideas about how social behavior worked.

To: <teresopolis@majordomo.srv.ualberta.ca>
From: "Robert Sussman" <rwsussma@wustl.edu>
. . . I think that we must remember there is a difference between what animals can do and what they do most of the time. We can and often do emphasize very rare behaviors because they are more exciting, spectacular, and of interest to a buying public but they might not represent the typical perhaps more boring everyday behavior of the animals. In the past, there was an attempt to change from collecting data by just ad lib notes because they tended to overemphasize these more noticeable behaviors. . . . I believe that our knowledge of primates has changed tremendously over the last thirty years but we have often "stacked the deck" in attempting to support our theories (many of which look amazingly like theories of the past with some new clothes). We often tend to dismiss much of the knowledge we have obtained these past thirty years if it does not support our theories . . .

If ideas about primate society have changed, why did they change? Our first approach to this question was to try "traditional" explanations: the role of factors like theory and methods.

Question 3: How Do Ideas Change? Do Specific Theories or Specific Methods Make a Difference?

To: <teresopolis@majordomo.srv.ualberta.ca>
From: "Alison Jolly" <ajolly@arachne.Princeton.edu>
OK . . . the reason this turnip isn't responding on shifts in theoretical stance is that it all seems too obvious (I said this before). Some shifts which come to mind are:

1. The importance of individuals (1960s) influenced by Jane Goodall, the early Japanese work, including Imanishi, the Stuart Altmann/Imanishi book and the first three red covered volumes of *Primates,* also Robert Hinde, and Thelma and Yvette Spencer-Booth.

2. The shift from male to female emphasis in social roles. I always think of Richard Wrangham's female-bonding article, but see Sarah Hrdy's writings, both original work and reviews of others (1970–1980s).

3. The shift to cognitively explicit explanations. Don Griffin, also Byrne and Whiten, and many of the rest of us (1980s–1990s).

4. A current shift which is influencing me—maybe many others, toward looking at society as a larger whole. Back to Imanishi, maybe! Influenced by Pam, Mariko Kiraiwa-Hasegawa, Hiro, but also Maynard-Smith and Szathmary's *Major Transitions in Evolution.* Also a course I have been teaching on the evolution of everything.

5. Notably absent: the influence of Donna [Haraway] and other "primatologist-ologists"! Donna has had a great influence on me but I still picture the social context as a prism though which we view the primates, more or less askew, more or less with rainbow edges . . .

To: <teresopolis@majordomo.srv.ualberta.ca>
From: "Pamela Asquith" <Pamela.Asquith@ualberta.ca>
I wonder how much WHO suggests a new idea affects changes in theory, etc. It seems a few doyennes of primatology publish an idea and many follow suit. . . . It is no coincidence that they are highly talented primatologists, of course, but I suspect that the same idea from a less well-known or well-respected person would die on the drawing boards.

It is interesting that within the Japanese science, though some people are more highly respected than others, everyone's ideas get a fair deal there because that is the way they expect to work. A prima donna tends to be resented by students and colleagues alike due to lack of a team spirit.

To: <teresopolis@majordomo.srv.ualberta.ca>
From: "Hiroyuki Takasaki" <takasaki@big.ous.ac.jp>
I make my one yen (about 1/1.2 cents these days) contribution while the rest of you are sound asleep because when all of you are awake, I am asleep. A prime example of theories, or conceptual means/framework for better perception of the world, which greatly influenced the genesis of the Kyoto school field primatology in Japan is Imanishi's "specia" theory. He regarded every biological species as having its "species society" (= specia, singular; speciae, plural) composed of its member individuals. Through comparison of the speciae of primate species, he proposed, the evolutionary pathway of human society would be elucidated.

In his view, each individual is regarded not like a dead museum specimen or a captive in a zoo, but as a live member of the respective specia to be studied in the field condition. Therefore, individual identification was, has been, and still is regarded as a sine qua non for collecting basic data for such a study. Provisioning is not a prerequisite, but was a methodological short cut devised in the early stages to facilitate habituation which provides observers with high-resolution individual identification by narrowing the distance between the observer and the observed (just like the setting in Teresopolis provided for primatologists).

Without Imanishi's view of the world of living things, no adherence to individual identification would have been pursued in the early stages; ecological studies feasible without provisioning are likely to have preceded sociological studies instead, or nobody might ever have seriously attempted to study Japanese macaques which live mostly in ecologically disturbed habitats. All Teresopolis participants are familiar with the discoveries that followed: basic rank, dependent rank, matrilines, male transfer, etc. to cite a few from Japanese macaque studies.

To: <teresopolis@majordomo.srv.ualberta.ca>
From: "Pamela Asquith" <Pamela.Asquith@ualberta.ca>
In response to Hiroyuki's note about Imanishi's "specia" theory, much of the Western primatology world has been sleeping while the Japanese wrote, due not so much to the earth's rotation as to linguistic, cultural, and other walls. Imanishi's ideas on "specia" have been readily available in English since 1960 and several times afterwards.

If you trace the Japanese influence on primatology, it is indeed the DISCOVERIES that Hiroyuki mentions—basic and dependent rank, matrilines, etc.—that have been picked up, not the underlying theory or methodology that sustained and continues to sustain this corpus of continuing discoveries . . .

SECTION 3

A Diversity of Primatologies: Other National Traditions

How do the histories vary across national traditions? We ask those who work in different traditions.

The extent of variation in history among primatologists trained in North America, Britain, and Continental Europe was surprising. However, from the start we had expected important differences between the development of primate studies in the "West" (Europe and North America) and the "East" (primarily Japan). For one thing, as graduate students in the late 1960s and early 1970s, we both had read the newly translated reports of Japanese field studies in their journal *Primates*. It was hard not to miss the differences in methods, ideas, and tone of presentation from what was standard practice in our own North American tradition. Furthermore, by the time of our project, we had between us extensive field experience in Latin America, Japan, and Kenya, and contact with foreign scientists in those countries and at international conferences. As noted in the history in section 1, primate studies is now an active and recognized science in many countries and almost all parts of the world. We were not able to sample scientists from all these countries. Instead we chose two national traditions where primate research is very well established and which have existed long enough to build up their own substantial histories. Indeed, Japanese scientists independently developed the practice of studying primates in their natural habitats in the same post–World War II period that Western scientists were fanning out into tropical countries to observe our primate relatives. It was particularly appropriate to examine the Japanese history of ideas about primate society. None of the other national traditions of primatology have been in existence quite as long as Japan's and Euroamerica's, but we chose Brazil as an example of a country which has a sizable cadre of indigenous primatologists and a recognizable body of literature with its own national flavor.

Because we wanted to compare other national traditions in primate field studies to our own, Pamela Asquith was a natural choice as a contributor. For many years, she has been studying the unusual phenomenon of a science (primatology) that was independently invented in two places. These two very different points on the globe had little cross-fertilization or communication until both traditions were well-established. We also wanted the perspective of a Japanese primate fieldworker, preferably from the renowned "Kyoto school." Hiro Takasaki was appropriate because of his training at Kyoto University and his extensive fieldwork with Japanese macaques, and with chimpanzees at the well-known Japanese site of Mahale Mountains in Tanzania. To maintain this pattern of presenting both a native and an "ambassadorial" perspective, we asked Maria Emília Yamamoto, Anuska Irene Alencar, and Karen Strier to provide us with their views of the history of ideas about primate society in Brazil. Karen Strier has worked in Brazil alongside her Brazilian colleagues for many years, studying the muriqui and helping to train Brazilian students. Emília Yamamoto and Irene Alencar are Brazilian primatologists who have worked extensively with callitrichids both in the field and in captivity.

Together the four chapters presented in this section give us a taste of the rich local histories of primatology as they have developed in different countries. As with the information from the "pioneers" in section 2, these histories of national traditions are not yet written down. These perspectives could be lost as practitioners retire and primatology becomes ever more a "melting pot" international science. Further, there are important details and insights in these chapters that go beyond what is generally known about the Japanese and Brazilian traditions. Hiro Takasaki, for example, in "Traditions of the Kyoto School of Field Primatology in Japan," traces the history of an important and little-known (to the West) tradition in primate studies. He focuses on seminal publications that have not been translated into English. For the sake of comparison with our history, he divides Japanese field primatology into similar stages, providing much information about who did what during those periods in Japan. Focusing on Kinji Imanishi, the founder of the Kyoto school, and Junichiro Itani, his first pupil, Takasaki shows us how their methods of studying primates in nature and their ideas about primate society came to influence so many Japanese scientists. He describes some of the characteristic features of this school—their strategic use of anthropomorphism, their belief in the lasting power of good descriptive data, and their dedication to longitudinal, holistic, collaborative research that has led to discoveries such as the fission-fusion "communities" of chimpanzees. He concludes that the members of the Kyoto school of Japanese primatology have a very distinctive worldview, first expounded by Imanishi and now shared by almost all who trained under him or Itani—a view of the world of living things as created neither by an "Almighty God nor any omnipotent theory of evolution." He shows how Japanese primatologists often assume the inverse of Western scientists (e.g., complexity versus parsimony) and notes that there is pressure for Japanese scientists to produce papers more similar to those published by others. He also argues that the diversity of different schools of thought is good for science–that there are always some parts of the world which are better "viewed upside down."

This theme of the counteracting forces of powerful global science versus rich local traditions in science is also taken up by Pamela Asquith in her chapter, "Negotiating Science: Internationalization and Japanese Primatology." Asquith uses a set of recent interviews with both older and younger Japanese primatologists to address some questions frequently asked about Japanese primatology by Western scientists (and indeed asked during our workshop). Why do they continue to publish descriptive papers and what do they see as the value of their methods and approaches? She conceptualizes Japanese scientists as negotiating their way through a series of formal and informal rules and expectations that they must meet and follow in order to both adhere to local tradition and successfully publish in international journals. Her chapter shows us the world of primate science from the Japanese perspective—how their history leads them to deal with issues of fieldwork and descriptive data, language, publications, conferences, internal critiques, and Western theories such as sociobiology. Like Takasaki, she concludes that in spite of differences among individual Japanese primatologists, there is a strong sense among practitioners of a distinctive Japanese tradition. And that while there is pressure to become more like Western scientists in order to meet "international standards," there is also a recognition of the power of their own research paradigm. The negotiation she describes is fundamentally about whether "international science," dominated by North American and European primatologists, will allow divergent paradigms such as that of Japanese primatology to count as valuable contributions to the discipline.

Although the origin and development of primatology in Brazil was quite different from that in Japan, some similar issues arise. In their chapter, "Some Characteristics of Scientific Literature in Brazilian Primatology," Maria Emília Yamamoto and Anuska Irene Alencar trace the history of Brazilian primatology over the past two decades. They argue that because Portuguese colonization was not characterized by an interest in natural history, the study of animals in nature came late to Brazil. When studies of primates did begin to flourish in the 1980s, they originated from two unusual sources: biomedical interest in the callitrichids (marmosets and tamarins) that reproduce relatively cheaply and easily in captivity, and as part of the conservation movement to study and preserve the forests of Brazil. Yamamoto and Alencar employ a quantitative analysis of publication patterns in Brazilian primatology to elucidate general national trends in topical, species, and geographical interests. They conclude that research is very patchily concentrated on only certain topics and regions of the country due mainly to financial constraints. Because the study of callitrichids has dominated Brazilian primatology, Yamamoto and Alencar use this research as a case study to demonstrate how the characteristics of the species we choose to study may affect our views of primate society. They suggest that many Brazilian primatologists view females as powerful and conspicuous in primate society because the role of the dominant female in callitrichid groups is quite notable. Furthermore, much research has been conducted on the complexities and ramifications of this pattern of female behavior. They also discuss issues of gender and power among scientists. Their chapters suggest an intriguing analogy between the struggle for recognition of female scientists in competition with male scientists and the struggles of scientists from developing nations, such as Brazil, competing for recognition on the international stage with scientists from developed countries. They conclude that Brazilian primatology has not yet had much of an impact on international science in part because of low numbers of publications, but also because of similar language, publication, and negotiating issues that are faced by Japanese, and indeed all, primatologists outside North America and Europe.

In "An American Primatologist Abroad in Brazil," Karen Strier describes some aspects of her long-term field project at the biological field station of Caratinga. She has collected data on the endangered muriqui monkey, lived alongside local farming people, collaborated with Brazilian colleagues, and trained both Brazilian and American graduate students. Strier describes how she was first attracted to muriqui monkeys for theoretical reasons (e.g., their diet and social system) but was almost immediately drawn into practical, conservation-motivated research. Muriquis are a "flagship" species for conservation efforts in South America, and conservationists have been active in all aspects of the research at Caratinga. Strier reflects on the similarities and differences between Brazilian and American primatology in terms of graduate training, attitudes toward nature, science and conservation, and the interaction between local and foreign scientists in Brazil. She also discusses how working in Brazil has changed her research approaches, the questions she asks, and the implications that this has for our understanding of primates, particularly the comparison of neotropical monkeys to Old World monkeys and apes.

The chapters found in this section serve several important functions. They document many aspects of regional histories of primate studies that are otherwise unknown outside their local circles. They show us how differently the field of primatology began in various places and how divergent have been the changing perceptions of primate society. And not least, they cause us to reflect on the unspoken assumption that our own history and our own view is universal. Each of the authors in this section makes a plea for the value of diversity in science. As international researchers operating in a global village, we usually take for granted that we are all working from the set of assumptions, speaking the same language, competing on a level playing field and seeking the same outcome. These chapters help to document how diverse histories can set implicit agendas, maintain asymmetries, and act as barriers to cross-national understandings of scientific practice. Some of our contributors believe that these and other national traditions of primate studies are all converging into one international science. If so, is this a case of natural convergent evolution, or is it an example of North Americans and Europeans imposing their views on the rest of the world? The perspectives of our contributors on attitudes toward the Japanese paradigm of primate studies *and* on "international science" are explored in the e-mail exchanges at the end of this section.

7 Traditions of the Kyoto School of Field Primatology in Japan

Hiroyuki Takasaki

Introduction

Each primatologist is a distinct individual, differing in ideas and background. At the same time, since researchers are social beings, they form schools. Among Japanese primatologists, too, there are schools—Kyoto, Osaka, etc. This plurality of primatological schools in Japan has been clarified for the rest of the world only recently (Asquith 1991), although some may still lump them together because of common traits derived from the same cultural background (e.g., Kitahara-Frisch 1991). This article limits its scope to a particular school formed by Kyoto University–linked researchers—those who taught or were educated there, their students, and disciplinary sympathizers. This school has been one of the most powerful driving forces in the community of primatologists, at least in Japan, and their field sites are now scattered all over the natural primate habitats.

By definition, the school boundary is vague; however, any Japanese field primatologist knows whether he/she belongs to this school or not. Being native to this school, I can hardly describe it equidistantly in comparison with other schools. As such, this limits the scope of this article. Unfortunately, I have only witnessed the development of the Kyoto school in the latter half of its existence, well after the bifurcation of the major institutions—the initial Laboratory of Physical Anthropology (Department of Zoology, Faculty of Science) in Kyoto and the Primate Research Institute at Inuyama. Also, I am afraid that I may be biased by the subculture of the Kyoto Campus, as I spent many of my undergraduate through postdoctoral years (1976–1991) there. For the historical accounts on the earlier development, I depend on various articles written by predecessors (e.g., Imanishi 1970; Itani 1975, 1983, 1985b, 1993; Sugiyama 1965b, 1985) and hearsay.

This article concentrates on covering pivotal publications in the Japanese language little known outside Japan, since there are already several excellent reviews of the body of Japanese research that has been published in English (e.g., Asquith 1984, 1986a, this volume; Huffman 1991a; Kitahara-Frisch 1991).

History

The Kyoto school was undoubtedly one of the centers where primatology passed through its early developmental stages. The stage divisions proposed by Strum and Fedigan (this volume) are useful as rough descriptions of the development of this school too. As the early "Monkey Studies" in the school were largely based on the Japanese macaque, *Macaca fuscata,* historical reviews of research on this species (e.g., Itani 1983; Huffman 1991a; Kawai 1964, 1969; Watanabe 1993) provide overviews of the early development of primatology in this school. Similarly, historical reviews (e.g., Itani 1977c, 1993; Nishida 1990a, 1994) of the later "Ape Studies" most heavily conducted on the longhaired chimpanzee, *Pan troglodytes schweinfurthii,* depict the subsequent development of primatology at Kyoto. The following is intended to be a rough sketch of the overall history of the school.

PRE-STAGE 1: EARLY STUDIES

The study of primates in Japan started in October 1948 at Toimisaki in Kyushu, when S. Kawamura and J. Itani encountered a troop of monkeys during their fieldwork on semiwild horses (Itani, personal communication). The study of Japanese primates, then, began as a natural expansion of the ongoing studies of animal societies organized by Imanishi, then lecturer at the Department of Zoology at Kyoto University (see Imanishi 1955, Kawamura 1957, and Kawai 1955 for other animals studied at the time).

Imanishi, who received his D.Sc. in 1940 based on his systematic, biogeographical, and ecological study of mayflies, was a frontier scholar with a rare personality, talent, and a pioneering spirit. During World War II, he explored the Inner Mongolian region and became familiar with grassland mammals, such as gazelles, and livestock animals, such as horses and sheep, as well as with the Mongol people. He developed his idea of animal sociology as a natural outgrowth of his *World of Living Things* (1941b), which he had written to leave behind in the event of his death during the war. Legend has it that he was reading C. R. Carpenter's monographs (1934, 1940) on a *tatami* straw mat laid on the concrete floor of the annex building of the Department of Zoology, when Japan was still unrecovered from the war damages. He subsequently became the founding father of primatology in Japan.

STAGE 1: 1950–1965

In 1951, the Primate Research Group was organized by D. Miyadi, professor of ecology at the Department of Zoology, and Imanishi, who had been transferred in 1950 to the Research Institute for Humanistic Studies (RIHS) at Kyoto University. Imanishi published *Prehuman Societies* (1951), which influenced Japanese primatology thereafter for a long time. His book and subsequent papers (e.g., Imanishi 1965) suggested that the elucidation of the evolution of human society be the main goal of animal studies and he established a comparative study of nonhuman primates and social mammals.

In August 1951, Itani and K. Tokuda succeeded in provisioning the Koshima troop with sweet potatoes (Itani and Tokuda 1958). This method dramatically decreased the distance between the monkeys and the observer and thereby facilitated habituation. At the same time, Itani and Tokuda adopted the individual identification method, which they had long used by then in studies of other mammals, such as horses and deer. The same set of methods were applied successfully to many other Japanese macaque troops and later to various other primate species. This led to the initial burst of findings (Watanabe 1993), such as concentric troop formation (Itani 1954), basic and dependent rank systems (Kawai 1958), matrilineal troop structure and youngest ascendancy rule in rank (Kawamura 1958a), paternal care (Itani 1959), culture (Kawamura 1959), and troop fission (Furuya 1960; Sugiyama 1960). Since then, long-term studies have been continued at many sites.

In 1956, the Japan Monkey Centre (JMC), a zoo and museum owned by the Nagoya Railway Company, was opened at Inuyama, Aichi Prefecture. Itani, and later Kawai, were posted there, and in 1957, the JMC began publication of *Primates*. This primatology journal began to publish in English two years later and *Primates* gradually became an international journal, bringing fame to Japanese field primatologists. Japanese primatologists used this journal as the main media for announcing their discoveries to the rest of the world.

In 1958, the members of the First Gorilla Expedition (Imanishi and Itani) were dispatched from the JMC. They traveled across Africa from Tanganyika to Cameroon, and then throughout Europe and North America. During this preliminary survey, they looked for sites suitable for studying great apes (Imanishi 1960a). In Europe and North America, they met with colleagues in primatology and anthropology and absorbed the primatological information available at the time. In this same year, S. Kawamura went to Thailand for a preliminary survey of gibbons (Kawamura 1958b). The JMC dispatched the Second and Third Gorilla Expeditions, in 1959 (Kawai and Mizuhara 1959; Kawai 1961) and 1960 (Itani 1961).

In 1959, Imanishi became the head of the newly created Division of Social Anthropology in the RIHS. He published in the field of social anthropology also (e.g., Imanishi 1954) and collaborated with social and cultural anthropologists as well as many scholars in various disciplines. Imanishi's (1960b) paper in *Current Anthropology* and the collection of early findings on the Japanese macaque (see Imanishi and S. A. Altmann 1965) brought international attention to the development of primatology in Japan.

In 1961–1963, Kawamura, K. Yoshiba, and Y. Sugiyama, in a team headed by Miyadi, studied the Hanuman langur in India by habituation without provisioning (Sugiyama 1965a), and Sugiyama (1965b, 1980) reported their first observed infanticide of nonhuman primates. In 1962, the Laboratory of Physical Anthropology (LPA) was opened as part of the Zoology Department and headed by Imanishi. Itani returned to Kyoto to support Imanishi at the LPA together with J. Ikeda, a physical anthropologist, and Sugiyama. The first generation of graduate students from the LPA became the manpower for the next stage of Japanese primatology and various articles and books—including those by Imanishi (1960a), Itani (1954, 1961), and Kawai (1961)—announced anticipation for the emerging discipline.

After Itani's return from the Third JMC Gorilla Expedition, the focus of research was shifted to chimpanzees, because of the observational and political difficulties for long-term research on gorillas in the region following decolonization (Imanishi 1966). On his way back from this solo expedition, Itani briefly explored the Kibale and Budongo Forests of Uganda, and the Gombe Stream in Tanganyika (Itani 1961, 1977a, 1977c, 1993). Thus, the Kyoto University Africa Primatological Expedition (KUAPE) was organized. The KUAPE lasted from the first (1961–1962) to the sixth (1967–1968) expedition. Up to the third team (1964–1965), Imanishi, Itani, and their graduate students partook in the primatological research. However, the chimpanzees in the savanna woodland on the eastern shore of Lake Tanganyika—at Kabogo, Kasakati, and Filabanga—were not easily approached. No chimpanzee group was successfully habituated in spite of all patient efforts, and long-term studies were not feasible at any site.

STAGE 2: 1965–1975

In 1965, K. Izawa, during his study at Kasakati, made a preliminary survey of the Mahale (=Mahali) Mountains area. The local people recounted stories of chimpanzees raiding sugarcane fields, and he found Kasoje (=Kasoge) to be a promising site for provisioning and habituating chimpanzees (Izawa 1977). T. Nishida, who was a new member of the KUAPE's fourth team, was sent to Mahale. In an inland survey on foot in the same year, Itani and Suzuki (1967) witnessed a multimale-multifemale procession of chimpanzees across a clear hill crest in Filabanga. They subsequently suggested the presence of bisexual unit groups in chimpanzee society (Itani 1977a, 1977c, 1993). By the end of 1966, chimpanzees at Kasoje began to feed on sugarcane and bananas, and the presence of the unit group was confirmed (Nishida 1968, 1970, 1973b, 1981; Itani 1977c).

In 1967, the Primate Research Institute (PRI) of Kyoto University was opened in Inuyama. Kawamura and Kawai became professors there and Sugiyama was an associate professor, with Kawamura organizing research mainly in Southeast Asia and Kawai leading studies of cercopithecoids in Africa.

Kawamura trained S. Kawabe, M. Kawabe (née M. Yamada), N. Koyama, and K. Norikoshi at Osaka City University before he came to the PRI. S. Kawabe raised Japanese macaque infants in order to study their development (Kawabe 1964). K. Wada's (1964), Hazama's (1965), and Yamada's (1966) observations confirmed the transfer of male Japanese macaques between troops. Koyama (1967, 1970) conducted a detailed study of dominance rank order with the Arashiyama troop and, together with Norikoshi, described the process of a troop fission along matrilineally grouped subdivisions within the troop (Norikoshi and Koyama 1975). It was made clear that the matrilineal structure remained unchanged by fission within the newly born daughter troops. Based on these and numerous other findings, Kawai published *Ecology and Society of Japanese Monkeys* (1964, 1969).

Sugiyama and Suzuki studied chimpanzees in the Budongo Forest in Uganda for a while before the Amin administration (Sugiyama 1981; Suzuki 1977, 1992), and they habituated the chimpanzees without provisioning. During this period Suzuki (1971) recorded the first instance of infanticide by chimpanzees. Izawa, posted first at the JMC and later at the Miyagi University of Education, and A. Nishimura (=Toyoshima), first at the PRI and later at Doshisha University, looked for new frontiers upstream the Amazon. They had been studying New World monkeys since 1973 (Izawa 1979, 1991; Nishimura 1991), subsequent to earlier research in Stage 1 by K. Tokuda and K. Wada (Tokuda 1962).

Nishida's success in provisioning and habituating chimpanzees in the Mahale Mountains opened a new era of chimpanzee research. After the confirmation of multimale-multifemale unit group structure, Nishida and K. Kawanaka found that females transfer between unit groups (Kawanaka and Nishida 1975; Nishida and Kawanaka 1972). (Before finding this, they had studied intertroop relationships in Japanese macaque society [Kawanaka 1973; Nishida 1966].) Nishida got a post at the Department of Anthropology at the University of Tokyo, where he organized teams to continue the chimpanzee research at Mahale (Nishida 1973b, 1981, 1990b, 1994).

T. Kano, who joined the KUAPE's fourth team, walked extensively throughout the hinterlands of western Tanzania in an effort to understand the distribution and density of nonhuman primates (Kano 1971, 1972). In retrospect, this study should have been undertaken prior to the other fixed-site chimpanzee habituation attempts. His study revealed that the highest population density of chimpanzees was in the Mahale Mountains area and that the lowest density was in the most arid habitat of Ugalla. Later, this experience led him to conduct an extensive survey of the distribution of bonobos before choosing any long-term study site. After Nishida's preliminary survey of bonobos in Zaire in 1971, Kano carried out his search by bicycle in 1973 for suitable sites and located Wamba (Kano 1986). S. Kuroda (1982) underwent his initiation into fieldwork in Africa by studying bonobos.

In 1972, Itani published *Primate Social Structure*. He reviewed the nonhuman primate societies known at the time (more than half of all primate species) and attempted to trace the evolutionary path of human society. By comparing "diachronic" social structures, he concluded that the matrilineal group structure found commonly among the cercopithecoids, in contrast to the anthropoid apes with their apparent absence of matrilineal structure, is unlikely to be linked to the origin of human society (Itani 1972, 1977b, 1980, 1985a).

STAGE 3: 1975–1985

By the mid-1970s, the use of statistical tests and quantified observational methods (J. Altmann 1974) was widespread. The Japanese macaque studies in this stage were "characterized by resolutions to several important problems proposed by preceding studies" (Itani 1983): T. Enomoto (1978) found that copulation rarely occurs between closely related individuals; A. Mori (1975) found intertroop variability and flexibility in the use of a vocal signal; K. Watanabe (1979) studied alliance formation; K. Kitamura (1977) recognized a persistent male-female "peculiar proximate relationship" (PPR); Y. Takahata (1982a, 1982b) confirmed the presence of PPR and found statistically significant avoidance of copulation in PPR dyads as well as within the third degree of consanguinity; and K. Sugawara's (1980) study of encounters of solitary males at Koshima was also unique. Preceded by U. Mori (1974), who focused on interindividual relationships observed in play, H. Hayaki (1983) analytically described the senior play partner's self-handicapping. A sociobiological approach was gradually attempted during this stage, first with regard to development of infants (e.g., Hasegawa and Hiraiwa 1980; Hiraiwa 1981; Hiraiwa-Hasegawa 1983).

By this stage, ecological aspects of the Japanese macaque also became a focus of attention. T. Iwamoto (1974) from Kyushu University initiated research on the dietary preferences of Japanese macaques, and K. Masui (1976) compiled demographic data over their entire distribution range. The long-term record of the Takasakiyama population was also analyzed (Masui et al. 1975) and, in the mid-1970s, a younger generation of primatologists began to habituate the Japanese macaques without provisioning. The site chosen was Yakushima (Maruhashi et al. 1986). First, long-neglected ecological topics were studied (e.g., Maruhashi 1980), giving way to sociological and behavioral studies (e.g., Furuichi 1983; Mitani 1986). Accumulated ecological data allowed statistical analyses to construct general dynamics of some ecological parameters of the species (Takasaki 1981; Takasaki and Masui 1984). In these analyses, troops were treated as ecological units as well as social units (Maruhashi and Takasaki 1996). At the end of this stage, a research manual of the Japanese macaque was written (Takahata 1985).

Waves of the global explosion of human population eventually reached the shores of Lake Tanganyika and, even in the vicinities of the Mahale Mountains, threats to the wildlife became a concern. In 1975, the Japan International Cooperation Agency (JICA) became involved, in response to requests by the Tanzanian government, in creating a national park (Itani 1993; JICA 1980). In 1985, the Mahale Mountains area was finally announced as the eleventh national park in Tanzania, and the habitat and the estimated population of about 700 chimpanzees became better protected. A number of fieldworkers have served as long-term JICA experts there, and their basic research and the parallel continuation of studies by Nishida and his associates have produced findings too numerous to cite in this review (see Nishida 1990b). However, two of the most striking findings were the repeated incidents of infanticide (e.g., Kawanaka 1981; Norikoshi 1982; Takahata 1985b) and the extinction of a unit group (Nishida et al. 1985).

In 1981, the Laboratory of Human Evolution Studies (LHES) was newly created out of the LPA, although it continued to be housed with the LPA. Itani headed the LHES, and continued training students of ecological anthropology and primatology. Well established by this stage, most of his primatology students did their master's research on the Japanese macaque and their doctorates on either the chimpanzees at Mahale or the bonobos at Wamba. Fieldwork was strongly encouraged. The most physically fit among his students were dispatched to try to study the gorilla again (Yamagiwa 1984, 1993).

By this stage at the PRI, ecological, sociological, and behavioral fieldwork was established in Cameroon (chimpanzees, gorillas, mangabeys, guenons, mandrills, drills, patas monkeys), Ethiopia (gelada baboons; see Kawai 1979b), Guinea (chimpanzees), Indonesia (macaques, orangutans), Malaysia (macaques), and Zaire (bonobos)—and, of course, in Japan (Japanese macaques). The study sites covered all vegetation types of primate habitats. The title of a book by Kawai (1979a), *Forest Produced the Primates,* may conveniently summarize the divergence of field primatology in this institute. Two of the early members of the Kyoto school, Sugiyama (1978, 1981) returned to the chimpanzee again, now in Guinea, and Suzuki (1985, 1992) started to study the orangutan in Borneo.

STAGE 4: 1985 TO PRESENT

In Stages 3 and 4, the journals in which the Kyoto school primatologists publish their papers became more international. During the same period, collaborative efforts between foreign researchers and Japanese primatologists were apparent. D. S. Sprague from Santa Cruz, M. A. Huffman from Colorado, P. J. Asquith from Oxford, D. A. Hill from Cambridge, and L. A. Turner from San Diego all underwent a successful period of participant observation in the Kyoto school. They have become ambassadors between the West and Japanese primatologists.

In 1985, the Center for African Area Studies (CAAS), Kyoto University, was established, with Itani as the director. Moving there from the PRI, Koyama began to organize lemur studies in Madagascar (Koyama 1991). In 1988, Nishida moved to the LHES.

In July 1985, only in the very last stage of development, the Primate Society of Japan was established. Incidentally, this society is one of the most concerned with conservation activities of all academic societies in Japan. A workshop on primate conservation is almost always held at the annual meeting of the PSJ. Japanese macaque conservation has become a particular concern since these monkeys have recently caused problems for some farmers.

Kawai and Kawamura retired from the PRI in 1987 and Sugiyama and Kano filled their niches. Kawai edited *Prehuman Sociology* (1990), which was to commemorate his own retirement. This voracious reader, editor, translator, and writer also published a book having the classic title *The Descent of Man* (1992). Itani retired from the CAAS in 1990 and *The Chimpanzees of the Mahale Mountains* (Nishida 1990b), *Cultural Historiography of Monkeys* (Nishida et al. 1991), and *Natural Historiography of Man* (Tanaka and Kakeya 1991) commemorated his retirement. These collections of articles illustrated the diversification of primatological and anthropological research in the Kyoto school.

Results of long-term studies continued to be published—from the Japanese macaque at various sites (e.g., Agetsuma 1995; Hill and Okayasu 1996; Koyama et al. 1992; Mori and Kudo 1986; Muroyama 1991; Nakagawa 1994; Oi 1988; Soumah and Yokota 1991; Sprague 1989; Takahata et al. 1994; Tanaka 1995; for more recent findings, see *Primates* 39[1], 1998, devoted to this species), to the chimpanzees at Mahale (Hamai et al. 1992; Hayaki 1990; Tsukahara 1993), and the bonobos at Wamba (Furuichi 1988; Hashimoto 1997; Idani 1990; Ihobe 1992). Some of the formerly poorly known species were also moved to the list of better studied primates (e.g., the mandrill in Cameroon by Hoshino 1985; the Sulawesi macaques by Watanabe and Matsumura 1991; and the Tibetan macaque by Ogawa 1995). In 1988 Mitani (1996), who traveled on foot into the swamp forest upstream the Ndoki in Congo, located a new site where the western lowland gorilla (*Gorilla gorilla gorilla*) and the Tchego chimpanzee (*Pan troglodytes troglodytes*) live sympatrically. Thus, another long-term ape study site was found. T. Nishihara spent more than a year there, and published preliminary data of the Ndoki gorilla's feeding ecology.

Another notable aspect of this stage is the fusion of field primatology with other primatological disciplines and sciences. T. Matsuzawa, a PRI-based cognitive psychologist studying chimpanzees, and his students came to be involved in the chimpanzee field research at Bossou (Matsuzawa 1995). M. Huffman, in collaboration with researchers in various disciplines, has probed into the medicinal plant use by chimpanzees (Huffman 1993). Also advances in molecular biology brought a formerly undreamed-of method to field primatology. O. Takenaka's molecular biology team at the PRI, in collaboration with monkey watchers, applied minisatellite DNA fingerprinting to a Japanese macaque troop caged at the institute, and they detected no correlation between a male's rank and his number of offspring (Inoue et al. 1990). In a wild patas monkey population in Cameroon, DNA extracted from blood samples revealed a fair chance for nontroop males to sire offspring (Ohsawa et al. 1993). Even minute tissue debris (e.g., single hairs and buccal cells in wadges) were used for microsatellite DNA typing (Takasaki and Takenaka 1991). Using this method, Sugiyama and colleagues (1993) revealed a case of probable paternity by a nongroup male in the Bossou chimpanzee population, and C. Hashimoto (1996) reconstructed a missing part of maternal kin relations in a group of bonobos at Wamba by sequencing the D-loop region of mitochondrial DNA.

Imanishi, the founding father of the school, died in June 1992. He left behind fourteen volumes of his work (Imanishi 1974–1975, 1993) on ecology, anthropology, primatology, and mountaineering. J. Yamagiwa's (1994) *Origin of Human Family* has its undeniable roots in Imanishi's classics.

Characteristic Features

Many, if not all, of the Kyoto school fieldworkers regard themselves primarily as anthropologists and not primatologists. In particular, those in the Kyoto Campus have always been housed together with physical, cultural, social, and ecological anthropologists. Even some of the field primatologists have been cultural, social, or ecological anthropologists too. *The Forest of Gorillas and the Pygmies* (1961) and *In Search of Chimpanzees in the Wilderness* (1977a) by Itani are not only narratives of a primatological expedition but also ethnographies. So are *The Children of the Mountain Spirits* by Nishida (1973b) and *The People and Chimpanzees of the Bossou Village* by Sugiyama (1978). Kawanaka translated R. Fox's (1977) *Kinship and Marriage. The Fire of Elya* (1987) and *One Man's Tales of the Forest* by Kano (1997) are collections of folktales of the Mongo in Zaire. With *Forest Storytellers in Cameroon* (1992), A. Mori proved he is also a folklorist. Kitamura (1990) and Sugawara (1991), both of whom studied nonhuman primates for their doctorates, now watch human primates. The very reason these researchers have chosen primates as their study subjects, then, is to gain some insight about *human* nature. Thus, Hayaki wrote *The Human within the Chimpanzee* (1990).

Anthropomorphism has often been cited as one of the characteristic features of Japanese primatology (e.g., Asquith 1981, 1986a; Kitahara-Fritsch 1991). In the Kyoto school, this was evident from the very beginning. It was not simply because of the man-animal closeness rooted in the Japanese culture, in which man is not regarded as a special being.

Imanishi advocated strategic use of anthropomorphism as a means of studying nonhuman primate and mammal societies, and the underlying reason that justified anthropomorphism is man's relative phylogenetic closeness to primates. These ideas implicitly assume evolution and man's place in nature. This unhesitating adoption of anthropomorphism can be traced back to Imanishi's view of the world of living things (Imanishi 1941b). He repeatedly emphasized this idea in his writings (e.g., Imanishi 1952), and the boldness of his reasoning, naive but neither careless nor baseless, was the secret for the seemingly easy discovery of nonhuman primate culture by the Kyoto school primatologists (Itani and Nishimura 1973). It was also the reason for remarks on monkey personalities from the very early stage (Itani 1954, 1957; see also his editorial notes of *Primates* reviewed in Asquith 1996).

Less frequently cited, but probably more important to an understanding of the development of the Kyoto school, may be the Japanese way of perceiving primate societies. Let us focus on this point for the rest of this chapter, with particular reference to their approach to understanding chimpanzee society (Takasaki 1984).

Wild chimpanzees repeatedly fission and fusion in grouping, which confused primatologists greatly in the beginning. Suppose a primatologist is faithful to the reductionist dictum: "As long as a phenomenon explainable by a simple model with lower level elements, no more complicated model assuming a higher level structure is necessary." If scientists carry out only short-term observations of wild chimpanzees, they will conclude, "Among chimpanzees there is no stable association other than that between the mother and offspring." Indeed, there were several researchers in the West who reached this conclusion (e.g., even Goodall 1965b; but see also Goodall 1990).

By contrast, the Kyoto school researchers continued to study chimpanzees for many difficult years in the bush, as if believing that the highly social chimpanzee was unlikely to have an uncomplicated social grouping above the individual level (Itani 1977a, 1977c, 1993). At last, Itani and Suzuki (1967) observed a procession of chimpanzees with a presumed multimale-multifemale composition of the unit group. Later, at the foot of the Mahale mountains, two neighboring unit groups, K and M, were found, and the discovery of female transfer between unit groups soon followed (Kawanaka and Nishida 1975; Nishida and Kawanaka 1972). Studies of various interindividual relationships within each unit group, as well as interunit-group relationships above the unit group level, have continued since then. Here, it is interesting to note that, although in recent English primatological literature the chimpanzee "community" and "unit group" are used interchangeably, the Kyoto school researchers originally used "unit group" so as to reserve "community" for a possible higher level structure comprising two or more unit groups (Itani 1980).

In the West, on the other hand, a different explanation still prevails, which emphasizes the differences in ranging behavior between males and females. To paraphrase Wrangham (1979), "Although male chimpanzees associate to form communities, females do not." This depicts the chimpanzee society such that females forage in quest for food, while allied males secure their community territory against others and mate with females who happen to roam in their community range. The Kyoto school chimpanzee researchers have presented evidence to the contrary (Hasegawa 1990; Kawanaka 1984; Uehara 1981). In the pictures they drew from data, both males and females associate to form the unit group. Incidentally, the bonobo society has also been found to fit with this bisexual unit group model (Idani 1990; Kitamura 1983).

Obviously, the sociobiological approach did not win much support within the Kyoto school. However, this does not mean they were ignorant of it. During my undergraduate years (1976–1980), many students in zoology courses were reading E. O. Wilson's (1975) *Sociobiology* and R. Dawkins's (1976a) *The Selfish Gene.* Most of those who might have been fascinated by sociobiology or behavioral ecology, however, went into disciplines other than primatology—entomology, ecology, or ethology. The presumably omnipotent theory of sociobiology was not to the taste of many of the Kyoto school primatologists.

Imanishi's (1941b, 1949b) influence during the formative years of the school was profound. He proposed the concept of "species society" (*specia* in Imanishi's terminology) after recognizing the "habitat segregation" in space and time between phylogenetically close species. As a means of probing into the internal structure of each species society, studies of animal societies with individual identification were initiated. In an attempt to trace the evolution of human society, Imanishi wrote *Prehuman Societies* (1951), which compared nonhuman animal societies.

By admitting the entity of "species society," it becomes possible to assume its subunits. This conceptual framework has been particularly useful for studying highly social primates, as well represented in Itani's (1972) *Primate Social Structure* and subsequent papers (Itani 1977b, 1980, 1985a). This framework assumes that a species society is composed of its member individuals at the lowest level, and postulates that congregating individuals may form a unit at an upper level above the individuals. It allows us, as necessity arises for interpreting observations, to assume further suprastructures. As the species society and its member individuals are simultaneously assumed, assumption of another level of social structure readily follows at an intermediate level in the pathway of thinking. By observing a Japanese macaque troop, anyone can see it is not unreasonable to regard the troop as a social unit above the individual level. It is not much different from perceiving the body as a highly organized multicellular organism consisting of cells that form tissues, organs, etc. in intermediate subindividual levels.

The above illustrates the sociological approach typical in the Kyoto school, which portrays animal societies as an extension of Imanishi's "species society" concept. Even some ecological studies of primates in this school may be better interpreted in this framework (e.g., Maruhashi and Takasaki 1996). Today, many Japanese primatologists, even in the Kyoto school, shape their papers deliberately so as to make them compatible or interfaceable with the international majority. However, the school's discipline has been developed with a different character. The Kyoto school and the sociobiological approach differ in paradigm. Favorably speaking, they are complimentary to each other. Unfavorably speaking, they are in a relationship to cross each other, or even never cross but pass each other without meeting.

Conclusion

Traditions in the Kyoto school—strategic anthropomorphism, structural view of primate societies, highly valued long-term fieldwork, etc.—can undeniably be traced back to Imanishi. However, the shared worldview of many of the Kyoto researchers seems to be the driving force behind the distinctive character of this school of primatology.

Imanishi's worldview was clearly expressed in his *World of Living Things* (1941b). Those who find it attractive are likely to share a similar view of the world of living things, created by neither the Almighty God nor any omnipotent theory of evolution. Imanishi envisioned a world or "society" of living things, and, as an extension of it, he attempted to picture its evolutionary path. Evolution, i.e., the unfolding of the self-development of the world of living things, is implicitly assumed in this worldview. Imanishi's view of evolution, or "history," is inseparable from his view that the society of living things is composed of elements existing in space. In this view, an evolutionary pathway is simply an extension of the "society of living things" existing across time. In comparison with sociobiology or behavioral ecology, which attempt to explain the world of living things in the tradition of Darwinian theory of evolution, "Imanishian biosociology" has an upside-down paradigm structure.

Language has been and still is the biggest obstacle for Japanese primatologists when communicating with colleagues overseas. Perception of the world is, to a considerable degree, bound by the cognitive structure of the language ("language" here might be better rephrased as "culture-language complex"). In the Japanese language, my address, for example, is arranged in the order—Japan, 700-0005, Okayama, Ridai-cho 1-1, Okayama Science University, Anthropology Laboratory, Takasaki Hiroyuki—completely the reverse of its order in English. This illustrates how a person's spatial position relative to the rest of the world is perceived in Japanese. The above upside-down paradigm structure may not be unrelated to the perception of the spatiotemporal world structure in the language.

Although science is supposed to be universal, each scientist is a cultural being, affected by his or her own cultural background. The Kyoto school primatologists have tried, during the past three decades, to communicate with those who do not share the same school culture, and at the same time have tried to absorb the theories and findings presented by others. The gap, however, may seem to have hardly been narrowed. This is not necessarily, however, a bad situation. As long as the primatologists of the Kyoto school continue to shed light on aspects neglected by others, their contribution to primatology and human knowledge remains distinctive. In other words, the Kyoto school will only vanish when its researchers start to produce "good, perfectly compatible" papers that are little different from those presented by the rest of the scientific community.

Intercultural and school differences, as well as individual idiosyncrasies, are essential to the advancement of science. Such differences are comparable to the biological diversity and the gene stock that are necessary for the global ecosystem's stability and for evolution of new diversity. There are always some parts of the world that may be viewed better upside down.

Acknowledgements

Besides the symposium participants, organizers, and editors, the following people read and commented on earlier versions, which greatly contributed to the revision: H. Ihobe, J. Itani, K. Izawa, M. Kawai, K. Kawanaka, M. Mitani, T. Nishida, K. Sugawara, Y. Sugiyama, S. Uehara, J. Yamagiwa, and K. Takasaki. To these people, I make grateful acknowledgment.

8 Negotiating Science: Internationalization and Japanese Primatology

Pamela Asquith

. . . the scientific paper is a kind of fraud, for its neat format bears no relation to the way in which scientists actually work: imagination, confusion, determination, passion—all the features associated with scientific creativity have been purged from it. *(Wolpert 1992, 101)*

Introduction

An additional "reformatting" of scientific practice to that noted above by Wolpert (1992) is the standardization and translation of science—not between languages, but between scientific traditions—by journal editors and reviewers, as well as by scientists themselves, mindful of dominant idioms. Our "Teresopolis" discussions about national traditions in primatology, and whether national traditions even exist, despite the past forty years of efforts to standardize primatological publications, raised questions with regard to Japanese primatology that are pertinent to understanding *inter*-national contributions to the discipline. These questions devolved to the following three issues. One, particularly for those with experience of reviewing Japanese papers for publication in international journals, was "why" do many Japanese continue to write very descriptive papers—what are they accomplishing by this? Secondly, Teresopolis participants observed that among those Westerners who have worked with Japanese, or in field sites adjacent to Japanese researchers, some perceive no particular differences between Western and Japanese practices of primatology, whereas others "see, but cannot understand" differences. These same two impressions have been expressed by Japanese researchers as well. Why is there this difference of perception, and what are the "differences" that are so difficult to articulate and to understand? The third issue was: "Is there not more than one Japanese 'approach' to primatology, one being in the Western (internationalized) mode, and another based on a 'traditional' approach, which Westerners probably would not understand?"

This chapter addresses these questions as a series of *negotiations* (about terminology and the meaning of data, about "air time" in which to express findings and ideas, and about scientific and national identities) that take place within and between primatological traditions. And I do believe these traditions exist. I think it is helpful to perceive Japanese scientists as negotiating their way through a set of formal and informal rules and expectations in preparing research for publication. Evidence for underlying differences between traditions appears at junctures of attempting and failing to put something into words for the international discourse; when editors and reviewers make decisions about publication, and during informal occasions when Japanese feel freer to speak their mind. This chapter explores the negotiations in which Japanese primatologists must engage in order to participate in international science. As an anthropologist studying this Japanese scientific community, my intent is to act as a purveyor of Japanese perspectives on these issues, and to put those perspectives into context for those who are outside the Japanese tradition.

The material in this chapter is based on my participant observation of Japanese primatologists from 1981 to 1984, a recent set of interviews of a cross-section of the same and new primatologists in 1997, and occasional visits to Japanese field sites abroad and in Japan in the intervening years. Additionally, the views of current Japanese graduate students are incorporated for their perception of the discipline and a glance at the future. I will try to use the words of individual Japanese scientists[1] to address the issues that were raised at Teresopolis and the issue of negotiating international science. In this way, they themselves are allowed to discuss and assess, to evaluate primatology within Japan, and to give their views of foreign primatology.

Background on Japanese Primate Studies

What do Japanese primatologists mean by what they say? Pertinent to an understanding of their current views, and to the first question of why some Japanese primatologists continue to submit descriptive papers for publication, is a brief review of the historical goals, concepts, and methodologies of Japanese primatology.

The study of primate behavior in Japan was undertaken with two major goals: one was to contribute to an understanding of the "whole life" of the "whole species" or species society (Asquith 1991; Imanishi 1957; Sibatani 1983); another was to trace the evolution of human society (Nishida 1990a; Takasaki this volume). Studies of other Old World and New World primate species were undertaken from early on, but the initial training of primatologists was based on the endemic Japanese macaque (*Macaca fuscata*), and initial methodologies and concepts were carried over from these studies. Kinji Imanishi (1941a), founder of Japanese primatology, recorded his view of nature and nature study in his seminal book *The World of Living Things,* which was written in Japanese. The book was published years before any Japanese research on primates, but some of the ideas expressed therein were foundational to Imanishi's and his students' approach to the study of primates. In this book Imanishi introduced the concept of the "species society," or *specia.*[2] He referred to the specia for the first time in English in a paper written for *Current Anthropology* (Imanishi 1960) in which he summarized results of the first decade of Japanese primate research. Few Japanese primatologists mentioned the species society in English-language papers written between 1950 and 1970, but it underlay an entire approach.

The study of the species society was a study of social relations among the members of the group, and among other groups of the same species. The concept was developed to fill the lacuna that Imanishi perceived between the study of whole communities of species and that of the individual animal.[3] To that end, the researchers gathered details on intraspecific variation in behavior and group structure, historical change in groups, individual life histories, and so forth. They believed that all the variability over time and place of these cultural, individualistic animals must be identified before one could understand the overall structure—hence, their long-term studies of thirty and more years. This concept determined, in part, the descriptive nature of Japanese primatology during its first two decades, as did the logistics of long-term studies, involving a group effort covering decades and generations of fieldworkers, each with his or her own project, as well as the larger research question to which they contributed data. Who was to say what details of behavior might not be useful in the future for determining the larger picture?

Early Japanese studies of primates were described as sociological (Imanishi 1957) as they sought the structure, at several different levels, of the species as a whole, not only of a particular group. Intraspecific variations in social structure and behavior were rarely discussed in relation to ecological variation, but were considered rather from the viewpoint of cultural differences[4] (see Yamagiwa and Hill 1998 for a recent review). This is characteristic of what came to be called the "Kyoto school" of primatology. Indeed, as Takasaki (this volume) points out, many of these primatologists regard themselves primarily as anthropologists. This was the largest and most influential nexus for primatology in Japan. However, while those trained at Kyoto and their students constituted the majority of primatologists in Japan for at least two decades after primatology began, individuals later followed their own research directions and preferences as they migrated to other institutions throughout Japan. In particular, there was a division between those who preferred a biological approach to primatology and those who retained a sociological or anthropological approach. Furthermore, the Osaka-trained primatologists had a separate history altogether, stemming from ethology and comparative psychology, though they too use long-term studies and make similar observations about Japanese and Western primatology as do the other primatologists described below. Some maintain that the Kyoto school no longer exists (but cf. Takasaki, this vol.), and others contend that neither Imanishi nor Junichiro Itani (of whom more below) ever intended to make a "school." In fact, for those who think the latter, the most important characteristic of the Kyoto pioneers was the emphasis that was placed on originality, new ideas, and doing work that had not been attempted before. They were quite prepared to throw out old ideas if better ones became available.[5]

Nonetheless, the goals of Japanese primatology as stated earlier exerted an influence over new generations of primatologists in Japan. Junichiro Itani has been especially influential in consolidating Japanese primate research and facilities for it, in training new generations of primatologists, and in promoting some of Imanishi's views. Of relevance to our question about descriptive reports is the fact that Itani headed the Kyoto school for many years and provided the intellectual milieu described above for the majority of Japanese primatology students. Although Itani's enormous abilities and contributions have been recognized by the international academy,[6] I was told that many of his ideas were not introduced to the West. Itani "wrote for his students," according to one informant. Many students contributed data to the underlying goal of understanding the species society, without giving the context. Thus, Western readers saw only the data and not the purpose for it. The apparent absence of a context or rationale[7] for descriptive contributions was perhaps a result of such factors as working within a particular research paradigm (in this case, the species society and the search for the evolution of human society). The larger context was tacitly understood by members working within the paradigm, and often on the same groups of primates, lending a descriptive, open-ended character to reportage.

Negotiating Description

Current reflections on the descriptive nature of Japanese reports help to dispel some of the exoticism that perhaps surrounded the specia concept for Westerners. Most primatologists say that the concept is no longer used at all. Yet, an examination of currently stated reasons for descriptive accounts reveals that they are not an entire departure from prior rationales. This is true for comments from primatologists of all generations and all schools and institutes in which interviews were carried out.

There has been, since the mid-1970s, a concerted effort to elucidate ecological factors influencing the social organization of Japanese macaques and other species. With recognition of structural and behavioral disparities between provisioned and unprovisioned populations of Japanese macaques, researchers have gathered long-term comparative data on unprovisioned groups at Kinkazan and Yakushima (Yamagiwa and Hill 1998). Yet, despite twenty years of data from these sites and forty years from provisioned sites, Yamagiwa and Hill (1998, 268) remark in their concluding paragraph that "[i]t is (still) too early to elucidate the evolutionary processes that led to the social organization or ecology of the Japanese macaques . . . ," and that "[t]he tentative nature of many of the conclusions reflect the need for more data for these two sites." Further, "[i]deally additional sites should be established between these two, to provide data for conditions between the two extremes." It is safe to say that contemporary Japanese primatologists are as committed to long-term study as were the pioneers, despite the development of new perspectives. Indeed, Yamagiwa and Hill (1998) remark that because current fieldworkers' primary research goals often differ, there is a need to coordinate the collection of basic data by a series of fieldworkers working at different sites and times so that data can be applied to the larger question of evolutionary processes that led to the social organization or ecology of Japanese macaques.

Another question that has been held over from pioneering days for some primatologists is "How did hominization occur?" According to one informant, we first need a full description for primate species of nutrition, ecology, behavior, structure, and so forth, all of which will feed into an explanation of the evolution of human society. Hypotheses and interim conclusions about hominization have not yet been written about much, as this is thought to be too speculative, since not enough is known. The fact that North American researchers have not been as reticent to speculate about hominization is readily apparent. These differences are reminiscent of the views held by American and European anthropologists of each other in the 1960s (and, it seems to me, today as well). While Americans described European ethnographic publications as being "like [those of] Japanese, descriptive and cumulative," from the European perspective, American approaches, though fresh, dashed through layers that European ethnographers found vital to their understanding of the society (Hofer 1968).

Several primatologists expressed the view that papers written for Japanese readership (published in the Japanese-language professional journal *Reichôruikenkyû* [*Primate Research*]) are far more descriptive than those written for international readership. They suggested that lengthy "objective" description leaves the data open to other interpretations and thus that a reader can "feel closer" to the animals. One researcher explained that each person experiences the monkeys for him- or herself through reading detailed descriptions of place and activity.[8] Examples were cited of papers containing lengthy descriptions of behavior that were prepared in Japanese for *Primate Research,* but were later reduced to two lines for publication in the *International Journal of Primatology.*

According to several primatologists, one thing that is missed in nondescriptive reports is the anecdotal input. Some informants noted that Japanese recognize the importance of anecdotal material.[9] A great many students who were embarking on their first field research were told by the pioneering mentors to simply "go and look" and to "get an impression of the animals and their lives." Many wrote everything down. One informant noted that many first-generation primatologists felt that writing impressions was important and that long descriptions often resulted in monographs. Yet the "gaining of impressions" about the animals' lives was also seen by informants to be an important and powerful tool to identify what may be important (to test), one which enabled the observer to take many factors into account in forming hypotheses, and to remain free of prevailing opinions and views about animal behavior. I will return to this last point in the section on Japanese views of Western primatology.

On the negative side, this has led to a great many boxes of unused (and probably unusable, because unreadable, by any other than the original authors) data. Additionally, at those Japanese primate sites where data were not continuously collected by keepers (or others) between researchers' visits, no firm conclusions can be drawn about many aspects of group structure, such as change in group membership. Certainly, not everyone thinks that rampant description is a good thing. In one informant's estimation, during the 1970s, the journal *Primates,* which was the major outlet for Japanese papers at the time, became rather "thick" (with long descriptive papers) and its quality went down for a time. Another remarked that he thought many of the descriptive papers produced in Japan were mainly a result of a lack of knowledge of Western scientific methods.

Negotiating Language

Closely associated with misunderstandings about description is the "problem" of language. Almost without exception, primatologists said that they could not express what they wanted to say in their English-language reports. This partially accounted for their avoidance of speculation and their omission of the larger context for their descriptions in published papers. Graduate students remarked that new ideas may not be published because of their inability to defend them comfortably in English. They also noted that foreign language is rarely the strong point of science students and that they feel they are rather poor in English. Several remarked that writing the discussion part of the paper is the most difficult. Yet, this constitutes a real conundrum: Japanese graduate students in primatology are told that they may as well not bother to publish in Japanese as this will not get them anywhere professionally. They must publish in English, and at some institutions, such as Kyoto University, publication of papers in English has been one of the requirements for the doctoral degree in science.

Those who do attempt to say more often find passages cut by journal reviewers or editors. I have heard many, many complaints by Japanese researchers that English-speaking reviewers and editors cut much of the detail from their submitted articles and always ask that the paper be modified to address a specific topic if it does not appear to do so. Two comments, made seventeen years apart, reveal how alarmingly inert the international community has been toward this problem of language. In 1980 a Japanese primatologist related to British primatologist Vernon Reynolds that "[t]he biggest problem which we can't write in our scientific articles in English is the basic way of thinking by the language barrier" (V. Reynolds, pers. comm., Mar. 21, 1980). In 1997, Naosuke Itoigawa (pers.

comm., Feb. 24, 1997) commented to me that "[w]hat is written in English is not what we want to say or what we are thinking. English writing is somewhere in between, and it is not descriptive enough."

Negotiating "Air Time"

Any scientist knows that publication in respected, accessible venues is essential for exposure of their ideas and findings in the scientific academy. Scientists outside the European and North American centers have often claimed that they are marginalized and not taken seriously. This claim is open to debate (Gibbs 1995), but it is a perception that has been shared by Japanese scientists in some fields, even though clearly, they are recognized as world leaders in others.

JOURNALS

Historically, the assessment of Japanese contributions to primatology is revealing in terms of where papers have been published. The first international journal of primate behavior was *Primates,*[10] begun in 1957 by the Japanese. Since 1959, the journal has been published in English. Other international Western-based journals in which primate behavior articles regularly appear, such as the *American Journal of Physical Anthropology, Folia Primatologica, Journal of Human Evolution,* and the *International Journal of Primatology,* as well as nontaxon-based journals such as *Nature, Animal Behaviour, Behaviour,* and *Zeitschrift für Tierpsychologie,* have been perceived to rank higher than *Primates* by some Western and Japanese primatologists. As for Western-based specialist primatology journals, the first of which began in 1962,[11] it was not until 1976 that a Japanese-authored paper appeared in one (i.e., Nishida 1976). The dates for first publication of a Japanese-authored paper on primate behavior in some of the other nontaxon-based Western journals range from 1973–1988.[12]

A common observation among scientists who perceive themselves to be marginalized is that their contributions are ignored or refused publication in mainstream journals (Gibbs 1995). If we consider contemporary primatology, however, in terms of citation of findings in international journals, Japanese research over the last twenty-five years appears to be wholly accepted. After all, the Japanese must publish in English and, since the late 1970s, they have published in Western-based journals as soon as their papers have been put into a suitable form for those journals. During the same interval, there has been a steady increase in the number of Western-authored papers that appear in the journal *Primates.* And, despite the lingering perception in some quarters (of both Japanese and Western primatologists) that it is less desirable to publish in *Primates,* according to the journal's "impact factor," *Primates* has occupied the second and third rank among the specialist primatology journals for 1994–96.[13] Rather, the point of marginalization is in how their papers reach an "internationally standard" form for publication.

Western primatologists and reviewers have told me that they often cannot discern the main point of the Japanese-authored articles. Some reviewers work quite hard to "improve" the papers to make them acceptable to the dominant discourse. At one level, it is a simple problem of English fluency or lack thereof. However, it is more fundamentally the case that Japanese authors often cannot express their own paradigm adequately in English. The problem here is whether members of one tradition understand what members of the other mean to say. Western primatologists appear to have not fully understood what underlies Japanese approaches to primate studies and they also appear to be largely unaware of different paradigms, or to think that these have no repercussions for the discipline anyway. Evidence for this comes from early Western responses to Imanishi's description of the Japanese approach (1960, and see comments following his paper) and, more recently, in the comments arising from both Japanese and Westerners in dialogues such as those described throughout this paper.

TRANSLATION AND DISCUSSION

Over and above citation, the willingness of the international community to give full voice to the Japanese contributions can be measured by how much non-Japanese speakers wish to hear the discussions that Japanese would be more comfortable holding in Japanese. In other words, will we provide translation facilities for even a couple of symposia at international meetings; for instance, those with topics that are likely to generate a lot of debate? Meetings of the International Primatological Society (IPS) never have done so; the 1994 Wenner-Gren-hosted meeting of great ape researchers held in Mexico, which included several Japanese primatologists,[14] did not (even though the entire week was spent in discussion, not in giving carefully rehearsed papers); nor have any other international meetings and workshops that have been held in the West and included a large component of Japanese contributors, with but one exception.[15] There is an additional loss here: we do not learn the diversity of Japanese opinion from these anglicized contributions to primatology. Of course, this is not restricted to primatology. Scientists (and social scientists) throughout the world must publish in English for maximum exposure.

Doubtless, some of the same kinds of losses are occurring in other fields as well.

Negotiating Identities

JAPANESE VIEWS OF THEMSELVES

Internal critiques of historical aspects of Japanese primatology center on several factors: on Imanishi's anti-Darwinian views (e.g., Hiraiwa-Hasegawa 1992; Sakura et al. 1986); on provisioning (especially once ecological studies got underway; see Yamagiwa and Hill 1998 for a review); on the undue influence of old ideas from cultural rather than biological perspectives in the study of primates; and, related to this last kind of influence, on the culture of science in Japan, in which open criticism outside their own schools is rarely voiced, certainly not by younger to senior scientists.

It should be stated that anti-Darwinian views were not taken up by Japanese primatologists (Asquith 1986b; Sakura et al. 1986; see also Sibatani 1983), even though they were by other segments of Japanese society. Imanishi was against the idea of competition as a driving force in the natural world (which he identified with Darwinian evolution), and this did feed into his "species society" idea. Mariko Hiraiwa-Hasegawa is one of the very few Japanese who has commented on the history and methods of their discipline to a Western audience. Her critique of Imanishi was directed to two aspects of his work: his idea of the relationship between an individual and its society (based, Hiraiwa-Hasegawa felt, on a personal rather than a scientific view of the organic world), and his evolutionary theory (Hiraiwa-Hasegawa 1992). The appearance of Richard Dawkins's (1976a) *The Selfish Gene* constituted a paradigm shift for Hiraiwa-Hasegawa (the book was later translated into Japanese, in 1980). She had read Lorenz, Tinbergen, Hinde, and so forth, but had been unable to put their ideas together as a result of the paradigm in which she had been trained. In her understanding, the pioneers of primatology were not oriented to biological explanations for behavior. Indeed, three of my informants from separate schools noted that, during the 1970s, primatology students were discouraged from introducing material into seminars from papers by Hamilton, Trivers, Wilson, or Dawkins, who were, of course, forging the sociobiological paradigm.

On the other hand, it should be noted, as many do, that it is Imanishi's early publications and approach to the study of nature that were influential. More than one primatologist has said that he would take Imanishi's

(1941a) first book for desert island reading. Several primatologists who were among the first and second "generations" of Japanese primatologists have remarked that this book, and his *Logic of Living Biosocieties* (Imanishi 1949a) were among his most influential writings for them. Perhaps more importantly, Imanishi introduced and critiqued Western theories in a way that most other professors at the time did not. There was much stimulating discussion at Imanishi's seminars, and students could learn to attend these discussions without fully accepting Imanishi's own views.

As for the current situation growing out of this history, there are, as would be expected, varying points of view, including some that contradict each other. Some primatologists felt that Imanishi and Itani made a canopy under which many worked, and that although that paradigm has had negligible influence for many years, the Japanese have not subsequently developed their own method and theory. Contemporary graduate students state that, of the pioneers, Junichiro Itani's work is read most. Several (students and professors) have said they feel that Japanese copy from the West, yet the Western paradigms are not wholly satisfactory. Japanese researchers study ecology, behavior, and a wide range of problems, and they write in English, but, some say, they do not have a larger picture; one commented that there is "no perspective now" on Japanese studies, and no new ideas. Self-critical comments varied among those who said the current research questions are too easy; that people are belaboring small points; that they cannot express new ideas convincingly in English; or that there is a lot of imitation of Western methods.

Many problems were discussed with respect to description. Professors say that as graduate students they were told just to gather lots of data, and then were asked what they thought the data meant. More recently, graduate research projects sometimes are not worked out in advance and some find it difficult, therefore, to write up results. A graduate student commenting on the experience of several fellow students remarked that this lack of planning before fieldwork (the "go and see" approach) led to a sort of "data mining" after the return from the field as the student struggled to organize a mound of descriptive notes on behavior, did statistical tests, and tried to find a theory into which it would all fit. One professor remarked that it used to be easy to use data for behavioral description, but now the emphasis on ecology means that they cannot use descriptive behavioral data so much. In fact, across the schools and institutes, there is as much variability in graduate student supervision as one finds in Western universities. Students either decide themselves on a topic, have an idea assigned, or just go and observe for a time, and then decide. However, as was pointed out by a professor, the idea of "go and look" is difficult as experience is needed in order to "see things" and to interpret behavior. Another professor noted that, contrary to what he identified as the Kyoto school's natural history approach to field study, some young researchers now scarcely want to see the monkeys' habitat or behavior in the field. "They mainly work to try to get data to fit a sociobiological hypothesis. Then, their 'atheoretical' [*sic.*] papers are easily accepted in the major journals."

On the positive side, several observations were made to the effect that there has been no shortage of good ideas on the part of Japanese researchers, such as grooming as social currency, deception, or female impact on male hierarchy, among many others, but that these were originally discussed or written in Japanese. There often was not the confidence to state these ideas in English until they had become accepted ideas in the West. Hence, the appearance of "following" the Western lead. Despite the self-deprecatory remarks cited above, readers should be cautioned that when asked to reflect on one's discipline, people tend to assess reasons for problems rather than successes. Primatology in Japan is thriving and its practitioners are dedicated to and, above all, excited about their work. Overall, in this very brief snapshot of current thinking about the state of primatology in Japan, it seems that while most research perspectives[16] have changed since about 1980 from the earlier paradigm, while there is a lot of variability in fundamental research aims among individuals and schools, and while it is recognized that good contributions are made by Japanese to international primatology, there is nonetheless a lingering perspective of "us" and "them" vis à vis Western primatology. The next section addresses this perspective.

JAPANESE VIEWS OF WESTERN PRIMATOLOGY

Although primatologists in Europe and North America may object to being lumped together as one "Western" approach, this is how my Japanese informants referred to them, so I will use their terminology. Their comments about Western primatology can be grouped under ideas about theory, method, and attitude toward (and treatment of) Japanese primatology.

Sociobiology is recognized as a "big theory" of the sort that comes once in many years. As in the West, many Japanese researchers want to clarify the mechanisms involved. At the same time, some of those researchers, and others not working on sociobiological problems, state that sociobiological theory is not a sufficient explanatory framework for what they want to know about the evolution of behavior and social structure. They explain that formerly, Japanese researchers looked at the group, whereas sociobiology concentrated on individuals' strategies that were not amenable to the specia concept. In addition, they note that today Japanese researchers recognize that individual (animals) may cause changes, but that Western researchers are not so interested in research on individuals (which often appears as anecdotal reporting), but in generalizations. One informant stated that the Japanese are interested in describing a particular monkey, or particular things; and that they collect "facts," not explanations. Japanese researchers, I was told, see very detailed things and think that they have "better eyes" than their Western colleagues. This does not mean that they fail to extrapolate general conclusions; rather, the "natural history" stage characteristic of early primatology continues in the Japanese tradition *alongside of* the more quantitative and sophisticated analyses of behavior that have become possible.

As against the earlier comments about creativity, or lack thereof, is the observation by Suehisa Kuroda that, for him, Imanishi's and Itani's influence was precisely the opposite of following a set paradigm. In their encouragement of doing work that had never been attempted before, and in their encouragement of originality and new ideas, based on a thorough and broad acquaintance with the animals, they freed young researchers from feeling the necessity to work only within the confines of prevailing theories. In fact, Kuroda felt, Japanese researchers are far less bound by prevailing (and often short-lived) perspectives than their Western colleagues.

Many of the Japanese scientists I interviewed complained that Western editors cut description or anything that does not fit the prevailing model out of submitted manuscripts. A graduate student said that she feels there may be some different viewpoint she perceives when talking to Western primatologists, but is unable to specify what it may be. There remains, still, a pronounced feeling in Japan that projects are ultimately a contribution to a group effort. Informants remarked that they do not appreciate those foreign researchers who come to Japan to collect their own data based on the Japanese habituation and kin-recognition record and immediately leave Japan without much discussion with them.

From the other side, some North American primatologists even today advise their graduate students not to publish in *Primates;* some current teachers who were students in the 1970s were advised against even citing research published in *Primates* in their doctoral thesis. Of course, many, and perhaps most, Western primatologists do not subscribe to these views. As mentioned at the beginning of this paper, among both Western and Japanese primatologists, there are those who perceive no difference between the traditions, and among Western primatologists are some who suggest that there may be two traditions within Japanese primatology—an "international" tradition and another which they do not think they would understand. The difference in these impressions among Western primatologists is perhaps due to factors such as with whom they worked, or what they have read of Japanese work. Among the Japanese scientists, these different perceptions may be a factor of their generation, the school in which they were trained, or, again, their awareness of their own traditions and what may be residual paradigms in current approaches.

Some of these factors can be illustrated in the views of current heads of Japanese institutes. Their opinions and observations do not necessarily represent those working in their institutes, but they have a certain role in the interface with the international science community. Thus, for instance, comments by Toshisada Nishida, current professor of the Laboratory of Human Evolution Studies at Kyoto University, president of the International Primatological Society, and well known among Western chimpanzee researchers for his work at the Mahale Mountains chimpanzee project (begun in 1965), illustrate two of the above factors very well. The first factor is the impression by some Western primatologists that they perceive little difference between the two traditions. This has been stated on different occasions by Westerners who have worked at Mahale or met Professor Nishida (or others) in the field or at conferences. The second factor is with regard to the Japanese conception of what their own tradition is. These come together in the following comments by Nishida. He feels that "modern science" is Western in origin and is of Western culture. As a champion of the "Western scientific method," Nishida sees characteristics such as "descriptive" reports as "the result of a lack of knowledge of Western scientific methods as part of Western culture" (e-mail corr., May 6, 1998). Japanese reports that are, for instance, descriptive, have simply failed to meet "Western" standards. He related that he thought that, ideally, Japanese researchers would work for a couple of years in a Western environment to gain the context of Western methods (and relates that Tokyo University, where he formerly taught, requires this of young scientists). The Kyoto school, he says, has never done this since Imanishi's time, nor has Nishida himself. It appears, then, that while Nishida feels that Japanese have made many good contributions to primatology, they should ideally be gained and reported according to accepted international standards and style. This sounds eminently reasonable, yet many of his colleagues would, I suppose from their comments, disagree, since they wish to report things that the "international standard" does not allow. Yet Nishida's views about scientific method are perhaps better known in the West than those of many of his Japanese colleagues. Hence, perhaps, the perception on the part of some Western primatologists that there is little difference between the two traditions. Also, Nishida's observation that Western science *is* modern science, and is embedded in Western culture, prompts the question, "What, then, of Japanese culture and what may it contribute to primatology?"

Summary and Concluding Thoughts

The dynamics of research and human relations within and between science communities are complex. There is a real risk of giving an over-simplified and sanitized view. The research in Japanese primatology contributes to "international primatology," of course. There is, as well, a community of Japanese primatology that has its own Japanese language journal and holds its own professional meetings. Just as one could argue for "schools" within the European and North American contexts, there are also within Japan different "schools" of primatologists. I have been concerned in this paper to (a) give a Japanese response to the three questions raised at the beginning of the paper which hint at the existence and character of their primatological tradition, and (b) to articulate where the nodes of negotiation between Western and Japanese traditions can arise and why they mostly have not yet done so.

The idea that there may be an international Japanese and a traditional Japanese primatology is, I think, illustrative of how national traditions are disregarded or modified to fit the dominant idiom by those who decide what gets published in respected journals or by respected presses. At the same time, most, or perhaps all, Japanese primatologists intending their work for English-language publication "play the game" and aim for an international "scientific language" in their papers. Of course, many Japanese primatologists are, or perceive themselves to be, trained practitioners of international science only—it is only at junctures such as failures to have something accepted or stated as one chooses that the other paradigmatic potentials appear.

What, frankly, has surprised me is that while there are almost as many different views as individual interviewees, I came away in 1984 and again in 1997 with a strong sense of a unity of outlook on certain questions—indeed a Japanese tradition in primatology. That tradition is comprised of a belief that the explanation of behavior and social structure should be constituted of a complex of interacting factors (rather than single explanatory factors, as in sociobiology); of the importance of knowing the whole context of behavior; of having a holistic view—with the perception that the descriptive and paradigmatic expression of those complex and interacting factors has yet to be fully voiced on the international stage. This lack of an international voice will persist as long as those working in the dominant idiom continue to attribute differences only to one or another of (a) linguistic idiosyncrasies, (b) descriptive rather than theoretical reports, or (c) differences in peoples' research problems. The awareness that there are different underlying research goals is as close to a good starting point for understanding each other as we may get. But I am talking about differences that are more profound than those that arise simply from focusing on one or another of Tinbergen's four "whys," or that result from different background disciplinary foci (psychology, anthropology, ecology, and so forth). Rather, it means understanding the research paradigms and the different means of expressing them. Ultimately, this is about negotiating *what* will count as contributions to the discipline and be accorded an open-minded scrutiny on the international scientific stage.

Perhaps it will only be through practicing primatologists who work in yet other national traditions (such as African or South American), and who identify such things as different paradigms, epistemologies, relations with the object of study, and so forth, that we may become more sensitized to what these traditions could add to the discipline. A Teresopolis participant, Emília Yamamoto, who is a Brazilian primatologist, remarked that the *Neotropical Primates Newsletter* is used by many Latin Americans to publish papers that do not get accepted in the more traditional journals. She noted that "[t]hat does not mean that they are bad science, but that they do not conform." In her opinion, changes in paradigms in primatology come from the countries at the center of science. That suggests to Yamamoto that "Brazilians, Japanese, Mexicans, etc., cannot dare, because we would be discredited. The alternative is to follow the mainstream" (Teresopolis e-mail discussion, Yamamoto, May 29, 1997).

As a primatologist who is knowledgeable about and appreciates the classics in Japanese primatology, yet who is firmly supportive of biological and ecological explanations of behavior, the following comment by Yukimaru Sugiyama is, I think, very interesting. It is one of the most telling and, in its way, most moving remarks I have heard, and it echoes Yamamoto's observation. Sugiyama stated: "At present[17] I am the Editor-in-Chief of *Primates* and the Director of (the) Primate Research Institute. In such a position my main work is to find and show our own way of science which is also accepted by Western scientists. Dark and long way" (e-mail corr., June 20, 1996).

Acknowledgments

An earlier version of this paper was prepared for the Wenner-Gren conference "Changing Images of Primate Societies: The Role of Theory, Method, and Gender," held June 15–22, 1996, in Teresopolis, Brazil. Thanks to

Linda Fedigan, Shirley Strum, Sydel Silverman, and Wenner-Gren staff for organizing these meetings, and to fellow participants for most interesting discussions.

I am grateful to the following Japanese colleagues and graduate students for reflections on their discipline in 1997: at the Primate Research Institute in Inuyama, M. Huffman, T. Matsuzawa, A. Mori, H. Ohsawa, Y. Sugiyama, O. Takenaka, I. Tanaka, K. Watanabe, and graduate students, K. Hashiya, S. Hayakawa, H. S. Kim, M. Matsubara, C. Oku, and M. Shimada; at the Japan Monkey Center in Inuyama, graduate student M. Hamai; at the Laboratory for Human Evolution Studies, Kyoto University, T. Nishida and graduate student L. Turner; in Kyoto, J. Itani, and S. Kuroda of the University of Shiga Prefecture; in the Department of Ethology, Osaka University, N. Itoigawa and M. Nakamichi; in Tokyo, M. Hiraiwa-Hasegawa of Senshu University. Thanks to J. Yamagiwa of Kyoto University for providing unpublished material. I am also indebted to H. Takasaki of Okayama University who, since 1981, has provided a ready ear and thoughtful response to my reflections about Japanese primatology. I am further indebted to L. Fedigan, M. Hiraiwa-Hasegawa, M. Huffman, N. Itoigawa, S. Kuroda, T. Matsuzawa, T. Nishida, Y. Sugiyama, L. Turner, and J. Yamagiwa for their thoughtful responses to a draft of this paper. Together, they have greatly enhanced my understanding of the perspectives reported in the paper. Responsibility for the interpretations and any errors expressed in the paper rests entirely with the author.

I also wish to express my gratitude to K. Nakatsuka and S. Kuroda in Kyoto, and to A. and M. Murase in Osaka, for kindly giving me a home away from home during my stay in Japan. Funding for the interviews was provided by the University of Alberta's Support for the Advancement of Scholarship (SAS) Fund, to which I make grateful acknowledgement.

NOTES TO CHAPTER EIGHT

1. My informants in the 1997 interviews are listed in the Acknowledgments. In the text of the chapter, I do not write every informant's name after each piece of information if several people said almost the same thing, or if an informant has requested anonymity. Specific attributions are made in instances where the informant represents a unique viewpoint, or has stated that he or she wishes attribution to be made for a particular point. Sources are also identified where I wish to represent the particular views of the current heads of the main primate centers and institutes in Japan. I will have made my point successfully if readers are made aware

that there is quite a large variety of perceptions among Japanese about Japanese primate studies. At the same time, some commonality in goals, methods, and reportage of their research, stemming from what I contend to be a Japanese tradition in primatology, will also be apparent.

2. The Japanese term is *shushakai* (species society).

3. Takasaki (this volume) provides examples of the insights that this perspective lent to their understanding of the structure of Japanese macaque and chimpanzee societies.

4. This simple statement belies the complexity of the underlying theory that Imanishi had developed. It may seem odd that although the founder of primatology in Japan was himself an ecologist, Imanishi's own ideas resulted in an early emphasis on the "culture" rather than the ecology of primates. This is related to his notion of "species identity" or a built-in ability to identify fellow members of the same species. Imanishi considered the species as more important than the individual in their organismic existence in a system of nature. Individuals obeyed the rules set by the species and not vice versa. He developed a new discipline that he called "animal sociology." Sibatani (1983, 27) remarked that "[a] generalized concept of 'biosociology' subsequently came to occupy a firmly-entrenched position in the episteme of scientific and intellectual communities in Japan, and successfully resisted the invasion of sociobiology . . . until quite recently. . . ." For a translation of passages of Imanishi's own explication of species society, see Sibatani (1983, 29–30) and Imanishi (1984, 361). A translation of Imanishi's (1941) classic is forthcoming: Imanishi Kinji, *The World of Living Things,* trans. P. Asquith, H. Kawakatsu, H. Takasaki, and S. Yagi.

5. Suehisa Kuroda (e-mail corr., May 27, 1998) and Tetsuro Matsuzawa (e-mail corr., May 19, 1998).

6. Itani received the Thomas Henry Huxley Memorial Award in 1984 for his contributions to anthropology.

7. Here, it is important to understand that although a topic or rationale might be given for any particular research report (such as "bark eating behavior" or "male transfer"), these provided contributions to the larger goal of understanding the "whole life" of the species rather than being tests of particular hypotheses only. Thus, when sociobiological theory became so prominent in Western reports in the 1970s, for many Japanese such "single factor" explanations were thought to be too narrow to gain an understanding of the animal society. And although most Japanese reports have incorporated sociobiological theory since the mid-1980s, many primatologists still regard it as only a partial explanation of behavior (as do some Western researchers, for that matter).

8. An example of the "borrowing of experience" that may be possible through someone else's detailed description was given recently by Sterck (1998, 250) when she acknowledged the influence of Y. Sugiyama's descriptions of male Hanuman langur takeovers in Dharwar that "made her realize how much [their] behavior resembled that of the Thomas langurs [she] studied and also where they seemed to differ. This helped [her] in predicting the difference between male takeover and female split-merger."

9. Likewise, we may recall the general call by Western researchers for anecdotal material at the beginning of work on "deception" in nonhuman primates (Byrne and Whiten 1990, Whiten and Byrne 1986). Also see Byrne (1997).

10. The journal *Primatologia,* first published in 1956, was a forum for papers on morphological studies of primates. Earlier, Western-based journals in which papers on primate behavior occasionally appeared were more general animal behavior and psychology journals.

11. *Bibliotheca Primatologica,* followed by *Folia Primatologica* in 1963.

12. Kano (1983); Masataka (1983, 1988); Matsuzawa (1985); Matsuzawa et al. (1983); Nakamichi et al. (1983); Nishida (1973); Takahata et al. (1984). Two earlier articles published in Western journals by Imanishi (1960) and Miyadi (1964) were overviews of the state of research on Japanese macaques, rather than research papers.

13. Y. Sugiyama (e-mail corr., May 25, 1998).

14. One of whom remarked to me that although the meetings were interesting, Japanese participants were not able to say all that they would have liked to say due to the language barrier.

15. A meeting of Japanese and North American primatologists who had conducted research on the Arashiyama (Kyoto) and Arashiyama West (Texas) Japanese monkey populations, held in Canada in 1987 (Fedigan and Asquith 1991). The Wenner-Gren Foundation provided financial assistance for this meeting.

16. I heard little self-reflection about the effect of gender on research, except occasionally in the suggestion of research topics by a supervisor "because you are a woman" (for instance, the suggestion to observe mother/infant behavior). However, whereas during the 1980s there were fewer than a half-dozen female graduates in primatology, today, at Inuyama at least, almost all the primatology graduate students are women. What effect, if any, this has on the discipline remains to be seen.

17. Sugiyama retired from Kyoto University and the PRI in 1999.

9 Some Characteristics of Scientific Literature in Brazilian Primatology

Maria Emília Yamamoto and Anuska Irene Alencar

Brazil is a country of immense natural diversity. This is especially true with respect to primates, since Brazil is home to the largest number of species in the world. This impressive biological diversity has not produced, as one would expect, a similarly impressive body of scientific literature in primatology. This is a very recent discipline in Brazil, and up until the early 1980s, only a few pioneers had put their research efforts into the study of primate behavior and ecology. Portuguese colonization was not characterized by an interest in nature since natural resources were viewed mainly in terms of profit, and early naturalistic descriptions of Brazilian flora and fauna were carried out mostly by other Europeans. It is noteworthy that the first publication on Brazilian fauna and flora, *Historia Naturalis Brasiliae,* by Piso and Markgraf, was sponsored by the Dutch, who for a short period (1637–1644) occupied part of Northeastern Brazil (Menezes 1992).

The study of animal behavior, in general, appeared rather late in Brazil and followed the classical ethological approach of Lorenz and Tinbergen. The first M.Sc. theses and Ph.D. dissertations on the subject were awarded in the beginning of the 1970s and were focussed mostly on invertebrates and rodents (Fuchs 1995). Primates were much later subjects of investigation.

The interest in New World primates, especially callitrichids, followed the early papers by Epple (1973a, 1973b, 1975), Kleiman (1977a, 1977b), and Box (1975a, 1975b, 1977), and research interest in these species heightened due to the low cost of reproducing these animals in captivity (Stellar 1960) and the political stabilization of South America, and more specifically, of Brazil. Very soon, many laboratories in the United States and Europe had established new callitrichid colonies and many foreign scientists came to Brazil to observe these animals in their natural habitats. This resulted in an increasing number of Brazilian scientists and students becoming interested in the study of our primates. Some were informally trained by the visiting scientists, while others pursued formal degrees abroad and returned to Brazil to start their careers. In addition, some of the scientists coming from other countries remained here. This set of scientists formed the research core that helped to establish a true Brazilian primatology. Accordingly, from the middle of the 1980s on, we can speak of a Brazilian primatology, rather than the scientific papers of a few Brazilian primatologists.

The aim of this chapter is to analyze Brazilian primatology by first presenting some general quantitative trends in the discipline and then moving to a specific case study. By examining the amount and breadth of publication, the preferred areas of study, the geographical distribution of laboratories working with primates, and the international impact of Brazilian primatology, the establishment of primatology in Brazil may be explored. A "case study" review of the research on callitrichids in Brazil will illustrate the nature of Brazilian primatology and will provide an interesting example of the role of females in one group of primates.

The Scientific Literature on Primatology in Brazil: 1985–1994

To assess the scientific literature on primatology in Brazil, we reviewed *Current Primates References* (CPR) from 1985 to 1994, and all publications that indicated a Brazilian address were noted. Data regarding the subject (the same list as that found in CPR), the species studied (callitrichids, cebids, unknown, and other primates), and the kind of publication used (international journals, Brazilian journals, international books, Brazilian books, informal publications), as well as the name of the authors, laboratories, and institutions to which they belonged, were recorded.

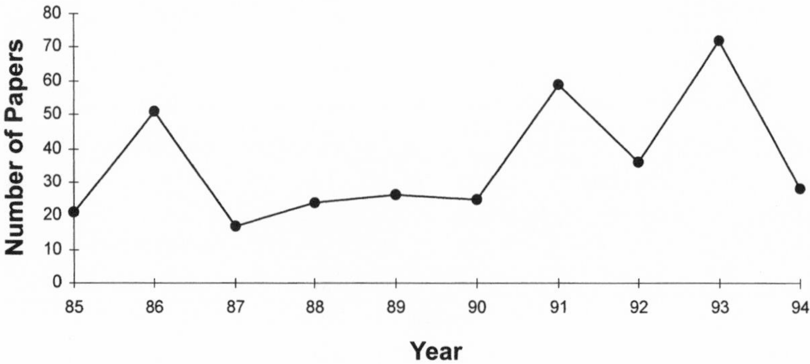

Fig. 1. The number of papers published per year by scientists living in Brazil, from 1985 to 1994. Source: *Current Primate References*

The publication of Brazilian articles on primatology showed three peaks, in 1986, 1991, and 1993. These years coincide with the annals of the meetings of the Brazilian Primatological Society (*A Primatologia no Brasil*). The mean number of publications per year was 38.8. Since 1991 there has been an increase in the number of publications, with a mean of 51.8 articles in 1991–1994 (fig. 1).

The choice of the type of publication indicates a change over the period studied. During these ten years, papers in Brazilian books, especially *A Primatologia no Brasil,* became progressively less frequent, and international periodicals and informal publications became the preferred kind of publication.

Only a few subjects have been investigated by Brazilian primatologists, and, of the 39 possible subjects present in CPR, more than 50 percent were not studied by Brazilian primatologists. The most intensely investigated topics were ecology and conservation, behavior, and primatology (general). Coincidentally, these are subjects that require only small financial investment (fig. 2). The sum of the papers published on these three subjects is larger than the sum of papers published on all other 36 topics. South American primates were the most studied group and were the subject of 68.6 percent of all publications.

Overall, 56 laboratories were identified that are involved in primatological research. The most productive were those from federal state universities in all five Brazilian geographic regions (North, Northeast, Western Central, Southeast, and South). The most active regions are the Southeast, due to a concentration of universities and research institutions,

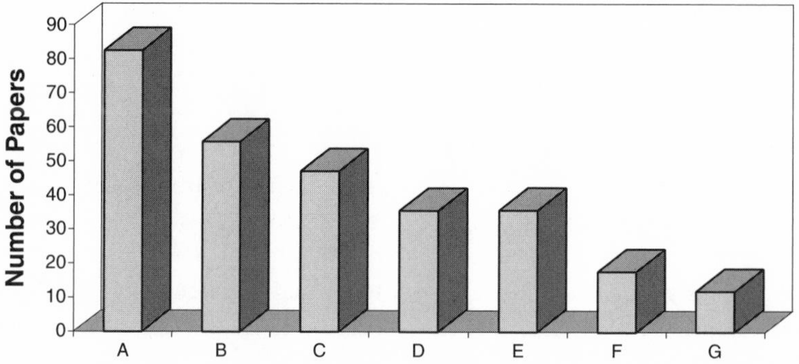

Fig. 2. The number of papers published by subject. Only subjects that had ten or more papers during 1985 to 1994 were included. A—Ecology and Conservation; B—Behavior; C—Primatology (General); D—Genetics; E—Parasitology; F—Nervous System; G—Management of Colonies.

and the North, probably due to its natural resources from the Amazonian forest. The Western Central and Northeast regions showed a more discrete production of primate studies, with only a few laboratories engaged in ongoing research.

In summary, the trends observed in Brazilian primatology during 1985–1994 suggest that it is increasing and becoming less dependent on national publications. Nevertheless, the focus on only some topics and in only some regions of Brazil mirrors what has happened in scientific research in Brazil in general, and this limited range of topics, regions, and institutions is a problem that is worrying research agencies. We expect that any actions taken by these agencies and the government would equally affect primatology. The choice of subjects for study is certainly affected by the availability of financial support, and it is not surprising that the preferred areas, like ecology and behavior, are those that require less investment, both in laboratories and in the recruitment and training of technical personnel.

Unfortunately, the impact of publications by Brazilian primatologists on the scientific literature is negligible. Brazilian scientists, and by that I mean native-born Brazilians or those based in the country, are, with few exceptions, scarcely known abroad. There are several reasons for this. First of all, the number of papers published is still very low. Fifty-one articles per year is, by any standard, a low amount. Besides, many articles are published in periodicals and other publications that have a limited distribution, and therefore, are not easily accessible to most scientists around the world. This is true in Brazil, not only for primatology, but for most other areas of knowledge. Gibbs (1995), in a recent paper in *Scientific American,* showed that Brazil shares only approximately 0.65 percent of mainstream journal articles, placing it twenty-sixth among all nations, a characteristic of scientific production in most of the underdeveloped world. The second reason is that most of our production is published in Portuguese, making it inaccessible for many scientists around the world. Brazilian scientists are struggling for recognition. Perhaps because all Brazilian primatologists, men and women alike, struggle for recognition, questions of gender have not represented a major concern for Brazilian primatologists as they may have for North American and European scientists.

During the discussions at the Wenner-Gren symposium in Brazil, and in Strum and Fedigan's position paper (1996), much was said about the difficulties that a female scientist faces when competing with male scientists. We suggest that these difficulties parallel those encountered when scientists from an underdeveloped country compete with Western scientists. Gibbs (1995) showed that even when scientists from underdeveloped countries publish in mainstream journals, their papers have less impact (are less cited) than papers in the same journal from North American or European scientists. We believe that in both contexts a power issue is at stake: women competing to have the same recognition given to men and underdeveloped world scientists competing to receive the same credit given to their colleagues from North America and Europe. We suggest that it is always the dominant group that sets the standards, oftentimes in such a way that it is extremely difficult for the other group to meet those standards, creating what Gibbs (1995) refers to as "the vicious circle." As such, one must either comply with these standards or accept a lack of recognition. Compliance reinforces the dominant position, perpetuating the vicious circle.

In the case of Brazilian primatology, the transformation of the vicious circle to a virtuous one could be made by providing more opportunities for Brazilian scientists to present their ideas and to collaborate with colleagues from richer countries. Most likely, new procedures would have to be developed to establish unprejudiced environments, which would be the only way to have a productive outcome. This is the challenge that must be faced.

The Social System of Callitrichids and the Role of Females

Possibly, the best studied group of primates in Brazil are the callitrichids, many of them endemic to this country. This family comprises two general groups, the marmosets and tamarins, each with two genera: *Cebuella,* the pygmy marmosets, and *Callithrix,* the marmosets; *Saguinus,* the tamarins, and *Leontopithecus,* the lion-tamarins (Mittermeier, Rylands, and Coimbra-Filho 1988). This family was initially studied in captivity, and the first publications, relying on their behavior in captivity and on the lack of sexual dimorphism, proclaimed them monogamous, living in extended family groups (Epple 1975b; Evans 1983; Kleiman 1977a). Later papers suggested that these were oversimplifications and that callitrichids presented social and behavioral complexities far beyond the features first proposed. The role of the callitrichid female became more and more notable as knowledge about the social behavior and organization of this group advanced. The aim of this section is to examine how these changes took place.

Sussman and Kinzey's (1984) paper marked the beginning of a change in the way callitrichids were regarded. Although they did not present any new data, their ideas ultimately changed the direction of later research on callitrichids. They proposed, in a more incisive manner than ever before, that callitrichids do not live in monogamous, nuclear families or in extended family groups. Instead, they suggested that callitrichids are communal breeders living in multimale groups with a mostly polyandrous mating system. These two features, polyandry and a communal rearing system, emphasized the role of the reproductive female. As the literature would later show, their conclusions were based primarily on tamarin data. Nevertheless, these ideas marked a change in our understanding of callitrichid social behavior.

Papers by Abbott and Hearn (1978) and Abbott (1984) stated that subordinate *C. jacchus* females are physiologically suppressed when in the presence of a dominant female. Approximately half of the adult female offspring of the dominant female, however, are exceptions to this rule since they cycle but do not reproduce. The logical conclusion, therefore, was that only one female is able to reproduce in any group. Field data on group composition, showing that groups usually contain more than one adult male (Garber, Moya, and Malaga 1984; Rylands 1986), and occasional observations of copulations between the dominant female and nondominant males (Baker, Dietz, and Kleiman 1993; Goldizen 1987; Terborgh and Goldizen 1985) were to shed a new light on the relationship between females in callitrichid groups, substantiating the idea of polyandry. Many papers, some of them by Brazilian scientists (Rylands 1986, 1989; Yamamoto and Araújo 1991), explored this possibility both in captivity and in natural groups. However, the picture was still incomplete. Ferrari and Diego (1989), Snowdon (1990b), and Abbott and colleagues (1993) showed that the variability between genera in Callitrichidae is greater than first imagined, and Ferrari and Diego (1989) in particular showed that there are extensive ecological differences between marmosets and tamarins, with consequent differences in behavior and, possibly, mating systems. These authors suggested that groups of marmosets, being more stable and maintaining a higher level of relatedness, can be characterized as nuclear or extended family groups. They also proposed that a monogamous mating system is the norm in this subgroup in which polyandry only occurs fraternally. Tamarins, on the other hand, form unstable groups with low levels of relatedness and a high degree of intrasexual competition, allowing for a more variable mating system.

The occurrence of a single reproductive female is still a key feature in the discussion of callitrichid mating systems, and the reasons for deviations from this pattern are highly controversial. Snowdon's 1990b paper, although entitled "Mechanisms Maintaining Monogamy in Monkeys," reviewed data that opened the door for understanding polygyny, already suspected from field data, but only clearly observed in *Saguinus fuscicollis* (Goldizen 1987). The paper by Snowdon (see also Abbot et al. 1993) reviewed the data on the suppression of ovulation in subordinate females in different callitrichid species. The idea of suppression of ovulation in subordinate females does not hold for all callitrichid species. In *Saguinus,* specifically in *S. fuscicollis* (Epple and Katz 1984) and *S. oedipus* (Ziegler et al. 1987), subordinate females experience total suppression of ovulation. On the other hand, *Leontopithecus rosalia* females show no sign of suppression in the presence of their mothers, and the two ovulate in synchrony (French and Stribley 1987). In *C. jacchus,* 50 percent of the daughters living with their mothers ovulate, but only one female will ovulate in peer groups (Abbott 1984). These are, of course, captive data. But the differences between species suggested that in very specific circumstances subordinate females might reproduce in the presence of a dominant female: polygynous groups can indeed occur in callitrichids. Reports about polygynous groups subsequently increased (Dietz and Baker 1993; Digby and Ferrari 1994; Goldizen 1987; Roda 1989), and the results from these, as well as the earlier studies, raised increasingly complex questions.

One central question relates to the reproductive strategies of dominant and subordinate females. The advantage of the dominant female strategy is clear: by physiologically and/or behaviorally inhibiting subordinate females, the dominant females monopolize resources and helpers (Digby 1995). A subordinate female is faced with less promising alternatives: (1) wait for a reproductive vacancy in her natal group; (2) disperse to an incipient social unit; (3) migrate into a breeding position in a nonnatal group; or (4) reproduce as a subordinate female (Digby 1995; Digby and Ferrari 1994; Ferrari and Diego 1992; Rylands 1989). The first three alternatives may be more advantageous in terms of reproductive output (Digby 1995), but reproductive or territorial vacancies may be scarce, especially for some species (Rylands 1996). The last alternative, although not necessarily permitting a high reproductive output, may produce offspring with good chances of survival when avoidance of simultaneous births with the dominant female is achieved (Digby 1995; Arruda pers. comm.).

This last alternative poses a new question: Why should a dominant female allow a subordinate to reproduce in her group, thereby using the resources and helpers that she could monopolize for the sake of her own infants? In all reported cases of polygyny where familial relationships were known, breeding females were kin (Alonso and Porfírio 1993; Dietz and Baker 1993; Goldizen et al. 1996; Rothe and Koenig 1991). One could argue that the dominant female could benefit, through kin selection, from her daughter's, sister's, or mother's reproduction, especially if it did not occur in synchrony with her own.

Data from our laboratory and field studies again showed that this is a more complex phenomenon. Data from three free-ranging groups of *C. jacchus* followed for four years showed that two groups had two reproductive females and that one had a single reproductive female. Differences in the number of breeding females could not be explained by differences in size of home range or group, or in the number of adult males and females in each group. The association between breeding females and two incidents of infanticide suggested that even when two females from the same group breed, one is dominant over the other (Araújo 1996; Arruda pers. comm.; Digby 1995; Digby and Ferrari 1994; Yamamoto et al. 1996).

In our laboratories, we observed the behavior and assessed the hormonal profiles (progesterone levels indicative of ovulation) of five pairs of *C. jacchus* females in competition for one male over a twenty-week period (Alencar 1995; Alencar et al. 1995; Cirne et al. 1995). The pairs differed markedly in behavior and hormonal status. Two of the pairs showed a well-established behavioral dominance from the beginning of the observation period with only one female of the pair ovulating (Group 1). In the three other pairs, both females behaved similarly and showed similar levels of plasma progesterone—either both or none showed post-ovulatory patterns of plasma progesterone (Group 2).

Dominant females presented similar behavior in both types of pairs, but subordinate females differed statistically in the levels of agonism displayed. Group 2 subordinate females showed higher levels of agonistic behavior than their counterparts in Group 1 (fig. 3). These results suggest that there are two kinds of subordinate females in *C. jacchus:* those that submit and those that resist subordination. The occurrence of groups in the wild with one or two breeding females, then, may be related to the

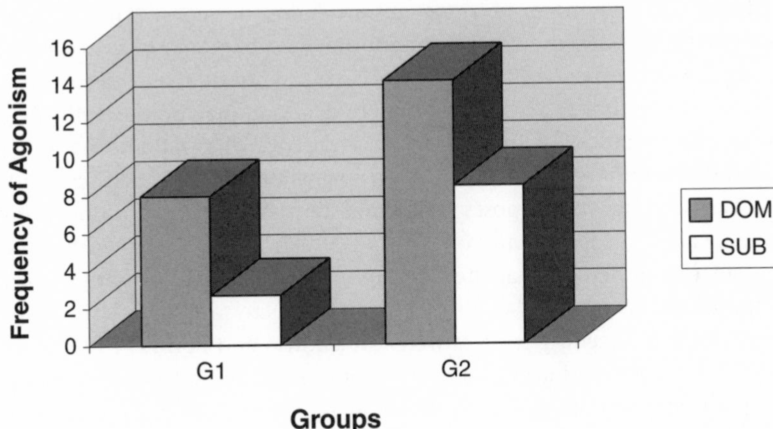

Fig. 3. Frequency of agonism in dominant and subordinate females of Groups 1 (clear dominance) and 2 (contested dominance) when competing for one male. Statistical test: MANCOVA, df = 3,36; F = 10,22; p = 0.0029 for differences between dominants and subordinates. MANCOVA, df = 7,11; F = 8,37; p = 0.0023 for differences between subordinates of Group 1 and Group 2.

ability of the dominant female to maintain reproductive suppression in subordinates. The social and reproductive relations between females, in addition to ecological and species-specific variables, should therefore be considered when assessing the options faced by a subordinate female.

How does this discussion on callitrichids help us to understand the questions posed by Strum and Fedigan's "Theory, Method, and Gender: What Changed Our Views of Primate Society?"? We believe it brings a fresh view to the discussion since we are talking about a primate group that is, in many ways, different from the intensively studied Old World monkeys. In callitrichids, males are not conspicuous or are, at most, only as conspicuous as females. The questions asked by Strum and Fedigan can be at least partially answered through the analysis of the behavior and reproduction of dominant females and through the analysis of the relationship between dominant and subordinate females in the group. The complexity of the relationships within a callitrichid group, the diversity of behaviors and physiological mechanisms, and the ecological variability of this primate family allows for different kinds of relations between males and females. Relationships can be exploitative, cooperative, or complementary depending on what aspect of behavior is analyzed and from what point of view this analysis is done. Further investigation is required to understand the relationships between and within male and females in

Callitrichidae, and more needs to be discovered about these scarcely studied species in order to paint a behavioral picture of this fascinating group of primates.

The discussion of the scientific literature in primatology from Brazil gives us some insight into the difficulties of producing first-rate science in an underdeveloped country: the lack of resources and good laboratories are the least of our concerns. Brazilian scientists are struggling for recognition and acceptance in the global scientific community, and questions of gender, so important in developed countries, are accordingly less important here. Although there are some signs that this may be changing, it is also necessary to change the way that scientists from underdeveloped nations are regarded by journal editors and referees. In a field such as primatology, where source countries are principally part of the underdeveloped world, efforts should be made to open the doors for scientists native to these countries. Finally, scientific collaboration should be encouraged, and in Brazil there are good examples that collaborations work (see Strier this volume).

Acknowledgements

M. E. Y. wishes to thank the Wenner-Gren Foundation, L. M. Fedigan and S. Strum for the invitation to participate in the symposium that preceded this book. Both authors wish to thank L. Fedigan and S. Strum for many valuable suggestions for the improvement of earlier versions of this paper. Also, we wish to thank I. T. Jarreta for helping with the collection of data for the first part of this paper, and CNPq for grants to both authors.

10 An American Primatologist Abroad in Brazil

Karen B. Strier

Introduction

Brazil is a country larger in area than the continental United States. Within its borders lies much of the Amazon, the world's largest tropical rainforest, as well as the entire Atlantic forest, now one of the most threatened ecosystems on earth. Together, the Amazon and the Atlantic forest support more than twice as many primate species as any other Latin American country (Rylands et al. 1995). The Atlantic forest alone, which has been isolated from the Amazon since the Quaternary, supports six primate genera, including two endemics. Habitat destruction during the last half of the century has reduced the Atlantic forest to less than 5 percent of its original extent (Fonseca 1985), and as a result, all of the endemic primate species and subspecies in this region are now threatened with extinction (Rylands et al. 1995).

Considering its high species diversity and the opportunity its primates represent for comparisons across genera that have less restricted distributions, Brazil's Atlantic forest should have long been a magnet attracting field primatologists from around the world. Yet, this was not the case in 1982, when I first visited the southeastern corner of the Atlantic forest to explore the possibilities of conducting a field study on the largest of its endemic primates, the muriqui monkey (*Brachyteles arachnoides*) for my doctoral dissertation. I was not the first foreign primatologist to pursue a field study in southeastern Brazil, nor was I the first researcher interested in muriquis. I was, however, extremely fortunate to arrive at a time when Brazilian scientists were already actively engaged in conservation efforts, which included a strong commitment to obtaining basic information about the behavioral ecology of endangered primates in the region.

Biologists and conservationists at the Universidade Federal de Minas Gerais, in particular, were generous with their support as I initiated my original study at the Estação Biológica de Caratinga (EBC), a privately owned forest fragment where they had found one of the only remaining populations of muriquis in the state (Aguirre 1971). Continued support from them, and from other faculty and students affiliated with other Brazilian universities, has been instrumental throughout the past seventeen years as the ongoing project at the EBC expanded in scope and as we launched two comparative studies on muriquis at other sites. At this writing, twenty-nine Brazilian undergraduate and graduate students from twelve universities have participated on these field studies, and nearly all have gone on to pursue careers in the Brazilian research and conservation communities.

Brazilian interests in the potential applications of basic research to muriqui conservation profoundly influenced the ways in which the project has developed. More indirectly, but perhaps no less significantly, are the ways in which these conservation priorities interacted with, and changed, the comparative perspectives on primate behavioral ecology that I have come to hold (Small 1995). It is impossible to distinguish between the effects of the intellectual impact that Brazilian colleagues have had on my research, the insights that the muriquis themselves have suggested, or the roles that shifts in methods, theory, and an expanding literature based on the work of other scientists, including many of the contributors to this volume, have played in the process. Similarly, it is awkwardly presumptuous to speculate about what, if any, impact these altered views of primates may, or might yet, have on primate studies. What follows, then, is more along the lines of a case study description of how one primatologist, coming from an American anthropologist's perspective of baboons-as-"typical"-primates (Strier 1994a), was converted, during a seventeen-year period of field research on endangered muriquis in Brazil, to regard New World monkeys as central to understanding the social evolution and behavioral variation within the primate order.

Theoretical Influences

BACKGROUND

American field primatology developed within a strong anthropocentric tradition in which an understanding of human behavioral evolution played a major role in dictating which species were most relevant to study and how the studies were approached (Richard 1981). Semiterrestrial monkeys, such as baboons, provided ecological comparisons for extrapolations about the lives of hominid ancestors, who also exchanged the safety of trees for the open African woodlands and savannas. The apes, and particularly chimpanzees, provided phylogenetic comparisons with humans as our closest living primate relatives. With such special status, it was not surprising that they tended to fall outside the general perceptions of how other, more "typical" primates should behave (Strier 1994a).

The New World monkeys, with their arboreal lifestyles and more distant ancestry to humans, were often relegated to the status of second-class citizens in comparative analyses of primate behavioral analyses. Considering that only one neotropical genus, *Alouatta,* ranked among the ten primate genera accounting for 60 percent of the published primate literature from 1931 to 1981 (Southwick and Smith 1986), it is not surprising that New World primates were so poorly represented in early comparative models of primate behavioral ecology (Strier 1990, 1994a).

The disproportionate absence of data on many New World primates has meant that the perspectives of many American anthropologists interested in nonhuman primates for their links to humans were strongly skewed by the better-studied Old World taxa. Even now, New World primates are still difficult to situate in comparative analyses of primate social behaviors ranging from male sexual coercion of females (Smuts and Smuts 1993) to heterosexual consortships (Manson 1997). Our ignorance about them was even more pronounced in the late 1970s and early 1980s, when primatologists were adeptly incorporating evolutionary and ecological theory into their models.

These comparative models generated explicit, testable hypotheses about how primates should behave under particular conditions. In the process, they facilitated a shift in emphasis away from particular species as models for the evolution of human social behavior toward understanding the underlying principles affecting primate behavior (Fedigan 1982; Tooby and DeVore 1987). Marginalized primates, including many New World monkeys, became prime targets for field research. Because many of these monkeys had not previously been studied in any depth, if at all, in the wild, they could serve as independent controls for testing predictions based on the models constructed without them.

Muriquis were particularly well-suited subjects for such an independent test of Richard Wrangham's (1980) synthetic ecological model of female-bonded primates, which based its major behavioral dichotomy on whether primates relied on patchy, defensible foods such as fruits, or on evenly dispersed, abundant foods such as leaves. Muriquis were known to exhibit anatomical traits consistent with both frugivorous and folivorous specializations, so predictions about their social behavior could be framed in terms of neat sets of alternatives depending on what they actually ate and how these foods were distributed (Strier 1992a).

The fact that muriquis were known to be endangered also figured into my original desire to study them, but in those early days, it was their pivotal position on this fundamental dietary continuum that made them so attractive compared to other, equally endangered, primates. It took seeing a wild muriqui that first time, and interacting with knowledgeable conservationists, mostly in Brazil, for me to begin to appreciate the complementarity and compromises between basic, theoretically driven *versus* applied, conservation-motivated research.

OVERLAPPING PERSPECTIVES

In the beginning, the dual objectives of the basic and applied research, and the means to achieving them, were identical. Systematic data on muriqui feeding, ranging, grouping, and social behavior were needed, on the one hand, to evaluate ecological predictions about muriqui society, and on the other hand, to begin to describe their basic habitat requirements and social structure for management purposes. The fact that these data would come from one of the two muriqui groups then comprising an isolated population inhabiting the small (860 ha), disturbed forest fragment at the EBC was only a minor concern to me at the time. After all, few primate field sites that were accessible to American researchers in those days could claim to retain truly unaltered, pristine conditions. Furthermore, my plan was to directly compare dependent behavioral variables to what were largely regarded as independent, quantifiable ecological variables, and it was the ways in which the different facets of muriqui society mapped on to these ecological variables, however changed they might be from their natural state, that were most important.

The Brazilian conservationists who took me into their care recognized the importance of all remaining forests still known to support muriquis. But, for them, the EBC was a special site, as it was not only through their proximity to the area, but also through their subsequent cooperative agreements with national and international NGOs and the owner of the forest and the surrounding farmland, that the protection of one of the last remaining strongholds for muriquis was being secured. Installing field researchers at the site was consistent with their overall plans to establish a scientific presence that would emphasize the value of the site and facilitate the development of conservation education programs in the region. Whatever progress I could make in habituating the muriquis for others to observe, photograph, and film, and whatever information I could obtain about them, would contribute to the more far-reaching conservation efforts underway on their behalf.

SHIFTING PRIORITIES

Among the most important sets of findings to emerge from that first year of intensive study were that male muriquis maintained strong, highly affiliative relationships with one another and that, although adolescent females immigrated into the study group, male group membership remained stable. These findings implied that the primary social and kinship bonds among muriquis involved males, by contrast to the female bonds expected from the "typical" primate model. At the same time, the indication of female-biased dispersal in muriquis implied that the much-discussed, and still controversial, possibility of managing isolated populations by translocating individuals to maintain gene flow would be more likely to succeed if females, rather than males, were targeted.

These findings were a clear example of how basic research could provide insights for informed conservation planning and for an increased probability of success for future management endeavors. But, reconciling the ecological correlates of the muriquis' patrilineal society was an entirely different matter because their overall diet and feeding, ranging, and grouping patterns were more consistent with those exhibited by other female-bonded primates. Thus, while the question of dispersal and bonding was considered more or less resolved from a practical standpoint, it was still a wide-open question within the theoretical framework of primate behavioral ecology.

Other findings that appeared to contradict theoretical expectations were equally puzzling. For example, it was not at all obvious why muriqui society was so strongly egalitarian, or why group members so rarely employed aggression in negotiating potentially competitive situations. The absence of any discernible agonistic-based hierarchies among and between males and females attracted a great deal of attention among primate behavioral ecologists, but it did not seem to stimulate much excitement within the conservation community.

If anything, the near absence of overt, agonistic competition, either among females for food, among males for mates, or between males and females in any context, raised questions about the limitations imposed by the unique conditions at the study site. The muriquis and other primates in this forest were protected from hunting, but many of the largest of their natural predators had also been eliminated from the area. Furthermore, not only has the study group more than tripled in size over the years due to high birth and survivorship rates relative to mortality, but its adult-biased age structure at the start of the study implied that the population may have suffered from some demographic or reproductive catastrophe in the recent past (Strier 1996a).

It wasn't clear whether the low levels of overt competition observed were a consequence of a population living well below the carrying capacity of the forest at the time or a highly unusual feature of this species' society. And, it was difficult to assess just how representative this and, by inference, other findings about the EBC muriquis really were. Data from other muriqui study sites at the time offered few comparative perspectives, in part because each site was so ecologically distinct and the populations of muriquis and sympatric primates and their potential predators were so different from one another in their sizes, compositions, and densities.

Still more confounding were the possible taxonomic implications of a long history of isolation between northern and southern muriqui populations (Rylands et al. 1995), which may have persisted since the Pleistocene glaciations (Kinzey 1982). Indeed, despite additional studies launched by other researchers at new sites in the last decade, all that can be said with any confidence at this point is that northern and southern populations differ in morphology (Lemos de Sá and Glander 1993) and that, independent of their geographic locations, muriquis occur at higher densities in small, disturbed habitats than in large, primary forests (Strier and Fonseca 1996/97).

Back in the mid-1980s, while puzzling over whether it was legitimate to generalize findings on the EBC muriquis to other populations and to the genus at large, publications nonetheless began to appear in which all available data on muriquis, independent of study duration or field conditions, were being treated as equivalent and, in some cases, were compressed into single species averages for comparative purposes. The initial satisfaction that any scientist feels when they see their work cited in the literature began to give way to increasing discomfort whenever qualifications about the field conditions were not also mentioned.

Generalizing from a single study to a species or genus in comparative models has the effect of reducing the many potential sources of behavioral variation to the level of systematics (Strier 1997). Such generalizations can also produce highly misleading perceptions about the most appropriate courses of action for conservation, particularly in the case of a primate like the muriqui, whose geographic distribution is so severely fragmented. Legitimate concerns with the applied sides of systematics have led many conservationists to encompass observable regional or population variation into new subspecies, or even separate species, designations (Rylands et al. 1995). Generalizing about muriqui behavioral ecology within a comparative primate framework, and applying these generalizations to what should be tailored regional or population-specific conservation plans that accommodate local historical and ecological conditions, seemed ironically at odds (Strier and Fonseca 1996/97).

Methodological Influences

POPULATION VARIATION

For conservation biologists, the population is the unit of analyses and it is subject to stochastic demographic and environmental processes. This perspective contrasts sharply with more traditional approaches in primate behavioral ecology, in which behavioral data are recorded from individuals and summarized for the purposes of species-level comparisons, and ecological data are treated as independent determinants of behavior (Strier 1997). In 1986, when I was teaching the first Brazilian student who participated on the project how to recognize the members of the study group by their natural markings, I was still thinking in terms of individuals and the life histories we were going to be monitoring. It was not until later, as the muriqui group was steadily increasing in size, first to 50 percent, then to twice, and ultimately to more than triple what it was at the outset, and the sex ratio began to shift to one in which sexually active females increasingly outnumbered males, that the implications of the conservationists' population-level phenomena for understanding behavioral ecology began to make sense.

With the increasing group size and female-biased sex ratio, the previously cohesive study group began to split up into temporary mixed-sex or all-male parties, and influxes of males from another muriqui group in the forest began to periodically travel and copulate with study group females (Strier 1994b). As the population continued to grow, a third muriqui group formed and the study group extended its home range to include parts of the forest that were previously unexploited and had been undergoing natural regeneration from prior disturbances (Strier et al. 1993).

Focusing on individuals and their life histories, or regarding ecological variables, such as food availability and distribution, as static (or even predictably seasonal) determinants of these muriquis' behavior, could explain only some of these changes. And, if such demographic and ecological fluctuations could alter the lives of the EBC muriquis to such a degree over a relatively short period of time, how reliable were the published descriptions of any species' behavior when its range of responses under diverse demographic and ecological conditions were not known?

CONSERVATIVE TRAITS

Despite the dynamic changes in many aspects of the muriquis' behavioral ecology over the years, other patterns in their lifestyles have remained remarkably constant. In particular, the original depiction of muriqui society, with its female-biased dispersal and strongly affiliative, nonaggressive bonds among philopatric males, has persisted. Muriqui grouping patterns evidently respond facultatively to ecological and demographic variables, but their dispersal and bonding patterns appear to be more resistant to such external factors.

A similar configuration of labile behavioral responses to fluctuating conditions and conservative, phylogenetically inert patterns of dispersal was supported by comparisons with other muriqui populations (e.g., Lemos de Sá 1991), other primates closely related to muriquis (Strier 1990, 1992b), other New World primates (Strier 1999), and other primates across the order (e.g., DiFiore and Rendall 1994; Strier 1994a; Rendall and DiFiore 1996). While ecological and demographic conditions might influence rates of dispersal, phylogeny nonetheless appears to be a better predictor of sex biases in dispersal patterns and their corresponding implications for social relationships and affiliative bonding.

Decoupling what ecology and demography versus phylogeny can explain about primate behavior remains a major challenge in primate behavioral ecology (Strier 1994a). The populationist approaches employed in conservation biology, which emphasize local variation and the far-reaching impact of stochastic demographic and environmental events, offer an additional level of analysis that has rarely been explored from a comparative perspective (Strier 1997). Merging such population-level approaches with the individual and species-level approaches of behavioral ecology provides a way to tease apart the various sources of behavioral variation, as well as making the links between basic scientific and applied conservation research more explicit.

Many of these ideas had been advanced in the literature long before I began to consider them myself, but the influence of the conservation movement that I encountered in Brazil no doubt played a role in my efforts to put the EBC muriquis into a comparative perspective. In the process, other primates began to look different to me as well. For example, female-biased dispersal or dispersal by both sexes is more common among New World primates, and indeed, all other anthropoids, than the "typical" baboon model implied. Indeed, among New World primates, female bonding is only prevalent in the genus *Cebus,* and even then, many of the characteristics associated with female bonding in Old World monkeys are expressed in somewhat different ways (Strier 1999). For example, although *Cebus apella* females tend to remain in their natal groups with other members of their matrilines, the alpha male appears to hold a distinctly prominent position in determining female access to resources and in mediating female social relationships compared to alpha males in most other female-bonded primate societies.

The fact that New World primates resemble apes more closely than other Old World taxa in their life history parameters (Ross 1991), as well as in their dispersal and bonding patterns, raises additional questions about the potential importance of reproductive biology in comparative models of behavior. For example, male reproductive strategies can be predicted, at least in part, by the degree to which male competition and affiliation can influence female reproductive rates (Strier 1996b). In particular, reproductive seasonality or slow intrinsic rates of reproduction can limit the effects, and thus the selective benefits, of intrasexual competition among males when compared to primates with fewer seasonal constraints on reproduction or higher intrinsic reproductive rates. Factoring female reproductive biology into behavioral equations extends comparative perspectives on primates in new ways.

Knowledge of reproductive biology is also of obvious relevance to conserving endangered species. Thus, as in the beginning, both the theoretical and applied sides of the muriqui research have converged once again. As the focus of the long-term study has expanded to include noninvasive fecal steroid analyses of the hormonal correlates of muriqui behavioral ecology (e.g., Strier and Ziegler 1994), it has also succeeded in obtaining basic life history data on gestation and cycle lengths (Strier and Ziegler 1997), and in providing noninvasive tools for assessing reproductive viability and failure in this and possibly other wild muriqui populations.

The Human Dimension

THE PUBLIC APPEAL OF THE SPECIES

Studying an endangered primate in the early 1980s in Brazil meant that everyone there—from students, to colleagues, to the media—paid attention. The fact that muriquis are large, furry monkeys with pleasant faces and graceful bodies contributed to their original and persisting charismatic appeal. International interests in conservation and biodiversity were also gaining momentum at the time, and muriquis, along with other cuddly mammals such as China's endangered pandas, were proving to be effective advocates for the conservation cause.

As early as 1984, conservation education campaigns put drawings of muriquis on Brazilian postage stamps and on the covers of local telephone directories. News about muriquis continues to make headlines in major Brazilian newspapers and on Brazilian television. Local and foreign journalists regularly visit the field site to film the muriquis and to interview both the students who work on the project and the owner of the forest, who, until 89 years of age, was still an articulate spokesman for conservation in general and muriquis in particular.

That muriquis also happen to be such interesting primates has often emerged in these reports as an afterthought to the overarching conservation theme. The unusual aspects of their society, including the peaceful affiliations among males and the egalitarian relationships among males and independent females, who move between groups and freely assert decisive preferences in their sexual liaisons, have certainly added to their media success (e.g., Brownlee 1987). Outside academic circles, however, it has been the larger story, of a photogenic primate pushed to the verge of extinction and a landowner who set aside forest to protect them long before environmental-protection legislation or tax-relief laws were enacted, that has made the muriqui story such a compelling one.

ATTRACTION TO STUDENTS

The effect of the conservation movement on Brazilian university students was evident from the time that I first visited Brazil as a student myself. While conducting my dissertation research, I lived and worked among Brazilian students who, like me, were embarking on their research careers and studying the flora and fauna, including other primates, at the EBC. The primary difference between us was that I came to Brazil armed with generous research support, while most of my contemporaries there were financing their studies on graduate student stipends. A few of them also had some funding from conservation organizations, but it was barely enough to make ends meet. Yet, despite these financial difficulties, they were every bit as committed to their research as I was and far more informed about the implications of their research to conservation.

The difference in our access to resources affected how I wrote the budget for the first NSF grant, and all subsequent grants, to support the long-term study. It also affected how I viewed the long-term study from the start. Instead of hiring paid field assistants to work for me and turn over their data, I wanted to collaborate with Brazilian colleagues to provide research opportunities and training for interested students. The goal was to train new students annually to help maintain the long-term demographic data, while at the same time helping them to develop independent studies that would allow them to continue on in basic or conservation-related research.

Unlike some primate habitat countries in the 1980s, it was relatively easy to find interested students who qualified for the project in southeastern Brazil. Conservation education programs and NGOs were expanding in this region, and endangered primates, like the muriqui, inhabiting protected forests, like the EBC, figured prominently in their missions. Faculty colleagues at Brazilian universities recommended the opportunity of participating in the muriqui project to their students, and the students who joined the project recommended it to their friends. These referrals, and the students' critical evaluations of prospective participants, have been fundamental to the project's successful continuation.

The fact that current students on the project play such a critical role in recruiting and helping to train their successors has also contributed to the development of a sense of community and deep loyalties among us. The students have been uniformly committed to the project's long-term objectives, and all have the opportunity to leave the project with their own set of data for their own graduate theses. Most of the students return to the forest, and the muriquis, years later on visits, and they make efforts to attend one another's public theses defenses. For example, at one of the muriqui master's thesis defenses at the Universidade de São Paulo in the early 1990s, twelve of the fourteen students present at the celebratory lunch had been or were currently participants on the project.

Many of the students join the project after completing their bachelor's degrees, and then use the data they collect for their independent studies to strengthen their application to graduate programs. In Brazil, as in most other countries, the top graduate programs are inundated with more applicants than the faculty can admit, particularly when fellowship funds are limited. The problem of limited admissions is compounded by the impressive fact that so many of the potential faculty advisors with expertise in primate field studies also devote considerable time and energy to conservation activities.

One solution to this dilemma for the muriqui students was recently suggested by colleagues at the Universidade Federal de Minas Gerais. By making me an honorary member of their graduate faculty, those students who had worked on the project and who passed the exams admitting them to the program could continue to work on their theses with me. This voluntary affiliation also provides me with the gratifying opportunity to see the products of their research, and the multifaceted goals of the long-term project, develop.

LOCAL COMMUNITY INTERACTIONS

Like me, the students who participate on the muriqui project at the EBC are required to integrate into the local farming community in which the forest is situated. In 1983, the owner of the forest donated a vacant house at the edge of the forest for researchers, and funds from Brazilian and international conservation groups paid for its inaugural and subsequent renovations.

The field site is administered by a Brazilian NGO, which maintains a housekeeper and an extremely competent handyman. Researchers pay a daily fee determined by nationality (and availability of funds) and length of residency, which covers food and maintenance supplies for the field station. It is a rustic setup. Bunk beds line the three small bedrooms, and until 1992, when electricity was brought in, we cooked over a gas-powered stove, studied under gas lanterns, and heated water by burning dead wood. Even the muriqui fecal samples, which we had begun to collect for the hormone study, had to be stored in an electric freezer at a local bar ten kilometers away.

The dirt road that leads to the research house continues on through the forest to coffee fields and pasture on the other side. Local farmers pass by the house, usually on foot, on their way to and from their homes and fields. The family that owns the farm and forest also pass by almost daily, and still stop to inquire about the muriquis and our well-being.

The farm was more active when the research began than it is today, but at the time, none of us had a car at the field site, so our only way of leaving was by a bus that stopped two kilometers away. Nowadays, the students maintain a small Volkswagen, which has increased both our ability to seek help in case of emergency and our contact with residents of the surrounding towns. At their own initiative, several of the muriqui students began teaching evening high-school-level science courses in one of these towns. Others joined in the farm's Sunday evening soccer games.

Still others formed close and enduring friendships with families living on the farm.

All of us recognize and appreciate the fact that the project depends on our having permission to live and work on private property. The benefits of such a dependent situation include the security and protection that the owner provides. The authority that he and his family maintain in the region is unchallenged. Local farmers and town residents respect his property and the umbrella protection that he confers on us. Prohibitions against hunting in the main area of the forest have not, to my knowledge, ever had to be enforced since the project began, nor do the local people cut live hardwoods from the protected forest for personal use.

These prohibitions against hunting primates and clearing the forest were widely respected prior to the initiation of the muriqui project, but the year-round presence of researchers there has undoubtedly been a reinforcing influence. The local farmers have witnessed countless visitors, from international film crews and ecotourists, to busloads of grade-school children, who come to the forest for the chance to see the muriquis from our study group swinging peacefully through the canopy and posing unabashedly for photographers. Both directly and indirectly, the vision of the Brazilian conservationists who first recognized the accessibility of this site and the value of the muriquis for increasing public conservation education and local conservation awareness is being realized.

Concluding Remarks

Each time I renew my visa, whether for a brief visit to meet with colleagues and students or for an extended field research mission, I am acutely reminded of the precariousness of my status as a foreign primatologist abroad in Brazil. Even after seventeen years, the opportunity to study such extraordinary primates as muriquis still feels like a rare, and easily revocable, privilege. Approval of my research permits depends entirely upon sponsorship and endorsement from Brazilian colleagues, and daily monitoring of the muriquis depends entirely upon the students who participate on the project. So, although I raise the funding to support the project, it is my colleagues and students there who are ultimately responsible for the project's continuation.

These same Brazilians have also been responsible for articulating the explicit links between basic behavioral ecology research and applied conservation. The field study has always been one component of a wider network of conservation activities, and as a result, it has been influenced by the populationist perspectives of conservation biology.

Few primates can compare to the muriqui, either for its charismatic appeal in generating interest in conservation, or for its example of the importance of New World monkeys to understanding primate behavioral diversity. The striking behavioral contrasts between muriquis and the more familiar Old World monkeys have helped stimulate comparisons among other New World primates, and ultimately, between the New World primates and Old World monkeys and apes. Reconsidering the phylogenetic distribution of primate dispersal patterns, which in turn influence, and are influenced by, kinship and social relationships, is just one of the ways in which New World primates have contributed to fundamental shifts in our perspectives on primate societies, including anthropocentric distinctions between humans and other primates. New World monkeys, including, but not limited to, muriquis, serve an essential role as out-groups for understanding the extent of variation in primate behavioral adaptations. At the same time, the behavioral diversity they represent has become central to comparative considerations of primate behavioral ecology.

Acknowledgements

I thank Drs. Linda Fedigan and Shirley Strum for inviting me to participate in their conference and suggesting this topic for their volume and the Wenner-Gren Foundation and the other conference participants for providing this opportunity to reconsider the processes involved in changing images of primate societies. M. E. Yamamoto generously read and commented on an earlier version of this manuscript.

E-Mail Exchanges

There are well-documented differences between the development and practice of primate studies in Europe/North America and in Japan. During our workshop in Brazil and our subsequent e-mail exchanges, we explored the implications of these distinctions, expanding the cross-cultural/cross-national comparison to include some of the more recent primatological traditions from other countries. The asymmetrical and problematic interaction of Western science and Japanese science captured everyone's interest, leading us to ask: *Why do Western scientists accept Japanese data but not their theory and practice?* The diversity of traditions, the variety of practice, and what emerged as the marginalization of non-Euro-American primate studies raised another question: *Are there many primatologies, or an international science?* Primatologists from "other" national traditions increased our awareness of the difficulties of being part of an international science.

Question 4: Why Do Western Primatologists Reject the Japanese Way of Doing Primatology While at the Same Time Accepting Their Data? This Seems Particularly Strange Since They, at Least the Kyoto School of Primatology, Were Right about Some Important Aspects of Primate Society Long Before Western Primatologists Understood the Issues.

To: <teresopolis@majordomo.srv.ualberta.ca>
From: "Thelma Rowell" <thelma@ingleton.demon.co.uk>
Again I have to ask, who is we? I don't reject the Japanese way of studying primates, although I have to confess I haven't found it quite as theory-

free in its approach as has been implied. . . . (What I reject vehemently is the hypothesis-testing approach insisted on by grant-giving agencies, which if people actually did what they promise to do would prevent us ever discovering anything really new, ever . . .)

To: <teresopolis@majordomo.srv.ualberta.ca>
From: "Pamela Asquith" <Pamela.Asquith@ualberta.ca>
. . . What is striking is that some of what Japanese reported early on (e.g., female support for alpha monkeys—reported in the early 1960s before chimpanzee politics became apparent, etc.) was ignored, whereas it received huge press (as it represented a marked change from earlier ideas) when reported by Westerners. More than one North American primatologist has told me that s/he was advised not to cite Japanese findings in thesis and papers in the 1970s at least. Some teachers apparently still feel this way too. And certainly some people are advised not to submit to *Primates* except as a last resort. . . . Imanishi's specia is very relevant to why the Japanese "way" of doing primatology has been rejected:

a) just what their long-term goal is in gathering so much descriptive data appears to be misunderstood;

b) the embellishments in their reports that contribute to the long-term goal (understanding the "specia" or "species society") get deleted by international journal editors and reviewers. I heard this complaint over and over again in my interviews in Japan in February 1997. Yet one can see why seemingly irrelevant materials appear merely prolix to Western readers. So we need to begin by understanding (a) . . . yet solving (a) will indeed be only a beginning.

To: <teresopolis@majordomo.srv.ualberta.ca>
From: "Dick Byrne" <rwb@st-andrews.ac.uk>
I am getting puzzled that everyone is talking about "them" and "the" Japanese way. Is it not possible that there are TWO: two groups of Japanese primatologists with two different approaches, one of them much more like "the" Western approach? My contact has mainly been with researchers associated with Mahale and especially with Toshisada Nishida, and I have never felt that their aims and methods were greatly different to mine. This is confirmed by their publications (although of course their seemingly irrelevant descriptions may have been deleted by Western editors), and I'd rather assumed that the more traditional Japanese approach to primatology would be much more different and hard for me to understand.

To: <teresopolis@majordomo.srv.ualberta.ca>
From: "Pamela Asquith" <Pamela.Asquith@ualberta.ca>
. . . As to Dick's good suggestion about there being more than one Japanese "approach" to primatology, and that some Japanese primatologists are in the Western mold, I had begun to think that might be so too until returning with different questions this year. Indeed, I come away with more of a sense of a unity among Japanese primatologists than ever before, despite different styles among Japanese. For instance, Toshi Nishida said that he did

not think there was much, if any, difference between Western and Japanese primatology. He related, however, that data gathering might be different. For example, a Western primatologist, Nishida said, would go out, find his target animal at Mahale, observe for an hour and return to camp. Nishida, on the other hand, would follow his focal animal (or animals, he didn't say which) through the entire day as . . . he wanted to understand "process": in this case, what happened before and after and around the particular behavior he was observing. Do methodological differences constitute a different science? If methods reflect in part a seeking of different kinds of information, at the very least we may expect some different results (or opinions) . . .

To: <teresopolis@majordomo.srv.ualberta.ca>
From: "Brian Noble" <brian.noble@utoronto.ca>
. . . Pam's comment on discursive exchange between Japanese and non-Japanese researchers of Japan in several disciplines . . . points to how such exchanges sustain an inside/outside relation, where the inside "ideas" are somehow discursively insulated from the outside "historical/intellectual" conditions of the disciplines. . . . it would be challenging to try and get a grasp on what's going on here. Pam's research [is] on the collapsing together of Japanese "culture" into primatological practice and so into what counts as "natural" behavior in primate society. Can you [Pam] comment on why this interpermeability inside Japanese primatology is so unimpeded and contrast this with the flow between Japanese and say American/Canadian and British primatology? . . . Flows between Japanese and non-Japanese primatology are ostensibly high in ideas (results from one primatological domain get used in others), and ostensibly lower in practice (which seems to include propositions about theory, method, gender). Does this mean that method/theory and situation of the practice are dismissible in scientific exchange, while results are all that counts, so long as they are useful to the researcher at hand?

Finally, the other flow that equally intrigues me is the public cultural one, where as Hiro has pointed out, Jane Goodall stands as important a public science figure in Japan as she has been in Britain and North America. If Jane is part of Japanese knowing by whatever means, and Japanese knowing collapses into primatology in some way, is not Jane Goodall's practice through scientific publication or mass media eventually flowed into Japanese primatological knowing, hence results that are in turn shared in and between primatologies?

To: <teresopolis@majordomo.srv.ualberta.ca>
From: "Pamela Asquith" <Pamela.Asquith@ualberta.ca>
. . . Donna has shown very effectively that "American" primatology also collapses culture and primatological practice. The interpermeability is as unimpeded as it is in Japan, surely. Whether that holds for the European and South American primatologies remains to be studied there (effects of political/social/economic/intellectual background/religion, etc.).

(on the second question) I don't think that theory and situation of the practice are dismissible in fully understanding the results that are reported. Besides not knowing what the author intends by the report, the "naive" reader may misrepresent the implications of the results entirely. This has perhaps happened with the "specia" concept and with the separate goal of tracing "hominization" (evolution of human behavioral patterns) as many Japanese primatologists call it.

(on the last question) Results have certainly been shared in and between primatologies, regardless of authorship. But Jane's "practice" raises an interesting phenomenon in contrasting Japanese and Western proclivities. Japanese were heavily influenced by Jane (Kuroda got into primatology after reading *In the Shadow of Man;* he trusted Goodall's results more than other Westerners for the very reasons that Westerners said were unscientific in her approach, etc.). Jane is highly respected as a persevering, patient observer who lets the chimps "speak" for themselves: Japanese have "trusted" her findings more than did Westerners at first.

Perhaps for individual Japanese we can say that method/theory and situation of practice was MORE important than results, whereas for Westerners, results, provided they are obtained by legitimate scientific methods, are more important. That is, for the young researchers who always got their initial training (master's level) on Japanese macaques, their work was a contribution to a large corpus of data collected within a certain framework to answer certain larger questions posed by, especially, Junichiro Itani, as Imanishi's premier student. In some ways, what they found did not matter that much—they needed simply to be trained in the mold. Today, I am told that graduate students want to begin on ape studies or monkeys species overseas right from the beginning of their training. The long-term local macaque records now have large gaps.

By the same token, many Japanese have jumped aboard new hypotheses or lines of inquiry generated by Westerners and directed research reports toward them. How far their history underlies current practice requires careful examination because the younger primatologists are unaware of that history. Yet they are taught by those trained in an earlier tradition.

The diversity across traditions as well as within traditions of primatology, across generations and institutions, made us reconsider the term "primatology" and even question whether primatology exists. What does this diversity mean for primatology as a science?

Question 5: Are There Many Primatologies, or an International Science?

To: <teresopolis@majordomo.srv.ualberta.ca>

From: "Linda Fedigan" <Linda.Fedigan@ualberta.ca>

. . . I have always assumed that primatology exists and that it is an international science. I found it somewhat disconcerting when one reviewer claimed that primatology is a taxonomic identifier rather than a science. When I began my training in primatology in the late 1960s, early 1970s, I had read everything I could get my hands on by Rowell and Hinde and Kummer and van Hooff and Itani, as well as Jolly and Dolhinow and DeVore and Bernstein. These are the people who shaped my thinking, and over the years, although the primate literature has grown much larger, I continue to follow it without "prioritizing" the researchers by their national origins. Of course I know that there are differences between Japanese and Brazilian and European and British and Mexican and American primatologists, and that there are differences between Berkeley-trained and Harvard-trained primatologists . . . but I always assumed that we still share a common endeavor to better understand primates—their behavior, ecology, evolution, morphology, genetics, conservation. I do not agree with Dick that everyone shares the same motivation for studying primates (i.e., to better understand humans), although I think that many do have this motivation. Shirley and I outlined ten key questions about primate social behavior [for the workshop]. Although these may not be all-encompassing or worded exactly as others would word them, I do think that these enduring issues are international in primatology—the Japanese and Brazilian primatologists may approach the questions somewhat differently but are fundamentally just as interested as North American primatologists in why primates live in social groups, in what the relationship is between the group and its environment, in the patterns of ontogeny, learning, cognition, in the relations between the sexes, etc. . . . When I am told that there is, in fact, no international scientific community I feel a sense of identity crisis. . . . What does it take to qualify as a science, an international science? . . . What about entomology or ornithology or anthropology for that matter? Primatologists have their own thriving professional associations and international journals but they do not have independent departments in universities. . . . Primatology has a corpus of knowledge and a fundamental set of methods and a group of theories relevant to interpreting behavior/ecology/evolution/morphology although there is clearly much variation in methods and theory. At what point is a science considered to have come of age? . . .

To: <teresopolis@majordomo.srv.ualberta.ca>
From: "Robert Hinde" <rah@joh.cam.ac.uk>
. . . About differences versus commonalities in primatology—it is important to add that all the individuals involved had their own inclinations and idiosyncrasies. . . . But the publications of these workers, who are now scattered in a number of research centers in different parts of the world, have become part of primatology. It makes one wonder whether, in view of the forces that draw primatologists nearer to the facts "out there," the differences beloved by science studies specialists really matter for long. It is relatively easy to document differences in the approaches, backgrounds, theoretical orientations, and so on of those who study nonhuman primates. But it is unbalanced to do that without placing them against a background of the commonalities. If that is done, they may be seen as relatively trivial. . . . many of the differences that do arise come . . . simply because the researchers are studying different problems . . .

To: <teresopolis@majordomo.srv.ualberta.ca>
From: "Emília Yamamoto" <fyamamoto@digi.com.br >
. . . I would like to put in a few words about native primatologies. I believe there is one primatology because when we say "primatology" we refer to that from mainstream journals

and books. And that is the one that derives from the British/American tradition. Even in underdeveloped countries we are urged to publish in the "right" journals, and that means those that are included in international indexes, such as ISI. These do not allow for diversity. One has to conform to the mold. If anyone of you is familiar with *Neotropical Primates Newsletter* you may have noticed that this has been used by many Latin Americans to publish their papers that by format or content do not get accepted in the more traditional journals. That does not mean that they are bad science, but that they do not conform.

. . . As Gibbs put it very well in his 1995 paper in *Scientific American,* scientists from underdeveloped countries are less acknowledged (meaning less cited) than their American or European colleagues that publish in the same journal. That suggests to me that we, Brazilians, Japanese, Mexican, etc., cannot dare [to initiate changes in ideas], because we would be discredited. The alternative is to follow the mainstream.

SECTION 4

Enlarging the Lens: Closely Related Disciplines

How do ideas and histories vary across disciplines? We ask experts in other disciplines.

Primatology not only developed in many different parts of the world with resultant cross-cultural variation, but also has a complex ancestry, budding off from anthropology, psychology, and zoology (ethology/animal behavior). Others might identify additional ancestral sources (ecology or anatomy for example), but these three disciplines seem to us to play the major parental roles—that is, behavioral primatologists almost invariably receive their degrees and occupy academic positions in departments of zoology, psychology, or anthropology. At the same time, these disciplines obviously continue to follow their own intellectual channels, and therefore they are not only parental, but also collateral to our field. Whatever the flow of ideas, methods, and theories between the disciplines, each has its own configuration. Thus, it occurred to us that we might find some answers to our question about how ideas change by a comparative examination of the influences of theory, method, gender, and culture on ideas in these related fields. This comparison would be one way of holding some factors constant (e.g., the preference for sociobiological explanation in both primatology and zoology) while rotating the others (e.g., proportions of women practitioners). In any case, we felt that it would be illuminating to have experts from these related fields tell us how and why they felt that scientific ideas about sociality had changed over time in their disciplines.

We decided to sample two of the other subdisciplines of anthropology (sociocultural anthropology and archaeology), as well as zoology (animal behavior) and comparative and cognitive psychology. We sought contributors who were active in their own fields of research and who also had a reflexive stance towards research practice. We invited Naomi Quinn to participate because she had already published on the role of women, gender, and feminism in cultural anthropology drawing comparisons to primatology. Philosopher Alison Wylie has background training in archaeological fieldwork and has for some time been examining the nature of archaeological research and the impact of feminism and feminist theory on archaeology. Zuleyma Tang-Martinez was trained in animal behavior, and she teaches not only in the field of animal sociality but also in women and science and biological determinism. Stephen Glickman is well known for his research on that sexually anomalous creature the hyena (where females sport erectile clitorises and ubiquitously dominate the males). He has also published on the history of his discipline of comparative psychology. Richard Byrne's long-standing interests in questions of primate mind, his background training in psychology and in evolution, his current work in cognitive ethology, and his field research on baboons, gorillas, and chimpanzees, made him a natural choice to help us understand the differences and parallels between psychology and our own field.

The contributors to the chapters in this section each take one or more aspects of their discipline—some important theory or concept or some assumption about gender and females—and trace its winding path over time in order to elucidate the variables that play a role in changing science. For example, in her chapter, "The Divergent Case of Cultural Anthropology," Naomi Quinn charts the route that women anthropologists studying gender have taken through the recent history of cultural anthropology. Covering both academic politics and theoretical developments, she offers this history of gender studies in cultural anthropology as a cautionary tale for behavioral primatologists. Although the two related fields were swimming in the same intellectual currents in the 1970s, they soon diverged. Evans-Pritchard's views on the "position of women in primitive societies" is used as an example of the masculinist gender myths prevalent in anthropology prior to the 1970s. Quinn then describes how feminist anthropologists initially countered androcentric depictions with their own universal and evolutionary gender origin myths. Soon, however, they recognized the overgeneralizations in their models. These moves stimulated a surge of field studies on gender and the collection of a veritable mountain of ethnographic data. This was, as well, a productive period of theorizing about gender. Up to this point, the trends in studies of male-female relations in both cultural anthropology and primatology show some arresting parallels. However, Quinn argues that for reasons both internal to cultural anthropology (e.g., cultural relativizing and tendencies to many small factions) and external (the decontructionist wave), the anthropological study of gender stalled at the level of ethnographic case studies in the late 1980s. Poststructuralism tainted feminist theory with the brush of Western imperialism since all theory is seen as a form of culture-bound ideology. Because poststructuralism makes a virtue of particularism, it ghettoized feminist anthropological theory and eclipsed its accomplishments. Quinn also presents cultural anthropology as a type of academic circus with only one center ring (since, unfortunately, anthropologists do not prize multiple theoretical agendas), a ring from which women and women's issues have been marginalized. She gives specific examples of where this has occurred. The chapter ends with a question first articulated by the feminist philosopher Nancy Harstock: Why is it, just when we are forming our own theories about the world, uncertainty emerges about whether the world can be adequately theorized?

Alison Wylie traces the history of gender studies in another subfield of anthropology in her chapter, "Standpoint Matters, in Archaeology for Example." Outlining the pivotal papers, books, and conferences that brought gender issues to the attention of archaeologists, she argues that the rapid emergence of archaeological interest in women and gender since 1984 is a strikingly late development compared to the other subdisciplines of anthropology, in particular sociocultural anthropology and primatology. Wylie suggests that the gendered dimensions of past cultural systems were just one casualty of the positivist or "processual" paradigm that reigned in archaeology until the early 1980s. The processual paradigm ignored or devalued most ethnographic aspects of the past. By contrast the postprocessual challenge within archaeology facilitated conditions under which it was possible to raise issues in the archaeology of gender, although the postprocessualists did not do so themselves. Focusing on survey data gathered from participants in the first large conference on the "Archaeology of Gender," Wylie tries to determine whether feminist issues in archaeology arose after a critical mass of women entered the field, some of whom had been politicized by the women's movement. She concludes that the majority of those involved articulate not an explicitly feminist standpoint but one of sensitivity to gender issues. There are some parallels to primatology, where "gender sensitivity" has also been attributed to those scientists who took up questions of what it means to be female. Wylie explores the value of standpoint theory in understanding why some archaeologists were motivated to take up issues of gender when they did. Briefly introducing us to the long-standing debate over standpoint theory among feminists, Wylie suggests how the marginalized position of women in archaeology may have played a role in the emergence of the archaeology of gender.

In "Paradigms and Primates: Bateman's Principle, Passive Females, and Perspectives from other Taxa," Zuleyma Tang-Martinez broadens our comparison to the discipline of animal behavior. She argues that sometimes theories come to be accepted as an item of faith with little or no testing. They then function as blinders that limit creative inquiry and innovative scientific insights. The example she chooses to explore is Bateman's Principle—essentially the argument that females invest more than males in reproduction because of their initial greater investment in costly eggs rather than cheap sperm. This hypothesis, she argues, has reached paradigmatic proportions without being tested or validated, and anisogamy has been the cornerstone of many classic theoretical papers in the evolutionary biology of behavior. Tang-Martinez suggests that the assumption of anisogamy has greatly influenced the way we conceptualize primate society because it has dominated the study of male-female differences and mating systems of all animals. Sarah Hrdy had earlier challenged the application of this model to primates and Tang-Martinez sets out to do the same for a larger array of animal species. She documents much recent research to show that, contrary to Bateman's Principle, sperm are not cheap and males may become exhausted and refuse female invitations to mate. After establishing that male reproduction can indeed be quite costly, Tang-Martinez next documents that rather than passively, or even selectively, accepting the "best" male, females in many species actively solicit a lot of matings from many different males. She argues that while the current paradigmatic view of male-female interactions places conflict at the center (the "war of the sexes"), we could just as appropriately focus on mutual benefit and selective cooperation between the sexes. Her take-home message is that theory is obviously essential to science, but there is danger in accepting theories without question or without recognizing their limits—thus allowing them to become dogma.

Stephen Glickman, in "Culture, Disciplinary Tradition, and the Study of Behavior: Sex, (Rats), and Spotted Hyenas," is concerned with the multiple influences on comparative psychology between 1920 and the present. To illustrate how such influences operate, he uses as case studies two lines of research in which he has been extensively involved: sexual behavior and differentiation in the laboratory rat and the spotted hyena. To set the background, he provides a short history of the roots of modern academic psychology—from the turn-of-the-century battle between functionalists and structuralists, to the rise and fall of behaviorism and the cognitive revolution of the 1960s. Noting ubiquitous assumptions regarding female sexual passivity in the psychological literature, Glickman shows how his mentor, the renowned comparative psychologist Frank Beach, came to change his view of the female role in sexual behavior. Starting with the common assumption that female animals are sexually passive, Beach was influenced by his research on female choice in dogs, by his graduate students' research, and by cultural events to recognize the extent to which females are active in choosing sexual partners and shaping sexual interactions. Ultimately, Beach coined the word "proceptivity" in reference to female sexual initiatives, a term that is still in use today. Drawing parallels to stages in primate research, Glickman traces the history of research on laboratory rats and hyenas. He concludes that gender-related cultural expectations as well as historic disciplinary traditions and the reward system of academia all influence changing ideas in psychology about sexual behavior and sexual differentiation.

In his chapter, Richard Byrne also pursues the theme of "changing ideas." In "Changing Views on Imitation in Primates," he traces the history of the concept of imitation and its application to primates. He argues that the concept of imitation was originally a rather fragile notion among primatologists, and one close to our everyday folk-understanding of the term (sham, counterfeit). However, it soon developed theoretical sophistication and in the 1990s has become a hot topic for primatology. The concept of imitation has been modified by two strong outside traditions, experimental and developmental psychology, and Byrne explores alternative approaches to primate imitation. He suggests that the dominance of the experimental approach with its hidebound insistence on laboratory-style control, rather than naturalistic observations, may have adverse consequences for the understanding of imitation in primatology. The case study of imitation offers up insights about how scientific ideas change, how primatology has changed, and how changing preferences in methods and theories influence our concepts of animal mind.

All of the chapters in this section address the question of how ideas change in science and all of them conclude that such change is not a simple, linear matter. The multifactorial nature of change in science may be the one conclusion that these authors hold in common. Focusing mainly on the rise and fall of influential theories (e.g., in respective order: Chodorow's psychoanalytic theory of mothering as the prime mover of male dominance, processual archaeology, Bateman's Principle in animal behavior, behaviorism in psychology, Piagetian models of development), these authors show how internal debates, accumulating data, and external forces work together to bring about shifts in scientific practice and preferred models of behavior and sociality. Four of the five chapters also describe how their discipline came to question traditional assumptions about sex, gender, and the role of females. But apart from these very basic similarities, the local histories of the disciplines that are presented are too variable and idiosyncratic for us to be able to use this small sample in our comparative design of holding some factors constant while rotating the others. We realize now that there are many local contingencies (e.g., the impact of postmodernism on cultural anthropology, postprocessualism on archaeology, the cognitive revolution on psychology) that play key roles in diverting the histories of any field along its own trajectory. Even subfields of the same discipline, whose practitioners rub shoulders in the hallways everyday, may proceed along quite different intellectual paths with little traffic between them. Certainly members of the same field trained at different institutions may have quite different perceptions of the impact and value of the same theory or method. Two prime examples that came to light during our workshop were the different reception and impact that sociobiological theory had on scientists from different backgrounds and institutions and the taking up of gender issues in different decades by cultural, primatological, and archaeological anthropologists. Why and how this happens in science is pursued in the e-mail exchanges presented at the end of this section.

11

The Divergent Case of Cultural Anthropology

Naomi Quinn

A woman sees herself being seen. Clutching her pencil, she wonders how "the discipline" will view the writing she wants to do. Will it be seen as too derivative of male work? Or too feminine? Too safe? Or too risky? Too serious? Or not serious enough? Many eyes bore in on her, looking to see if she will do better or worse than men, or at least as well as other women.

Ruth Behar (1995, 2)

In this paper I trace the path that women studying gender, as a sometimes more- and sometimes less-identifiable group, have taken through the recent history of one discipline, cultural anthropology. As will become apparent, the cultural anthropology story is as much about academic politics as about theoretical developments, implicating the one deeply in the other. In this respect, mine is a cautionary tale for behavioral primatologists. Marilyn Strathern (1987) has argued that the social sciences contrast with the natural sciences simply in being fractious and fractured by nature, but I do not believe any academic discipline to be immune to the kind of academic politics I will describe. On the other hand, we cannot assume parallel developments in the two disciplines just because they happen to be relatives and happened in the 1970s to be swimming in the same intellectual currents when, along with other behavioral and social sciences, both embarked on the renewed study of male-female differences and relationships. That distinct characteristics and histories can lead to different places is the second lesson. In fact, circumstances set behavioral primatology and cultural anthropology on profoundly divergent courses. My explanation for the divergent course taken by cultural anthropology looks to two factors that I will take up in turn in this paper—one intrinsic to its subject matter and approach, and the other visiting it from outside.

Let me characterize these briefly to begin. First, even though both disciplines are field sciences, what we might call the data-to-theory ratio has been, up to now, much greater in cultural anthropology than in behavioral primatology. I do not mean to detract for a minute from the meticulous care and intensive labor required of cultural anthropological field research. Yet nothing impressed me more about the brief but memorable trip that some of us got to take, following the Teresopolis conference, to Karen Strier's field site among the muriqui monkeys of Brazil, than how hard-won was field primatological data. So many questions, so few answers, I thought to myself at the time. By contrast, as I will describe in this paper, cultural anthropology reached a point, sometime in the late 1970s, at which existing theory was overwhelmed by the findings with which it was being inundated. This is not just an observation about the quantity of these findings; the intrinsic complexity of the human gender relations story that they told also distinguishes cultural anthropology from behavioral primatology.

The second factor was a well-known late-twentieth-century intellectual development in the humanities that has had considerable influence on several humanities-near social sciences, most notably cultural anthropology. This intellectual current has not reached behavioral primatology at this writing (or, to be more accurate, has arrived in the form of practitioners of science studies studying them, rather than as an influence on their own studies of the nonhuman primates). As was observed not too long ago in a *Science* (Sept. 1993, 1798) piece describing the growing separation between biological and cultural anthropology:

> The divide has become much more pronounced as biological anthropologists have become deeply involved with the latest tools of molecular biology and theories of evolutionary ecology, while many cultural anthropologists have been caught up in the wave of deconstructionist thinking that has been sweeping the humanities.

I am tempted to bend this metaphor to say that this is the story of the women cultural anthropologists who have ridden that wave and of those who have fallen away from it or been wiped out by it.

In considering how these two factors have played themselves out, setting cultural anthropology on its path, I focus on one train of events in the recent past of my discipline. The objects of my attention are the highly visible events in the center ring—what, in a different metaphor, Mascia-Lees and colleagues (1991) have dubbed the "prestige discourse" in the discipline. A central tenet of the argument I want to make is the effect on theory, and on the position of women vis-à-vis theory, of the show going on in that center ring of the academic circus that is cultural anthropology. To contextualize current events properly, however, especially for nonanthropologists as well as for a younger generation of anthropologists just coming on the scene, I must go back briefly to the beginnings of "the anthropology of women" (as it was then called) in the 1970s.

Feminist Myths

Before the 1970s cultural anthropology, like biological anthropology and many other disciplines, had been dominated by accounts so decidedly and unselfconsciously biased by academic male assumptions about the natural order of things that they seem grotesque today and can most usefully be understood as tellings of a masculinist gender origin myth that was widely accepted then. It is a gendered picture of social life that will be familiar to primatologists because it differs hardly at all from the versions of it that primatologists were spinning during the same period. The British social anthropologist E. E. Evans-Pritchard (1965) offered an unusually complete rendition of this picture in his paper "The Position of Women in Primitive Societies and in Our Own," first delivered as the 1955 Fawcett Lecture at the University of London, one in a series of lectures established in memory of British suffragette Millicent Garrett Fawcett. Evans-Pritchard opened his lecture by noting that "Mrs. Fawcett" had been a leading feminist who might not have accepted his conclusions or sympathized with his "way of examining and commenting on the facts" (1965, 38), and he acknowledged his own lack of understanding of and enthusiasm for the feminist movement. He delivered the opinion, however, that "circumstances, and therefore the climate of opinion, have so changed, that were Mrs. Fawcett alive today she might well have accepted some of the points I would like to make." "The facts" were that, in so-called "primitive" societies, "every woman finds a husband," and "running a home is a whole-time occupation, to which is added the care of small children," of whom "parents have as many as possible" (Evans-

Pritchard, 1965, 45–46). Since it is also the case that "the primitive woman has no choice" about marrying, it follows that "given the duties that go with marriage, [she] is therefore seldom able to take as much part in public life." At no pains to disguise the contemporary social issue that motivated him to address this subject, Evans-Pritchard (1965, 50) went on to observe that therefore men and women in primitive society, unlike our own, do not intrude on or compete with each other, and he recommended such clearly demarcated spheres of activity in favor of the "blurring of social differentiation between the sexes" that had arisen in his own England. Separate spheres were for women's own good, he said, affording them their own domain of control as well as the protection they needed, because "in primitive societies men invariably hold the authority."

Evans-Pritchard (1965, 51–52) went on to observe that the "primitive" family, too, and all its members, benefit from women's restriction to the domestic sphere, since "arguments, bickerings, and other domestic unpleasantness" are reduced when there is no "fundamental challenge to man's position in the home" and fewer joint activities over which to disagree. "Primitive woman," he concluded, does not resent her situation, which only seems inferior to us, looking in. "She sees herself as different from man and as having a social status different from him; but if we may state her attitude in our own way of speaking, it is for her less a matter of level, than of difference, of status. Primitive women do not see themselves as an under-privileged class as against a class of men with whom they seek to gain social equality. They have never heard of social equality; and also they do not want to be like men."

Beginning in the mid-1970s, under the impetus of second-wave feminism and with the (mutual) reinforcement of parallel critiques emerging in neighboring fields such as behavioral primatology, women cultural anthropologists (with a relative abundance of whom, for historical reasons, our field was blessed) gave revisionist attention to these biased accounts. Two books published at that time, *Woman, Culture, and Society* (1974), edited by Michelle Rosaldo and Louise Lamphere, and *Toward an Anthropology of Women* (1975), edited by Rayna Reiter (now Rapp), gained landmark status in this regard. Articles in each book became classics that were and still are as widely cited outside as within cultural anthropology. As Faye Ginsburg and Anna Tsing were later to observe in the introduction to their 1990 edited volume *Uncertain Terms,* "Early second wave feminists posed questions for anthropologists regarding the world-wide origins and scope of gender inequality, male dominance, and patriarchy. In response, feminist anthropologists provided global answers" (1990, 4). What these classic articles proffered were new, feminist versions of the gender origin myth.

If the point of the masculinist myth was to explain and justify as natural the gendered order of things with which British and American academic men were comfortable, the point of these feminist origin myths was to explain and justify why that order was unnatural. If, by way of grounds for his mythological account, Evans-Pritchard had only to speak for primitive woman and refer to the facts about her situation, the feminists who wished to revise the entrenched cultural myth had more explaining to do than that. There seem to have been, as Bell has noted (1984, 245–246, quoted in Strathern 1987, 284–285; see also Quinn 1977), two strategies for constructing satisfying feminist myths: universalist and evolutionist. In universalist accounts, male dominance was variously attributed to the control men gain over women due to compulsory heterosexuality (Rubin 1975); to men's association with culture and women's to nature, due to the inescapable role of the former in childbearing and nursing (Ortner 1974); or to men's association with the public and women's with the domestic (Rosaldo 1974), due to the psychodynamics of mothering, argued to affect boys and girls differently (Chodorow 1974). Boys, in this latter view, must separate from their mothers in order to attain their gender identities, and hence learn their own status as an abstract set of rights and duties associated with formal authority and formal roles, and achieved in peer groups outside of domestic units; girls, who do not have to so separate, learn relational, one-on-one skills that are required and valued in the domestic world.

Evolutionist myths harked back to a golden age of gender equality in the supposedly "egalitarian" foraging societies that organized the human way of life for the first 99 percent of human history (Leacock 1978). In these stories, various gender inequalities arose thereafter with the institution of brideservice marriage (Collier and Rosaldo 1981); sedentarization (Draper 1975), the rise of private property, class, and indigenous states (e.g., Leacock 1972; Sacks 1974; Silverblatt 1978, 1987; Muller 1977, 1985; Gailey 1987); or the spread of colonial rule (Etienne and Leacock 1980). While the universalist versions, especially those of Rosaldo and Ortner, got into especial trouble for the implication that gender inequality was biologically based and hence natural and unchangeable, all these accounts were ultimately to be contested on empirical grounds.

Research Agenda

In referring to these as gender origin myths, I mean to call attention to their ideological function. At the same time, as academics' accounts, these new "myths" provoked scrutiny of both their explanatory adequacy and their empirical validity. That the scrutiny was so instant, energetic, and extensive attests to the political urgency women anthropologists felt about the issues involved.

I believe I may have been the first, in a 1977 review of the literature on "women's status," to take Rosaldo, in particular, to task for the overgeneralized and underanalyzed nature of such claims about women's position cross-culturally; but I was far from the last. The arguments in *Woman, Culture, and Society* and *Toward an Anthropology of Women* and their critiques also invited fresh empirical investigation, and in this, more than in their service as feminist myths, lay their enduring significance. The two books together—including some of the articles in them that never achieved "myth" status because their arguments were already more causally complex (e.g., Sanday 1974; Sacks 1974) or less sweeping (e.g., Collier 1974 and Leis 1974; Draper 1975; Faithorn 1975; Brown 1975)—set a substantial research agenda on which numbers of women anthropologists then commenced. Continuing to parallel developments in behavioral primatology to this point, cultural anthropology initiated an extraordinary period of field research on gender. An army of women (and a smattering of men) cultural anthropologists went out into the field to examine the position of women, gender roles and relations, and (increasingly a topic of interest) ideologies about gender, in all kinds of societies. This flood of empirical research continues today without signs of abatement (see Miller 1993, 10–13). The result, as I will detail in the next section (see also Ginsburg and Tsing 1990, 4–5), has been the discovery of new variations defying the monolithic explanations for gender inequality that had been favored at first and adding a much-needed complexity to the picture.

Research at this new stage was no less theoretical for its liberatory impulse and its empirical bent. Profitable debates and questions revolved, for example, around the emergence of gender inequality in "simple" societies (Collier 1988; Kelly 1993); the position of women in emerging states (Coontz and Henderson 1986; Sacks 1979); and the role of ideology in promoting gender inequality in class societies (Martinez-Alier 1974; Stolcke 1981). Comprehensive review essays summarizing and assessing this literature reflected a growing concern for theory that could address and explain the newfound complexity. In the most recent general review of which I am aware (and what could turn out to be the last comprehensive assessment of this outsized and exponentially growing literature that anyone is willing to attempt), Carol Mukhopadhyay and Patricia Higgins (1988) covered anthropological studies of gender over the prior ten years. This impressive review of what the authors now referred to as "feminist anthropology" (after its beginning as "the anthropology of women" and an interim period as "the anthropology of gender") attested to the robust ongoing research tradition I have described, even though it identified a number of issues on which crucial evidence was still lacking, obvious questions had not been pursued, proffered explanations were flawed, or overarching syntheses were badly needed.

In retrospect, it is easy to see why the collected volumes by Rosaldo and Lamphere and by Reiter were written when they were and had the impact that they did. One has only to remember back to those heady feminist times in the academy (di Leonardo 1991; Gordon 1995). The critique, research, and revision that came on the heels of their publication are no less understandable; not only were they paralleled in behavioral primatology, they were familiar phases in disciplinary-based gender studies of all kinds. A new generation of students were poised and motivated to undertake the fresh anthropological fieldwork and the theoretical rethinking that the new anthropology of women spurred.

At the same time, more than one irony came of these disciplinary developments. Mid-1970s cultural anthropology provided "global answers" to the feminist question about why women were subordinate to men. However, as the field produced new and conflicting ethnographic evidence that raised many uneasy questions about those easy answers, feminists outside cultural anthropology seemed to become disinterested in women anthropologists' struggles to square general explanation with empirical evidence. Cultural anthropology lost its special place as mythmaker to feminism. It did so, we shall see, at the very time when it was providing new and better answers.

Within cultural anthropology itself, further ironies were to come. Mukhopadhyay and Higgins (1988) wrote forcefully, to be sure, of both the difficulties of explaining gender asymmetry in all of its new complexity and of the need for new empirical studies directed at further unraveling the complexities that had emerged. If a realist impulse made them and other authors (see, e.g., Miller 1993; Pasternak et al. 1997; Silverblatt 1988) emphasize how much was left to do over and above what had already been accomplished, there was nothing defeatist in the tone of their writing. Indeed, there was no reason for discouragement; the theoretical contribution of this anthropological enterprise, in its twenty-year history and under its successive names, has been truly impressive. It is remarkable, then, that feminist anthropologists have largely failed to claim, or even comprehend, their own theoretical achievement. The reasons for this, I will argue, are both intellectual and political. But first I must convey, however summarily, the nature of this little-appreciated theoretical achievement itself.

Theoretical Discoveries

First of all, feminist anthropology has definitively answered the pressing question of the 1970s with which it began: male dominance is not universal. Not only did the differences between men and women that had initially been treated as indicators of "male dominance" fall apart into multiple independent variables, but few of these separate measures of women's position, status, or role relative to men proved universal, either. These points were appreciated quite early on in the enterprise (Quinn 1977; Whyte 1978; see endnote 3) and have become axiomatic. Second, a number of very widespread differences between men and women, some of these amounting to inequalities between them, have been documented. Yet, for every near-universal, insistent exceptions have cropped up, exceptions that have proven to be invaluable clues to the more usual patterns that obtain. As these exceptions have shrunk the list of "men and women everywhere," the third achievement of the anthropology of women has been to document the ample and illuminating contexts and ways, cross-culturally, in which women exercise autonomy and authority, claim respect and influence, and live in complementarity and, yes, equality with men. The fourth achievement of this anthropology has been the identification and documentation of a medium-sized list of variables that together appear to pattern gender roles, relations, and beliefs cross-culturally, accounting for gender differences and inequalities when these occur. These variables range from the biological (notably, but not limited to, women's role in reproduction, especially lactation) to the social (for example, the gendered system of property ownership and distribution) to the ideational (for instance, regionally distinctive, historically durable ideological complexes in which gender figures). The fifth and final finding is that these variables hardly ever operate independently. Now we are in a position to appreciate, not only that there are multiple causes of the differences and inequalities between men and women, but that, in interaction, these lead to highly complex and variable outcomes.[1]

Nancy Chodorow, reconsidering her original thesis about the effects of mothering cross-culturally and in our own society, brings our attention to this last point, and characterizes the kind of feminist anthropological theory needed to account for these multiple, interacting processes:

> In my current view, feminist understanding requires a multiplex account—perhaps not as acausal as thick description, but yet not necessarily claiming causal explanatory status—of the dynamics of gender, sexuality, sexual inequality, and domination. It is the focus on relations among elements, or dynamics, along with an analysis and critique of male domi-

> nance, which define an understanding of sex and gender as feminist, and not just the exclusive focus on male dominance itself. I no longer think that one factor, or one dynamic, can explain male dominance (even if I still have my own predilections for particular theoretical contenders). An open web of social, psychological, and cultural relations, dynamics, practices, identities, beliefs, in which I would privilege neither society, psyche, nor culture, comes to constitute gender as a social, cultural, and psychological phenomenon. This multiplex web composes sexual inequality. . . .
>
> My early writing . . . implied that women's mothering was *the* cause or prime mover of male dominance. I would now argue that these writings document and delineate one extremely important, and previously largely unexamined, aspect of the relations of gender and the psychology of gender. (1989, 5–6)

Yet, as obvious as Chodorow's argument may seem and as logically as it may seem to follow from the research findings of feminist anthropologists, it is a conclusion that many feminist anthropologists themselves would no longer embrace, or perhaps even be willing to endorse. And even though some continue to call for the pursuit of explanatory generalizations along the lines that Chodorow would seem to be advocating,[2] few any longer venture it. The collection of ever new ethnographic cases of gender systems continues but, as many of the introductions to the newer edited collections and the vast preponderance of case studies over more general theoretical essays in these volumes reflect, this ethnographic evidence accumulates in a veritable explanatory vacuum.

In place of general explanation, a new climate of particularism has overtaken feminist anthropology. The rationale for this particularism appears to be that each ethnographic case can only be understood in its own terms, the unique convergence of multiple, interacting causes of difference and inequality between men and women. I would hardly disagree with a call for attention to the concrete, contextualized particularities of cases; that, after all, is what got feminist anthropological theory this far. I take issue, however, with the many feminist anthropologists today who would leave it at that—who assume that in the particular cases of women in given societies, no universal processes are at work, that no patterns, however complex and dynamic, repeat, and that no generalizations can be sought.

Theoretical Impasse

Why has feminist anthropology failed not only to pursue, but even to recognize, a successful theoretical program? Why has anthropological research on difference and inequality between men and women cross-culturally largely stalled at the level of particular descriptions of ethnographic cases?

The next logical task would seem to have been the construction of theories that address the dynamic interactions among variables—Chodorow's "relations among elements." Such a venture might employ multivariate modeling to capture the results of these complex interactions, setting these complexities in motion to show how the more general, recurring "social, psychological, and cultural relations, dynamics, practices, identities, beliefs" interact and, in interaction, produce particular cases or types. Even more usefully, perhaps, it might untangle the complexities, relying on past findings to identify critical variables and their interactions, and then returning to these for closer and more informed investigation and more sustained and thoughtful explanation. These next steps—ordinary ones in the practice of science—were ones that, as I have described in the previous section, some cultural anthropologists who studied gender were beginning to take.

If most feminist anthropologists turned their backs on this endeavor, it may have been that they found themselves over their theoretical heads. For their parent discipline offered no established way of thinking about the complex picture that was emerging from their research. To the contrary, as a research tradition, cultural anthropology is exceptionally unprepared to address multicausality. The discipline is built on ethnographic case studies of individual societies (see Strathern 1988, 5–6, for this point); comparison has been a weaker impulse and only a sporadic undertaking. Cultural anthropology ingrains in its practitioners a related habit of cultural relativizing that denies human commonalities (see Brown 1991, 154–156) and that has been most recently authorized, though it was not invented, by the prominent anthropologist Clifford Geertz (see Spiro 1997). Finally, the field has long been characterized by a disciplinary politics of explanatory schools disposed to square off against one another rather than seeking to synthesize their diverse explanatory insights (see Strauss and Quinn 1997, 18). Steeped in these three traditions, most cultural anthropologists are indisposed to multicausal thinking and unexposed to ways of incorporating it into theory.

A widely cited 1980 article by Michelle Rosaldo—the tone of which Mukhopadhyay and Higgins (1988, 481) aptly characterize as "soul-searching"—was written by an author at just such a loss for multicausal thinking about gender inequality. The article marks an intellectual turning point for feminist anthropology, when universalism gave way to particularism, at the same time that it exposes the illogic that led one of the subfield's most prominent theorists from one to the other:

> Gender in all human groups must, then, be understood in political and social terms, with reference not to biological constraints but instead to local and specific forms of social relationship and, in particular, of social inequality. (Rosaldo 1980, 400)

Between biological universals and "the local and specific forms of social relationship and . . . inequality," there seemed to occur to Rosaldo no alternative. Biological universals having been excluded from consideration, no cross-cultural patterns can be expected, nor general explanations sought.[3]

Not Chodorow's acknowledgment of causal complexity, but Rosaldo's throwing up of the explanatory hands and acceptance of causal particularity, was the move that, overwhelmingly, feminist anthropology went on to make. As in one recently published instance I have run across, the word "explain" itself now warrants quotation marks (Abu-Lughod 1993b, 11, who describes how Chodorow and others "try to 'explain' the differences" between men and women).[4] These new scare quotes, as well as a facile but prominent new language for talking about gender as wholly particular because wholly cultural, signal a further development. At this point, we will next see, the "deconstructionist wave" sweeps up all tendencies against complex explanation of cross-cultural differences between men and women in anthropological thinking, and carries them forward. A strategy born of theoretical frustration, particularism now becomes a virtue.

The turn to particularism has had an amplifying effect on the abandonment of cultural anthropological theory about gender. It has obscured the impressive theoretical accomplishments of feminist anthropology that, so far, I have delineated. The anthropology of gender exchanged its initial overgeneralizations for new generalizations that were empirically better supported and analytically more useful. Particularism recast this theoretical advance as a failure to find generalizations at all. In so doing, it can hardly have encouraged further explanations of this sort. Of course, poststructuralism, with its outright rejection of general explanations, was to reinforce this message heavily.

Academic Politics

In a wide-ranging summary of the influences on contemporary feminist anthropology, Micaela di Leonardo (1991, 30) has advocated: "Although social constructionism can shade into poststructuralism, it cannot, when it is located inside historical and social scientific analysis, degenerate into a nihilistic stance holding either that there is no truth or that, in Foucauldian logic, we are all trapped in the prisonhouse of language." Yet, social constructionism that is not so located frequently does degenerate into antitheory or at least theoretical paralysis on the part of cultural anthropologists (see, for an example in the realm of gender, Moore 1994).

Several poststructuralist currents converge on cultural anthropology today. Their influence on the anthropology of gender has been mixed. At its best, poststructuralism has made us more "alert to how androcentric and Eurocentric folk views influence scholarship" (Mukhopadhyay and Higgins 1988, 487), and it has led us to "approaches that favor specific histories, debunk essentializing categories, and are alive to the voices that politics and academics have muffled" (Silverblatt 1988, 429). At its worst, it makes suspect, because Western, the entire enterprise of searching for explanatory generalizations about gender cross-culturally, and theorizing about these (Wolf 1992). New theory turns out to be old rather than contributing to any helpful revision of the old. As I will support with the history of feminist anthropology, I believe this unnecessary theoretical extremism to be more politically than conceptually motivated.

The most well-defined channel by which poststructuralism has entered into American cultural anthropology is the "new ethnography," or what di Leonardo (1991, 22) characterizes as the "'ethnography-as-text' school." These cultural anthropologists radically reconceptualized ethnography, which is no longer to be thought of as the collection and accounting of facts about other worlds but is now to be viewed as a genre of writing, past authors of which have employed various rhetorical strategies for asserting the ethnographic authority to tell their accounts of those worlds. Just so, we saw Evans-Pritchard invoke a universalizing discourse about "the primitive woman," his access to the ethnographic "facts" about her, his right to speak for her, and a distanced, decidedly masculine, objectivity on the subject. This new theory of ethnography, however, like poststructuralism more generally, had consequences that were not altogether positive.

I have argued above that the 1970s research agenda for the cross-cultural study of gender was a fruitful one, up to the point at which it became stalled for lack of a theoretical framework equal to explaining its findings, caving in to an inclination to particularism. I will go on to argue that poststructuralism next brought with it the marginalization of this successful anthropological effort and the eclipse of its accomplishments. The true nature of these efforts and accomplishments—the continuous, cumulative record of research that I have indicated—was not just obscured but obfuscated, by being subsumed within the new framework of "ethnography-as-text," and reinterpreted as a genre shift. Thus, Deborah Gordon (1995, 438) comments, "If we can call feminist anthropology a genre, clearly it has changed over time from a bastard form of the expository essay into feminist ethnography." Gordon explains the change:

> For feminist anthropologists who must meet the professional demand to write ethnographies, textual experimentation has transformed the "personal is political" into a new form of academic capital. As university presses experience greater pressure to sell books and as a competitive job market drives up publication rates, anthropologists can now become writers. (Gordon 1995, 438)

There is more to this story than genre politics, however. What Gordon describes is not simply a new form of academic capital that just happened to come along as something else happened to go out of style. As I have indicated, what she describes is the displacement of the old anthropology of gender from the center ring by a new, poststructuralist-inspired, feminist ethnography. In the discipline, explanatory work begun by Lamphere and Rapp and Rosaldo and summarized by Mukhopadhyay and Higgins is no longer in the center ring. Behar makes this displacement clear when she comments about the 1986 book *Writing Culture,* which chartered "ethnography-as-text" and its impact on women's theory in cultural anthropology:

> [T]he classic texts of that historical moment—*Woman, Culture, and Society* and *Toward an Anthropology of Women*—were perceived as original and ground breaking, offering a major paradigm shift in the theorizing of anthropology as an intellectual, political, and cultural practice. But the *Writing Culture* critique showed that the mark of theory, as Lutz [1995] argues, is ultimately male controlled. Feminist anthropologists may have carried the theoretical day, but by the standard of the avant-garde textual theory promoted by *Writing Culture* they wrote in terms of a notion of grand theory that was outdated, even conservative. (Behar 1995, 14)

In a discipline that exhibits the "center ring" phenomenon as markedly as does contemporary cultural anthropology, some might want to take the position that it is an ugly academic politics best sidestepped and ignored; that other good work of all kinds should proceed and can even flourish outside the center ring (as indeed the anthropology of gender has managed, to some extent, to do), alongside the work that happens to be occupying the center at a given time, and without regard for the differential prestige granted each.

However, as the case of feminist anthropology demonstrates, these are serious politics, having the object of nothing less than displacement. The ever-realist Behar (1995, 11) quotes Pels and Nencel (1992, 17) as saying, "To be taken seriously in the academy, we also have to write ourselves *in* the history of the discipline and, consequently, write *off* rival academic currents" (italics theirs). Being in or out of the center ring, then, can have serious consequences for the fate of one's intellectual position. As we have seen in the case of feminist anthropology, the show in the center has a way, if not of proscribing other kinds of work, then of affecting the seriousness with which they are taken and even suppressing their circulation.

My own experience and observations of how the prestige system works in cultural anthropology give me no reason to doubt anthropologists Mascia-Lees and colleagues when they assert that

> the significant power relations for many of these new postmodernist anthropologists are not global but parochial, those that are played out in the halls of anthropology departments, those that are embedded in the patriarchal social order of the academy in which male and female scholars maneuver for status, tenure, and power. (1989, 16)

Or, they might have put it that most of these scholars maneuver not to be left out of the center ring where status and power reside. The prohibition on the discredited theoretical approach has not been most widely spread and longest perpetuated by the leaders of the new theoretical movement itself, who, as in other center-ring takeovers, may be few and are likely to be the first to move on to new pastures. The prohibition is now most diligently promoted by their followers, those who imitate in a desire to be included or in fear of being left out. In any event, as Mascia-Lees and colleagues point out, the least explicitly articulated but most sweeping prohibition against explanations of difference and inequality between men and women—which come from research traditions that no longer carry what Behar calls "the mark of theory"—does not appear on the page. Rather, it is spoken in the corridor and the search committee and, they might have added, in the classroom and the tutorial and the margins of term papers, where it is less public and hence more difficult to address. The devaluation of this "outdated" body of research is most readily communicated, thus, to the most eager followers of all—our graduate students. It is through their exclusion from these students' curricula, reading lists, and critical consideration that the hard-won findings and explanations of a twenty-year tradition of gender studies are being most effectively lost, while the new poststructuralist intellectual tradition of gender studies that has supplanted it lengthens its life.

Exclusion from the system of prestige that operates so prominently in disciplines like mine marginalizes by dictating, for a given period of time, not only what is on the tip of people's tongues and on their reading lists, but who is getting real resources: graduate students, job offers, invitations to conferences. In this particular ousting, a signal event was the conference on experimental ethnography held at the School for American Research in Santa Fe in the summer of 1983, which led to the publication of *Writing Culture* (Clifford and Marcus 1986). Only one woman, and she a nonanthropologist and the partner of one of the other participants, invited at his behest, was included in the conference and the book. The world of women anthropologists went berserk. Large amounts of ink (to which I now add) were spilled over the exclusion of women from the conference and the book, and over ill-considered remarks made in the book as to why women and their writing were excluded, and over the book's cover. The question was pursued whether the conference organizers were justified in excluding women's writing because, as was claimed in the book, women did not really do self-consciously experimental ethnography; or whether women, as some women anthropologists claimed, did indeed do their own brand of experimental ethnography and had perhaps even pioneered it. Noted author bell hooks (1990, 123–131) weighed in to the controversy. Last year, finally, a group of women published the 457-page answer to *Writing Culture* from which I have already quoted, as well as drawn this paper's epigraph. *Women Writing Culture* (Behar and Gordon 1995) is a volume the irony of which, coeditor Ruth Behar (1995, 5) observes in its introduction, is that it "might never have come about if not for the absence of women in *Writing Culture*."[5]

The 1983 conference is telltale not only about men's interests, which have already been dissected in print (see Sangren 1988; Mascia-Lees et al. 1989), but about women's. After all, women anthropologists have been excluded time and time again from conferences and edited volumes, as we continue to be; and many of these exclusions have been egregious. Indeed, in *Women Writing Culture* Catherine Lutz (1995, 252) draws attention to the spate of books about "grand theory" in anthropology and related social sciences, published during the last decade, that systematically exclude women and feminist topics. Why did this one exclusion from this particular conference and this particular book cause such enormous anger? Why would someone say, as Behar (1995, 5) does, that "*Writing Culture* took a stab at the heart of feminist anthropology, which was devalued as a dreary, hopelessly tautological, fact-finding mission." The answer is that women presumed, this time, that they had already won a place in the center ring. Women then learned that men, as usual, were utterly oblivious to women's achievements and aspirations. The men were simple enough, on this occasion, to reveal, in print, how truly oblivious they had been.

However briefly in terms of disciplinary history and however inadvertently on the crest of second-wave feminism, women anthropologists like Shelly Rosaldo, Louise Lamphere, Rayna Rapp, and Shelly Ortner were in the middle of things. They were widely read and appreciated by feminists outside of anthropology, avidly followed and taken up by other women anthropologists, foremothers overnight to a vigorous research tradition, and, perhaps most significant of all, discussed and acknowledged (if not actually widely cited) by male anthropologists. For the first time, they were performing, not one fluke woman at a time, but *en masse,* in the center ring. This group of women anthropologists had its expectations raised spectacularly. It was possible to imagine that women were finally full-fledged members of the discipline.

At the same time, however, the anthropology of women and of gender had become ghettoized within cultural anthropology. This occurred in spite of hopes that the study of gender would be incorporated with the pursuit of more traditional anthropological questions, and explicit efforts on the part of women anthropologists, like the 1986 conference that resulted in the volume *Gender and Kinship* (Collier and Yanagisako 1987), to make this happen. (Significantly, these overtures came from the women's side.) In spite of such efforts at joining theories, the women studying the gender side of the equation remained separate as a group. This came about when virtually all of the women but virtually none of the men in cultural anthropology were drawn to the new anthropology of women. And this concentration of women in a subfield that was so intellectually energized and energizing for them was also a political liability. Men then coming along to make their careers in anthropology, as new generations continually do, could hardly claim the center, as some men were bound to try and do, by doing feminist anthropology. But, what they could do, the only strategy open to them to claim the center ring for themselves once again, was to ignore what the women were doing. Ghettoized as they were in their own subfield, the women anthropologists were targets for mass replacement. The men would have done the same (and have done many times) to any competing group of anthropologists, men and women alike, whose school was susceptible to being overturned; gender was only an incidental factor in this case.

Coincidentally in 1982/83, the academic year during which the *Writing Culture* conference was being planned, I occupied an office at the Institute for Advanced Study in Princeton, New Jersey, opposite that of George Marcus, another visiting member of the Institute that year and one of the two conference co-organizers. Immersed as I was in my cognitive anthropological work, I first heard of plans for the conference when I was made aware of a controversy. Another, younger woman anthropologist on our floor at the Institute, Judith Goldstein, had challenged George on the exclusion of women from the list of conference invitees. Since, as Judy says, "I always run my mouth," the argument continued off and on. On the one hand, Judy remembers being bemused. She wanted to know, not only how George could possibly leave women out, but also how anything that was supposed to reframe anthropological theory and transform the discipline as George hoped it would do, could leave out feminist anthropology, which had already reframed so much theory and had been so transformative. Successive women anthropologists, assuming, as Judy did, that women had already arrived, repeatedly displayed the same incredulity when they learned about our exclusion from this conference. On the other hand, George was not hostile to Judy's blandishments or defensive toward her, she says, but simply uncomprehending. He did not think that the examples she gave of feminist anthropology fit the conference and he did not see the problem, since a woman had already been invited to attend. His reaction fits my sense of a men's agenda being totally blind to women's concerns. It also coincides with di Leonardo's (1991, 23–24) explanation that "[e]thnography-as-text writers simply fail to subject their own deeply held representations to the same operations they perform on feminism" when they treat the latter as "a culture-bound ideology to be held at a distance and analyzed critically."

The story plays itself out. In explaining why professional women anthropologists may not have experimented as much with ethnographic form as they might have in the period before men burst on the scene with experimental ethnography, Abu-Lughod guesses:

> If feminist anthropologists have not pushed as hard as they might on epistemological issues nor experimented much with form, it is perhaps because they preferred to establish their credibility, gain acceptance, and further their intellectual and political aims. (Abu-Lughod 1993a, 19; see also Moore 1994, 125)

The irony here, of course, is that now women are producing experimental forms of ethnographic writing for the same reason. Women not only want to be accepted, they want to be, and have come to expect to be, in the center ring. And why should we not, after all? This aspiration becomes a problem because women almost always have to choose between being ghettoized or being camp followers, between being judged, in Behar's words, "not serious enough" or "derivative."

We have to choose because we are both devalued and powerless in academia as in other of our institutions, in ways—and this is the deeper irony—that the displaced anthropology of gender has made a good start at exploring and explaining. As Margaret Mead (1949) projected onto women everywhere, we in academia are devalued whatever we contribute; moreover, we are powerless to change the terms of our participation. George Marcus and James Clifford did not need to exclude women deliberately, out of a fear of being eclipsed by feminist claims to having already pioneered the experimental ethnography they wanted to tout as their own. Far more simply, these men knew who were the important people to include if they wanted their conference to set a trend. It should not have been a surprise to anyone that those people were not women. Moreover, Marcus and Clifford were in a position to exercise their selectivity: senior male anthropologist Doug Schwartz, president of the School of American Research, was an unlikely person to ask, "Hey, George, how come no women on the list of conference participants?"[6] Feisty Judith Goldstein raised the question, but had no clout.

It is regrettable when anthropologists cannot find a way, in our academic practice, to prize multiple theoretical agendas, in this case theory about differences and inequalities between men and women cross-culturally, alongside theory about the writing of ethnography. If there is one lesson I could wish that primatologists would learn from this story about their sister discipline and carry home to their own, it is not to repeat our experience of letting academic politics license and curtail anthropological theory. These politics prevent us from recognizing and confronting the real theoretical predicaments that should preoccupy us, like the one confronted by cultural anthropologists who study gender. That politics have impeded theory in the case of my own discipline is all the more regrettable, because cultural anthropological explanations for gender inequality bear on the very gendered academic practices that, I have argued, devalue this latter body of explanation.

Writing about the exclusion of women from *Writing Culture,* Mascia-Lees and colleagues (1989, 15) and Abu-Lughod (1993a, 17) acknowledge the suspicion that many feminist anthropologists have about "ethnography-as-text," a suspicion that they found well articulated in a set of remarks by political philosopher Nancy Hartsock (1987, 196). "Why is it," says Abu-Lughod, paraphrasing Hartsock, "that just when subject or marginalized peoples like blacks, the colonized and women have begun to have and demand a voice, they are told by the white boys that there can be no authoritative speaker or subject?" Mascia-Lees and co-workers append Hartsock's next observation that it is not only "precisely when women and non-Western people have begun to speak for themselves," but, moreover, when they have begun "to speak about global systems of power differentials." Curiously, however, both stop short of including the final question Hartsock (1987) raises. Brandishing the *t*-word that they, as new-style feminist anthropologists, are now prohibited from using, but which Hartsock sees as vitally informative to the voice that speaks about global systems of power differentials, she asks: "[Why is it, j]ust when we are forming our own theories about the world, uncertainty emerges about whether the world can be adequately theorized?" Cultural anthropology offers an object lesson. Just as women cultural anthropologists had deconstructed the masculinist gender origin myth, reconstructed it in their own terms, and used this as the springboard for theories of their own difference from and inequality with men, they were told by their male colleagues that there could be no further theory on that subject. And some of us not only capitulated to this prohibition, but now—and this is the deepest irony of all—rearticulate it in the name of feminist anthropology.

NOTES TO CHAPTER ELEVEN

Revisions of this paper have benefited from the helpful comments of Judith Goldstein, Jennifer Hirsch, and Wendy Lutterell, as well as the good suggestions of Shirley Strum and Linda Fedigan, organizers of the Wenner-Gren conferences, *Changing Images of Primate Societies: The Role of Theory, Method, and Gender,* for which it was written.

1. Parenthetically speaking, it is in large part because of the number of these variables that have been uncovered and our understanding of their complex interplay, that cultural anthropologists have little patience with the monocausal, oversimplifying explanations for male-female differences in terms of human mating and reproductive strategies that have been visited on human societies by some sociobiologists and evolutionary psychologists. Cultural anthropologists would also want to raise methodological issues and questions of Western bias.

2. Barbara Miller (1993, 4), for instance, while advocating a greater attention to variation cross-culturally and intraculturally, cautions, "But do not think that I mean that the wider generalizations should be abandoned; instead they should be tested against, reformulated, and enriched by more localized insights."

3. The very work that Rosaldo footnoted (along with my own 1997 review) as having posed a challenge to her thinking and having prompted the 1980 article offered a multivariate approach to gender inequality. This was a statistical analysis published in 1978 by sociologist Martin King Whyte that tested the correlations among all the then-accepted indicators of "women's status" cross-culturally, using ethnographic evidence available to Whyte at that time. Whyte found clusters that were only weakly correlated with one another (within-cluster correlations of items not being especially high, either), a result strongly suggesting that there was no unitary "woman's status" or any single universal cause for the "subordination" of women. Less widely cited than the Rosaldo article has been, this outsider's book can nevertheless be said to have had a decisive effect, not only on Rosaldo's thinking but, partly though her article, on feminist anthropology more generally. Its

negative finding regarding women's status was widely assimilated. However, its positive model of how the newly evident complexity of gender inequality could be theorized was not.

4. Similarly, Abu-Lughod (1993b) puts quotes around the word "describe," saying, "Some [cultural feminists] attempt to 'describe' the cultural differences between men and women."

5. Behar (1995, p. 6) adds, "Fortunately, although *Women Writing* Culture began as a feminist response to *Writing Culture,* it grew into something much larger."

6. In fact, during the time when he was co-organizing the conference, George reported to me his disquiet with the fact that one of the conferees insisted on bringing his young children, since Schwartz did not like having children on the premises and the conference center was not set up to accommodate them.

12

Standpoint Matters—In Archaeology, for Example

Alison Wylie

Introduction

Unlike primatology, archaeology has never been described as a "feminized" field, much less embraced as a prospectively feminist science.[1] Indeed, popular images of archaeologists and their practice tend in the opposite direction. From the pages of *National Geographic* to *Raiders of the Lost Ark,* the archaeologist of media fame is represented in distinctly manly terms; whether "hairy chested" or "hairy chinned" (Ascher 1960, 400), whether beset by challenges physical or intellectual, stereotypically he is engaged in a heroic quest for treasure or for truth (or both) that is highly masculinized. This attention is not altogether unwelcome among practitioners. Reports circulate of bumper stickers sighted (apocryphally) in the southwestern U.S. and Alaska that declare: "Archaeologists are the cowboys of science" (Gero 1983, 51; 1985). But perhaps the myth is attractive precisely because, in most respects, it systematically misrepresents the more mundane and complicated realities of archaeological research.

Certainly archaeology has had its share of influential women practitioners, some of them legendary for flouting the conventions of their time, many others responsible in less dramatic ways for significant contributions to the field in all areas of practice; a number of recent histories celebrate the famous foremothers and reclaim a good many others from undeserved obscurity (Claassen 1994; Babcock and Parezo 1988). But despite the prominence of some women in the field, until the mid-1970s they did not account for more than 13 percent of the membership of any of the major North American societies of archaeology (e.g., the Society for American Archaeology; the Archaeological Institute of America); their representation jumped to 18 percent in 1973 and then to 30 percent in 1976, stabilizing at 36 percent from the late 1980s to the present (Patterson 1995, 81–82). And despite these recent gains, a good deal of detailed empirical analysis of the status of women in archaeology has appeared in the last twenty years that documents entrenched patterns of gender inequity and workplace segregation evident in the training, employment, funding, publication, and recognition of women, patterns which suggest that women continue to be marginalized even as they gain strength in numbers and in visibility within the field (Nelson, Nelson, and Wylie 1994). Perhaps, then, there is some truth in the popular myth; archaeologists may not be heroes, exactly, but they have mainly been men, and by all accounts the disciplinary culture associated with fieldwork in archaeology is highly masculinized, more like that of geology than primatology (for the comparison with geology, see Moser 1996, 1998).

Perhaps, too, it is not surprising that until quite recently Anglo-American archaeologists have shown little interest in the questions about women and gender that were first brought to prominence, in the 1960s and early 1970s, by the growing number of feminist practitioners working in such closely affiliated disciplines as sociocultural anthropology and history. It was not until 1984 that a first, exploratory paper appeared in which the case was made that archaeology both needed and could sustain a systematic program of "gender research." The authors, Conkey and Spector, argued the need for a systematic critique of the assumptions about gender that routinely inform archaeological interpretation—assumptions about gendered divisions of labor, gender roles and relations, even gendered identities, that look suspiciously like those familiar from 1950s sit-coms—and they identified a number of questions archaeologists might fruitfully take up about the gendered dimensions of the prehistoric and historic cultures they study. In this they drew inspiration from the rich traditions of feminist research that were already well established in the social and life sciences. Feminist sociocultural anthropology, they observed, had already developed through several stages of critique, remediation, and autocritique; it offered substantial resources, empirical and theoretical, for recasting archaeological questions and reinterpreting archaeological data. They also noted the constructive influence of feminist thinking on "reconstructions of earliest hominid life" (Conkey and Spector 1984, 6–8), unseating "man-the-hunter" models of human evolution and opening up a number of hitherto unexplored interpretive possibilities (e.g., as represented in Hager 1997b). Although they did not discuss primatology directly, its role in expanding the range of models that inform paleontological theorizing has clearly been pivotal in the development of feminist research programs in both paleontology and anthropology.

Conkey and Spector's article generated intense discussion as soon as it appeared, but it was another seven years before the first major publications began to appear that took up the challenge they had posed. The earliest of these was a collection of essays, *Engendering Archaeology: Women and Prehistory* (Gero and Conkey 1991), which was the outcome of a small working conference convened in April 1988 specifically for the purpose of encouraging colleagues to consider questions about women and gender in concrete empirical terms. This meeting was soon followed by an open-invitation conference on "the archaeology of gender" at the University of Calgary in November 1989 (the 22nd annual Chacmool Conference), and then by a growing number of conference symposia, workshops, and autonomous, special-topic meetings of various scales organized in Australia, the U.K., and the U.S., many of which have produced edited volumes, special issues of established journals, and published proceedings. In a bibliography of archaeological work on gender compiled in 1992 (a listing of 284 conference papers and publications), less than 10 percent of the entries predate 1988 and the vast majority of these were conference papers that never appeared in print; within two years of the initial conferences, the number of conference presentations more than trebled, and the publication rate showed a fourfold increase (Claassen 1992). The literature in this area has grown exponentially since 1992, generating a number of extensive bibliographies (e.g., Bacus et al. 1993; Hays-Gilpin 1996), several new edited volumes (e.g., Seifert 1991; Wright 1995), monographs (Spector 1993; Gilchrist 1993; Wall 1994), and, most recently, a reader (Hays-Gilpin and Whitley 1998) and an ambitious thematic overview (Nelson 1997). So despite the fact that little more than Conkey and Spector's 1984 paper was in print by the late 1980s when the first archaeological conferences on "gender" were organized, there seems to have been widespread latent interest in the topic that has since given rise to an enormously rich and rapidly expanding research program, one that is bearing fruit in virtually all areas of archaeological interest.[2]

From the time these "gender research" initiatives began to appear in archaeology, I have been intrigued by a number of interrelated questions which complement several of those posed by Strum and Fedigan for the Teresopolis conference (1996, 58–65). Why, I asked, had research on women and gender been so slow to emerge in archaeology, and why did it take hold with such vigor at the turn of the 1990s, some twenty years after comparable developments in neighboring fields and at a time when, by popular account, we were well into a "postfeminist" era? Was the growing representation of women in the field responsible for this late-breaking development? Certainly the majority of those involved in the new work on gender were women, but what was the nature of their interest? Were they motivated to take up questions about women and gender because of preexisting feminist commitments, scholarly and political? Finally, given that the very success of this research program throws into relief the androcentrism inherent in the research agenda and orienting assumptions of much contemporary archaeology, I was interested in its implications for the ideals of scientific objectivity and neutrality that are still highly prized by most North American archaeologists.

My answers to these questions have been elaborated elsewhere (Wylie 1991a, 1991b, 1996, 1997). I summarize them here for the purpose of providing what I hope will be a reinforcing comparison with two critical insights that emerged prominently in the Teresopolis discussions. The first is a general presumption against reductive, single-factor explanations in science studies: given the complexity of scientific practice, it should be assumed that any explanation of the emergence and trajectory of distinctive research programs (e.g., recent developments in primatology; the emergence of an archaeology of gender) must take into account an expansive network of factors which are contingent, emergent, and fundamentally interdependent. More to the point, it must be treated as an open, empirical question which factors are relevant in a given context and what role they play. This undercuts any sharp distinction between "internal" (broadly epistemic, cognitive) and "external" (sociopolitical) factors, and any asymmetrical presumption in favor of the explanatory relevance of factors on one side of the divide, over against factors on the other.[3] The second critical insight is that essentialist appeals to "gender"—for example, to distinctive women's "ways of knowing" or styles of inquiry—invariably underestimate the complexity not just of scientific practice but, more generally, of gender roles and relations, gendered identities, and gender symbolism wherever they play a role in our lives.

A further thesis follows by extension of these two points: that the gendered dimensions of scientific practice take the form they do only in interaction with myriad other factors, and that they play a role in shaping the course and outcomes of inquiry contingently, as activated or made salient by specific features of context and circumstance. There is little sense to be made of claims that "gender" functions autonomously; like science itself, it is always situated.

While I fully endorse these cautions against reductive and simplistic forms of explanation in science studies, especially where they involve untenable forms of gender essentialism, I also want to resist a tendency, evident at several points in the Teresopolis discussions, to conclude that considerations of gender can be set aside, that more important or, as Latour put it (in the e-mail exchange), manifestly more "interesting" factors are at work shaping scientific practice than the contingent structures of gender relations, gendered identities, and gender symbolism that are so often reified and essentialized in popular discussion. There are certainly more and less interesting and productive ways of conceptualizing the gendered dimensions of scientific inquiry, and it cannot be assumed in advance that gender, in any of the senses that may be useful to science studies, will necessarily prove to be relevant, much less fundamental, to our understanding of a particular constellation of scientific practices. But far from providing a rationale for dismissing questions about "gender and science," I will argue, the considerations of complexity that continuously resurfaced at Teresopolis make it clear that a thoroughly reflexive, symmetrical, and empirically grounded program of science studies must take seriously the gendered dimensions of scientific practice (see, e.g., Lloyd 1995; Longino 1994).

Archaeological Comparisons

From the outset, the questions about "the archaeology of gender" that concern me have been widely and actively discussed by those who pioneered work in this area. Their reflexive explanations take into account virtually all the types of conditions and factors that Strum and Fedigan identify in connection with primatology (1996), with one possible exception: that of changes in standard fieldwork protocols. Some cite a significant shift, in the early 1980s, in the theoretical commitments that had dominated archaeology through the 1960s and 1970s, while others note a coincident broadening of the interpretive conventions that determine what kinds of questions archaeologists can reasonably ask of their data. Many point to the influx of women into archaeology in the late 1970s and 1980s, and related changes in their role and status in the field; and some cite the influence of feminist scholarship in neighboring disciplines and of the women's movement more generally. What emerges on closer examination is that all of these factors are in some sense necessary but none are sufficient to account for the late but dramatic growth of archaeological interest in questions about women and gender. Perhaps more to the point, the interaction between these factors is crucial for understanding how and why any one thread in the network could play an enabling or catalyzing role in the context of the whole.

An early explanation for the dearth of work in archaeology on questions about women and gender, set out by Conkey and Spector when they first argued the case for feminist initiatives (1984, 22–24), was that, until the early 1980s, Anglo-American archaeology had been dominated by the New Archaeology, a self-consciously positivist research program whose advocates emphasized the need for strictly scientific modes of practice and, consistent with this, endorsed a reductive "ecosystem" conception of culture. On this "processual" paradigm, all aspects of cultural life and the development of cultural systems were to be explained in broadly functionalist terms, in terms of the adaptive responses of cultural systems to their external environments. The gendered dimensions of these systems (whether conceived as patterns of social organization, cultural conventions, or "ideational" constructs) were just one casualty of a predisposition to ignore all "internal," ethnographic aspects of the cultural past. These were ruled out of consideration both because they were taken to be inaccessible (on positivist criteria of adequacy) and because, on a strict ecosystem model, they were strictly dependent variables and therefore could be presumed to be explanatorily irrelevant (this argument is discussed in Wylie 1991a).

By the time Conkey and Spector were writing their formative paper on the prospects for gender research in archaeology, a strong offensive against the New Archaeology (processual archaeology) had been launched by "postprocessualists," who challenged both its stringent epistemic ideals and its substantive claims about the cultural subject. Postprocessualists argued, by appeal to the results of ethnohistoric research, that the form and dynamics of cultural systems (and the resulting archaeological record) are often significantly shaped by "internal" factors, sometimes even to the point of rendering them nonadaptive; archaeological accounts of the cultural past will be incomplete, at best, if these factors are systematically ignored. They also argued, by appeal to philosophical critiques, that the positivist ideals embraced by the New Archaeology are inherently untenable and profoundly limiting for archaeology. They undertook to recuperate various forms of interpretive inference that the New Archaeologists had rejected out of hand, exploring a range of strategies for getting at the dimensions of cultural life that, in their zeal to set archaeology on a firm scientific (positivist) foundation, processualists had ruled out of consideration; these include not only community organization and social relations, but also the symbolic "lifeworld" of past cultural agents. Conkey and Spector presupposed the earliest of these developments when they argued, in 1984, that the time was right for archaeologists to explore questions about gender. In effect, they suggested that postprocessual critiques had made it possible to raise questions that were, in a sense, unthinkable in the terms of strict processual archaeology; they had opened up conceptual space for an archaeology of gender at the same time as they brought into being such things as a "critical archaeology" (Preucel 1991), an archaeology of "symbolism and structuralism" (Hodder 1982), an archaeology of mind (Leone 1982), and various forms of "interpretive archaeology" (Tilley 1993).

While postprocessualism does seem to be a crucial enabling condition for archaeological research on gender and has profoundly shaped its development, there are two problems with this account. The first is that most New Archaeologists quite sensibly practiced a rather different archaeology than they preached. Despite the rhetoric of commitment to a strict ecosystem theory and stringently positivist modes of practice, many were intensely interested in the internal dynamics of the cultures they studied, and took great initiative in devising strategies for documenting, in archaeological terms, such inscrutables as cultural interaction networks, patterns of social stratification, diverse forms of community and household organization, and processes of change in all of these over time. Given this, the real question is, Why did these more expansive processualists not make questions about the *gender* dimensions of prehistoric social organization and divisions of labor an explicit focus of investigation?

The second problem with an internalist account of this sort is that, as I have indicated, postprocessualists had decisively challenged the most restrictive commitments of the New Archaeology by the early 1980s, but questions about gender continued to be marginal in archaeology for most of a decade. This lacuna is particularly striking when you consider that, on occasion, postprocessualists had identified feminist initiatives as just the sort of politically self-conscious archaeology they hoped their critiques of positivism might inspire. In the event, the main advocates of postprocessualism did little to develop feminist archaeology themselves and, when explicit critiques of sexism and androcentrism were articulated in archaeology (in the early 1990s), postprocessualists provided as much grist for the mill as did processualists.[4]

Although they differ on most other issues, it seems that postprocessual archaeologists share with processualists a number of (largely implicit) presuppositions about gender familiar from popular culture and common sense; for the most part they treat gender as a stable, unchanging, biological given and on this basis set it aside as a proper subject for archaeological investigation. The socio-logic here seems to be that if the social roles and psychological dispositions associated with men and women are essentially, "naturally" theirs, a function of genetic or hormonal or psychoanalytic makeup, they can be assumed to be stable across time and cultural context. In this case gender is not, strictly speaking, a cultural variable and is not relevant for explaining cultural change. For the same reasons, common wisdom about what women and men are capable of or are likely to do (what social roles they are likely to occupy, what contributions they are likely to make to community life) can be used as the inferential basis for building more challenging interpretive arguments, but does not require investigation in its own right. So long as such assumptions formed part of the common foundation of archaeological thinking—unacknowledged and largely unquestioned—postprocessual critiques could not have given rise to an "archaeology of gender" on their own, much less to a feminist archaeology. The question, then, is not "Why so late?" but "Why ever?" What drew attention to these assumptions in archaeology, raised them to consciousness, and opened them to critical scrutiny, in the late 1980s?

To address this question, I have argued, it is crucial to consider not just changes in the intellectual history and methodological commitments of Anglo-American archaeology, but also a range of sociopolitical and cultural features of this research community and its practice (Wylie 1996, 1997). I expected that the emergence of research on gender in archaeology would follow roughly the same course as had the development of comparable research programs elsewhere a decade or two earlier. In sociocultural anthropology and in many other social sciences, for example, the key factors include the influx of women to the social sciences and humanities in the 1970s, and the influence on many of these women of the second-wave women's movement; as feminists they had the conceptual tools and the political inclination to question the kinds of assumptions about gender that were taken for granted in archaeology until quite recently. A review of early conference programs did confirm my impression that it was primarily women who were attracted by the initial conference announcements and calls for papers on "gender." Their representation at these meetings more than inverts the ratio of women to men in the discipline; women routinely contributed over 80 percent of the papers listed at a time when their representation in the major archaeological societies stood at 36 percent (I take 1988 as the baseline for this assessment). Moreover, the various studies published by archaeologists on the status of women in the field make it clear that this level of representation was a recent development; the first cohorts in which women accounted for more than a third of the members of the major archaeological societies in North America entered the field in the late 1970s and early 1980s.[5] And, indeed, the first stirrings of interest in questions about gender became apparent at just the time when members of these cohorts would have come of professional age in archaeology, in the mid to late 1980s.

In order to learn more about the motivations and interests of those who were drawn to, or responsible for, the first conferences and publications on gender, I undertook a survey of everyone who participated in the 1989 Chacmool conference, and I interviewed a number of the "catalysts" who played a role organizing this and other initiatives at a time when there was virtually nothing yet in print on "the archaeology of gender." In the process I gathered some demographic data that refines the general picture I have just described. I learned that, although the average age of Chacmool participants is very similar for men and women (43 and 40 years, respectively), they show quite different age distributions; the men range widely across age grades and are four times as well represented in the most senior cohorts as are women, while altogether 60 percent of the women (twice the proportion of men) are clustered in the 26-to-40-year age range. The women who attended the Chacmool conference were drawn primarily from the first cohorts in which their representation in archaeology rose above 30 percent; they were relatively young professionals who would have completed their graduate training and begun to establish themselves as professional archaeologists by the late 1980s. So, in these respects the broad outlines of the "critical mass" thesis seems to be born out for archaeology.

I also learned that the vast majority of women who responded to the Chacmool survey attended the conference specifically because of its topic; 96 percent of the women (compared to 75 percent of the men) cite this as their main reason for attending, and, perhaps more telling, only a fifth of the women say they had ever previously attended a Chacmool conference, while over half the men report that they had been regular Chacmool attenders. And although 60 percent of all participants say that the Chacmool conference opened up a new area of research interest for them, a substantial number (75 percent) also report a preexisting interest in questions about gender; a quarter indicate that they had taught or taken women's studies courses. At the same time, however, I learned that this avowed interest in women and gender as a research topic does not necessarily reflect an explicitly feminist standpoint. A striking number of those who responded to the survey made it clear that they are deeply ambivalent about identifying as feminists; altogether 40 percent of women disavowed the label, and many of the 60 percent who did accept it registered serious misgivings about what it might mean. Even those who clearly identified as feminists indicated that they had had little involvement with women's groups or feminist activism; the most common sort of involvement reported was that of "being on a mailing list" or "sending money," usually to women's shelters and reproductive rights groups. While these levels of involvement in women's studies and activist groups is no doubt higher than is typical for North American archaeologists, taken as a whole the results of the survey decisively undermine my expectation that the majority of participants in the Chacmool conference on "the archaeology of gender" would prove to have had previous involvement in the women's movement and had welcomed the Chacmool conference as a first public opportunity to integrate their activist commitments with their interests in archaeology.

By contrast, most of the "catalysts" I interviewed did prove to have been independently politicized as feminists, often around workplace issues within archaeology or around sexuality issues, and most said that they had taken the initiative of organizing one or another of the early conferences or publications on "the archaeology of gender" because they wanted to bring a feminist perspective to bear on the research questions that interested them as archaeologists. Many were quite explicit on the point that it was, in fact, the direct experience of gender inequity in everyday life (especially their working lives) that had alerted them to androcentric or sexist assumptions which they then recognized as implicit in archaeological thinking about the cultural past. Often these practical and conceptual incongruities led them to feminist literature in other fields which, in turn, sharpened their critical perspective on androcentrism and sexism in archaeology, providing the conceptual tools for developing feminist research initiatives in their own field. So my initial hypothesis was born out by those who created the opportunities for exploring archaeological questions about gender, but these opportunities proved to be a powerful magnet for many others who had no such clearly defined interest in feminist research or activism. The Chacmool respondents themselves support this conclusion; when asked why "an explicit interest in research on gender and/or feminist approaches . . . [had] not been visible in archaeology until the last few years [and] why is it emerging now," only 16 percent cited the influence of feminist activism or scholarship. It is also supported by content analyses of the early conference proceedings; Hanen and Kelley (1992) note a dearth of feminist references in the Chacmool abstracts, which is consistent with the tenor of the majority of survey responses.

What emerges, then, is a situation in which a majority of those involved in the first conferences and publications on "the archaeology of gender" seem to articulate (in their survey responses and conference presentations), not an explicitly feminist standpoint, but what I have described elsewhere as a standpoint of sensitivity to gender issues (Wylie 1997); this characterization is also developed in some detail by Hanen and Kelley (1992, 203). I am struck by parallels with the cautious attribution, to (some) women primatologists, of a particular "sensitivity" to social relations (Hinde, this volume), or "gender consciousness" (Jolly, this volume), and by similarities with Haraway's account of the diffuse, indirect effects of an "atmosphere of feminism" (Haraway, this volume), which inclined both men and women working in primatology in the 1970s and early 1980s to "take females more seriously," an atmosphere which established for Western women, in particular, a "higher motivation to reconsider what it meant to be female" (Haraway, this volume).

In archaeology these motivations and sensibilities are evident in a growing awareness, particularly among women who entered the field in the late 1970s and early 1980s, of the gendered nature of their own experience, perhaps provoked by the fact that they were members of the first professional cohort to substantially disrupt the demographic status quo, a microcosm of "gender trouble" in the larger society. As amorphous and ill-defined as this standpoint is, it seems to have been sufficient to incline some members of this cohort to greater incredulity about the androcentrism inherent in extant research programs. Although they were clearly cautious about adopting an overtly political (feminist) stance—working, as they did, in a field that was still much influenced by positivist sensibilities, and at a time when postfeminist sensibilities were on the rise—they make it clear that they were aware of the contested and contestable nature of gender roles in their own personal and working lives. And this self-consciousness about gender relations was enough to put them in a position—to motivate them—to question taken-for-granted assumptions about women and gender, to identify gaps in analysis and to envision a range of alternatives for inquiry, that simply had not occurred to their older, predominantly male colleagues whose gender privilege includes an unquestioning fit between their gendered experience and the androcentrism that is, in part, constitutive of the research traditions in which they participate.

On Matters of Standpoint

Much remains to be done to specify the content of this "grass roots" standpoint of gender sensitivity (as Hanen and Kelley describe it, 1992), and to determine how it shapes archaeological practice in diverse contexts and how it relates to the gender politics of the larger society. But it should be clear from the example itself that the attribution of a gendered (or gender-sensitive or feminist) standpoint need not involve any essentialist assumptions about those to whom it is attributed, nor need it involve any claim to the effect that the standpoints in question confer automatic epistemic privilege on those who occupy them. The claim is that the gender institutions and conventions that define, in part, the standpoint of an epistemic agent (or epistemic community) make a *contingent* difference to what they are in position to learn or to know, what assumptions they will take for granted or be inclined to question, what research problems will draw their attention, and what opportunities they will have to pursue them within a field of practice in which their capacities for action are both enabled and constrained in many other ways.

In urging this formulation of how a gendered standpoint can be said to matter, I draw on a long-standing tradition of debate among feminist social scientists and science studies scholars about the prospects for developing a viable feminist standpoint theory. As originally formulated, in the early 1980s, feminist standpoint theory was inspired by Marxist theories of knowledge and psychoanalytic theory (e.g., Hartsock 1983; Harding 1983; and, in some respects, Keller 1985), reframed in feminist terms to make sense of how gender socialization and social differentiation along gender lines might confer epistemic advantage on those who occupy positions that are subordinate in the terms set by these social structures. By extension of familiar arguments about the epistemic implications of class position, these standpoint theorists suggested that women may be in a position to understand a great many things differently and, indeed, better, precisely because of their subordination as women. Although these early accounts are more nuanced than subsequent critiques might suggest, they were largely discredited when empirical and intellectual challenges to essentialism took hold in the later 1980s and early 1990s (see, e.g., Harding's critique of such accounts including, implicitly, her own earlier work, in Harding 1986; and Keller's later formulation of her position, 1992a). In fact, sociologist Dorothy Smith had long argued the need to formulate standpoint theory in rigorously contingent and empirical terms (1974, 1987). She argued, for example, that given the kind of work women typically do, the kinds of expectations they internalize, and the various ideological contradictions they negotiate as women, they might well be less inclined to accept the gender stereotypes and associated ideology that legitimate their marginalization than those who are privileged by the sex/gender systems that this dominant ideology supports. Indeed, Smith suggests, their successful performance of feminine roles and identities requires women both to understand in concrete detail the mechanisms by which the illusions of sexist ideology are created and maintained, and to obscure the role that they play in this process of construction.

This kind of analysis has been elaborated in especially powerful and insightful detail by feminists who integrate the insights of women who are "insider-outsiders" to privilege in many other respects than gender alone, in particular, women of color and women living in poverty whose survival may depend on an intimate knowledge of the world of race and class privilege in which they often work (a world in which they are marked as outsiders), as well as of the subdominant communities in which they are insiders (see, e.g., Collins 1991a, 1991b). The advocates of a standpoint epistemology construed in these terms are careful to acknowledge that positions of subdominance and enforced boundary crossing do not necessarily confer epistemic advantage. Narayan makes this point when she observes that conditions of oppression are often maintained in part by ensuring that those in subordinate positions have little access to the kinds of information and training, the conceptual tools and explanatory frameworks, necessary for them to understand the systemic causes of their oppression (1988, 36). Being an insider to an oppressed community often means being an outsider to certain kinds of information networks and educational privilege. At the same time, however, Narayan (among others) makes a compelling case for recognizing a range of advantages that may accrue to epistemic agents who are marginal to entrenched systems of privilege.

In some cases, the epistemic advantage contingently conferred by disadvantage in other respects is as straightforward as having access to certain kinds of information not available to those in positions of privilege. To cite an example that figures prominently in this literature, black women domestics in the U.S. frequently come to know more about the private lives and psychological dynamics of the families they work for than do many members of these families (e.g., Collins 1991a; for a powerful fictional illustration of this point, see Neely 1993). Often, too, there is epistemic advantage in having had to develop a capacity to recognize behavioral patterns, discern motivations, and make connections that others can safely ignore in order to negotiate a context in which you are comparatively powerless (a point elaborated by Narayan 1988, and illustrated, again, by Neely 1993). And, crucially, the advantage of an "insider-outsider" position may be evident in a sensitivity, born of divergent background experience, to anomalies, contradictions, and implausibilities that have gone unnoticed in the assumptions or explanatory models taken for granted by dominant insiders to a powerful epistemic community. As Collins describes her own experience, the misrepresentations of black family structure inherent in post-Moynihan sociology were patently obvious to one who brought to her professional training as a sociologist a grounding in the culture and history and experience of the black community (1991b, 52–53).[6]

The central tenet of feminist standpoint theory as it emerges in these accounts is that those who are marginal to established structures of privilege for any number of socioeconomic, political, or cultural reasons—gender being one dimension on which such differentiation is articulated—may well prove to be better positioned to understand a given subject domain than those who are comparatively privileged precisely because of their social-political location. What counts as compromising baggage on standard objectivist accounts may confer crucial advantage in maximizing quite standard epistemic virtues: empirical adequacy, internal coherence, consistency with established knowledge in other areas, explanatory power, and also perhaps simplicity, heuristic manipulability, extendibility across domains, and fruitfulness. Where the emergence of an "archaeology of gender" is concerned, it would seem that a gender-sensitive standpoint has made just this kind of contingent and localized difference to inquiry. Such a claim is enormously difficult to establish; like the sciences it is meant to describe, it requires painstaking empirical and conceptual work, the mobilization of resources—conceptual, methodological, and empirical; material, technical, and instrumental; political, economic, and social—that are dispersed through a wide network of fields concerned with the study of science, as well as those of feminist scholarship, and those that are themselves the subjects of study. But the sheer difficulty of delineating the gendered dimensions of such a standpoint—the fact that you almost always learn something you did not know when you started, that you often discover you were wrong—makes it clear that neither the sciences under study nor the enterprise of studying science in this way are a matter of arbitrary fabrication, at least not inherently or necessarily. If we use the best empirical and conceptual tools we have to understand, in detail, what conditions make a difference to our ability to conduct specific sorts of inquiry, we stand to improve the disciplines we valorize as scientific, to provide a basis for systematically assessing the limitations and scope of the knowledge they produce. This may not hold much appeal for those who can take for granted ready access to the epistemic privileges associated with science as it stands, but there is a great deal at stake for those who have been marginalized by the existing institutions of science. I am reminded, in this connection, of Glickman's account of maze-running setups in studies of the sexual behavior of rats: "[I]t wasn't until the 1960s that female rats were given an opportunity to express their interest in sexual contact with males" (this volume).

Conclusions

As Strum and Fedigan described the mandate of the Teresopolis conference, it was to "extend . . . previous single factor analyses and expand . . . Haraway's multifaceted socio-cultural account" (1996, 3); we were to consider as wide a range as possible of both "intrinsic" and "external" factors, a distinction that was itself problematized as discussion proceeded. In the event, much of the discussion focused not so much on specific explanatory or descriptive claims about primatology, but on the adequacy of a particular kind of explanatory claim about science. Consensus quickly emerged that it is just too simple to describe primatology as a "feminized" science, much less to claim that its recent developments can be explained by the strong representation of women in the field and the stereotypically "feminine" styles of inquiry and interests sometimes attributed to them as women. It is not clear who has seriously held such a view; certainly Strum and Fedigan's review of work by feminists who have taken an interest in primatology suggests a much more nuanced appreciation of the complexity of ways in which the practice of primatology might be said to have been shaped by the gender politics, roles, identities, and symbolism associated with the field (1996, 58–65). But consensus there was that such views must be decisively repudiated.

Quite apart from the specifics of primatology, any assumption that the fortunes of an enterprise as complex as primatology can be explained by appeal to attributes typical of the gender of its practitioners ignores twenty years of feminist scholarship which has decisively undermined such essentialist construals of "gender." If anything has been accomplished by feminists in the academy and outside it since the advent of the second-wave women's movement, it is an increasingly nuanced appreciation of how diverse women are, and what different things it can mean to be gendered male or female. The gendered dimensions of our identities, and of the cultural norms and social institutions that structure our practice, are always—contingently, situationally—a function of innumerable other conditions that structure our lives and opportunities. In short, essentialist construals are bound to be inadequate; the particular form that gender relations, norms, and identities take must be specified with care, and always in relation to a great many other factors that, together, determine whether gender is, in fact, a salient dimension of our practice.

Essentialist appeals to gender—explanations that posit a "way of knowing" distinctive of women in general—also ignore significant parallel developments in postpositivist science studies which make it clear that scientific practice and the knowledge it produces cannot be understood in reductive terms, whether these be epistemic or sociological. What counts as evidence, rigorous method, powerful explanation, sound scientific reasoning—the "intrinsic" epistemic factors privileged by objectivist accounts of scientific inquiry—is in many respects a conventional matter; this is the central insight to be drawn, not only from the expansive literature of recent sociology of science (cf. Cussins, this volume), but from philosophical analysis of the ways in which different "styles of reasoning" delimit what scientists can consider investigable (e.g., Hacking 1985), and of the interplay between "constitutive" and "contextual" values (Longino 1990). Moreover, none of these epistemic virtues is unambiguous. For example, what counts as empirical adequacy is notoriously slippery, subject to interpretation that depends as much on the purpose at hand—on pragmatic considerations and cultural conventions—as on the way the world is; sociohistorical studies of episodes in which evidential standards are forming or changing make this point with particular clarity (e.g., Shapin and Schaffer 1985), but it is central, as well, to philosophical analyses of cases in which evidence is controversial (e.g., Mayo and Hollander 1991) or underdetermines competing descriptive schemes and their metaphysical presuppositions (e.g., Dupré 1993, section 1). Finally, rarely can these virtues be maximized together (see Longino 1994, 479). For example, demands for explanatory power or generalizability routinely require the strategic compromise of empirical adequacy, compromises that are warranted by disciplinary history and the local dynamic of debate, technical and theoretical resources, the influence of neighboring fields and their conventions, the prospects for (or repudiation of) technical application, and so on; these tensions are especially clear in debates about criteria of adequacy for models in science (see, e.g., Wimsatt 1987; Levins 1966) and for laws in physics (Cartwright 1983).

At the same time, science is hard work, and it is work made hard, in part, by the (often) intractable nature of the things scientists study (Bhaskar 1978, 57; see also an elaboration of this argument in Wylie 1996). To varying degrees, the social-natural worlds we negotiate for practical and scientific purposes have a powerful capacity to subvert even our most deeply held convictions; this is a capacity that, at their best, the routines of empirical inquiry exploit to good effect, forcing us to redraw the limits of our knowledge and rethink what we have assumed even about such enigmatic subjects as the distant cultural past. In itself, then, the commitment to a thoroughly empirical and "symmetrical" approach to the study of science—a commitment that precludes any easy exemption of apparently successful science from sociohistorical scrutiny—establishes no presumption in favor of exclusively sociological explanations. As Pickering puts the point, scientific practice "is situated and evolves right on the boundary, at the point of intersection, of the material, social, conceptual (and so on) worlds" (1990, 710). Increasingly, as science studies scholars of all disciplinary backgrounds attend to specifics of the disciplines they investigate, they move toward an uneasy consensus that none of the positions defined by the long-standing opposition between broadly epistemic ("internalist") and sociohistorical ("externalist") studies of science have the resources, on their own, to make sense of all the dimensions and forms of practice that constitute the disciplines we identify as scientific. Moreover, as they attend to the dynamic practice of science (rather than its products), there is growing recognition among them that all the elements constitutive of science are historically contingent and emerging; the form they take and the effects they have in any given context are a function of their mutual "mangling" (Pickering 1995), as elements of extended networks of "actants" of various kinds (see Latour; Cussins, this volume), constellations of "ideas, things, and marks" that take shape as they stabilize into context-specific configurations (Hacking 1992, 44).

In short, given insights emerging from science studies—including philosophy of science—it would be remarkable if recent developments in primatology and archaeology could be explained in terms of a single, detachable factor, whether social or epistemic. And it would be especially remarkable that "gender," conceived in essentialist and reified terms, could play such an explanatory role, given the results of well over a decade of feminist scholarship that has systematically challenged gender essentialism in all its forms.

As I indicated at the outset, however, I do not believe that these considerations establish a brief for dismissing questions about the relevance of gender constructs and conventions to the practice and products of science. To conclude that these questions are in some sense distractions from the main issue(s), an "unrewarding" avenue to pursue, is to ignore, or dismiss, not only the constructive implications of arguments against gender essentialism but, more specifically, the substantial body of research produced in the last several decades by feminist scientists and science studies scholars who draw inspiration both from the growing sophistication of feminist theory (the elaboration of an increasingly rich and nuanced understanding of "sex/gender systems"), and from the hybrid traditions emerging in science studies. As diverse as they are, feminists who have closely studied a variety of sciences—like Collins, Haraway, Hanen and Kelley, Harding, Keller, Lloyd, Longino, and Smith, just to name those who have been cited here—make it clear that there are a great many contingent and localized senses in which the institutions and goals, assumptions and practices of science are "gendered," sometimes with considerable consequence for those who find themselves systematically excluded or trivialized, either as subjects or as practitioners of sciences whose modes of practice are structured, in part, by stereotypic assumptions about what they can (or should) do as women. If we are committed to understanding the sciences in all their rich contingency and complexity, I conclude that we should be suspicious of any research program that is structured, in advance, so as to systematically "disappear" gender (Longino 1994).

NOTES TO CHAPTER TWELVE

1. See Fedigan's response to the popular perception that primatology boasts a disproportionate representation of women (1994), and Fedigan's (1997) and Strum and Fedigan's (1996) discussion of both popular media and scholarly feminist interest in the possibility that primatology may be a distinctively "feminine" or "feminist" science.

2. These developments are described in more detail in Wylie 1997.

3. By epistemic factors I have in mind the considerations that philosophers have tended to privilege in their "rational constructions" of scientific practice: factors that are assumed uniquely relevant to the content and evaluation of knowledge claims such as the bearing of evidence, the logical structure of arguments for and against particular claims, requirements of internal coherence, and so on. See Longino for a useful account of "epistemic values" (1990, 77–78) and of the distinction between broadly internal and external factors; she refers to these as "'constitutive" and "contextual" factors (e.g., 1990, 4–7). The symmetry in explanatory accounts of science at issue here is one according to which "good" science is properly explained by appeal to the compulsions of evidence and rational argument, while failed science is to be explained by appeal to external factors (intrusive bias, distortion).

4. In one especially notorious passage, Shanks likens archaeology, specifically excavation, to striptease—each "discovery is a little release of gratification"—pretty clearly reaffirming dominant assumptions that the subject position of the archaeologist is normatively gendered male (Shanks 1992, as discussed by Engelstad 1991 and Gilchrist 1992).

5. Compare these figures to those published by the American Anthropological Association; women archeologists showed an increase of nearly 40 percent through the 1970s, more than twice the rate of increase for anthropology as a whole (for the details, see Wylie 1994, 1995).

6. For a recent discussion of these developments, see a forum on standpoint theory recently published in *Signs:* Heckman (1997), with Hartsock, Collins, Harding, and Smith contributing as respondents.

13 Paradigms and Primates: Bateman's Principle, Passive Females, and Perspectives from Other Taxa

Zuleyma Tang-Martinez

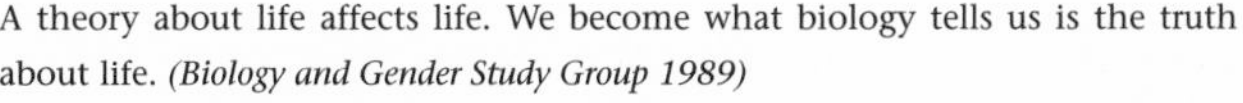

A theory about life affects life. We become what biology tells us is the truth about life. *(Biology and Gender Study Group 1989)*

To be believed, scientific facts must fit the world-view of the times . . . our facts come to be accepted on faith and large numbers of people believe them even though they are in no position to say why what we put out are facts rather than fiction. *(Ruth Hubbard 1989)*

Thomas Kuhn (1970) suggested that periods of "normal" science are governed by the dominant paradigm of the field, while revolutions in science occur only as a result of paradigm shifts. Paradigms are beliefs that, at any particular time in the practice of science, are accepted as unquestionably true and self-evident. Paradigms govern the practice of science by providing a worldview or framework in which science is done. This framework determines what questions are considered important, how questions are framed, and how data are interpreted. Although there is a positive aspect to paradigms, because they help to guide and provide coherence to scientific research, there is also a downside, as paradigms may come to be regarded as so self-evident that scientific evidence in support of the paradigm is considered unnecessary. In such cases, paradigms come to be accepted almost as items of faith and they function as blinders that limit creative inquiry and innovative scientific insights.

Although not everyone agrees with all aspects of Kuhn's treatise, there is no question that certain untested hypotheses are, at times, accepted as patently obvious and then used to frame questions and interpret data, with relatively little concern about the validity of the hypothesis. The unfortunate consequence of accepting such "self-evident" hypotheses or beliefs is that they can come to dominate a field by becoming dogma that defines and restricts the acceptable boundaries for scientific research, whether theoretical or empirical. Furthermore, such a dogmatic belief often serves as the fundamental assumption on which other theorizing and scientific research is based. Then, despite the lack of evidence in support of the initial assumption, results of the investigation are further interpreted within the framework of the assumption, presumably adding additional "credibility" to the untested hypothesis.

Bateman's principle (1948) is one example of such an untested hypothesis that has reached paradigmatic proportions (somewhat similar arguments have been made previously by Dewsbury 1982 and Hrdy 1986). Bateman's postulate that eggs are more costly than sperm appears self-evident from the much larger size and higher nutrient content (e.g., yolk) of an egg, as compared to the tiny packets of chromosomes that constitute a spermatozoon. The fact that eggs are larger than spermatozoa has been used to argue (1) that females necessarily invest more in offspring (beginning with the production of gametes) than do males; (2) that females will, therefore, be selective and sexually conservative (often signifying "passive") because they risk the loss of a large investment if they mate with the wrong male; and (3) that males will be promiscuous (usually meaning "sexually active") because they have nothing to lose by mating with as many females as possible and "the word excess has no meaning for a male" (Dawkins 1976a). Bateman's principle has influenced the ways we think and theorize about primate societies, because of the dominant role it has played in the study of male-female differences, and the evolution of mating systems and sexual selection (e.g., Daly and Wilson 1983; Krebs and Davies 1981; Wittenberger 1981). Orians (1969) begins his classic paper on the evolution of mating systems by using the principle of anisogamy as the cornerstone of his ideas. Trivers (1972) bases his hypothesis on the role of parental investment in sexual selection and the evolution of mating systems solidly on the idea that eggs are costly and spermatozoa are cheap. Despite Dewsbury's (1982) and Hrdy's (1986) papers, many evolutionary biologists and behavioral ecologists continue to assume, in a variety of contexts, that sperm are exceedingly less expensive than eggs, that males have unlimited quantities of sperm, and that male reproductive success depends primarily on access to females. Moreover, since each male theoretically produces enough sperm in one ejaculation to fertilize all of a female's eggs, each female should only need to mate with one male. Therefore, a female's reproductive success depends, not on her access to numerous mates, but rather on her finding the best male, and on her access to resources she can utilize for herself and her offspring. Thus, the assumption of costly eggs and cheap sperm becomes the starting point of many modern studies in behavioral ecology and sociobiology (see also Fisher 1930; Williams 1975).

Dewsbury (1982), in a prescient paper entitled "Ejaculate Cost and Male Choice," was one of the first to critique the dominance of the principle of anisogamy. More recently, the focus on sperm competition (e.g., Birkhead and Moller 1992; Dewsbury 1984; Gomendio and Roldan 1993) and on the role of females in controlling their own reproduction (e.g., Eberhard 1996; Rosenqvist and Berglund 1992), including their role in seeking extra-pair copulations (e.g., Gowaty and Bridges 1991; Smith 1988), has contributed to the slow but perceptible erosion of the exaggerated reliance on the principle of anisogamy.

Recent studies on sperm competition and the role of females in sexual behavior have shown that the description of sperm as cheap and eggs as costly is an invalid oversimplification. This realization has the potential to dramatically alter our current understanding of male-female differences.

Costs of Male Reproduction: The Sky Is Not the Limit

Many evolutionary biologists assume that because males produce millions of sperm on a regular basis, "the sky is the limit" with regards to the number of females that each male can inseminate. The production of millions of tiny sperm has been accepted as evidence that sperm are cheaper than eggs. While this is probably true if one compares one spermatozoon with one egg, Dewsbury (1982) points out that this is not a valid comparison. Males produce millions of sperm for each egg produced by a female—and males *need* millions of sperm to effectively fertilize one egg. Males also produce various costly accessory gland secretions contained in semen, and in some species they also produce spermatophores. Thus, the relevant comparison is between the cost of one egg and the cost of one ejaculate or one spermatophore. Even this comparison may not be valid because, even though the first ejaculate contains sufficient sperm to fertilize the egg, some females require repeated ejaculations to become pregnant (Dewsbury 1978, 1982; Eaton 1978; Lanier et al. 1975). In such species, it would be most accurate to compare the cost of one egg to that of the minimum number of ejaculations needed to fertilize each egg.

Other recent evidence dramatically demonstrates the high cost of sperm production. Van Voorhies (1992) reports that male nematodes (*Caenorhabditis elegans*) that produce normal levels of sperm have significantly shorter life spans than do males with reduced numbers of sperm. Furthermore, it was the production of sperm, rather than sexual behavior, that resulted in the shortened life span. Mutant male worms that mated normally, fed normally, and produced fewer sperm, had life spans that were 65 percent longer than worms that produced normal numbers of sperm and engaged in the same amounts of sexual behavior. Since *C. elegans* males produce relatively small numbers of sperm (about 3,000 to 5,000), the effect of sperm production on life span is not the result of producing inordinately large numbers of sperm. In contrast, increasing egg production up to three times normal levels (to approximately 800 to 1,200 eggs) did not result in shorter life spans, although each egg is equivalent in volume to about 500 spermatozoa. Based on his results, Van Voorhies (1992) concludes:

> . . . spermatogenesis, rather than oogenesis or the physical act of mating, is a major factor reducing lifespan in *C. elegans. This contradicts the traditional biological assumption that large oocytes are much costlier to produce than small sperm.* (emphasis mine)

Unlike *C. elegans* males, which produce relatively small numbers of sperm, males of most species produce millions or hundreds of millions of sperm every day. Why must males in most species produce so many sperm to inseminate a very small number of eggs? The answer seems to lie in two interrelated areas: (1) sperm competition: in many species, including most mammals, the male introducing the largest number of sperm into the female's reproductive tract has the highest probability of fertilizing her eggs (Gomendio and Roldan 1993); and (2) the hostility of the female reproductive tract to male sperm (see section below).

Sperm competition has resulted in selection for increased numbers of sperm and/or for more "efficient" sperm. Although millions of spermatozoa may be introduced into the female reproductive tract during each copulation, only a very small percentage of these sperm ever have the opportunity to fertilize an egg. For example, in both the domestic rooster (Brillard and Antoine 1990, cited in Birkhead and Moller 1993), and the rat (Gaddum-Rosse 1981), of the many millions of sperm transferred to the female, only about 100 (in the rat) to a few thousand (rooster) ever reach the site where fertilization can occur. In most mammals, of the sperm that reach the oviduct, only about 20 percent are "capacitated" and capable of actually fertilizing an egg (Bedford 1983).

Another suggestion is that males produce so many sperm because not all sperm are capable of fertilizing eggs. According to this hypothesis, some spermatozoa are specialized to form copulatory plugs or to attack or block rival sperm in various ways (Bellis et al. 1990; Birkhead et al. 1997).

Despite the millions of spermatozoa that are regularly produced by each male, sperm are sufficiently costly that sperm production is limited (Dewsbury 1982) and, after repeated ejaculations, sperm counts are reduced or depleted in a variety of species (e.g., Birkhead and Fletcher 1995; Halliday 1976; Huber et al. 1980; Sims 1979). In some species, it takes as long as a week for recovery to occur (Almquist and Hale 1956). In humans, for example, after complete depletion of sperm supply, predepletion levels were reached only after 156 days of recovery (Freund 1963). Males of some species may be frugal in sperm production, reducing the number of sperm produced when there is little opportunity for the female to mate with another male (Baker and Bellis 1989; Gage 1991; Shapiro et al. 1994). The production of the substances that constitute semen or spermatophores also is limited and costly (Dewsbury 1982). Following depletion, a period of recovery is necessary before these substances are again produced in adequate amounts for copulation to be successful.

In addition to these costs, all males at some point reach satiety and are physically incapable of continuing to mate, even if the male's sperm supply has not been completely depleted. After mating to exhaustion, male rats showed little recovery of sexual behavior after three days of rest, and complete recovery of all measures of sexual behavior did not occur until the rats had rested for eleven days (Jackson and Dewsbury 1979). A similar decline in sexual behavior was reported in newts that had mated repeatedly (Halliday 1976).

Perhaps due to the assumption that males have unlimited numbers of cheap sperm and will always mate with as many females as possible, relatively few studies have addressed male choice (Dewsbury 1982). Even fewer studies have asked whether males that are still capable of copulating will ever refuse to mate with willing females. Recent evidence suggests that they may. Male European starlings (*Sturnus vulgaris*) refuse 27 percent of copulations solicited by females (research by Eens and Pixten, reviewed by Gowaty 1994).

It could be argued that a female's cost of reproduction is still higher than a male's because the female carries the eggs, gestates, and may have primary responsibility in caring for the young. However, the belief that females invest inordinately more than males in parental care may reflect a mammalian bias. Mammals are only a tiny minority of the earth's fauna, and in other nonmammalian vertebrate taxa, there is either little parental care, or care is shared by both males and females (e.g., most birds, cichlid fishes) or may be primarily the responsibility of *either* females *or* males (some fishes and amphibians). Among invertebrates, in the relatively few species that have parental care, it is either the responsibility of the female (e.g., crustaceans, hymenoptera) or it is shared by males and females (e.g., termites, some beetles). Thus, we should be wary of all-encompassing hypotheses that assume that high levels of female care are universal. Also, even if some females have high costs of reproduction because of gestation and nursing (as in mammals), it is not clear that these costs are higher than the costs incurred by males in searching for females, courting, territorial defense, mate-guarding, or direct male-male competition. If these activities are necessary and essential for successful reproduction, they should be included in the costs of male reproduction.

Big Eggs Do Not Equal Passive Females

One corollary of the principle of anisogamy is that females have much to lose by mating with the wrong male. This led to the prediction that a female should be sexually coy and guarded, being careful to mate only with the "best male" who would fertilize all her eggs and provide her with the genes and resources she needs to produce superior offspring. According to this formulation, females have nothing to gain by mating with more than one male, and female copulations with multiple males should not typically occur.

Although Darwin (1859, 1871) recognized the significance of female choice, most early evolutionary biologists downplayed its importance and directed more attention to male-male competition (Birkhead and Moller 1993). In recent years, however, an interest in the ways that females control their own reproduction has been accompanied by a dramatic increase in the number of both theoretical and empirical studies of female choice. The realization that females are active participants in evolution is reflected in the appearance of books that focus specifically on female roles (e.g., Bjorkqvist and Niemela 1992; Eberhard 1996; Wasser 1983a).

In this section I examine the pre- and postcopulatory mechanisms by which females influence which sperm will fertilize their eggs. I also review some of the ways that females can influence the genotype and/or phenotype of their offspring, independent of the male's genetic contribution.

FEMALE TACTICS PRIOR TO COPULATION

Precopulatory female choice is not limited to choosing the *one* "best" male. Instead, modern evidence has demonstrated that females may exert choices by mating prolifically with one or multiple males, by playing an active role in soliciting these copulations, and by engaging in a surprising number of "extra-pair copulations" with males other than their supposedly monogamous mates. Additionally, it has recently been suggested that females may also exercise indirect or "cryptic" choice (Eberhard 1996; Wiley and Poston 1996) through a variety of mechanisms.

Cox and LeBoeuf's (1970) classic study demonstrating "female incitation of male competition" in elephant seals (*Mirounga angustirostris*) was an early demonstration of an indirect way in which females can influence the outcome of male-male competition; by doing so, they determine which male is most likely to mate with them. Wiley and Poston (1996) suggest that females inciting male-male competition may be only one example of the many ways in which females exercise "indirect female choice." According to them, direct female choice involves preferences for particular mates; indirect female choice occurs any time that a female, by her behavior (other than direct preferences) or morphology, influences which member(s) of the opposite sex are most likely to fertilize her eggs.

Indirect female choice includes all instances of females inciting intrasexual competition among males, thereby promoting sperm competition and/or increasing the likelihood that they will mate with the most dominant male or the male in best physical condition. It has been suggested that, in many species, when females advertise or broadcast their estrous condition, they may be inciting male-male competition by attracting numerous males. Examples include the copulatory calls of some female birds (Montgomerie and Thornhill 1989) and chacma baboons (O'Connel and Colishlaw 1994; but see also critique by Henzi 1996, and response by Colishlaw and O'Connel 1996), the calls of estrous chimpanzees (Hauser 1990), and the sexual swellings of Old World primates (Clutton-Brock and Harvey 1976; Hrdy and Whitten 1987). Copulatory chases solicited by females (as in bearded tits, *Pannurus biarmicus,* research by Hoi, reviewed in Gowaty 1994) may also be a mechanism by which females promote sperm competition by inciting males to compete in the chase and mating with males that are the swiftest or show the most endurance.

Other possible examples of females inciting male competition include situations in which females display synchronous estrus and gather in particular locations for mating, thus forcing all males in the population to aggregate in one place (Wiley and Poston 1996). One hypothesis for the evolution of leks, for example, relies precisely on such female preferences

(Bradbury et al. 1986; Queller 1987). Likewise, when females demonstrate strong mating site preferences, males are forced to compete for these sites in order to mate. Examples of such female preferences have been reported in some fishes and in butterflies (e.g., Davies 1978; Shields 1967; Warner 1987).

FEMALE SOLICITATIONS AND MULTIPLE MATINGS

Bateman's principle predicts that females will be highly selective of their mates, surveying the available males and mating only with the "best" male. Recent evidence suggests that rather than being sexually "coy," females of many species take the initiative in sexual behavior, solicit copulations (sometimes with extra-pair males—e.g., Birkhead and Moller 1992), and mate prolifically (sometimes hundreds of times within very short periods) either with a single male (Hunter et al. 1993) or with multiple males (Hrdy 1986; Keller and Reeve 1995).

For example, female felids typically mate with several males, and estrous lionesses may mate up to 100 times per day with multiple males (Eaton 1976, cited in Hrdy 1986; Eaton 1978), and may engage in copulations even when pregnant or lactating (Smuts et al. 1978). Female primates in a variety of species may actively solicit males both inside and outside their social group and may mate with several different males in quick succession. Moreover, in many primates, female sexual activity is not limited to the period when they are in estrus (Hrdy 1986; also reviewed in Eberhard 1996). Even in the highly monogamous gibbons (*Hylobates syndactylus* and *H. lar*), females have been reported to engage in extra-pair copulations (Palombit 1994; Reichard 1995).

Why are multiple copulations with multiple males so common? Hypothesized advantages to the female include (1) insuring against infertility of the female's first or primary partner; (2) increasing genetic diversity in the offspring; (3) fomenting sperm competition; (4) deceiving males into believing they have fathered the female's offspring, thereby increasing the likelihood of male parental care and decreasing infanticide against the female's young; and (5) increasing kinship relationships for the offspring (Davies and Boersma 1984). While all these hypotheses are to some extent controversial, what is strikingly clear is that the prediction that females should be sexually passive, except for active choice of one "best" mate, is not supported by the data and appears to be nothing more than a time-honored myth (see also Hrdy and Williams 1983; Hrdy 1986). Females of many species not only solicit males, they also regulate the pacing of male sexual behavior and control the order in which they mate with males (e.g., McClintock 1984; Huck et al. 1986a, 1986b). By doing so, a female that mates with multiple males may be able to control which male (or males) fertilizes her eggs. Thus, a female may influence paternity by mating only with a particular male during the time that she is ovulating, or by controlling the number and duration of mating bouts with each male, or the duration of intervals between bouts (Birkhead and Moller 1993; Dewsbury 1982). All of these factors are related to mating order effects that determine which male is most likely to fertilize a female's eggs (e.g., first or last male advantage). In some species, females also may be able to interrupt copulations prior to ejaculation, or may control how long ejaculations or spermatophore attachment lasts (Birkhead and Moller 1992, 1993).

POSSIBLE MECHANISMS OF FEMALE CHOICE AFTER COPULATION

Postcopulatory physiological adaptations recently have received attention as possible mechanisms of female choice. The female reproductive tract is "hostile" to sperm but, although some sperm are killed, others are spared (Cohen 1984). The current interpretation is that this is one mechanism by which females "select" certain sperm, while killing or incapacitating others.

Female effects on sperm survivorship and fertility are mediated by behavioral mechanisms, the chemical characteristics of the reproductive tract, immunological responses, capacitation of sperm, and physical barriers to sperm passage (Birkhead and Moller 1993; Birkhead et al. 1993; Roldan et al. 1992). In some species of birds and mammals, females are able to eject sperm immediately after copulation with certain males, thereby minimizing the chances that those males will fertilize their eggs (Birkhead and Moller 1992; Ginsberg and Rubenstein 1990). For example, female *Macaca mulatta* may discard sperm from some males, while *M. radiata* may sometimes terminate copulations prematurely (reviewed in Eberhard 1996). On the other hand, some insect females can store sperm in special organs and select the sperm that they subsequently use for fertilization (Eberhard 1996).

The acidity typical of the female reproductive tract is detrimental to sperm survival. Furthermore, shortly after sperm enter the female tract, there is an invasion of leukocytes that react immunologically with the sperm, killing many of them (Ausin 1957; Barrat et al. 1990; Birkhead and Moller 1993). Additionally, in both birds and mammals, the narrowing of the female tract by the utero-tubal junction, just prior to the isthmus of the oviduct, may limit the numbers of sperm available for fertilization. This may be further enhanced by agglutination of sperm at or near the utero-tubal junction as a result of other biochemical and immunological responses of the female. Thus, of the many millions of sperm that enter the female tract during copulation, only a tiny proportion even make it to the portion of the reproductive tract where fertilization can occur.

Selection of sperm by the female is not limited to ejecting, killing, or otherwise destroying sperm. Secretions from the female reproductive tract may also be responsible for "capacitation," which allows sperm to achieve "hyperactivated mobility" and gain the ability to undergo the acrosome reaction, both necessary conditions for successful fertilization (Bedford 1983).

The ability of females to affect the genotype and phenotype of their offspring, independently of the genetic contribution of males, has received only limited attention as mechanisms by which females exert control over their own reproductive efforts. Such mechanisms might include sex determination of offspring (as in hymenopteran females), maternal *inheritance* (the offspring always has the phenotype of the mother because *extranuclear* genes in the egg determine the phenotype of the progeny), maternal *effects* (the phenotype of the offspring is determined by the mother's *nuclear* genome; Russell 1996), gynogenesis (male genes are discarded immediately following fertilization; Schultz 1967), and hybridogenesis (paternal genes are incorporated in the offspring but discarded during subsequent gamete formation; McKay 1971).

I have reviewed only some of the primary ways in which females are sexually active and influence their own reproductive success. Other relevant phenomena are not reviewed because of a lack of space. For example, female-female competition for mates and "courtship-role reversals" may provide important insights into the evolution of female sexual strategies (Gwynne 1991).

Paradigms, Primates, and Culture

I began this paper by suggesting that Bateman's principle is a paradigm that has had an immeasurable impact on the way that we envision the interactions of males and females in all animal groups, including primates. Paradigms do not arise in a vacuum; they are shaped not only by existing scientific doctrines, but also by cultural values. Bateman's principle, for example, was influenced by nineteenth- and early twentieth-century evolutionary ideas about the nature of men and women. The prevailing attitude in the Victorian society to which Darwin and most other (almost exclusively male and aristocratic) evolutionary scientists of the period belonged viewed women as dependent, submissive, nurturing, deficient in energy, and sexually passive, as compared to men (Russet 1989).

These value-laden assumptions of Victorian society were then transferred to all animals, as well as to biological processes. These ideas were so pervasive that earlier in this century unicellular organisms were arbitrarily classified as female or male depending solely on whether they were large (nurturing) and inactive or small and active (see Sonneborn's critique of Hartmann 1929).

Bateman's principle and its corollaries have influenced our views of primates in many different ways, not all of which are directly related to reproduction. For example, the assumption that females are passively nurturing and interested only in childbearing undoubtedly contributed to the early primatological focus on male behaviors. Only recently has the critical role of females as core participants in primate societies been appreciated (e.g., Fedigan 1992; Hrdy and Williams 1983). Likewise, the assumption that females are sexually passive and interested in mating only with one male probably affected our ability to recognize that females in many primate species actively solicit multiple copulations (Dixson 1977; Hrdy 1986) and that they have probably evolved mechanisms of cryptic female choice (Eberhard 1996). It may also have delayed the realization that some primates have polyandrous mating systems (e.g., Sussman and Garber 1987). Lastly, Bateman's principle, by emphasizing female dependence and passivity, has contributed to the assumption that females always rely on males for resources. An increasingly accepted alternative proposed by Wrangham (1979) is that females control resources, while males distribute themselves so as to have access to the groups of females that defend the resources.

Although the recent attention paid to the role of females in controlling their own reproduction has changed the terms of the debate, much of the new emphasis on female "control" still reflects a metaphor based on contemporary human culture in Western societies. Females are no longer regarded as passive and dependent on the dynamic initiative of males, but rather as engaged in a "war between the sexes" (e.g., see reports by Gowaty 1994, and by Pitnick and Karr 1996). That is, females are now fighting back and having at least a limited say in their own reproductive destinies. These ideas reflect another dominant paradigm in modern behavioral ecology—one that places conflict at the center of male-female interactions and virtually all social behaviors (e.g., Dawkins and Krebs 1981; Trivers 1974).

Perhaps because our generation of animal behaviorists has inevitably been affected profoundly by the current climate of behavioral ecology, it is difficult to envision how some of the behaviors and phenomena discussed previously could be interpreted as the result of anything other than conflict. However, Eberhard (1966) emphasizes that conflict between the sexes must have limits and that the real issue in "female control" is one of "selective cooperation" by females, rather than all-out conflict. Others (e.g., Biology and Gender Study Group 1989) suggest that, in some cases, male-female interactions should be viewed as mutually beneficial exchanges.

Are there examples of cooperative behaviors between males and females? Once a female has selected a male or males for mating, complex, mutually beneficial interactions may occur that lead to successful fertilization. For example, in some species, penile stimulation and pre-ejaculatory intromissions are necessary for ovulation and/or successful inseminations (Dewsbury 1981; Eberhard 1996). Thus, male behavioral or physiological stimulation may be essential for successful female reproduction. Moreover, ejaculates in many species contain nutrients that are absorbed by females and may benefit the female, perhaps serving a purpose similar to the "nuptial gifts" of food (often considered attempts to trick and manipulate females) used by some insects. Semen in some species may contain agents, including chemical secretions, sperm products, and leukocytes, that appear to stimulate and prime the female reproductive tract for successful implantation of fertilized eggs (examples reviewed in Eberhard 1996). On the other hand, the female reproductive tract "capacitates" sperm, enabling it to fertilize eggs, and then protects spermatozoa by storing them in the isthmus of the oviduct prior to fertilization. During fertilization itself, the fact that there are a series of interactions between the sperm and the egg, including the grasping and drawing in of the sperm by the microvilli produced by the egg, suggests an interactive process. Behaviorally, females may cooperate with males by providing morphological structures (e.g., grooves in some invertebrates) that are grasped by the male during copulation. Female primates often interact with copulating males, by reaching back to touch or steady the male, maintaining eye contact, or using communicatory signals that may facilitate the interaction (e.g., Kanagawa et al. 1972; also discussed in Eberhard 1996). All of these interactions suggest that during copulation there are also ongoing "cooperative" or mutually beneficial interactions between females and the males they select. This emphasis on cooperation does not negate the importance of female control, nor does it imply that there is no conflict between the reproductive interests of males and females. It does, however, provide a different framework—one that acknowledges the conditional and complex interactive nature of male-female relationships. It also emphasizes that, while males have an interest in successfully mating with the female, the female shares this interest once she has selected the male or males with whom she wants to mate (reviewed in Pitnick and Karr 1996).

Other aspects of male-female sexual and courtship interactions may also be interpreted in a framework that highlights the mutually beneficial outcomes of the different strategies used by males and females. For example, a major point made by Wylie and Poston (1996) is that male competition for mates cannot be logically separated from indirect female choice because the former usually depends on conditions determined by females. By setting up these conditions, females are indirectly selecting a mating partner. Females may incite male-male competition by advertising their sexual condition through vocalizations, visual signals, or odors. This could be viewed as mutually beneficial to both sexes: the female alerts the dominant males to the fact that she is receptive and, simultaneously, she increases her chances of mating with the most dominant males. Evasive behaviors of females, which result in group chases by males, may also promote male-male competition, resulting in the female mating with the male or males that are most persistent in the chase, have the most stamina, or are best able to monitor and follow the female's movements. Possible examples of this type of indirect female choice may include the courtship chases of certain birds (e.g., Moller 1994) and squirrels (e.g., Koprowski 1993; Schwagmeyer and Wootner 1986). Likewise, preferences for certain mating sites, aggregations or synchronization by receptive females can encourage male-male competition. This is advantageous to the males in that it makes it easier for them to find females, while females also benefit by increasing their chances of mating with the most competitive males.

In summary, indirect female choice is mediated by female behaviors and/or physiological characteristics that indirectly limit the set of potential mates available to the female (Wylie and Poston 1996). As such, indirect female choice can be considered a form of female control and a female strategy for mating with superior males, yet it is also ultimately beneficial to those males that are indirectly selected by the females.

Conclusion

The main point I have attempted to make in this paper is that science does not function in a vacuum. Rather, it is guided profoundly by the paradigms of the field, and these paradigms are in turn affected by cultural values and biases. The study of primates, like behavioral ecology and evolutionary biology, has been influenced by the dominant paradigms of the field. But paradigms are not inherently harmful; indeed, by providing a shared framework for a field, they can lead to important discoveries and new insights. The danger lies in accepting paradigms without questioning, and without realizing that they offer us only a restricted view of the world. Theory must continue to be an integral part of primatological research, but theoretical worldviews must not be accepted unquestioningly simply because they have become dogma. Rather, theory must be balanced by attention to empirical data, by testing alternative hypotheses, and by an openness to alternative interpretations of data.

Acknowledgements

I am grateful to George T. Taylor and Amy McRae for suggesting references. Debbie Munro-Kienstra was invaluable in helping me to track down and photocopy relevant articles. I especially thank Linda Fedigan and Shirley Strum for inviting me to participate in this book and Steve Glickman for conversations that inspired me to write on this topic. Arlene Zarembka provided valuable comments on the original manuscript and revision.

14

Culture, Disciplinary Tradition, and the Study of Behavior: Sex, Rats, and Spotted Hyenas

Stephen E. Glickman

Shirley Strum and Linda Fedigan (this volume) have gathered substantial evidence in support of the thesis that general cultural and, particularly, gender-related attitudes influence scientific practice in the study of primate behavior. In Bruno Latour's paper (this volume), other sources of influence are added to the mix, and "good" science is shown to be characterized by exceptionally complex networks of interacting variables. The present chapter is concerned with the multiple influences that affected scientific decision making within the boundaries of biological psychology between 1920 and the present time. Two lines of research have been selected as "case studies": (1) a literature on sexual behavior in the laboratory rat, and (2) studies of sexual differentiation and sociosexual behavior in the spotted hyena.

These were not arbitrarily selected cases. First, the study of sexual behavior and sexual differentiation seemed a particularly culture-sensitive field (Fausto-Sterling 1985). Second, I spent several postdoctoral years working under the direction of Frank Beach, who was a pioneer in the scientific study of sexual behavior and used the laboratory rat as his primary subject.[1] I had done just enough work with rat sex behavior, using the standard paradigms of the 1960s, to feel a certain modest empathy in reading the work both before and after that time (e.g., Bermant et al. 1965). Similarly, for more than a decade, my research has focused on the hormonal substrates of sexual differentiation in the spotted hyena. When I embarked on the reading and reflection required for the present paper, I expected to find gender-related influences in both these research areas, and that expectation was fulfilled. The ubiquitous assumptions regarding feminine passivity were indeed located in both the literature on rat sexual behavior and in accounts of the process of sexual differentiation. However, there was a lot more to it than just gender. Historic disciplinary traditions emerged as powerful determinants of what psychologists observed and reported. Some of these traditions were enduring, while others shifted from decade to decade, but they always influenced (if not dictated), for example, the choice of species under study, the settings in which the selected subjects were observed, the questions that were asked, and whether the focus was on mental processes or overt behavior. In addition, more general processes in our academic culture were observed at work, particularly the rewards for carrying out experiments which have a reasonable probability of yielding publishable results and the resultant tendency to avoid lines of work with very uncertain outcome.

Psychology: The General Discipline and Its Subdisciplines

This section is concerned with some enduring disciplinary attitudes, some shifting disciplinary attitudes, and the impact of those attitudes on the subdiscipline biological psychology. That, in turn, necessitates a brief review of the roots of modern academic psychology, for it is in those roots that one finds the origins of the traditions that modulate psychological research to this day. In particular, I shall focus on two enduring attitudes that have shaped animal research by psychologists: (1) a strong preference for the control provided by the experimental laboratory, and (2) the belief that animals are relevant to psychology to the extent that they shed light on the human mind, or human behavior. However, the period from 1920 to 1990 was also marked by a set of dramatic shifts in opinion regarding the core questions of the discipline: from a psychology of mind, to a psychology of behavior, and back again to a preoccupation with mental life. These shifts had, and are having, a major impact on the course of animal behavior study in psychology.

ROOTS OF MODERN ACADEMIC PSYCHOLOGY

In the nineteenth century, lines of work from evolutionary biology, sensory physiology, and studies of the nervous system combined with developing philosophical perspectives to produce an "independent" discipline of psychology (Boring 1950; Leahey 1992). It was an invasion of territory traditionally assigned to philosophy. The threads came together in the experimental laboratory programs of Wilhelm Wundt in Leipzig. Only by a persistent commitment to a new *experimental* paradigm could college administrations in this country be induced to create new academic departments of psychology. During the final decades of the last century, many American psychologists traveled to Leipzig and returned to spread the gospel, including the "founder" of my department at Berkeley, George Malcolm Stratton. To be sure, there was a transitional period in which conscious "introspection" was appended to the emergent, quantitative experimentation. William James, the most eminent American psychologist of that era, insisted that introspection remain the primary route to understanding mental life. He viewed the rapid rise of the experimental tradition with alarm, commenting that "[t]his method taxes patience to the utmost, and could hardly have arisen in a country whose natives could be bored. Such Germans as Weber, Fechner, Vierordt and Wundt obviously cannot . . ." (James 1890, 192). However, even as he was relegating experimentation to a secondary place, his textbook of psychology (1890) was heavily laden with the new experimental data. Psychology was to change in many ways between 1890 and 1990, but a commitment to the control afforded by the experimental laboratory has been at the center of academic psychology in the United States. It has also placed serious boundaries on what would be "seen" by psychologists interested in animal behavior.

EMERGENCE OF BEHAVIORISM AS A DOMINANT PARADIGM

The starting date selected by Strum and Fedigan, 1920, represents a transitional point in the history of American psychology. Behaviorism had been announced by Watson some years earlier (1913), but it had not yet become the dominant paradigm of the field. The primary comparative psychologists, people like Margaret Floy Washburn and Robert Yerkes, were products of an earlier generation, when psychology was an eclectic discipline, with an experimental technology, but still relying on introspection as a primary route for understanding mental life. The major theoretical battle in the general discipline at the turn of the century was between the functionalist tradition, represented by William James and James Rowland Angell, and the structuralist tradition of William Bradford Titchener. All sides agreed that the study of mind and mental events was the central concern of psychology. However, James/Angell promoted a more naturalistic form of introspection and emphasized the adaptive functions of consciousness, while Titchener advocated a particularly rarefied form of introspection that relied on highly trained experimenters searching for the basic "elements" of consciousness. Using an analogy from biology, Titchener argued that the study of mental structures was the anatomy of the field, and that anatomical knowledge had to precede functional studies, which he compared to physiology (Titchener 1898). Margaret Floy Washburn (1908), as Titchener's student, was not only committed to inferences about consciousness, but also to the details of the sensory life of animals, for it was in the facts of sensory experience that Titchener sought his basic elemental "anatomy."

Watsonian behaviorism, with its discarding of consciousness and mental life, touched a responsive chord in American psychology. Born in 1878, receiving his Ph.D. in 1903, and publishing his classic announcement of behaviorism in 1913, Watson was elected president of the American Psychological Association in 1916. It was, if not a meteoric rise, an impressive one. Where the psychologies of the functionalists and structuralists had promoted "understanding," Watson promised the "prediction and control" of behavior. In the decades that followed, many psychologists were prepared to trade the abstract study of mental life for a muscular psychology that could effect change in a practical world. Behaviorism was an ultimately "American" product. Watson's emphasis on Pavlovian conditioning as a central explanatory device, and his use of animal data as directly applicable to predicting and modifying human behavior, were particularly relevant to issues raised at the Teresopolis conference. In this scheme, there are general laws of behavior whose basic processes transcend species variation: Ivan Pavlov's conditioned dogs, or B. F. Skinner's rats and pigeons, were seen as equally able to provide not just interesting information about the species under study but essential explanatory devices for predicting and controlling human behavior. The role of rewards and punishments ("reinforcement"), the conditions of efficient learning and performance, and the technology of eliminating previously learned habits ("extinction") were now available from studies of animals in controlled experimental settings.

As behaviorism developed during the decades between 1930 and 1960, there were significant variations on the basic theme. B. F. Skinner (1938, 1959) represented a Spartan elaboration of Watsonian behaviorism—rejecting mentalism, formal theorizing, and physiologizing. Edward Tolman (1932), although also rejecting physiological speculation, developed a cognitive behaviorism in which animals had a mental life. In this latter system, the behavior of rats (and people) was viewed as "purposive," and they were thought to navigate in their environments using "cognitive maps" rather than behaving as automatons running through a repertoire of conditioned reflexes.

For all their differences, Skinner and Tolman maintained two critical aspects of Watson's program: (1) animal work was central to the development of basic laws of behavior in people, and (2) the focus was on the universal characteristics of animals as learning organisms.

THE COGNITIVE REVOLUTION

By 1960, the stage was set for the decline of American behaviorism. In part, this was due to failures of delivery on various promissory notes. We had learned a good deal about learning in rats and pigeons without solving the most central problems of human existence. In addition, new traditions emerged which promised to shed light on human language, memory, and cognitive development. These were areas that often defied explanation in the language of a simplistic behaviorism and emphasized the value of making inferences about human cognitive processes. The so-called "cognitive revolution" was underway in psychology (Gardner 1985). There may be some special irony in the fact that this cognitive revolution, and the related decline of mechanistic behaviorism, was stimulated by the development of computer-inspired models in which human perception and thought were represented by charts that symbolized the flow of information in electronic devices (Gardner 1985).

THE STUDY OF ANIMALS IN AMERICAN PSYCHOLOGY

> From its inception, American psychology has been strongly anthropocentric. Human behavior has been accepted as the primary object of study and the reactions of other animals have been of interest only insofar as they seemed to throw light upon the psychology of our own species. There has been no concerted effort to establish a genuine comparative psychology in this country for the simple reason that with few exceptions American psychologists have no interest in animal behavior per se. (Beach 1950, 10)[2]

The preceding quote from Frank Beach was (and is) essentially correct. As outlined below, there was indeed a time when psychologists were deeply interested in animal work—but they were generic animals; furry, pliable, substitute humans, producing general laws that would be directly applicable to people. More than that. They were removed from nature, often bred for the convenience of laboratory scientists, and housed in caging systems where observations could be carefully "controlled." I believe that the lack of naturalistic reference points has been a persistent problem for psychology. Despite repeated calls for a more ecologically valid discipline by highly respected psychologists (e.g., Neisser 1976), many domains of the discipline remain bounded by traditional laboratory procedures.

IMPLICATIONS OF THE COGNITIVE REVOLUTION FOR ANIMAL RESEARCH: THE FLOW OF IDEAS WITHIN DISCIPLINES

On the positive side, the cognitive revolution encouraged psychologists working with animals to expand their studies to more complex aspects of mental life (although not necessarily more complex than those envisaged by Tolman). Much interesting work emerged (e.g., Ristau 1987). However, I believe that there are also some potential costs.

First, much of this work on animal cognition is, like the work reviewed by Margaret Floy Washburn many years ago (e.g., Washburn 1908), designed to determine whether this or that facet of the human mind can be located in the animal world (Roitblat 1987). Do rhesus monkeys display human-like routines for retrieving memories? Do pigeons recognize rotated objects in the same way as human subjects? Although this approach has a long and honorable history in the field, the studies are often removed from the natural lives of animals. To be sure, a field of cognitive ethology has developed which keeps in contact with the problems faced by animals in their real world, but the history of psychology is littered with dead papers on animals trained on arbitrary tasks for abstract reasons.

A second, more important concern involves the future of animal work in psychology departments. Now that psychologists are focused, once again, on human cognition, animal work once more recedes to a peripheral position. It is a curiosity, a luxury of a small subset of rich and/or particularly intellectual departments. During the heyday of behaviorism, with all of its flaws, there was a steady flow of ideas from animal work to the general discipline. The explanatory devices revealed by work in animal learning spread throughout psychology. Social learning, using principles derived from animal studies, became a primary explanatory device in social psychology (e.g., Bandura 1973). Behavior modification and desensitization became part of the armory of clinical psychologists (Favell 1973; Kazdin 1978). There were even attempts to "explain" the central tenets of psychoanalysis in terms of an animal-based learning theory (Dollard and Miller 1950).

The flow of ideas has now changed direction: it is from people to animals. Given the peripheral nature of animal work, the increasing expense of such work, and the sociopolitical concerns raised by animal rights activists (which find a sympathetic ear among many psychologists), it is difficult to be optimistic about the allocation of faculty appointments in this area in the future.

Another complicating factor is the role of physiology. In the past, physiological studies generally meant animal studies. But with the development of remarkable MRI, CAT, and PET technologies, work on the nervous system which once required animal subjects can now be studied in people. Cognitive neuroscience is indeed flourishing, but, to the extent that it is housed in psychology departments, it may be a nonanimal neuroscience. Personally, I think it is wonderful that many animals can be spared invasive procedures, but I question what will happen to physiological studies that are more specific to the lives of animals—studies of hibernation, daily torpor, or seasonal reproductive rhythms. The incentive to pursue these issues will not come from a cognitively oriented psychological discipline.

Sexual Behavior and the Laboratory Rat: 1920–1990

For the purposes of reproduction it is not essential that the female rat take the initiative in sexual relations with the male. . . . the female need only react with the assumption of the lordosis position in order to permit effective copulation. Her role can therefore be conceived as essentially passive. . . . (Beach 1943, 197)

Proceptive behavior is functionally as important as other patterns traditionally termed "receptive"; but the female's tendency to display appetitive responses finds little opportunity for expression in laboratory experiments which focus exclusively upon her receptive behavior. . . . The resulting concept of passive females receiving sexually aggressive males seriously misrepresents the normal mating sequence and encourages a biased concept of feminine sexuality. (Beach 1976, 115)

INTRODUCTION

Frank Beach always chose his words with care. In the citation from 1943, he took pains to emphasize that it was only for purposes of reproduction that the sexual behavior of the female rat could be regarded as "passive." In the same paper, he included a perceptive description of the active components of female behavior. However, he did not include any discussion of the functional role of female-initiated sexual activity. Studies of, for example, female choice of mates, were far in the future. Within the laboratory traditions of his time, there was certainly a bias toward focusing on the "passive," reflexive components of sexual behavior in female laboratory animals.

A great deal happened in the scientific world and the relevant nonscientific culture between 1943 and 1976. I am sure that Frank Beach was influenced by that culture, as well as his own later research on dogs, demonstrating very active female choice of mates (Beach and LeBoeuf 1967), and the writings of his graduate students (e.g., Tiefer 1978) and postdoctoral fellows (Bermant 1961; Doty 1974) on a broader view of female sexuality. Beach was as consciously committed to coming to grips with changing attitudes as anyone I have ever known, and his formulation of attractivity, proceptivity, and receptivity in the paper cited above was to have enduring impact on the conduct of sex research in rodents and other species.

A detailed account of the mating behavior of wild and laboratory rats in a spacious seminatural setting was provided for the first time by McClintock and Adler in 1978. This description was subsequently elaborated in a set of papers by Martha McClintock and her collaborators (e.g., McClintock 1984; McClintock 1987; McClintock et al. 1982). When a pair of rats is observed in a spacious environment, mating becomes a negotiated dyadic activity, in which the female signals her readiness to mate by placing herself in proximity to the male and then controls the pace of mating with appropriate mixtures of approach, avoidance, solicitation, and rejection. McClintock describes three components of female sexual solicitation: (a) initial approach and orientation within a body length; which may be followed by (b) grooming, crawling over the male's head, or presenting her hindquarters (in the case of an unresponsive male); and finally, (c) ". . . she turns away and runs directly away from the male either with a dart-hop gait, or the stiff-legged run of an estrous female" (McClintock 1984, 14).

The male requires a number of brief (approximately 0.5 second) intromissions in order to achieve ejaculation. The "prolonged" ejaculatory contact lasts perhaps 1.5 seconds. A female rat in a critical stage of her estrous cycle will display the lordosis reflex when palpated on her flanks by the forepaws of the male. During such intromissions the female remains motionless, with her rump elevated and her tail moved to one side, facilitating intromission of the penis of the male. Norman Adler, one of Frank Beach's graduate students, demonstrated that the sequence of male intromissions triggers a progestational reflex in the female rat which is essential for implantation of the fertilized ova (Adler 1969). "Negotiation" results from the fact that males, left to their own devices, will opt for short interintromission intervals, requiring a large number of intromissions prior to ejaculation. Females, given an opportunity to control contact by the male, prefer longer intervals between intromissions, reducing the number of intromissions required for ejaculation, as well as the number of intromissions required to trigger the crucial progestational reflex. McClintock (1987, 193–194) has also provided data from mating that occurs in socially complex, multimale, multifemale groups. Under these conditions, which presumably simulate common arrangements in nature, females mate with multiple males and participate in very different sequences of intromissions and ejaculation than those observed in paired mating tests.

A great deal is now known about rat mating behavior and there is more to be learned. We are always limited by our own sensory equipment; it was not until 1972 that Barfield and Geyer recognized that male rats emit a postejaculatory "song" that has an impact on subsequent mating behavior, and it was still later that scientists recognized that females were "singing" as well (Micek et al. 1991). The frequencies of these vocalizations were simply above the threshold for human hearing and their detection required use of an electronic transducing device. We are also limited by the broad and local scientific traditions of our fields, as well as those cultural biases noted by Strum and Fedigan (this volume).

DESCRIBING SEXUAL BEHAVIOR: SPACE, CONDITIONS OF TESTING, AND CULTURAL BAGGAGE

> While an active male and a receptive female will display copulatory activity in any type of cage, we have found one type especially well suited to our experimental purposes. . . . A circular cage 10 inches in diameter, constructed of 1/2-inch wire mesh which has been painted black, is an ideal structure for the observation of copulatory activity. The temporarily resistant female is deprived of corners in which she can crouch and prevent the male's mounting response. (Beach 1938, 358)

> . . . studying female solicitation and proceptive behavior in a small cage is akin to studying swimming behavior in a bathtub or ballroom dancing at a disco; it cannot be done. (McClintock 1984, 15)

It is a truism that the behavior one chooses to record and the conditions of observation determine the outcome of any experiment in this field. That is certainly the case for the copulatory behavior of male and female rats, as it is for studies of primates in captivity. Calvin Stone published an initial account of the sexual behavior of the male rat in 1922. He employed a bare, rectangular 8-by-12-inch metal cage. In Stone's account, all male sexual behavior was lumped under the category of "copulatory acts." By 1927, Stone provided a much more precise quantitative description of sexual behavior in the male rat, distinguishing between mounts without intromission, intromissions, and ejaculations. As Beach (1981) has indicated, it was this enriched behavioral analysis that permitted Stone to demonstrate that different components of the sexual repertoire were affected to varying degrees by castration, with ejaculatory capacity being the first to disappear, and mounting being the most persistent. However, it took a substantial period of time before Stone's detailed analysis became the standard approach. In early papers, Beach (1940) limits his quantitative analysis to "complete" and "incomplete" copulations. Then he moves to the use of rating scales which assess grades of copulatory activity (e.g., Beach 1942, 1946), and finally there is a return to the precise distinction between mounts, intromissions, and ejaculations, with still more detailed quantification of occurrence and timing (e.g., Beach and Jordan 1956).

Despite the limitations of his testing cage, Stone (1922) observed that, in addition to the lordosis reflex, female rats exhibited a darting/hopping gait and ear wiggling as characteristics of estrous. The focus of laboratory studies, however, became the "passive" lordosis reflex. Some initial enrichment was provided by the pioneering work of Josephine Ball, who entered the field as a graduate student in psychology at the University of California, Berkeley. She was recruited by Herbert Evans, a professor of biology housed in the same building on the Berkeley campus, to study changes in female sexual behavior which accompanied the stages of vaginal cytology he had identified in his work on ovarian cycling (Long and Evans 1922). Ball (1937) developed a complex ten-point rating scale of female receptivity which clearly emphasized that this was not an all-or-none phenomenon, and enabled delineation of the most likely timing of estrous as a function of both the four-to-five-day ovarian cycle and place in the day/night cycle.

Over the years, however, American researchers tended to focus on the lordosis quotient, or L.Q. (the percentage of instances, when "appropriately" stimulated, a female rat exhibited lordosis). The ease of recording this reflexive behavior and its obvious biological significance in the chain of reproductive behavior was undoubtedly part of the appeal for studying lordosis. The fact that you didn't even need a male rat also had some appeal. The experimenter's finger, applied to the rump of the female rat, would do. Finally, there was the (unfortunate) attraction of reducing a complex scientific problem to a single number (note the popularity of the I.Q.) and, it seems likely, that the resultant emphasis on a passive sexual role for the female rat was congruent with more general cultural attitudes.

INTELLIGENCE, INSTINCTIVE BEHAVIOR, AND THE CEREBRAL CORTEX

Several other aspects of Frank Beach's studies are particularly relevant in the context of the present conference. They were often driven by an interest in the common substrates of learning and instinctive behavior. First, Beach found that neocortical lesions resulted in serious disruption of mating behavior in male rats (Beach 1940). He then reported that similar lesions did not impede conception in female rats, that they even facilitated the lordosis reflex, although they did impair the integration of female copulatory activity (Beach 1943). Given the spatial limitations of his observational setting, his descriptions of female behavior were remarkably sensitive. He noted that ". . . extensive cortical removal reduced or eliminated the estrous female's tendency to orient her receptive reactions to the male and to solicit actively his attentions" (Beach 1943, 196). He also wrote that the individual components of female behavior often appeared at inappropriate times, for unnatural durations, in lesioned animals and that the rejecting "kicks" of the female were essentially eliminated by large cortical lesions (Beach 1944). In review articles, Beach, almost always, took care to note that the more complex, active components of female sexual activity were impaired by cortical lesions. However, occasionally (e.g., Beach 1961), the caveat regarding potential deleterious effects of lesions in females disappeared, and the reader of authoritative secondary sources was left with the message of a sex difference in the role of the neocortex.

As McClintock (1984) has observed, there were many cases in which a basic error was made in the rodent sex behavior literature; i.e., the appetitive, voluntary behavior of the male was compared with a single reflexive component of female behavior—lordosis. If, as McClintock recommended, those aspects of female sexual behavior representing more volitional activities—for example, solicitation—had been recorded, a very different outcome might have appeared. Nearly four decades after Beach's original studies, Sue Carter and her colleagues (1982) did reopen the question, and, with an enriched behavioral repertoire and with their animals observed in a somewhat larger enclosure, they found that the sexual behavior of female rats was indeed impaired by lesions of the neocortex.[3]

There is a second study that Beach (1939) carried out in which he attempted to determine whether there was a significant positive correlation between the vigor of "instinctive" behavior and measures of intellectual behavior. Male and female rats were accordingly tested for their learning ability in mazes, and then assessed for the vigor of their "instinctive" behavior. However, the instinctive behavior selected for male rats was sexual behavior, while the instinctive behavior selected for female rats was maternal behavior. It is important to remember that Beach (1937, 1938) had located profound effects of cortical lesions on maternal behavior in female rats (Beach 1937). From his point of view, one was surely more likely to locate a correlation between two patterns of behavior when both were dependent upon the integrity of the neocortex. Bias influenced decision-making in these studies at a much earlier stage, when research traditions were established that dictated the study of sexual behavior in small laboratory cages.

MORE ON ACTIVE MALE AND PASSIVE FEMALE RATS

On a number of occasions in the 1950s, researchers examined the extent to which opportunities for sexual behavior could function as a "reward," motivating the performance of a learned task. In every case that I have located, male rats were trained to run through mazes for access to females (Beach and Jordan 1956; Sheffield et al. 1951). It wasn't until the 1960s that papers appeared in which female rats were given the opportunity to express their interest in sexual contact with males (e.g., Bermant 1961).[4] In a result that anticipated the later work of McClintock and Adler (1978), Bermant's study revealed that the intervals at which females requested sexual contact with males were (a) dependent upon the nature of the last sexual contact (there were shorter latencies following "mounts" than intromissions and shorter latencies following intromissions than ejaculation), and (b) longer than the interval which the male would have preferred in a situation in which he had control.

Sexual Differentiation and Sociosexual Behavior in the Spotted Hyena

SOME BACKGROUND

Since 1985, I have been involved in studying a colony of spotted hyenas, maintained in the hills above the Berkeley campus. In nature, these animals are active hunters living in complex, female-dominated social groups (Kruuk 1972). The domination of adult males by adult females is not a small statistical effect. It is a ubiquitous, trans-situational characteristic, observed in feeding at kills as well as during transient interactions at the communal den. My colleague and collaborator Laurence Frank (who introduced me to hyenas) further demonstrated that hyena societies are organized through matrilines, with separate hierarchies among the males and females (Frank 1986a, 1986b).

However, it is the peculiar genital anatomy of female spotted hyenas that forms the basis of the Berkeley project. There is no external vagina, as the labia have fused to form a pseudoscrotum. The clitoris has hypertrophied until it is the approximate size and shape of the male penis, and females display erections similar to those of the male. This organ is traversed by a central urogenital canal through which the female spotted hyena urinates, copulates, and delivers her offspring. The existence of this uniquely "masculine" morphological profile in a female mammal challenges our understanding of the process of sexual differentiation, which relies on testosterone secreted by the fetal testes of the male to account for sex-typical variation in genital anatomy among existing mammals.

In her book *Myths of Gender* (1985), Anne Fausto-Sterling has documented the oversimplified portrait of sexual differentiation and development encouraged by popular accounts, including a remarkable neglect of the mechanisms of feminine development. Consideration of the spotted hyena forces attention to feminine development as an active process in all female mammals. Finally, because these hyenas are complex, long-lived mammals, with many features of social organization observed in Old World Cercopithecines, there is an opportunity to observe interactions between hormonal influences and natural social contexts, emphasizing a critical, if obvious, point: i.e., that hormonal influences on social behavior have no meaning apart from the situation in which the observations are carried out.

SEXUAL DIFFERENTIATION

> Clearly, in contrast to the body of work done on male development, the final word on the genetic control of female development has yet to be written. That it has not yet been fully researched is due both to technical difficulties and to the willingness of researchers to accept at face value the idea of passive female development. (Fausto-Sterling 1985, 81)

Sexual Differentiation: The General Theory Contemporary understanding of mammalian sexual differentiation relies on a genetic instruction that results in differentiation of the fetal gonads as ovaries or testes. Against this background of fetal bisexual potential, secretion of androgens by the fetal (or neonatal) testes results in development of the genital tubercle as a penis, fusion of the external labia of the vagina to form a pseudoscrotum, and maintenance of the internal Wollfian duct system, which develops into a set of structures associated with male reproductive function. If the fetal gonad develops as an ovary, the genital tubercle becomes a clitoris, the external vagina is maintained, and the Mullerian duct system forms the basis for the internal reproductive system of the female mammal.

There are also changes in the brain produced by the different hormonal states that characterize the developing female and male fetuses/neonates. These hormonally mediated dimorphisms in the brain provide one modulating influence on sexual behavior, although it is clear that in complex mammals, there is the potential for very significant interaction with the social environment.

Faced with the task of accounting for the presence of a "scrotum" and a penis-like clitoris in the spotted hyena, we turned to the theory described above, as it provided the only existing physiological explanation for the development of male-like genitalia. Indeed, over a span of years, we carefully tracked a mechanism that does deliver testosterone to the developing female fetus throughout pregnancy in the spotted hyena. Briefly, a cascade of events starts with the maternal ovary of the hyena secreting a hormone (androstenedione) that probably does nothing in its original form but is readily converted to testosterone, or estrogen, in the presence of appropriate enzymes (Licht et al. 1992). Such enzymes are present in the placenta of the spotted hyena, and this maternal androstenedione is then converted to testosterone by the placenta and transferred to the developing fetus (Yalcinkaya et al. 1993; Licht et al. 1998). In addition, several reports have now appeared which document "masculinization" of the genitalia in human female infants as the result of a defect in placental metabolism—which results in the human placenta behaving like a hyena placenta and converting inactive precursors into testosterone (Shozu et al. 1991; Conte et al. 1995). Our hyena studies have even been cited in the recent human literature as clarifying the mechanism of such "masculinization," a gratifying situation for any basic science group.

But now the story becomes interesting. As the result of a set of long-term experiments, designed to test the testosterone-based explanation of hyena masculinization, it appears that androgens like testosterone are not involved in the basic formation of male-like genitalia in either female or male spotted hyenas. Rather, they "tune" the system in both females and males. First, we were able to rule out the fetal ovary as the source of testosterone producing genital masculinization, as a female fetus was found to have a fully formed penile clitoris at a very early stage of gestation, prior to differentiation of the gonad (Licht et al. 1998). Then, we observed that administering antiandrogens to a pregnant female hyena, thereby blocking the actions of testosterone on the developing fetus, does not produce female offspring with a clitoris of reduced size and an ordinary external vagina, as current theory would predict. Such female infants still have a scrotum and a penile clitoris, but it is shorter and thicker and has a much larger opening at the tip (Drea et al. 1998). Finally, we found that both the clitoris and penis grow to adult dimensions following removal of the testes, or ovaries, in juvenile hyenas, which removes virtually all circulating testosterone from the system (Glickman et al. 1998). The differentiation and growth of the penile clitoris during fetal life appears to involve a new active principle of feminine development, which, in turn, has implications for male development. Examining the extreme case of the female spotted hyena calls attention, once more, to the wide variation in clitoral morphology in primates (e.g., Hrdy 1981, 166–167) and emphasizes how little has been done to explicate the underlying mechanisms.

This subsection began with a quotation from Anne Fausto-Sterling's *Myths of Gender.* I believe that she has correctly characterized the current state of theory/research in sexual differentiation, and our hyenas provide a dramatic example of the limitations of current understanding. The idea of "passive development" is simply nonsensical. If there is growth, there is an active process. There is no mechanistic theory of feminine genital development comparable to the theory of male development.

However, as someone working inside the field, I would also agree with Fausto-Sterling that there are other practical considerations that shape what scientists do. The existing model of male differentiation not only accounted for many features of sexual development, but, for several decades, it generated experiments that were easy to plan, technically feasible, and offered the prospect of reasonably definitive results.[5] Given the realities of academic life, and a limited scholarly lifespan, most researchers will opt for a line of work that promises answers-for-effort, publications, and extramural funding. If we were to discover a route to feminine development in hyenas that was easily extrapolated to other species, with technically feasible manipulations and a sound theoretical framework, I believe that recruiting investigators to the problem would pose no difficulties at this point in time.

Mating in Spotted Hyenas: Expectations and Realities Recently, with a set of hyenas in hand, and armed with the accumulated wisdom of nearly eighty years of systematic research on mammalian sexual behavior, Elizabeth Coscia and I set out to describe the mating pattern of the spotted hyena. We were determined to watch the animals under conditions that would give the females full opportunity to demonstrate their preferred, active, repertoire, while being careful to distinguish the early components of male behavior from those that (potentially) result in fertilization. At the time of writing, there have been no complete descriptions of mating behavior in this species from the field, and the available descriptions of mating in captive hyenas lack the detail required by modern understanding of mating patterns in other species. Our observations emphasize the ways in which morphology constrains and channels behavior.

Since the female hyena has no external vagina, and mating takes place through the urogenital canal of the clitoris, sexual behavior involves overcoming a set of problems (Drea et al. 1998b). The first obvious issue is anatomical, and some have speculated that the entire situation evolved to prevent rape (which, given the size, dominance, and aggressiveness of female hyenas, seems an unlikely occurrence). Fitting a penis into a second penis-like structure is solved, in part, by the female retracting her clitoris into the abdomen during copulation. In addition, the opening at the tip of the clitoris is much larger and more elastic in the female than the male (Glickman et al. 1992), and the glans of the male has a wedge-shaped contour, which may facilitate entry (Frank et al. 1990). There is a second anatomical difficulty. Due to the absence of an external vagina in the ordinary location, the male is forced to achieve intromission through a small opening housing the retracted clitoris that is situated in the rostral abdomen. Aiming for this limited target, in a relatively distal position, requires that the male "flip" his erect organ against the abdomen of the female. His posture is upright and (to the human eye) ungainly, while the female must remain virtually immobile while the male searches for the opening.

Finally, the whole process is made more complicated by the fear engendered in an adult male hyena by the proximity of an adult female. The problem is apparently solved during the first mating sequence by having the females abandon their normal aggressiveness toward males, assuming postures that range from stolid to subordinate.[6]

Elizabeth Coscia and I entered this study poised to record the active components of female behavior, and, given the otherwise boisterous behavior of female hyenas, we were initially disappointed at their passivity. In retrospect, this initial female "passivity" may be a prerequisite for convincing the male that it is safe to approach, while the immobility is necessary for successful intromission. However, having established their friendly intentions toward males during the first mating sequence, females do become much more proactive during subsequent sequences, regularly approaching males, grooming them (which we rarely see in any other context) and encouraging them to further sexual activity.

There are still no detailed accounts of female choice by spotted hyenas in nature. But we have certainly seen some clear instances of female preference for particular mates in captivity. It will be interesting to examine this question when more behavioral data are available and DNA studies have told us which males are successfully fathering offspring. However, it may be worth voicing a caveat at this time about embracing simple interpretations. Should it turn out that the "dominant" males are leaving more genes in succeeding generations, one might be tempted to think of this in terms of the usual linkages between aggression and dominance. However, it must be remembered that dominance in male hyenas is tied to length of residence in the group and that every male hyena enters a new clan at the lower rungs of the dominance hierarchy. Becoming the most dominant male is not based on barging in and taking over the male wing of the group. Rather, long-term residence requires great social flexibility—the capacity to appraise social situations and behave obsequiously when that was required, but to be opportunistically aggressive when circumstances permitted, slowly but surely ratcheting one's way up the male hierarchy. That scenario is in accord with my intuitions regarding female preference in our colony. If female hyenas turn out to mate preferentially with "dominant" males, what they are really choosing is the genetic complement that underlies sensitive social appraisal and effective action.

Female Spotted Hyenas: Affiliation, Aggression and Dominance The history of spotted hyena field studies is much more compressed than that of the primate studies reviewed by Strum and Fedigan. However, some similar trends certainly are apparent in both groups. For example, there has been a parallel move from descriptions of behavior in terms of age and sex roles to assessments of fitness that involve long-term studies of individually identifiable animals. I think one could also identify a shift to more cognitive interpretations of behavior. It has certainly affected our analysis of coalition formation (Zabel et al. 1992) and Christine Drea's studies of cooperative behavior in the Berkeley hyena colony (Drea et al. 1996).

Within our own research project, there has been a rather disorderly progression from initial papers that focused on dominance and aggression (Frank et al. 1989; Frank et al. 1991), to reports that emphasize the role of affiliative behavior in the maintenance of dominance-subordinance relationships (Zabel et al. 1992; Smale et al. 1995; Jenks et al. 1995), and on to the nature of play and other prosocial behavior patterns (Pedersen 1990; Drea, Hawk, and Glickman 1996). It should be noted that dominance hierarchies in hyena clans are much less ambiguous, or situation-dependent, than in many primate groups.[7] But it also seems likely that, in watching hyenas, we have all responded to the sometimes dramatic, aggressive confrontations that regularly occur between individuals, and attention to more subtle prosocial behavior patterns was deferred in the order of things.

Environment-Independent and Environment-Dependent Effects There are a number of aspects/implications of our research program which may be of interest to primate researchers. First, we were surprised to find that hyenas removed from Africa as infants, and reared in peer groups, recreated a natural matrilineal social system in Berkeley, California, complete with a dominance hierarchy among adult females, offspring who occupied ranks adjacent their mothers, and the complete subordinance of adult males to mothers and their juvenile offspring (Jenks et al. 1995). Second, we have reasonably clear evidence that the social context is a critical determinant of response to any hormonal predisposition. For example, in the case of play, Joanne Pedersen found that female hyenas engaged in more vigorous social play than males when tested in same-sex groups. However, when tested in more natural mixed-sex groups, the females stimulated the play of the males, while the presence of the males reduced the play of the females, eliminating differences between the sexes (Pedersen et al. 1990).[8]

In contrast, we believe that the natural social context accentuates sex differences in aggression and dominance (Frank et al. 1989). Female spotted hyenas normally spend their entire lives within a given clan in a familiar environment. Mothers, daughters, and sisters form alliances and coalitions which add to their power. The nature of male dispersal at puberty results in individual males entering the home territory of large, aggressive females with strong affiliative bonds. In the early years of our captive studies, without natural matrilineal bonds and without dispersal, several small males were able to forge bonds with females and use their larger, more powerful friends to temporarily raise their dominance status (Zabel et al. 1992). However, nature affords no such luxuries for immigrant males—who have enough trouble dealing with male residents.

In the course of discussing sex differences among people, Beach (1974) suggested that human societies often deal with overlapping distributions of traits by accentuating differences; in effect "creating" non-overlapping distributions. Hyena cultures also are capable of modulating sexual variation, either accentuating, or eliminating such differences through variation in the social context. Kim Wallen (1996) has recently provided an exceptionally thoughtful review of the complex, powerful interactions between hormones and social environments in rhesus monkeys. This is a ubiquitous phenomenon and scientists are well-advised to add cautionary notes to simplified accounts of the route from physiology to behavior.

The Loci and Ordinary Limits of Social Influence

One doesn't need to study primates in order to observe the operation of both disciplinary and cultural biases in scientific work. Such influences were fairly clear in the case of rat sex behavior, and they have operated, on occasion, in our hyena studies as well. In general, however, if you did what the rat sex researchers did in the 1920s and 1930s, you could see what they reported. More than that. They contributed to a cumulative literature. Cultural and disciplinary bias certainly appear, but Martha McClintock's liberating analysis of copulatory behavior in the female rat builds on the prior, painfully achieved, delineation of the major components of sexual behavior in the male rat by Stone and Beach and descriptions of the estrous cycle by Long and Evans (1922). Josephine Ball's richly detailed descriptions of sexual behavior in the female rat could go no further than they did until the gradual infiltration of ethological attitudes from a zoological tradition prepared the ground for more naturalistic studies in psychology.

In the histories recounted above, within still-undefined limits, the animals themselves appeared to control much of what was seen and described. Given the impoverished conditions of observation, Stone and Beach provided remarkably thorough descriptions of sexual behavior in female and male rats. It was later in the scientific sequence that social influences played out, and attention was directed to the "rich" active behavior of males and the "passive" lordotic behavior of females—a theme that was accentuated by a disciplinary tradition emphasizing experimental control, without regard to ecological validity. Similarly, although there are differences between observers in interpretation of hyena behavior, I cannot think of any major disagreements about what has been observed. No modern observer of spotted hyenas has ever claimed that males are more aggressive than females, or that they dominate females (except under highly circumscribed conditions; for example, the phenomenon of "baiting"). Characterizations of the affiliative matrilineal core of hyena societies appeared later in the process than descriptions of aggression and dominance, but that necessarily awaited the long-term study of identified individuals (Frank 1986b), and I don't believe it can be ascribed to any simple case of observational bias.

It strikes me that all of us engaged in behavioral studies bring a great deal of baggage into our scientific lives. In addition to the general culture of the 1990s, I am comfortably embedded in a liberal academic culture, where, for example, certain attitudes about behavior and its genetic, or physiological, substrates are much more acceptable than others. Given the long-standing antipathy to the politics of genetic determinism, it is no accident that our cluster of hyena researchers is alert to every interaction between hormones and the social context that we can find. We also have our individual life experiences, which can tilt the balance between caution and speculation, or change our attitudes toward what might be expected in the way of sex-linked variation in behavior. Finally, we have the established traditions of our disciplines and subdisciplines operating in our time. Daily rumination on these issues would leave little time for the practice of science. But, every now and then, it is worth reflecting on, or examining in detail, the unarticulated assumptions that underlie our scientific choices.

NOTES TO CHAPTER FOURTEEN

1. It should be noted that Beach personally worked with a wide range of animal species, and collaborated with the anthropologist Clellen Ford on a comprehensive comparative treatment of sexual behavior, including cultural variation among people (Ford and Beach 1951).

2. Frank Beach was known to a number of individuals attending this conference as a friend, colleague, and teacher. He was all those things to me, as well as a mentor and collaborator during the early years of our hyena studies. Perhaps more significantly, between 1937 and his death in 1988, Frank played a central role in the general development of comparative psychology in the United States and the specific emergence of research on hormonal substrates of sexual behavior. As an advocate of comparative perspectives in the general discipline of academic psychology, he was often swimming against the tide. Some of his attitudes were persistent and well ahead of his time in psychology. In other cases we can see his ideas change as the result of his own research and that of other scientists, as well as the impact of the general culture and personal interactions with colleagues and students. Leonore Tiefer, who carried out her graduate studies under Frank Beach's direction, has written that "Frank provided a powerful intellectual example of someone whose ideas and focus changed over the decades without his earlier work losing its value. He lived and worked in his time and within the intellectual paradigms of his time, but his focus on the future . . . kept him alive to the possibility, the reality, even the necessity of intellectual change" (Tiefer 1988, 442). I share her view.

3. They also reversed a long-standing tradition in the rat sex literature, in which male rats were habituated to the test enclosure and females (literally) were dropped from above, where they had been retained in a small compartment in the ceiling of the apparatus. Carter et al. (1982) allowed their females time to habituate to the apparatus prior to introduction of the males.

4. One of the oddities of this literature is that people seemed to have forgotten the implications of a still earlier set of publications on sexual motivation. In these studies, the sex "drive" was one of many behavioral dispositions assessed in an apparatus requiring the rat to cross an electrically charged grid to gain access to the reward (Warden 1931). In the course of such studies, Warner (1931) concluded that female rats in estrous were probably more motivated to gain access to males than male rats were to gain access to females.

5. Fausto-Sterling and others have suggested searching for a symmetrical role of the fetal/neonatal ovary in accounting for sexual differentiation of the female

mammal. Arnold Gerall, and his students Janet Dunlap and Shelton Hendricks, carried out a set of technically heroic experiments without uncovering any such processes in laboratory rats, at least as regards behavioral differentiation (Gerall et al. 1973). However, more recent studies suggest that there may be direct genetic effects that bypass the gonadal mechanism (Arnold 1996), and such mechanisms may well account for species variation in clitoral morphology, as well as sex differences in the brain.

6. In nature, Kruuk (1972) described "baiting," a pattern in which groups of males harass a single female, often with biting attacks, in the period preceding actual mating. This is the only time when male coalitions display aggression towards females.

7. In our captive hyenas, hierarchies based on dominance generally correlate 0.8 to 0.9 with hierarchies constructed from subordinate behavior. In addition, hierarchies constructed on the basis of interactions during competitive feeding show similar levels of relationship with those constructed during time-sampled recording of spontaneous nonsocial interactions.

8. A very different situation has been observed in rhesus monkeys. Goldfoot and Neff (1985) observed that adding a single male to a female group stimulated play, but adding a second male eliminated that effect because the males then played with each other.

15 Changing Views on Imitation in Primates

Richard W. Byrne

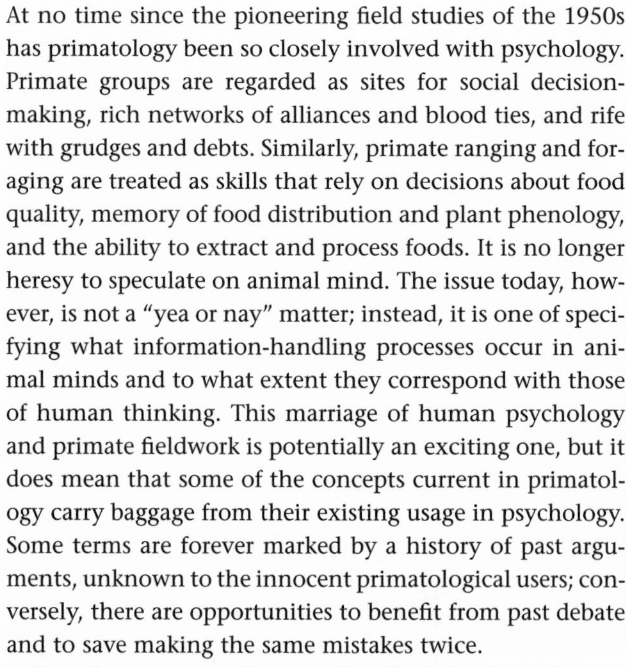

At no time since the pioneering field studies of the 1950s has primatology been so closely involved with psychology. Primate groups are regarded as sites for social decision-making, rich networks of alliances and blood ties, and rife with grudges and debts. Similarly, primate ranging and foraging are treated as skills that rely on decisions about food quality, memory of food distribution and plant phenology, and the ability to extract and process foods. It is no longer heresy to speculate on animal mind. The issue today, however, is not a "yea or nay" matter; instead, it is one of specifying what information-handling processes occur in animal minds and to what extent they correspond with those of human thinking. This marriage of human psychology and primate fieldwork is potentially an exciting one, but it does mean that some of the concepts current in primatology carry baggage from their existing usage in psychology. Some terms are forever marked by a history of past arguments, unknown to the innocent primatological users; conversely, there are opportunities to benefit from past debate and to save making the same mistakes twice.

In order to examine how another discipline's past history can influence the current state of primate research, in this chapter I shall take a single topic, *imitation,* which originates in psychology and has recently become important within primate research. As will become clear, imitation has been treated quite differently within developmental and experimental animal psychology, and both these traditions affect interpretations of primate behavior.

During the 1990s, imitation has been a regular topic for symposia at international primatological conferences; indeed, whole conferences have been convened on imitation alone, and several books have resulted which focus on the issue of primate (and other animal) imitation. Imitation is big news for primatology. (Sometimes this interest is lightly disguised under terms like "observational learning," "mimicking," or "social learning," but the question that sparks off the hottest debate is always, "Does the animal imitate?") I intend to show that the very idea of what "imitation" is, originally a rather fragile notion among primatologists and close to the everyday folk-understanding of the term, has changed in response to new data, but has also been buffeted by the two strong traditions within experimental psychology and developmental psychology. This essay will therefore take the form of a historical sketch, for whose brevity and overgeneralizations I apologize at the outset. The aim is to give a broad picture and not to be fair to every nuance of history or every person's viewpoint.

Innocent Days: Most Animals Imitate

Before 1975, the concept of imitation within primatology does not seem to have been much different from its everyday usage: and certainly the everyday understanding of imitation has not changed even now. This prescientific notion of imitation is that of a cheap trick, a sham of better things. Imitation is counterfeiting, taking an easy route to give a false appearance of real skill and ability. Not that imitators are regarded as dumb, exactly: they are evidently smart, in their sneaky, trickster way, and due some grudging respect as such. In certain formal contexts, imitation may be defined as illegal (plagiarism in school examinations or Ph.D. theses, for instance). But mostly, imitation is an allowable if not very respected shortcut to excellence; even art forgeries sometimes become valuable in themselves. The opposite pole to this sort of imitation is creativity, and that is what really demonstrates intelligence. Imitation counterfeits creativity. Imitation gets results, but they are derivative results, showing no more than a certain exploitative cunning; the higher processes of intellect produce creative novelty, demonstrating real insight and intelligence.

Applying these beliefs to primates, one must naturally expect monkeys and apes to imitate readily. Indeed, the everyday term for imitation is derived from the word for monkey or ape in many languages (Visalberghi and Fragaszy 1990). In English it happens to be "to ape," but we should not make anything of the distinction: the words monkey and ape were used interchangeably until recent times, as the "rock apes" of Gibraltar (Barbary monkeys) remind us today. Thus, when Japanese primatologists described the fascinating novel behaviors of sweet-potato washing and wheat placer-mining[1] they had observed among the monkeys of Koshima Island, the interest for most primatologists was in how these behaviors were innovated (by a young female, Imo), and how they spread as cultural traditions throughout the local population (Nishida 1987). That they spread *by imitation* was largely assumed: just such would be expected from quick, cunning animals like monkeys. Japanese monkey imitation fitted into a broad picture, in which any monkey, and indeed most other animals under the right circumstances, could imitate.

It was popularly believed, for instance, that blue tits in Britain had learned novel food-extraction techniques by imitation. In the late 1940s, it was noticed that some populations of tits had discovered how to get to the cream on top of milk bottles by tearing off strips from the cardboard tops while the bottles were on people's doorsteps. Ornithologists documented the spread of this habit from several foci of innovation. Later, cardboard tops were replaced by foil, and again the tits learned a novel technique to get the cream, pecking vertically down through the foil, and the habit spread. The researchers had not in fact described the leaning as "imitation" (Hinde and Fisher 1951)—but everyone knew animals could imitate, so naturally this was seen as a canonical case. In the last few years, I have many times listened to descriptions of the spread of the trick of hanging upside down on suspended bags of peanuts in Britain—from initially just the tits and siskins which naturally feed in this way, to greenfinches, sparrows, and most recently bramblings. In every case, the term "imitation" has been used to explain the spread. Birds only imitate a few useful tricks, but monkeys are seen as the great imitators of the animal kingdom, and apes—obviously—ape all and sundry.

The Assault on Animal Imitation

The first hint of a problem with this consensus came with the discovery of vocal dialects in Japanese macaques (Green 1975). Normally, primate vocalizations do not show dialects (Janik and Slater 1997), but dialect has been described in bird song. Many suboscine birds learn their songs by vocal imitation of adults while they are nestlings. Song-learning by vocal imitation is not error-free, and this allows variations to creep into future generations of songs; the site-fidelity of some species then tends to preserve differences, allowing a local dialect to develop. The small amount of work done on the acquisition of primate vocalizations suggests that the calls (if not their meanings) are tightly channeled by inheritance, and not subject to vocal learning. Green's discovery of dialects in Japanese monkeys was therefore surprising, but he was suspicious of his own results. Noting that it was only the "food call" which showed systematic interpopulation variation, and that all the populations he studied were artificially provisioned, he suggested an alternative explanation. Perhaps the people who fed the monkeys (not themselves scientists) were prone to feed monkeys more if they called louder. This very human tendency would then unintentionally reinforce the emission of whatever call was most striking when the monkeys were first fed, coincidentally a different call in each population. On this interpretation, the dialects are human products. The puzzle of the existence of only one single monkey call being learnt by vocal imitation, in one single species, is solved.

Green (1975, 309) also noticed that at one of his study sites, Koshima Island, the woman who had been provisioning the animals "regularly from the earliest days," "gave sweet potatoes differentially to those monkeys who washed them." This immediately suggested a completely different explanation for the cultural diffusion of the novel foodprocessing techniques. The maintenance and spread of the habits might be under (unintended) human control, based on the perceived "cuteness" or importance of special feeding techniques to the nonscientist feeders. In other words, the cultural tradition of washing sweet potatoes was part of the extended phenotype of the human provisioners, not the monkeys. Under this interpretation, there is no evidence for monkeys' ability to imitate actions, any more than vocalizations.

The subsequent attack on the idea of monkeys as imitators had two main proponents. Galef, an experimental psychologist from McMaster University, reexamined the published descriptions of the original spread of novel habits among Japanese macaques, and argued that their spread was too slow and too incomplete for imitation to be a plausible explanation (Galef 1988). Visalberghi, a comparative psychologist from Italy, examined the ability of naive monkeys to learn novel foodprocessing skills. She showed that the monkeys (both *Macaca* and *Cebus*) readily learned to wash dirty food after about two hours of individual experience. However, they gave no sign of looking for or benefiting from the sight of others who already showed the necessary behavior and, in general, their skill-acquisition showed all the signs of trial-and-error learning (Visalberghi and Fragaszy 1990). Even the blue tits did not escape revisionism: Sherry and Galef (1984) presented naive tits with the sight of a conspecific dealing efficiently with a milk bottle, with an opened milk bottle alone, or with normal unopened bottles. Both the sight of a skilled model working on a bottle, and the opportunity to explore an already-opened bottle, facilitated acquisition of the habit—and each was equally effective. Since, if nothing else, imitation has to involve copying of an observed action, the idea that an animal can imitate was once again called into question.

The apogee of this offensive against animal imitation came at approximately the same time in the publications by Tomasello, a developmental psychologist at Emory University. He questioned the assumption that even chimpanzees could "ape." An experiment, designed to test for imitation in a captive chimpanzee, showed that observers of a novel raking technique were more likely to rake for food afterwards, although they used their own idiosyncratic techniques and did not copy the particular actions demonstrated (Tomasello et al. 1987). Tomasello described this as "emulation" (copying the goal but not the methods of the observed individual) and went on to suggest that the traditions of tool-use among wild chimpanzees and the interpopulation variations in techniques might result from ordinary trial-and-error learning combined with a tendency to emulate others' goals (Tomasello 1990).

None of this work positively demonstrated that primates *cannot* imitate—unsurprisingly, since proving a negative is intrinsically difficult. But it raised sufficient doubts in people's minds that the onus came to be upon believers to prove that any animal (other than the human) could imitate.

Alternative Frameworks

If the everyday notion of imitation—a fairly easy way of counterfeiting creative intelligence, that one might expect animals to be able to do—was discredited, what alternative theoretical frameworks were possible? In principle, there were two candidates.

THE DEVELOPMENTAL PERSPECTIVE

One comes from child psychology. Developmental psychologists have long regarded imitation as an important component of normal development, either for its social implications (Vygotsky 1978) or its intellectual ones (Piaget 1951). The Piagetian framework, in particular, distinguishes several sorts of imitation that occur in an ordered *series,* one that is in step with other series of intellectual development. Each series is necessarily progressive and interlinked, because it reflects the unfolding of an underlying logical sequence of cumulative development. Across the various developmental series, when the child progresses up a step in one, it will also move up in all the others: at that point, the child enters a new "stage." To reach the (human, adult) endpoint, the *only* route is along the series of stages that Piaget mapped out. If so, then any other species that also attains some part of the human adult's mental skills must of necessity move along the same trajectory. This was used, initially it seemed with some success, to compare among primate species in their intellectual attainments (e.g., Parker 1977; Parker and Gibson 1977; Chevalier-Skolnikoff 1977). The developmental series in different skills were used as scales with which to measure intelligence. Great apes progress further along these scales than monkeys, and monkeys further than strepsirhines; none, however, reach full adult human capacity. (Few longitudinal developmental studies have actually been done; many researchers looked at the endpoint attained, assuming that this was reached in the expected way during development.)

Two sets of problems have beset this approach: from within primatology, and from child development itself. Most simply, when the data are examined, the pattern of primate abilities does not support the idea of cumulative, ordered series: instead, a species will often show some "higher" ability, yet lack the "lower" one that is supposed to be part of its development (see commentaries to Parker and Gibson 1979). If this were merely one item from one series, the failure need not be fatal, but it is more general than that; there is something wrong with the idea of an immutable series. Turning to child development, one finds that the whole approach is somewhat discredited. Numerous researchers, working in every area of children's competence, have failed to find clear Piagetian stages, in which every component ability is at the predicted level and they all change in step to the next stage (McShane 1980). Rather, mental development is much more modular than Piaget realized, with each skill developing to become semiindependent of others and not moving in rigid step with each other (Karmiloff-Smith 1993). There are admittedly some series showing progressive levels of a skill that are traversed by most children; but the simple picture as presented by Piaget is shattered and expecting it to apply across species is—with hindsight—certainly optimistic. Perhaps for these reasons, developmental notions of imitation have only found limited favor within primatology.

This is, I believe, unfortunate: rigid series of progression and discrete stages in development are unhelpful models, but developmental psychology has long articulated a sophisticated understanding of imitation that could benefit primate work. Developmentalists treat imitation not as a single unitary capacity, but as a complex skill that may develop, or be partially or incompletely present in an individual. Usually without stating so explicitly, the developmental idea of imitation often implies that *parts* or *levels* of a structurally complex whole are chosen for imitation, others discarded. For instance, Bauer and Mandler (1989) describe children's reproduction of sequences of actions demonstrated by an adult: children were much better at copying logically coherent sequences, like giving the teddy a bath, than disordered or illogical ones, and details that did not fit the main purpose tended to be omitted. Similarly, Kuhl and Meltzoff (1982, 1984), working with twelve-to-twenty-week-old children copying spoken words, found that they copied the pitch contour and prosody but not the exact fundamental frequency. In both these cases, the copying was described as "imitation," but it is clear that the children appreciated the structural organization of the behavior, copying only certain aspects and varying others. The process is very different to that which takes place in a videotape recorder or on a theatre stage.

However, a very different understanding of imitation has come to dominate primatological debate.

THE EXPERIMENTAL PSYCHOLOGY PERSPECTIVE

In recent years, those researchers who have been instrumental in questioning monkey and ape capacities to imitate have generally taken their simpler, "null hypothesis" model of learning from Experimental Psychology. The capitals are meant to connote more than simply the innocuous literal content of the words—to indicate that the subject has an attached philosophy. (There is a direct and related parallel with the meaning of Behaviorism, as distinct to the ordinary study of behavior that might be called "behaviorism.") Experimental Psychology (EP) espouses an essentially Behaviorist model of animal learning, while stopping short of the extreme Radical Behaviorism of B. F. Skinner, which lumps all forms of human learning into the same simple explanatory framework.

The basics of all EP explanations are the mechanisms of Classical Conditioning (or Association Learning), and Operant Conditioning (or Reinforcement Learning). Modern EP is not ignorant of the data of ethology, and now readily accepts the existence of evolved "constraints on learning" that so usefully guide many animals to learn precisely the things that benefit their Darwinian fitness (Garcia, Ervin, and Koelling 1966; Garcia and Koelling 1966). Some researchers even treat Association Learning as an evolved mechanism for learning the correlational structure of the world, and reinforcement learning as a mechanism that records the information gained from exploration (see, for instance, Dickinson 1980). There is indeed much truth in the view of animal learning as espoused within EP. However, there is not much place in it for *structure,* beyond the level of complexity of linear strings of associations. And the cardinal rule for deciding on a null hypothesis is that of simplicity, in the sense of Occam's razor or Lloyd-Morgan's canon.

In EP, imitation is seldom discussed, and this has been true for a long time. As a rough survey, I examined all experimental psychology textbooks—excluding books explicitly referring to "cognitive" psychology—held in the St. Andrews University library. Of the eleven major texts found, going back to the early 1900s, seven did not mention imitation; in the remainder, it received no more than a few lines of discussion. This should not come as a surprise, since EP workers typically study only single individuals, whether animals or people, and are seldom confronted with any signs that learning might be different in social contexts.

Social learning *can* be accounted for within EP by postulating additional principles that direct the individual towards particular aspects of the situation. Whereas constraints on learning make certain specific sorts of information easier or harder to acquire, these social principles give more general guidance. The most well-known and successful in "debunking" claims of imitation was called stimulus enhancement by Spence (1937) and local enhancement by Thorpe (1956). Seeing a conspecific engaged in a task makes the observer more likely to pay attention to, or explore, those aspects of the situation that engage the conspecific (the "stimulus configuration"). Once learning is focused in this way—on a specific place, at a specific time, and perhaps on a certain object there—trial and error learning can proceed much faster. Where the observer happens on the same successful actions as it observed, as a result of its individual exploration, we may be deluded into thinking that it has imitated the actions. Careful experiment would reveal that actually seeing the actions is not crucial and that "enhanced" individual learning is actually responsible.

EP researchers assert, in fact, that *only* careful experiments can reveal the true mechanisms of learning; this gives them something of a monopoly on virtue, since they are highly practiced experimentalists. What sort of experiment can reveal the ability to imitate? Heyes, an experimental psychologist at University College London, considers that only two experiments have in fact done so, neither employing primates (Heyes 1993). In one, Galef and his colleagues replicated an experiment (Dawson and Foss 1965) in which budgies observed a food reward being uncovered by one of two means, pecking downwards with the bill or tearing upwards with a foot (Galef, Manzig, and Field 1986). Afterwards, budgies were slightly more likely to use the means they had observed, although both methods were also used spontaneously. (Subsequently, this technique has been adapted for use with chimpanzees, with similar results: Whiten et al. 1995.) The other type of experiment is rather different. Observer rats watched a trained rat push a lever (that was close to the cage wall separating the two partitions) to the left or right. Since they watched from the front, the observed animal's right was the observer's left. When the observer rats were placed into the compartment with the lever, they tended to push more in the same direction as the demonstrator had done, left or right (Heyes, Dawson, and Nokes 1992). To avoid the criticism that stimulus enhancement would favor pushes towards the same point in space, the experimenters added a condition in which the lever was rotated clockwise 90 degrees, so it again lay against the cage wall; still the rats pushed in the direction observed. Heyes asserted that these experiments gave incontrovertible evidence of imitation in animals (Heyes 1993).

The Trouble with EP

The dominance of the EP treatment of imitation at first sight appears a wholly positive trend. The need for objective, replicable data is emphasized, and theories have a pared-down, parsimonious economy of mechanism. Most compellingly, instead of having to deal with the messiness of imitation as observed in everyday living, two simple experimental paradigms have been put forward as acid tests of true imitative ability.

Unfortunately, these experiments are also vulnerable to reinterpretation as tapping into cognitively simple, "enhancement" effects, instead of imitation. The argument is as follows. If stimuli can be facilitated by observation, then so can responses. In stimulus enhancement, brain records corresponding to environmental features are made more salient or more active when a conspecific is seen to interact with those features; in the corresponding "response facilitation" (Byrne 1994), brain records of actions in the observer's repertoire are made more salient or more active by seeing a conspecific doing some similar-looking action. This mechanism, a cognitively simple one, predicts the results of Dawson and Foss, Galef et al., and Whiten et al. Imitation is not proven by such experiments, because the actions are already in the test individual's repertoire, allowing a simpler explanation based on mere facilitation. This effect alone cannot explain Heyes et al.'s (1992) results, but they too have been challenged (Byrne and Tomasello 1995). When the rat is placed in the test section for the first time, it must orient itself relative to what it previously saw from the observing cage. Heyes and colleagues assume the rat uses the food hopper for orientation, treating the lever as moved. But perhaps the rat is confused, and imagines the (rotated) lever is in the *same* position as before, whereas the (fixed) food hopper has moved 90 degrees. By stimulus enhancement, the rat would then push to the place in space (it thinks) the lever was when the push was rewarded, and so *appear* to be imitating the motor movement it saw before.[2]

Accordingly, laboratory experiments are not necessarily a trouble-free route to salvation as has been claimed. And, if only lab experiments do hold the key, then we must grimly accept that no animal has yet been shown to imitate. One reaction to this is to try harder: design an ingeniously modified experiment that rules out even this rather technical critique. Instead, I suggest that it may be wise to back away from the minutiae and consider possible problems with the approach. The trouble may lie in EP's insistence on a formalism that can only deal with *simple* actions, as responses to stimuli. The apparent simplicity of this narrow focus may impede discovery of a learning mechanism as cognitively complex as imitation, and perhaps paradoxically we would do better to study a *complex* behavioral sequence, whether we use experiment or observation as our source of data.

Most behavior of neurally complex animals, such as primates and especially the great apes, is not well described as a set of unvarying "responses" in a countable "repertoire." As noted by Lashley (1951) and Dawkins (1976b), behavior is hierarchical, and this gives scope for a process of hierarchical assembly of arbitrarily complex patterns of action, limited perhaps by working memory capacity. Novelty of behavior, in a hierarchical system, is a matter of new arrangements of "old" actions; seldom, perhaps never, does an entirely new action need to be acquired. For instance, around six years old, children learn to tie their shoelaces. This behavior is "new," although all the component motor movements have been executed before, probably thousands of times. The novelty is in the carefully structured programming of these actions that leads to the novel consequence of tied laces: the order of the action sequence (e.g., half-hitch before bow), criteria for moving to each next action (e.g., the tightness of the half-hitch, before the bow is tied), and the necessary bimanual coordination (e.g., hold first loop tight with right hand while other loop made with left hand). Imitation in a hierarchical system becomes a similarly more open question: copying may not be at the "lowest" level of mimicking every detail, but instead the overall plan or program may be copied without the fine details (Byrne 1994; Byrne and Russon 1998). With complex actions, even "imperfect" imitations in which the observer chooses to copy only some aspects of the complex process may be identifiable in output behavior. The richness and complexity of the actions makes the a priori probability of their arising in spontaneous behavior as a result of individual learning so low, that their existence alone can give convincing evidence.

There is actually a wealth of observational data that great apes, at least, do replicate complex, hierarchically structured behavior they have observed. The many projects in which great apes have been home-reared all report numerous cases of copying of complex everyday actions—washing-up dishes, brushing teeth, combing head hair, (illegible) writing, and so forth; captive great apes even attempt to tie shoelaces, though with little success (e.g., Gardner and Gardner 1969; Linden 1974; Miles 1986; Patterson and Linden 1981; Russon and Galdikas 1996; Russon and Galdikas 1993; Savage-Rumbaugh 1986). Although these data are observational, experiments have confirmed apes' ability to copy arbitrary actions to order (at least for the common chimpanzee), by first teaching the apes to "do as I do" when they are given a certain signal (Custance, Whiten, and Bard 1995; Hayes and Hayes 1952). All of these cases rely on apes' copying human actions which only occur in the ape repertoire under captive conditions; obtaining evidence of imitation in natural behavior in the wild is intrinsically far more difficult. Nevertheless, the very rapid acquisition by gorillas of extended, highly structured, bimanually coordinated sequences of action in plant food-preparation may rely on imitation (Byrne and Byrne 1993, 1991). The gorillas acquire a highly *standardized* overall approach, but vary idiosyncratically in every detail of precise movements, the digits used, and the manual laterality; this striking contrast has led the researchers to suggest that "program-level" imitation guides acquisition of the skills.

None of these data, however, have convinced EP researchers. For instance, Heyes dismisses the gorilla data, not because it does not logically prove imitation (which it doesn't), but because the study is based on "anecdotal reports" (Heyes 1995, p. 1423); to field researchers, this is a strange description of 510 hours of focal-animal data.

Conclusion

This brief history suggests that the chief influence on the changing conceptualization of imitation within primatology has been that of another discipline, rather than any special characteristics of key protagonists. There is no association, for instance, of male researchers and experiments, females and reliance on observations; Heyes and Russon are women, Galef and Byrne are men. Nor is there any obvious geographical influence on changes in belief. Instead, the methods and theory of Experimental Psychology have been applied wholesale, and the other approach with a long history in psychology, the developmental one, has been sidelined.

Are there any lessons, for those simply hoping to advance knowledge most efficiently, from this analysis of the past? Firstly, it would be rash to reject all features of EP, just because—as I have argued—it has hindered understanding in certain ways. The discipline of evaluating evidence against Occam's razor—here, "Can the mechanisms of individual animal learning, already known in psychology, explain the data?"—has been wholly beneficial. Primatologists are now rightly cautious in their interpretations, and none would wish to return to the naïveté of the early days. And there is no doubt that the overly Piagetian nature of some developmental approaches to primates was a cause of the current relative lack of influence of developmental psychology on imitation in primates; the result may be regrettable, but the reason is straightforward.

However, hidebound insistence on laboratory-style experiments as the only route to scientific knowledge will hinder the progress of modern primatology, just as it did psychology in the past. Ethology avoided this pitfall, by proper organization of observations to avoid bias (J. Altmann 1974), by the development of quantitative methods that could be used in the field (Martin and Bateson 1986), and by judicious use of experiments sensitively designed to mesh with natural behavior. Thus, while ethology has benefited from some of EP's methods (the use of Skinnerian procedures in optimal foraging and other areas of behavioral ecology is a striking example), the intellectual traffic travelling the other way has been slow and hesitant.

A productive fusion of the best of each tradition will perhaps be made most likely by adopting a *cognitive* approach (as is often done now within developmental psychology). Cognitive psychology grew up in the 1960s as something of a dissident offshoot of EP, and now dominates the study of human mental function. The cognitive revolution, in psychology as in other disciplines, stems from the advent of computers and computational theory. In 1938, Turing had shown that a mechanical device, at that time imaginary but which he called a computer, could derive ("compute") most but not quite all numbers. The importance of this was that, as mathematicians had already realized, any definite process can be represented uniquely by a number. So Turing's proof was equivalent to showing that a computer could perform any process which could be clearly described. When first available as research tools, computers were mainly used for "number crunching." However, during the 1950s some scientists followed Turing's lead and began to attempt computer simulations of a wider range of processes. This realization of the power of the universal Turing machine to perform any fully specified process led to a search for proper and full specifications of human capabilities, formerly considered well beyond mechanistic understanding. The breakthrough for psychology was probably the demonstration that a computer program could solve problems of formal logic, going along solution paths that were much the same as those of beginning students of philosophy and making similar errors (Newell, Shaw, and Simon 1958). It was rapidly discovered that while computer models for "hard" tasks like logic and chess were easily feasible, it was much more difficult to give a computational specification of "easy" tasks, which children master with ease, such as using language or seeing a stable world of objects. Repeated failures to build computer models of "simple, everyday" processes showed the inadequacy of current psychological understanding of these phenomena. Thus psychologists realized that the disregarded tasks of everyday life were, after all, the great problems for a scientific understanding of mental function, and the whole subject of cognitive psychology has since built up in the effort to solve them.

Cognitive explanations are computational, in the sense that any model must be *definitely specified* not ambiguous, and must be *adequate* to perform the task. These prescriptions are difficult to meet and many putative explanations can be rejected as vague, or inadequate for the job in hand; those that are neither can then be tested in the normal ways of EP, using timing and error data to compare performance of model and human. Thus cognitive psychology uses many of the methods of EP, but augmented and sharpened with evidence from everyday life—models that fail to account for everyday behavior do not even merit serious testing. Like EP, cognitive psychology insists that models be in principle falsifiable by clearly specified observations, but unlike EP it does not restrict explanations to the relatively linear models generated by association and reinforcement.

The science of animal behavior is in a sense "pre-adapted" to a cognitive approach. The pioneering ethologists—Lorenz, Tinbergen, von Frisch, and others—all saw the power of naturalistic observations, properly and carefully described, to limit severely the range of possible models of the underlying mechanisms. Rather than rushing straight to experimentation, ethology has always urged a first stage of scientific description in order to rule out whole classes of explanation (as a result, when ethology does turn to experiments, they are often particularly effective as tools). Similarly, the ready acceptance of the brain-as-onboard-computer metaphor within animal behavior, and within primatology in particular, suggests that a the time may be ripe for a more explicitly cognitive approach to take hold. It would be nice to think so.[3]

NOTES TO CHAPTER FIFTEEN

1. Wheat placer-mining consists of separating wheat, which floats, from sand and mud, which do not, by throwing the mixture into water.

2. After this was written, the claim of imitation in rats was in fact withdrawn as insecure, but for a different reason (Mitchell et al. 1999). The researchers discovered that the rats were using odor cues on the swinging bar to govern their responses, so it may be that no other explanation—whether of imitation or disorientation—is required. This nicely illustrates how experimental data, just as much as observations, depend on prior assumptions and require careful interpretation.

3. And since these words were written, an independent but near-identical plea for a new synthesis of the cognitive and ethological traditions has been made, by several researchers who work on birds rather than nonhuman primates (Balda, Pepperberg, and Kamil 1998). They are coming at the problem from the opposite direction, working in fields where systematic ethological observation is routine but cognitive explanation is exotic. However, their conclusion is the same: the time is ripe for a properly Darwinian, cognitive science to be developed, in which the choice of experimental manipulation is always guided by prior ethological observations.

E-Mail Exchanges

We tried different approaches to assess what might have been important in changing our ideas about primate society. The historical and comparative framework of the situated histories from different national traditions unearthed a wealth of complexity that was destined to become part of the new context for interpretations. But another aspect of the "history" of primate studies was to be found in the "closely related disciplines" that produced modern primate studies. Comparisons with anthropology, psychology, and zoology should be doubly provocative since these are simultaneously parents—the "source" of primate studies—and siblings—sister disciplines with their own parallel development. The chapters in this section illustrate many insights that can be gained by looking across disciplines. However, both at the workshop and in the e-mails, the topics that generated the most interest (when we donned our cross-disciplinary eyeglasses) concerned the relative importance of two "factors": sociobiological theory and gender. This led us to ask: *Did sociobiology make a difference in our ideas about primate society?* and similarly, *Did women studying primates make a difference to our ideas about primate society?* The e-mail conversation shows not just what people think about the impact of theory or of gender on ideas. Listening in on the exchange between psychologists, cultural anthropologists, primatologists, animal behaviorists, and science analysts shows that both sociobiological theory and gender issues are problematic topics, whether it is their importance or the manner in which they could have had an impact on different disciplines.

Question 6: Did Sociobiology Make a Difference in Our Ideas about Primate Society?

To: <teresopolis@majordomo.srv.ualberta.ca>
From: "Robert Sussman" <rwsussma@wustl.edu>
Can something have such a bad influence that it negates any good influence it might have had? I think sociobiology has created an atmosphere in primatology that encourages sensational interpretation and broad-based theory without supportive data. It does this often at the expense of good attempts at data collection and often downplays the need for good data and the need to attempt to falsify hypotheses. This has led to an emphasis on adaptationist all-encompassing theories (e.g., the selfish gene—which is no more explanatory than was Bentham's pleasure and pain theory) at the expense of good data collection. There was a very good start in the 1970s on good community ecology, and the study of relationships between morphophysiology, behavior, and ecological preferences and community ecology. I believe we have lost a great deal of ground in these areas because of the emphasis on often nonsensical sociobiological nonquestions.

To: <teresopolis@majordomo.srv.ualberta.ca>
From: "Stephen Glickman" <glickman@socrates.berkeley.edu>
After following the various Teresopolis discussions, and occasionally writing dyadically to various participants, I find myself in the rather odd position of entering the lists as a defender of "Sociobiology." I have been critical of sociobiology in print with the usual array of thoughts about premature and/or inappropriate extrapolation to people, a sometimes poor ratio of data to speculation, bypassing mechanisms to focus on outcomes, one-tailed treatment of data, and ad hoc explanations of nearly everything. However, down in the experimental trenches, I think the ideas and attitudes of sociobiology served to change and energize the field. Behavioral ecology was certainly on its way before Wilson's book and, in a conversation at Teresopolis, Robert Hinde convinced me that the folk around David Lack were thinking about costs and benefits for individuals. On the other hand, for many of us, it was not until the appearance of G. C. Williams's book in 1966 that we really appreciated the implications of Hamilton's papers on kin selection. And although Trivers's extensions of kin-selection theory occasionally drove me up the wall (with, e.g., their pseudomathematical precision), they were very significant heuristic pieces that also provided new, alternative ways of looking at familiar data. There was also a clear shift in the literature from the 1950s to the 1970s, with analysis of social behavior moving from sociological demographic descriptions in terms of age and sex roles, to a contemporary focus on individuals and their genetic relationships. The approach also pressured us to think more sharply in terms of selective agents and has changed animal-watching to this day.

My initial, conservative (and incorrect) response to sociobiology was, "I've heard most of this before," and one can argue that the essential useful ideas would, inevitably, have filtered through the field–without the overclaiming, media-grabbing antics of some sociobiologists. However, Edwin Boring, the first devoted historian of psychology, expressed the following sentiments in 1920: "Founders are generally promoters, in science and elsewhere, and we therefore have 'to consider' the mechanisms of public attention."

In a conference devoted, in part, to the impact of "social" influences on the scientific process, it is important to recognize that there are things that we find really aggravating, that appear to be fairly common and even "useful" events in the history of behavioral studies.

To: <teresopolis@majordomo.srv.ualberta.ca>
From: "Dick Byrne" <rwb@st-andrews.ac.uk>
. . . Hear, hear Steve. The connotative meaning [to the question] seems to be "since sociobiology is clearly a Bad Thing, it'd be nice to see if it had any little bright sides." Yet I'm not the only one in Britain who would find that a bemusing suggestion and the question above much the same as if asked "if anyone cared to outline any positive influences of science."

For instance, a Tinbergen-trained swallow researcher I heard introducing a talk with an account of how, in the late '60s, he'd decided that ethology had pretty much run out of steam and (when he moved into developmental psychology) he only kept his swallow-ringing going because he liked the birding . . . and wow how wrong he was eh? Sociobiology totally transformed and revitalized the field, bringing it straight back to the heart of science, and lucky he kept up the ringing program so he could still contribute. And a left-wing anthropologist introduced a talk with an account of how, when he was heavily involved in fighting the heavy-handed police controls in the Isle of Dogs (London) in the early '70s, he thought sociobiology was near-Nazi filth, however once he actually read some of it he realized that it was in fact, awkwardly, er, right, and his subsequent career has been devoted to helping liberal left ideas come to terms with certain realities without falling apart. . . . Like those people I just see "sociobiology" as a mainstream part of orthodox science, crucial to understanding biology. Sure, I get annoyed when functional people fail to see the interest in causal explanation, or in cognitive organization (but causal types are at least as bad). Sure, it isn't always helpful when sociobiological types make too bold claims about which bits of human nature are already explained. But since when was a whole field damned by a few irritating enthusiasts?

To: <teresopolis@majordomo.srv.ualberta.ca>
From: "Bruno Latour" <latour@paris.ensmp.fr>
I have been shocked—for reasons different from Dick—by the very question on sociobiology, but since I have been shocked by all the previous questions I felt tired of being always dissenting and afraid of appearing opposed to the kind efforts of the organizers!!

The way sociobiology has been portrayed by social anthropologists has always appalled me. I take it as the most interesting way of bringing a body back to the society. If Dick is worried by some brands of naive causal reductionist mechanisms in sociobiology, then he should live among sociologists and anthropologists who believe in completely naive and entirely reductionist definitions of society, culture, symbolic order, discursivity, and such things—not things alas!

For me . . . sociobiology was the most illuminating access to a complete redefinition of what is a society. I have always liked the most fabulous consequence of it: that our very socialness antedates Hobbes and Durkheim by millions of years, that our very body, our very genes, have been shaped and selected inside some sort of social order. Far from reading sociobiology as the biologization of society, I have always read it in the opposite way: even the body, in all its details, has been shaped by a "secondary adaptation" to a social world.

Needless to say, it is complete anathema to the social sciences, who still believe the only way to do their work is to distinguish as much as possible nature from society. I take sociobiology as the least reductionist form of sociology now available—I say that with a grain of provocation, as usual of course. . . . In France such a position is tantamount to fascism since through an incredible twist of history, only the extreme right has read and translated Wilson and Dawkins. This is why I always publish in English!! In France social scientists believe people to be born out of "language." . . .

. . . we would have been better off . . . if we had tried to get at this question through the notion of well articulated or badly articulated bodies; the question is not to solve again the mind/body problem, but to see what sort of agencies we give to what sort of body parts—so to speak— . . . The question is not whether or not hormone level and aggressivity relate to one another, and if those two relate to human culture, but what type of agency carries hormone, aggressivity, and culture. Is it an interesting carrier or a boring repetitive one? Is it one that reduces the number of entities or one that increases them?—all these questions cut across the divide and seem to me much more productive than the reenacting of the "us" versus "them" debate of physical and social anthropologists.

In summary, I'd say that sociobiology reopened the question of the nature of society and social order that had been sealed off by cultural and symbolic anthropologists—this being said, there is of course an immense mass of nonsense in it, but Dick is right on that: which is the field who is going to throw the first stone? The main technical weakness of sociobiology, as I see it, is to not have enough biology and "too much economics"—now there is indeed a disproportionate amount of bad economics in economics! . . . you notice the irony; the main defect of sociobiology for me comes from another social science, economics! not from biology (Donna [Haraway]'s first book is an ideal antidote for those who believe that biology is reductionist).

One last remark . . . the advantage of science studies is to take scientists by their "practice" not their theory: thus reductionism in theory is an entirely different thing than practice. I have never met nor studied a scientist whose "practice" is reductionist!

To: <teresopolis@majordomo.srv.ualberta.ca>
From: "Naomi Quinn" <nquinn@duke.edu>

. . . just a query about your observation, Bruno, on the irony "that the main defect of sociobiology for me comes from another social science, economics!" Enlighten me if I'm wrong, but I always thought the real irony was that economics had originally gotten its conceptual framework from evolutionary biology, via social Darwinism; so that the two fields have passed their vision of the world back and forth, each one in turn validating its enterprise in terms of the other. I don't mean, of course, that there have been no distortions or misuses in the process.

To: <teresopolis@majordomo.srv.ualberta.ca>
From: "Alison Jolly" <ajolly@arachne.Princeton.edu>
Hooray for the good sense about sociobiology. . . . Economics is just as slippery a term as sociobiology, though . . . (sociobiology suits my worldview), but economics is just as multiple in form as biology.

To: <teresopolis@majordomo.srv.ualberta.ca>
From: "Donna Haraway" <donna_haraway@macmail.ucsd.edu>
Camille Limoges wrote a wonderful paper about the traffic between political economy and biology or natural economy from the late eighteenth century through the mid-nineteenth century. His focus was on the idea of division of labor—an absolutely crucial tool for doing physiology, factory studies, and much else. It was not a question of "naturalization" nor of "socialization," but of a thick trading and passage zone for conceptual development that helped shape the world we live in, complete with our modes of naturalsocial (one word) action. I think sociobiology, with its investment analyses and many other key conceptual tools, is much the same. We know the world in terms of information, energetics, cost-benefits, strategies, etc. That's not ideology, but a historically specific way of interacting in the world, including the doing of natural science. Sociobiological analysis is science at its best, and also just the kind of practice that configures squarely inside historyculture-nature . . .

I have long been fascinated with the near interchangeability of textbook material and journal articles in certain kinds of economic and evolutionary analysis—again, not as ideology or "reduction," but as a part of world making. . . . Evelyn Hutchinson's textbook in ecology, from his lecture notes, is full of amazing footnotes that are great examples of the fruitful trading zones that are part of the prehistory of sociobiology, or more generally of the kind of thing that gets the name "theory" in behavioral, evolutionary, ecological biology in the last few decades. Great stuff.

To: <teresopolis@majordomo.srv.ualberta.ca>
From: "Bruno Latour" <latour@paris.ensmp.fr>
. . . Reductionism is in fact hard to allocate and that is what Donna meant, I think. I propose a test: take any feature used by a scientist to account for a sequence of action—gene, behavior, neurotransmitter, myth—forget about which one pertains to which domain of the world—body or soul, nature or culture—and just itemize them. Now the first part of the test is to count "how many" there are in a given paper; already you might get some surprise. You can see that some sociologist, for instance, employs in his story three elements

like "power," "tradition" and "culture" and that a completely myopic neuroanatomist puts in action, to account for the piece of behavior of his "ant," let's say, twenty-five intermediary elementary paths from gene to kin to pheromones, etc. Don't bother that one deals with humans and the other with animals; just count three compared to twenty-five.

Now the question is who is the more reductionist? The sociologist who says that human power determines a lot of things or the neuroanatomist? With my first test, it is the sociologist even though he will claim he is antireductionist and not the neuroanatomist, even if he claims he "is" reductionist!

Now here is the second part of the test: relate each of the elements you counted by an arrow and take two color pens. Color in black those that are related by a strict relation of causality . . . and in red those that are related by a relation of "occasion" (A gives the occasion or may trigger B under some circumstances).

Now try the test again and compare the two, sociologist and neuroanatomist. . . . I say that the first one is reductionist and not the second, because the first one does not imagine any other type of causality (but the black arrow type) . . . while the second, although he lists twenty-five elements . . . says only four or five are causally active; the others are there for figuration or other roles.

Most of our dispute is now clarified. . . . there are simply "more" elements to be taken into account, thus the behavior is more interesting—in my sense. . . .

As with sociobiology, our e-mail discussions of gender showed the many different ways that people conceptualized the issues. Here too there was disagreement about whether gender was a positive force in changing science. Some of the dissension seems to reflect disciplinary influences.

Question 7: Did Women Studying Primates Make a Difference to Our Ideas about Primate Society?

To: <teresopolis@majordomo.srv.ualberta.ca>
From: "Bruno Latour" <latour@paris.ensmp.fr>
. . . the reiteration of the question, did women studying primates make a difference, proves an admirable persistence in asking a question, but is it a good one? Yes, of course women made a difference, but so did planes to access field sites, sampling methods, King Kong, Jane Goodall popularization, drugs to protect researchers, photographic cameras, gene testing, etc. Making a difference is relatively easy given the number of elements that participate in the construction of a field site to make relevant features that were invisible so far.

Now if the question is to order these differences . . . from important to nil, I don't think that is an answerable or important question. . . . The question is not, it seems to me, "does it make a difference?" but "does it make an 'interesting' difference?" that is, does it articulate in interesting new ways the phenomena of primates?

To: <teresopolis@majordomo.srv.ualberta.ca>
From: "Alison Wylie" <awylie@julian.uwo.ca>
How to formulate this: When is a question an interesting question? When is a difference an interesting difference? I suppose it must have to do, fundamentally, with what you want to know and, more to the point, what you want to do in the most general sense. . . . For one thing, it does seem unavoidable that being gendered male versus female makes quite a lot of difference to your prospects in the archaeological workplace. . . . But perhaps these sorts of questions about the role and contribution of women are not interesting because they don't concern the content of the science—they don't make an "interesting difference" to the science as such. Well, in fact they do, and pretty directly. The defining preoccupations/the trajectory/the results of long-running research programs are invariably shaped by community conventions that determine what sorts of questions, range of explanatory models, factors, data, methods are important—respectably "core" as opposed to marginal. And sometimes gendered patterns of support and recognition are constitutive of these conventions. . . . Is it the case that these questions are uninteresting where primatology is concerned? The fact that they've proven fruitful in collateral fields establishes only that they may be worth pursuing in primatology too. . . .

Whether these are questions that interest you enough to invest in them, to pursue them, must have to do with your situated interests—intellectual, political, idiosyncratic, whatever leads you to do and/or to study primatology. The growing number of women, and more specifically, feminists, has made at least one significant difference to science studies and to various fields (like archaeology) with a strong tradition of critical self-consciousness: questions about the relevance of gender to science (in any number of senses) are now being raised and systematically, fruitfully addressed in a number of contexts where, twenty years ago, they could hardly be broached.

To: <teresopolis@majordomo.srv.ualberta.ca>
From: "Zuleyma Tang-Martinez" <szthalp@umslvma.umsl.edu>
Have women scientists made a difference? In some ways I think this question is too simplistic. The real issue to me is that all of us have a certain way of looking at the world and certain assumptions we make about how the world is ordered and I believe those assumptions are based on our different personal histories and experiences. Because of the different socialization of men and women in our societies, I do think that men and women (as a generalization) do tend to look at the world differently and this may well influence what questions they ask, what their expectations are, what methods they are likely to use, and how they interpret their data. But let me ask another question: "Have American Indian scientists (male or female) made a difference?" or "Have any Papuan scientists made a difference?"

I ask these questions because I believe that it is not sex per se that is the interesting question here. Rather, it is the impact that any underrepresented group of scientists could have if it is allowed access to science. Moreover, the experience of oppression and/or colonization should also be a significant parameter, since such individuals will have experiences that are drastically and dramatically different from those of scientists belonging to the dominant culture, which in science . . . has been primarily white, Western European in origin, and upper class with the middle class gaining access only recently. The reason we ask whether women have made a difference is precisely because women have been historically grossly underrepresented in science and also have experienced various forms of oppression, from subtle to heinous. AND because in some fields women now make up a significant minority.

. . . Usually when we ask about "women scientists" in our tradition we are implicitly asking about white women scientists mostly (but not exclusively) belonging to a certain socioeconomic class, etc. A second point is that there are many other factors that shape our worldview in addition to our sex. These include our race, class, sexual orientation, disabilities, etc.

So, have women (I'm assuming white European and North American women) made a difference? You bet! I think all of us, on some level, would admit that they have. Has having more women in science affected the men around them and how they look at the world? You bet! However, any underrepresented group that suddenly is allowed into the inner sanctum would have had an effect and the end result is that science will be better in the long run because we will have different perspectives looking at the same questions.

. . . I also believe that there is nothing wrong per se with the way men do science. The issue for me is that all the science, all the perspectives, all the hypotheses and theories, all the interpretations, etc. not be the product of only one very limited segment of our society—white, middle- to upper-class men. If women had a monopoly on science, I hope that I would be in the vanguard urging that men be allowed in as full participants and suggesting that men are likely to have their own and different perspectives that would enrich science. And as soon as we allowed a few men in, we could start asking: "Have men scientists made a difference?"

To: <teresopolis@majordomo.srv.ualberta.ca>
From: "Alison Wylie" <awylie@julian.uwo.ca>
Zuleyma raises a number of points that have been central to the work on questions about gender in archaeology. . . . Questions about the difference women have made in archaeology—or to put it more generally, questions about the difference made by gendered standpoints, gender roles, gendered institutional structures, gender symbolism, or some combination of all of these—have invariably led to questions about who exactly these women are, and about how gender is inflected by the politics/standpoints of class and nationality and, more recently, by race/ethnicity and sexuality. The central point is that it's a conceptually and empirically open question whether or not gender (on some specific dimension) "makes a difference" (in some specified sense) to the development/practice of a particular research tradition. By extension, the very process of addressing these questions should be expected to transform them; indeed that is what makes them interesting.

To: <teresopolis@majordomo.srv.ualberta.ca>
From: "Brian Noble" <brian.noble@utoronto.ca>

I just want to add a point about dinosaur research nowadays. Most of you may already know or suspect that dinosaur research has been an exceptionally male-dominated, middle-class activity over almost all of its history. . . . But in relation to gendering of dinosaurs and vertebrate paleontology, a quick survey of visual and textual presentation in technical [literature], museums, pop TV, and movie-making will show that up until the 1980s, dinosaurs seemed to have no explicit sex/gender. For the most part, dinos always seemed to be doing what for so long has been assigned to masculine behaviors: fighting, hunting, being big and fierce. . . . But in the 1980s, a paleontologist by the name of Jack Horner assigned the very first feminine technical name to a dinosaur, that is the genus "Maiasaura"—"Good Mother Lizard." After he found skeletons of a particular adult dinosaur next to nests of eggs in a Montana locality, . . . after that girl dinosaurs started showing up all over the place. . . . After 150 years of dinosaur research, when sex was only a footnote . . . dinosaur researchers are no longer just doing the death of dinosaurs, they are doing dinosaur reproduction. . . . dinosaurs now have more than gender (and more than stereotyped male gender)—they have sex, they have sexually marked bodies, they procreate, etc.

Corresponding with that shift, and I'm not suggesting causality here only correspondence, the numbers of women coming into the study of dinosaur vertebrate paleontology has been on the rise. Interestingly, while almost all of the men working in dino palaeo do a lot of heavy-duty fossil collecting out in the field, of the few women now working on dinosaurs, only a couple do much field collecting. . . . more women are turning to highly technical laboratory work for a lot of different reasons. . . .

To: <teresopolis@majordomo.srv.ualberta.ca>
From: "Shirley Strum" <sstrum@ucsd.edu>

I've been silent but listening to the interesting responses on gender and gendered lives. . . . Brian's contributions suggest the real meaning of situatedness of careers and trajectories of ideas and disciplines that was mentioned continually in Brazil. I am struck by the multiplicity of factors, including ones that do not usually get entered in the inventory, but have much influence. . . . I still wonder how we go from situated lives, ideas, careers, disciplines to understanding something more about the importance of actors and their interaction. Do we take 100,000 life histories and do an analysis? Bruno objects to ordering interesting factors but not everything is of equal value and since you can't change everything, if you want an interventionist as well as an analytical framework, you need to go beyond the rich thick description that improves your sense of the complexity to a way of understanding that complexity. Am I wrong? Is this just another scientistic prejudice of a sometimes scientist?

. . . I think the politics of who can/cannot do fieldwork has been less relevant in primate field studies than in the other disciplines being mentioned. This too has multiple origins. First, although primate field studies are hard work, the tasks are not really physically biased towards one or the other sex, although possibly being gendered in some cultural settings. Thus technological advances have not liberated women in this field. In fact I think it is funding more than technology that has played a role, and in this sense the politics of funding has a surprising impact, allowing marginalized women to continue where centralized men might not be able to. . . .

Finally, I have been fascinated by the couples that I know who do primates (and animal behavior and ecology). My fascination is that they react almost identically and not in a gendered fashion. Of the couples [I can think of], I feel comfortable saying that if I had to rank "factors" (things that make interesting differences), theoretical framework seems the overarching one and there is not a gendered take on this. In fact, I did a blind test on reading articles [by these couples] and it is impossible to tell who wrote it, if you don't know in advance, because the topics, the language, and the analysis are identical. Now this is the situatedness issue in extreme and in microcosm. It tells me that no matter how complex the context, "interesting differences" do get ordered.

To: <teresopolis@majordomo.srv.ualberta.ca>
From: "Naomi Quinn" <nquinn@pc301.econ.duke.edu>
Having listened in on the various energetic debates going on at Teresopolis-mail, and having meant all along to participate, really, I sat down the other day to read the entire set of comments so far on gender/sex. . . . I agree with many individual points that have been made, but generally what I find missing from the discussion is a consideration of the role of men in gendering academia. . . . There are obvious ways in which men designed academia to be an idealized men's world, from its reward of an uninterrupted career trajectory to its reward of qualities (at risk, which I knowingly assume, of being called an essentialist) that academic women I know are, like myself, often not as good at—like bold assertiveness, selfish competitiveness, and individual glory-seeking. What is most interesting to me about this is the way in which our attention, in the present discussion, as elsewhere, is so often deflected from questions about how to change this reward structure, by questions about how women can be more successful in it.

. . . I believe Bruno to be politically misguided (sorry Bruno) in putting gender on a par with adoption of sample methods or the introduction of cameras. I do not share Zuleyma's belief (sorry Zuleyma) that there is nothing wrong per se with the way men do science; on the contrary I think there is something wrong often enough to be cause for concern. I think that Shirley is wrong (sorry Shirley) in her assertion that there is not a gendered take on theoretical framework—this is exactly what is often enough wrong with the way men do science.

I believe the following: there are psychological reasons, some deeper than others, why some men scientists, as men, produce male-biased theories. Theories that are biased in one or both of two ways: that replicate the sociopolitical status quo in these men's own society, one that privileges the men themselves and seems completely natural to them (and even to many women); and/or relatedly, that treat women themselves as inherently inferior, a difference that also seems entirely natural to the theory makers themselves. (These biases are paralleled and exemplified in academic life itself, in men's failure to see anything amiss in the picture of themselves assuming all the high positions and honors, and in men's tacit assumption that women do not make major theoretical contributions.) As obvious as some of my observations may seem to some of you, this bias, like race bias, is so deep, powerful, and naturalized that it is difficult for those who hold it to even see it, let alone critique and correct for it.

To: <teresopolis@majordomo.srv.ualberta.ca>
From: "Allison Jolly" <ajolly@princeton.edu>
Naomi: you are so right about male-bias, and the other biases we don't even recognize. I am sure that I have a whole lot that I don't recognize or admit to myself, being from a generation that just absorbed them. . . .

To: <teresopolis@majordomo.srv.ualberta.ca>
From: "Zuleyma Tang-Martinez" <szthalp@umslvma.umsl.edu>
Whoa! Naomi! Interestingly I disagree with very little of what you have to say. . . . We are socialized as scientists. This does not mean that we can't reach a point where we question, critique, and even reject at least a good portion of this "scientific socialization." But this is dangerous business and we are likely to be devalued and our ideas dismissed because by rejecting and critiquing, we have become "no longer real scientists." This, of course, has been the fate of more than one woman (and I would add some men), who after being trained as mainstream scientists, go out on a limb and become "science studies" folks and critics.

To: <teresopolis@majordomo.srv.ualberta.ca>
From: "Naomi Quinn" <nquinn@pc301.econ.duke.edu>
Zuleyma . . . the trouble with the real world is that the features of Western science you and I admire and wish to keep are all tied up with male-biased practices, in the same package. (Maybe this was your point.) . . . I think there are lots of styles of doing science, they are all valuable, especially when worked together, maybe some are more congenial to women by and large and others to men, or maybe not. What I was pointing out was a whole layer of male politics on top of scientific practice. Valuing scientific practices that men do more of and devaluing those that women do more of would be an example of such politics.

To: <teresopolis@majordomo.srv.ualberta.ca>
From: "Brian Noble" <brian.noble@utoronto.ca>
In relation to all this [discussion of gender and science], I see primatology as a place where the density of feminization (however that may be formulated by enrollment stats, by practices, by forms and styles, by language used, by questions focussed on, by affinity for different animals, by films getting produced, etc.) is greater than in many other locales in science (N. American science certainly) and in public culture overall.

SECTION 5

Models of Science and Society

Why and how do ideas change in science? We ask the science studies experts.

Why and how do ideas change in science? From the start we imagined this to be a complex process, so we included multiple factors in the "framework" designed specifically to understand changing ideas about primate society. However, the "potted history" made it clear that this framework by itself was insufficient. Specialized analytic resources were needed in order to make sense of even a minimalist history. The workshop added breadth, depth, and complexity to the "facts," putting even greater demands on the analysis. No matter how "interested" we were, it was soon obvious that as primatologists we lacked the necessary training and expertise to do a reliable and robust analysis.

For this reason, we turned elsewhere, to "science studies," for analytic tools, models, and insights. Science studies is a new, diverse, heterogeneous field drawing from the philosophy, sociology, history, and anthropology of science. We selected participants using a few principles. First, because the most provocative claims about changing ideas of primates asserted that women primatologists made an important difference, we sought the convergence of feminist concerns and science concerns in the field of feminist science studies. This made sense as well since the first extensive examination of primate studies as a science was done by a feminist science analyst, Donna Haraway (*Primate Visions,* 1989). Haraway, a historian of science, was an obvious choice for the workshop. She had a wide-ranging involvement with feminist theory and with historical and cultural studies of biological and social sciences. In addition, *Primate Visions* had captured broad attention and generated much controversy, eliciting very different reactions from scientists and from science analysts. Evelyn Fox Keller, also a historian of science, was our other candidate to address gender and science. Her work, spanning several decades, includes an interest in twentieth-century biology, the use of language by scientists, as well as changing feminist concerns over recent decades. She is probably best known for her award-winning book on the life and work of Barbara McClintock (*A Feeling for the Organism,* 1983).

We were not only concerned with "gender" and science. Bruno Latour seemed specially suited to represent a different perspective. Latour, trained as a philosopher and an anthropologist, is a well-known pioneer in science studies. After field studies in Africa and California, he specialized in the analysis of scientists and engineers at work. But he also had a long history of involvement with ideas about primates both as an analyst (he was the token science analyst in the 1978 Wenner-Gren international symposium that Strum convened on "Baboon Field Studies: Models and Muddles") and as a theorist (collaborating with Strum on the topics of "society and technology" and on models of socialness). Latour had even spent several weeks watching baboons in the wild!

To round out the list of science analysts we turned our attention to another major issue: the role of the media in conveying and molding scientific ideas about primate society. This required someone interested in the interface between science and popular culture. Gregg Mitman, trained as a historian of science, had recently worked on "cinematic nature," the way images of the environment have been constructed in twentieth-century American culture through the medium of Hollywood films and documentaries. This appeared a unique and valuable entry point into our workshop issues. He had an unusual ally in Brian Noble, the workshop rapporteur. Noble's interest in cultural studies of nature is combined with many years of experience in natural science/public media projects, including several on dinosaurs. Dinosaurs are not primates, but the public's fascination with them is intense and complicated. Noble offered insights from the vantage point of another "sexy" creature.

The discussion about science and society also benefited from participants "officially selected" to represent another field or area of interest. Alison Wylie and Pamela Asquith are both trained science analysts. We imposed on them to wear two hats. Although their chapters appear elsewhere in the book, they are relevant to this section as well. Many of the other participants were also selected with "science" criteria in mind. Some have worked on the history of their own or other disciplines or have strong interests in questions of science (Rowell, Jolly, Hinde, Sussman, Quinn, Glickman). Ideas relevant to "models of science and society" will also be found in their chapters.

After the workshop it became obvious that the readers of this book would benefit from some background on science studies, both in order to situate the science analysts who have made contributions and as the context for the "science questions." We asked Charis Thompson Cussins to provide this introduction. Thompson Cussins was trained in one of the recently created integrated programs of science studies. As part of the new generation of science analysts, she is able to "look back" on the short history of the field of science studies.

The chapters in section 5 are tantalizing windows onto important issues of science and of our project. They are not historical reviews or grand syntheses. Each chapter can stand alone; together they create a stimulating potpourri. It might even be that the juxtaposition of different topics, perspectives, and issues in this section replicates the approach advocated in many of the chapters, where complexity and diversity are not simplified but converge and are entangled, "bending and warping both our attention" and the objects of our interest.

The section opens with Charis Thompson Cussin's "Primate Suspect: Some Varieties of Science Studies." This chapter provides a very clear and useful introduction to the field of science studies, detailing some of its history as well as the variety of approaches. Thompson Cussins explores the "problem of scientific knowledge" from the different disciplinary perspectives that comprise science studies. She begins with sociology, describing the succession of perspectives starting with Merton and including subsequent critiques and elaborations such as the sociology of scientific knowledge, ethnomethodology and social worlds theory, and actor-network theory. The history of science is considered next with a review of key issues and controversies. This is followed by a quick look at the philosophy of science and why it sometimes seems the odd man out in science studies. The "politics of science and technology" studies, the anthropology of science, and women's studies and feminist studies of science are also included in the survey. In this way, Thompson Cussins helps locate the points of view represented by workshop participants, providing a kind of "meta" context for the reading of individual chapters and e-mail exchanges. She also discusses what science studies has to offer as viewed from the perspective of a science analyst rather than from that of a scientist.

In her chapter "Morphing in the Order: Flexible Strategies, Feminist Science Studies, and Primate Revisions," Haraway clarifies, extends, and elaborates her vision of science and of primatology presented earlier. She provides a guide to some of the less obvious "meanings" embedded in *Primate Visions,* opening with a confessional about her motivation for writing about primatology over the last two decades. This is followed by a defense of the value of science studies. Throughout, Haraway suggests that the complex intersecting worlds of primate lives and scientific practice both require and signal a different kind of approach to historical contingency, trope and narrative thickness, and situated interactions of subjects and objects (like simians and people). Primatology is not just a trading zone between nature and culture but a zone of implosion where "the technical, mythic, organic, cultural, textual, oneiric (dream-like), political economic, and formal lines of force converge and tangle, bending and warping both our attention and the objects that enter the gravity well." Imploded zones are interesting because that is where knowledge-making projects are emerging, at stake, alive. Entanglements thus become part of making primates, making science, and making society. Ideas of bias are discarded for a richer, more complex way of viewing process. Haraway uses the controversy over what is the proper social unit of chimpanzees to convince us that old ways of looking at science are inadequate to describe and interpret scientific practice.

Keller begins, in "Women, Gender, and Science: Some Parallels between Primatology and Developmental Biology," by clarifying the various meanings of gender and of "feminism" and the problematic relationship between the two, particularly in the analysis of cultural products like science. She illustrates these difficulties drawing on her research in the history of developmental biology. Christianne Nüsslein-Volhard, who was recently honored with a Nobel Prize for her contributions to developmental biology, provides an interesting case study with which to explore several questions. What does the presence of women in science have to do with gender and science? And, less directly, what has been the indirect impact of second-wave feminism on improving the access of women to science and on changing our collective assumptions about gender? Keller concludes that all scientific change needs to be understood in the context of a complex network of interaction: social, technical, and cognitive. She argues strongly that "gender" must be situated as well— that feminist concerns may or may not articulate with what happens in a science. The chapter ends with some conclusions from the analysis of developmental biology that could be relevant to understanding the history of primatology including gender and science.

"Primate Relativity: Reflections of a Fellow Traveler" demonstrates Latour's innovative approach to science as well as his familiarity with the way we do and think about the study of primates. Latour first offers a way to conceptualize the flow of scientific knowledge by diagramming five horizons of practice that are simultaneously necessary to make science productive. Each horizon transforms knowledge; reality circulates because of the changes that occur, not despite them. He then considers the inadequacies of some traditional "metaphors" about science, illustrating the advantage of replacing the metaphor of "gaze" with that of "proposition," and the implication that this has for notions of "theory" and "method." Latour's reconfiguration continues with a discussion of articulated and inarticulate propositions. Articulated propositions are "interesting"; they allow new or better "access" to entities, offering phenomena another occasion to speak. Good science is when both the definition of the observed and of the observers are made more articulated and articulate by the work. In this sense, primatology is exemplary as a vast building site where propositions about animals, men, women, primates, scientists, female scientists, theory, data are reformatted into something more articulate. Latour's innovative framework is relevant to interpretations of what primatology is and how ideas about primates have changed.

Mitman's "Life in the Field: The Nature of Popular Culture in 1950s America" discusses the rise of popular natural history in the context of American culture in the 1950s. Mitman tracks the growing distinction between professional and popular natural history, focusing on naturalists-photographers who appropriated the study of wildlife. The formal science of the postwar generation adopted a remote analytic view. Nature was transformed into a laboratory in order to increase scientific control and reliability. By contrast, Disney's "True-Life Adventures" and the individuals who filmed them serve as the representatives of the science of popular natural history. Their natural history emphasized experiential bodily knowledge, both sensory and emotional. Through this they often crafted knowledge that surpassed the understanding of trained biologists. Mitman discusses the resonance between the portrayal of nature in these films and important political meanings in Cold War American culture. The chapter also considers two popular accounts of primates that appeared in the 1960s: Schaller's book *The Year of the Gorilla* and the *National Geographic* documentary "Miss Goodall and the Wild Chimpanzees." Mitman suggests that their widespread appeal came from their connection to the tradition of popular natural history (i.e., intimacy, emotionality, individuality, freedom) and not by virtue of their scientific status. This provides a cautionary tale about the complexity of the interaction between popular and professional natural history which argues against a simplistic view that scientific knowledge is disseminated to a passive lay culture of consumers. This particular history also suggests that a methodological approach centered on individual experience, patience, and emotion predated and was not confined to the practices of women in the field. Instead it was part of a "public" science of nature that came to intersect with the revival of naturalistic field studies and which had important political meanings in American culture of that period.

Noble brings "popular natural history" into the 1990s, specifically the images of primates, women, and science, in "Politics, Gender and Worldly Primatology: The Goodall-Fossey Nexus." In this chapter he argues that primatology over the last four decades has become an exceptionally public science. The most public of the primatologists have been Jane Goodall and Dian Fossey. The most public primate species: chimpanzees and gorillas. A variety of matters are interwoven in this nexus, ranging from specific primate species to the importance of the English language in the cultural production of "nature." Noble deconstructs films and articles by and about Jane Goodall and Dian Fossey. He argues that both have acted as major influences on the public perception of primatology. His analysis suggests that Goodall has been represented over time as becoming increasingly culturally and politically situated—from the "girl" alone in nature to the powerful woman organizing international conservation efforts—while Fossey has been represented as moving from a culturalized woman to a naturalized being. The multiplex mediations that Noble traces outline the weaving of cultural narratives which influence popular perceptions of science, gender, and primates. They also suggest an increasing collapse of nature and culture via politics, including the politics of ape lives and the condition of ape habitats during the current biodiversity crisis. Noble argues that at least in "public" mediations, women in primatology have come to dominate the intersection of nature and culture.

The workshop papers which became the chapters for this section only hint at the resources that science studies and feminist studies provided during the week of discussions. The e-mails at the end of this section give a flavor of the exchange. It is impossible to summarize, in a few sentences, what we learned about models of science and society that is relevant to "changing ideas about primate society." Instead, two new synthetic chapters, "Science Encounters" and "Gender Encounters," have been added to the end of this book as a way to discuss the most important lessons.

16 Primate Suspect: Some Varieties of Science Studies

Charis M. Thompson Cussins

The June 1996 Wenner-Gren workshop set out to assess what had changed in primatologists' understanding of primate societies, and to isolate some of the elements responsible for those changes. As the workshop title—"Changing Images of Primate Societies: The Role of Theory, Method, and Gender"—indicated, it was felt that theoretical and methodological changes within the field and gender (whether of primatologists themselves, or as reflecting concerns in their science and in the wider culture) were both important, perhaps interacting, factors. The meeting was an encounter between primatologists and scholars in the humanities and social sciences who write about the intersections between science and other parts of society. The primatologists represented several national traditions of primatology, different generations of primatologists, and a balance of male and female primatologists.

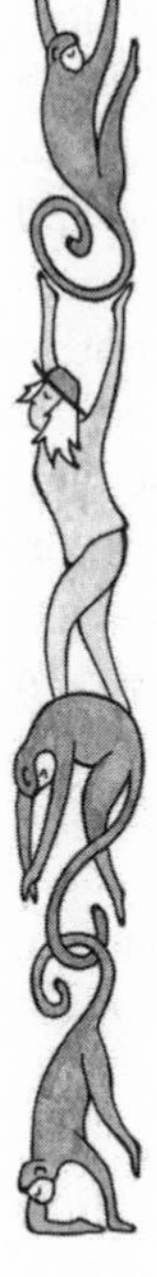

Shirley Strum and Linda Fedigan, the organizers of the workshop, are both known for the work they have done to relate scholarship in primatology to concerns among those who study science. Strum's collaboration over many years with the French sociologist/philosopher of science Bruno Latour, and Fedigan's active participation in feminist scholarship about science, stand as models of the kinds of interdisciplinary collaborations that are possible. In addition, both of the conference organizers and several other participants were featured in Donna Haraway's 1989 book *Primate Visions,* so that the feminist science studies community and primatologists have been in dialogue with one another for many years. These long-standing connections among the participants at the conference mean that the contributors to this volume are well placed to generate conversation reflecting the views of both those who practice science and those who study science. In an intersection not always noted for its fruitfulness, this is an important series of conversations (see Gross and Levitt 1994; Dickson 1997; Macilwain 1997a, 1997b; and Masood 1997).[1]

In this chapter I provide a road map to the field known as science and technology studies (sometimes simply called science studies). I present some of the motivations, controversies, and common explanatory aims of the field for nonspecialists. I draw distinctions among practitioners or schools of thought and pay particular attention to those present at the Wenner-Gren conference. I hope it goes without saying that the view presented here of the field is very much an idiosyncratic view, reflecting my background and tastes.[2]

Varieties of the Field of Science Studies

What motivates practitioners of science studies? The thing that unites much of science studies is an interest in scientific and technical knowledge: Who has it? What are its warrants? How is it acquired? What does it do? How does it order the world and the people in it? Likewise, much of science studies is united by archival and empirical methodologies for looking for answers to these questions. Topics such as visualization and the achievement of standardization in science and technology recur as loci for posing these questions. Many practitioners have an interest in both interpretive flexibility—showing that objects of scientific inquiry could have been captured and interrogated and described otherwise than they currently are—and the stability of scientific facts—contingency notwithstanding, facts tend to stick, to travel well through time and space, and to work. And many science studies folk have an abiding interest in the role that scientific truth plays as a resource for being able to talk authoritatively about, and draw boundaries between, nature, politics, and identity. Understanding any system of knowledge production is a fascinating and important intellectual project. The extent of, and changes in, science and technology at the end of the twentieth century suggests that scientific knowledge deserves particular attention at this time.

A useful taxonomic principle for science studies is one that divides the field up by "parental" discipline. For simplicity's sake, that is the main principle I shall use below. As with most classificatory systems, it doesn't perfectly capture the domain, and it is better thought of as a heuristic. The account below is biased to North American science studies, although science studies also flourishes in, among other places, the U.K., continental Europe, and India, and there are connections among them all. Science studies has a theoretically and institutionally important division over the question of whether it is an emerging discipline in its own right or whether it is important that it remain interdisciplinary.[3] This division affects how much emphasis is placed upon parental disciplines. Nonetheless, there are distinct parental disciplines in the academy from which the main strands of science studies have emerged, or where it currently thrives, and I thus find it a clear and helpful way to think of the field.[4]

Practitioners of science studies are frequently asked whether they are "for" or "against" science, and some people use this as a taxonomic principle of the field. I find it far more interesting to divide the field by intellectual content (although the two are not completely unrelated), so I shall not use "for" or "against" as an organizing principle. Because, however, the "for" or "against" question lurks beneath a good deal of misunderstanding, it is worth briefly addressing. I think that most people working in science studies would agree that one encounters scientists one admires more than others, findings that seem more compelling than others, and uses to which the sciences are put which seem more enlightened than others. In broad outlook, however, there is a spectrum from those who could be described as "apologists" for science or "hagiographers" of particular scientists, at one extreme, all the way to those who consider it their responsibility to be strongly critical of the sciences, at the other. In the majority of its variants, science studies is about understanding science and technology, not about analyzing individual scientists or about undermining science. There is a separate but related question as to whether those who study science consider it to be relevant whether or not scientists approve of their writings. Some practitioners care deeply for the opinions of scientists, especially those whose field they have studied. Others think that just as those in political science do not typically look to politicians to evaluate their work, or to give them the right to study that part of our shared culture, neither should science studies folk be beholden to those with an obvious interest in the outcome of their findings. It is the opinion of most in science studies that science should be studied like any large and well-funded part of society—be it religion, science, education, politics, sports, the courts, the media, the market—because outside scrutiny helps to keep the institutions in question honest.

Practitioners of science studies have an interest in the intimate connections between knowledge production and how we classify, order, and evaluate people, things, and nature. "Ordering, classifying, and evaluating people, things, and nature" is another way of describing "social order," which is the principal object of study of the social sciences. Most practitioners focus more on one or the other of these, stressing either knowledge production, or social order and the politics of knowledge. In fact, the disciplines where science studies flourishes can crudely be divided into those where the problem of knowledge is predominant, and those where social order and the politics of science and technology is predominant. I have divided the disciplinary classification offered below into the initial institutionally represented disciplines of science studies—history, sociology, and philosophy of science—grouped under the heading "The Problem of Scientific Knowledge," and other contributing disciplines grouped under the heading "The Politics of Science and Technology." It should be clear to the reader, however, that these are two sides of the same science studies coin, and that practitioners are united around the belief that politics and knowledge are interrelated (often they are coordinated in ways designed to keep them functioning as separate systems of authority) in interesting and important ways.

I start with, and pay the greatest attention to, the sociology of scientific knowledge, because it is the tradition within which I was predominantly educated and is thus the one I know best. It is also a central variant of science studies, echoing and incorporating concerns of practitioners in other areas of science studies. In the interests of brevity, I have not summarized case studies or findings of individual pieces of science studies scholarship, leaving that to the reader to pursue as interest dictates. I have referenced lightly, but, I hope, with care: I have listed only a few canonical works for each section, but in addition I have pointed to several review articles that the reader can turn to as he or she wishes for further references and original work in the relevant area.

I. The Problem of Scientific Knowledge.

SOCIOLOGY

a) Mertonianism American sociology of science has paradigmatically concerned itself with the relations between modern Western science and modern liberal democracies. Beginning with the work of Robert K. Merton, whose classic essays on the normative structure of science began to appear in the late 1930s, sociologists of science tried to show that the kinds of polities that are stable liberal democracies are also the kinds of societies that encourage and enable good science. It was argued that this was due to the similar normative structure of questioning, transparency, and disinterestedness possessed by both in their ideal forms. Free societies and good science are seen as going together on this view. A stable social order guarantees good science and vice versa. The totalitarian anti-intellectualism that derailed Nazi science can, according to Merton, be avoided by promoting the complex of moral attributes that ensures against abuses of both knowledge and power. These moral attributes, Merton believed, are the so-called KUDOS norms: communitarianism (or an attitude of sharing and openness), universalism (scientific knowledge should have the same content whether it is you or I or anyone else who possesses it), disinterestedness (science should be "pure" and not instrumental), and organized skepticism (an attitude of questioning should be directed to a focused area of empirical investigation). Collectively the norms function to determine the "normative structure of science." Scientists follow these norms, and expect sanctions of various kinds if they deviate from them. Merton proposed that these norms were absorbed by scientists during their scientific training and functioned thereafter as "internal policemen" keeping scientists from excessive competitiveness and interestedness (e.g., Merton 1942, 1973a, and 1973b). By following these norms, the scientific community would be protected from abuses of and by science. Similarly, fostering the equivalent norms in society at large would protect people from political excesses and abuses such as those evident in the totalitarian regimes of Europe.

Critiques of the foundations of Mertonian sociology of science began to appear in the U.K. from the mid-1970s, and spurred the development of what is called the sociology of scientific knowledge, or SSK. Scholars working more within the Mertonian paradigm of the sociology of science have concerned themselves with structural/institutional topics such as the funding of science, the professionalization of contemporary science, the way that peer review and citations work, access to the field by class, race, and gender, and so on. Although most science studies scholars within sociology have roots in criticisms of Mertonian sociology of science, a number of Mertonian scholars are actively contributing to science studies. These latter scholars are recognizable by their focus on the interactions between knowledge and the social structures of science they are analyzing. Examples of this work include studies of the interactions between scientific knowledge and the state, and studies of ways of delineating science from nonscience so as to be able to deploy it in institutions central to modern liberal democracies such as policy and the law. This scholarship is bringing to science studies a new dimension of the politics of scientific knowledge (e.g., Gieryn 1983; Gieryn and Figert 1990; Mukerji 1989).

b) Sociology of Scientific Knowledge (SSK) The sociology of scientific knowledge examines, in the words of Steven Shapin's title, *A Social History of Truth*. (Those interested in pursuing SSK further here should refer to Barnes and Edge 1982; Shapin 1984; Barnes 1985; Barnes and Shapin 1979; Shapin and Schaffer 1985; Shapin 1994. Steven Shapin is also a master of the analytical review essay: Shapin 1982, 1992. See also Mulkay 1979; Pinch 1985; Pinch and Bijker 1984; and Collins 1985). SSK developed partly as a critique of Mertonian sociology of science (for two early and oft-cited moves directly against Merton see, Mulkay 1976 and Barnes 1971). It shared with Merton the idea that ways of organizing society, and knowledge about the natural world, are deeply intertwined with each other even in modern scientific societies. But it departed from Merton by arguing that it wasn't the normative structure of science that mirrored society's mores, so much as the *technical content of science*. SSKers argued that scientific facts themselves are dependent upon equipment and methodologies that have a history, and that have to be agreed upon by scientists. Likewise, they pointed out that there was often considerable variability between practicing scientists as to how to follow agreed-upon methodologies. Similarly, scientific facts seemed to depend upon conventionally agreed-upon standards of dissemination, credibility, and proof. Thus, it seemed that there were things that were social about facts themselves. They began to raise questions such as the following, which are about both the technical content of science and social order: How do we know when we have a fact? Who is a reliable spokesperson about the natural world and when and why should we, or do we, believe him or her? What does it take to stabilize a scientific fact such that many different people in different places and in differing conditions can understand the same fact and use the fact in reliable and reproducible ways?

SSK built up its analytic tools by borrowing from an eclectic mix of disciplines. From the similarly named "sociology of knowledge" it took its interest in the social and cultural conditions for the construction and maintenance of facts, knowledge, and belief. SSK extended this inquiry into the systematic study of actual episodes of the production of *scientific* knowledge, previously considered out of bounds for sociological explanation (e.g., Fleck 1935; Mannheim 1936). Extending the sociology of knowledge into scientific knowledge was a move that was sociologically ambitious on two fronts, for it took on the two bastions of resistance to social explanation and insisted on "sociologizing" them. The first of these was scientific truth. Truth is supposed to be one, and to be the underlying reality on which social variants are played out, but SSKers insisted on the social and constructed nature of scientific truth, credibility, and authority, and denied truth a transcendent and ahistorical and foundational character. The second source of sociological ambition was to replace the great disembodied individual scientific minds of conventional historical narratives of science with collective accounts of scientific practice. In the accounts of SSK, great scientific minds emerge as one major product of scientific culture, rather than being the prerequisites for scientific discovery. This problematizing rather than simple valorizing of charismatic individuals in science is a shared theme among many practitioners of science studies, and is one of several strands that lend the field an anti-elitist and demystifying politics. Sociologists of scientific knowledge mostly have backgrounds in the natural sciences rather than in the social sciences or humanities, and became "sociologists" by the nature of their program of research rather than by their training. Sociology is thus less the "parent" discipline of SSK, and more of an adopted discipline.

From anthropology SSK borrowed and reinvigorated (and repatriated) the idea that social and natural orders are solved together (e.g., Douglas 1966, 1970). SSKers also followed anthropology's habit of looking for coherence in systems of (often primitive) belief without presupposing that those belief systems correspond to an independently existing standard of objective truth. SSKers used anthropology's methodology because anthropologists had worked out ways of examining the beliefs about the nature of different cultures, without presupposing that those beliefs simply reflected the state of nature. This anthropological approach allowed them to analyze scientific practice and discourse to see where and when and how facts about nature were established, without presupposing that modern science simply reflects true nature. This was important because it was the relation between scientific truth and nature that they were trying to explore. To explain what was radical about this SSK move, consider the following: "It is reasonable for Group X to do a rain dance to get it to rain." The function of rain dances and rain in Group X's life might be such as to make it rational to continue to link the rain dance and the rain, regardless of whether dancing is causally efficacious in the production of rain. Taking this same attitude to examining one's own scientifically informed belief systems is much harder, however, just because we do take for granted all sorts of external warrants such as causal efficaciousness. We think that if we do something to make it rain, it is only rational to continue doing it if indeed the method (sometimes) causes it to rain. At the end of the twentieth century, the regularities of the natural world provide us with the warrants for many of our beliefs, even if only implicitly, and even if the relevant science is not known by everybody. When scientific findings are proved wrong, we tend to give explanations for why one method or another didn't work, such as poor methodology, false starting assumptions, or faulty data analysis, and these explanations somewhat resemble anthropological explanations. But if we come upon a method of making it rain that (sometimes) works, such as seeding clouds, we think it works because we had the right theory that correctly represented the laws and workings of the natural world. There seems to be nothing more to be said than to appeal to the match between our theory and how the world really works. SSK took on the challenge to give anthropological-type explanations for true facts about nature within a scientific worldview.

SSK also drew on ideas from the later writings of philosopher Ludwig Wittgenstein to suggest that formal logical explanations have to come to an end, or hit "bedrock," somewhere. There is an infinite regress of ways to question knowledge. For example, if I say that the double helix contains cytosine, you can ask how I know. I can point to laboratory techniques for isolating and labeling components of biological samples. You can then ask why I believe those answers. I can reply by citing additional theories that implicate the same techniques, or by giving a history of the development of the techniques. You can then ask why I believe those elements of my story, and so on, ad infinitum. "Ways of going on" or "forms of life" including scientific practice, have boundaries shaped by shared and largely unquestioned assumptions, without which meaning and progress would be impossible because one would be compelled to question and doubt everything. SSKers took up Wittgenstein's antifoundationalism, agreeing that it is always possible to keep questioning and doubting any piece of knowledge. This led them to abandon the search for a scientific methodology that would provide a foundational level for scientific knowledge that was beyond doubt. They also incorporated Wittgenstein's alternative to a logical or methodological foundation for knowledge, namely, his turn to forms of life.

This antifoundationalism was combined with arguments found in various theoretical and empirical works in the history and philosophy of science. Arguments had been made in these disciplines for indeterminacy or underdetermination, and thus interpretive flexibility, as regards facts about the natural world (such as the historical studies of science by Thomas Kuhn [1970, 1977], or the philosophical writings of Willard V. O. Quine [1953, 1969]). Indeterminacy was evident in many famous examples, such as the claim that there is no way of deciding definitively which things are and which things are not heaps of sand. There is no definite number of grains of sand which is the dividing line between too few and just enough grains to constitute a heap of sand. For large numbers of grains of sand piled together, we know we have a heap, and for very small numbers we know that it would be incorrect to call the handful of grains a heap, but there is a large gray, or indeterminate, range in between.

Underdetermination pointed to the fact that there are different ways of perceiving and understanding the "same" natural phenomena, and nothing in nature itself that makes it the case that there is only one true classificatory scheme. For example, I could call all raised ground rising higher than one thousand meters above sea level a mountain, but there is no reason in nature why I should separate hills and mountains in this way. If we called those peaks between 700 and 2,500 meters mountains, it would be a different way of classifying nature, but it would be no less true to nature. Our choices of theories are underdetermined by nature, then. Combining indeterminacy and underdetermination with the anthropological view described above, one gets a methodological principle of relativism. Given that more than one true way of describing the world is possible, no one classificatory scheme, theory, or set of facts should be treated as being uniquely true to nature. The analyst of science should thus not treat explanations of accepted scientific knowledge any differently from the means used to describe and account for belief in unproven or alternative claims to knowledge. The reasons for the beliefs of expert Western scientists should be explained using the same resources as those used to describe prescientific or non-Western thought. This methodological principle became known as the principle of "symmetry," and was formalized in the so-called "Strong Programme" of the Edinburgh School of Science Studies (see Bloor 1991).[5]

For those who follow the realism/relativism debates, this methodological relativism enjoins practitioners to examine scientific belief systems without assuming that they have a special status because they are true. Instead, one examines how scientists constitute what is true and how they make truth work for them. This kind of relativism is thus agnostic as regards realism about the world because it has nothing to say about whether or not there is a real world that is systematically revealed and represented by our best scientific theories. An SSKer *could* believe that scientific knowledge was "socially constructed" in the sense of "made up to suit purposes other than the pursuit of truth," but it is much more common for SSKers to insist that scientific knowledge is constructed in the sense of requiring work and instruments and institutions and conventions to discover and sustain it. This second sense of constructed has little or nothing to do with whether or not one believes in an independently existing reality.

The forms of life and sociology of knowledge approaches of SSK combined to emphasize the importance of tacit knowledge and what scientists actually do over the kinds of explicit or formal renderings of the scientific method often assumed by historians of ideas and philosophers of science. This turn from studying science as the progressive accumulation of truths about the natural world divined by great minds, to the study of science and truth in the making, is a turn that unites many different strands of science studies.[6] SSKers shared with Merton the sociological perception that it was not primarily great minds that made great science (see Weber 1948). But they found little evidence in their empirical studies of scientific practice to support the idea that Mertonian norms were responsible for good science; indeed, for every norm for which they found evidence, such as openness, they found counternorms, such as secrecy. Instead they looked to the practice of science, and found technical standards and shared bodies of knowledge making up scientists' ways of going on. This was much as the historian and philosopher of science Thomas Kuhn (1970) had characterized periods of "normal science."

Given the SSK desire empirically to investigate rather than presuppose what makes something true, it makes a great deal of sense that SSK would concentrate on areas where the division between true and false, and between good and bad science, is not yet firm. Early-modern science is a favorite focus because it enables historians of an SSK bent to explore the rise of modern science as a system of the production of truth about the natural world. Likewise, the great scientific standardizations of the nineteenth century have been well studied (e.g., Dear 1985; Shapin 1994; Shapin and Schaffer 1985). Another common site, and for similar reasons, is the study of scientific controversy, where the truth is not yet broadly accepted as falling on one side or the other of the controversy. In social constructionist technology studies, which overlap to some extent with SSK, the predominant focus is on the contingency of the ways in which one or another technology design comes to be accepted (e.g., Bijker et al. 1987). Throughout all this work, the themes of credibility—who has it and how one gets it—, of trust—what must be unquestioned for any system of truth to be sustained—, and the authority of science—the ability of science to produce assent to its claims—, are crucial. These are the kinds of theoretical concepts that have been used in the place of appeals to transcendent truth in SSK to account for the success of modern experimental science.

c) Ethnomethodology and Social Worlds Theory Ethnomethodology is a branch of sociology that looks at the constitution of normative order.[7] Drawing on the writings of social theorists like Alfred Schutz and Harold Garfinkel, ethnomethodologists concern themselves with the phenomenological "life-world" of lived experience, and eschew a priori theoretical accounts of concepts like truth and rationality (Garfinkel 1967; Schultz 1967; Schultz and Luckman 1973). Classical ethnomethodological accounts have looked at legal reasoning, the performance of gender, and scientific practice (Garfinkel et al. 1981). Ethnomethodologists were dissatisfied with structuralist sociological accounts of action, which seemed to reduce all meaningful behavior to being the product of various structural social properties, such as one's gender, class, race, profession, and so on. These accounts seemed to render life as experienced by the person living it as merely epiphenomenal; to make people "judgmental dopes," as Garfinkel expressed it. Ethnomethodologists tried instead to find the things that structural sociological explanations picked up on—knowledge of and conformity with appropriate ways of behaving and assumptions about the nature of the world—enacted in the here and now, produced and reproduced in lived experience, rather than existing as a set of abstract rules of behavior and bodies of knowledge. Michael Lynch is the most prominent ethnomethodological scholar in science studies, and he has shown that science relies on the development and sustaining of specific immanent (present in the situation at hand) ways of experiencing and representing the world (e.g., Lynch 1985a, 1985b; Lynch and Woolgar 1990). Scientific rationality, like other forms of rationality, requires that scientists share a basic worldview (or at least repair disagreements where they occur), and that they trust and take for granted most of what each other considers to be appropriate ways of behaving, and rational ways of explaining and reconstructing events. Far from skepticism being the general attitude of scientific practice, the local building and reproduction of shared and unquestioned rationalities, with extremely focused areas of questioning, is the norm in science.

Social worlds theory is another significant sociological approach to the study of science, and it has some affiliations with ethnomethodology. In social worlds theory, analysts pick and delineate their site of study through following those people and things to which the group being studied has shared commitments. It draws on the tradition known as "symbolic interactionism," and has moved easily into studying scientific practice, inheriting an already well-established emphasis on institutions and professions (see Strauss 1991; Clark 1991; Knorr-Cetina 1992). In the scientific realm, a "social world" in this sense would include those people, places, objects, and the knowledge, complex of values, economics, and so on, that are found in fact to surround and create the piece of science of interest. Thus, if one wished to study reproductive medicine, as Adele Clarke, a social worlds theorist, has done, one would analyze those who practiced reproductive medicine, those they interact with, and the flow of people, instruments, capital, and theoretical and moral ideas making up the evolving field. Larger conceptual connections which link different social worlds—"arenas"—are also analyzed. Social worlds theory contributes to what has become another key concept of science studies, namely, the inseparability of moral, economic, and knowledge economies.

d) Actor-Network Theory Actor-network theory is a brand of science studies with which primatologists (or at least those who were participants at the Wenner-Gren symposium) will be familiar, as it is associated with the work of Bruno Latour and his colleagues at the Centre de Sociologie de l'Innovation at the École des Mines de Paris (e.g., Latour 1987, 1988, 1993, 1996c; Latour and Woolgar 1986; Callon 1986; also see Law 1986, 1991).[8] I have placed actor-network theory (ANT) under sociology here because of its institutional classification in both Paris and the U.S. as sociology, but ANT is as much philosophy as it is sociology, and additionally borrows from (as well as contributes to) anthropology and the history of science. I will restrict myself to making a few general points about ANT here as the reader has the pleasure of consulting Bruno Latour's contribution to this volume.

Actor-network theory is the most charismatic branch of science studies. It is infectiously celebratory about science and technology, portraying modern science as a hitherto unprecedented source of innovation not just for understanding the natural world, but as a means of connecting people and things over space and time. The appeal of ANT stems at least in part from this naive enthusiasm for the modern world, as accustomed as we are to technophobic portrayals of the modern condition. Alienation and deskilling, discipline and surveillance, colonialism and mass genocide, are all associated with the rise of modern science, and Continental European scholars, even more than their North American counterparts, are known for sociological and historical accounts that stress these elements. ANT differs from the Continental "critical tradition" in wanting to dissociate the possibility of critical understanding of science of technology from the necessity of being antiscience, or, as Bruno Latour calls it, antimodern.

ANT is remarkable for its theoretical boldness. I described a "principle of symmetry" above. That principle urged analysts to explain episodes of scientific discovery with the same anthropological approach used to explain discredited or superseded science. Scientists and their publics do not know "ahead of time" that they have truth on their side, and so to give the scholar of science the wisdom of hindsight would already be to misrepresent scientific practice. ANT theorists shared the interest of SSKers in making empirically visible and interesting all the questions about truth and objectivity that arise in the practice of science. ANT, then, also emphasizes "science in the making" rather than "science made." To distinguish between the two, Latour has advocated referring to the former as "research," and saving the word "science" for more formal renderings of science. ANT added a second dimension of symmetry to that proposed by SSK. They argued that analysts ought to treat "society" in the same conceptual way as "nature." SSKers rejected the appeal to transcendent notions of truth and objectivity, but did not similarly problematize using "social" explanatory categories such as trust, credibility, and authority. Likewise, ANTers argued that SSKers tended to see (contested) truths about the natural world as reflecting (battles over) political status quo. ANT maintained that it was just as problematic to take for granted an independently preexisting social or political order as it was to make appeal to transcendent truth or an independently preexisting nature. Instead, ANT theorists argued through case studies of scientific practice that it is networks of people and things together that make both new knowledge and technologies, and the social categories of identity and society. Pure social or natural facts are, according to ANT, end products of processes of separation and purification rather than preexisting distinct explanatory repertoires. The process whereby facts and technologies become stabilized and generally taken as true has been labeled "blackboxing." Scientific experiments yield results that can be expressed as facts within the appropriate theory. When these facts withstand tests and confirmation attempts, and are promulgated as truths in scientific journals or informal lore, they can be said to have become blackboxed.[9] The facts stand as a self-contained package, and are valid without having to make reference to the conditions under which they were produced. Truth is routinely precipitated out from science-in-the-making in this manner, and so should not be thought of as input to the process of research.

ANT points out that no scientist studies "pure" nature. Scientists and their equipment always intervene and interact in one way or another with the objects of study. Scientists set up scenarios or develop observational and recording skills against which the objects of study (whether nematodes or baboons or muons) manifest properties and behaviors. Dispelling the myth of pure nature adds credibility to ANT's refusal to presuppose clear-cut separations between the social and the natural or to provide explanations which reduce the social to the natural or vice versa. This has the immediate payoff that it enables people to talk to each other across disciplinary and conventional cognitive boundaries. But it is the generativity that this allows that gives the theory its excitement. ANT compellingly portrays science and technology as processes where genuinely new assemblages and alignments of things and people are brought about. Scientists allow the natural world to display properties which it captures. Once things are named, classified, and accorded certain properties or behaviors, they are well on the way to becoming "things unto themselves," or objective scientific facts (Latour 1994).[10] In ANT, then, the hybridity of social and natural things is the *condition* of producing objective facts about the natural world.

ANT has introduced a number of explanatory concepts that now have broad circulation. Among these are centers of calculation, obligatory passage points, enrollment, delegation, and networks. Centers of calculation refer to places or institutions at which particular kinds of facts are collected and disseminated, such as the CDC for infectious diseases, or SLAC and CERN for work in experimental physics. If certain kinds of facts have to pass through a particular site to become stable and standardized facts, that site is referred to as an "obligatory passage point." For example, it would not be unreasonable to argue that vectors or organisms involved in infectious diseases must pass through the CDC these days to be classified, standardized, and controlled. The CDC is thus an obligatory passage point for infectious diseases. "Enrollment" is used to describe the process whereby something or someone is allied to a particular view of the world, and comes to act or speak as evidence of that position. ANTers grant agency to both humans and nonhumans, so microbes are enrolled by Pasteur just as much as political and scientific allies are. "Delegation" describes the process whereby an assemblage or link in a network holds in place particular social relations, and a particular representation of part of the world. If I go to the South Pacific and bring back specimens from some remote island to Paris, I have connected a lot of people, places, and things together. I have displaced a tiny bit of one place (the remote island) to another place (Paris). Paris has become a "center of calculation," and if it is really successful, it will exert sufficient force on understanding specimens from the South Pacific that it will be an "obligatory passage point" for understanding future such specimens.

The "networks" of actor-network theory are much disputed: What are they made out of, and what work do they do? My understanding is that they are not supposed to be anything independently of the connections made between things, facts, and people that make up the scientific network being studied. The network metaphor gets its strength from its theoretical neutrality: it allows the analyst to consider each thing that goes into science in the making, without ruling out certain kinds of things ahead of time. The network metaphor is excellent for highlighting changes in scale—from local to global, as in the knowledge produced when the bit of the South Pacific island is brought to Paris; from weak to powerful, as when a physician at the CDC is able to classify and control an agent of infectious disease; and so on. These explanatory freedoms have proven fruitful for science studies because they seem to be a better empirical account of what actually happens in science in the making.

HISTORY OF SCIENCE

Recent developments in the historiography (theory of writing history) of science have meant that there is considerable affinity between the wider community of science studies and certain historians of science. Indeed, the history of science has provided the theoretical innovation for much of science studies. For example, a number of the founding contributors to the sociology of scientific knowledge are historians of science (Shapin 1992). The areas of the history of science that have received the most science studies attention include early modern European science, nineteenth-century field sciences, the history of modern biology, the history of medicine, twentieth-century physics, and the history of technology.[11] It is beyond my competence to comment extensively, so I shall just mention a few key aspects of contemporary science studies–affiliated history of science here. It is also worth noting that science studies has inherited from the history of science its widespread view that the rise of modern Western science is a phenomenon of special historical interest, and that the development of experimental science reflected and in turn shaped the societies where it developed (e.g., Zilsel 1942; Needham 1953; Merton 1970).

Historians of science who are also practitioners of science studies have reacted against what is called "Whig history," namely, the telling of history from some later point of view or later stage of development of society. This kind of historical narrative is accused of being "presentist," and appealing to events and outcomes, as well as to concepts and understandings, that were simply not available at the time to the actors being studied. This criticism has been raised against all kinds of historical narrative, but has a particular form when used against narratives in the history of science. For example, writing the history of genetics in a presentist way might involve giving an account of steps leading inexorably to modern-day genetics, and might involve interpreting earlier findings such as Mendelian inheritance as evidence of the presence of genes. By contrast to this presentism, a historicist approach that tried to use *actors'* categories would not use the modern-day concept of the gene in describing Mendel's work. Neither would it assume that what was important about Mendel's work was whatever aspects of it we can now translate into genetic theory. While a very presentist account might be a good way of writing other kinds of narratives about Mendel's science—say, high school science textbooks —it would be a bad way of writing history because it would not reveal what it was like to live in Mendel's time or think in Mendel's terms. An additional point connects very closely to the first principle of symmetry described above. If what the historian of science is trying to explain is the history of the acquisition of certified knowledge about the natural world, then to interpret the history of science in a presentist manner would be to render invisible the very processes of interest whereby the science in question changed (see Mitman, this volume), and/or became certified and accepted as true. It would also make the historical trajectory of science seem a lot more progressivist and deterministic than might in fact be the case, masking the contingencies that led up to the present divisions and theories of the sciences.

Some have argued that historical accounts of science predating science studies can be sorted into "internalist" and "externalist" narratives. Internalist historical narratives tell the history of ideas or intellectual history (sometimes called "talking heads history," for the excessive interest in the findings of a few "giants" like Newton and Einstein, whose ideas are portrayed as having had the power to change the history—idealism). Externalist histories of science relate such things as problem choice and interpretations of findings in science to events going on in social and political culture at the time and place of discovery. An externalist account of Newton's science would explain why he chose the problems he chose and why he brought to bear the resources that he did, in terms of the political context at the time, reading the science off the political order. On this picture, materialist factors, such as the economy, rather than ideas, become the wellsprings of change.

Embracing science studies has meant for some historians of science rejecting both internalism and externalism. The turn against disembodied intellectual history has included an emphasis on all the things that internalist accounts seemed to leave out, such as the actual practice of science, the instrumental and other material culture of science, and the various literary cultures of science. An additional part of the revolt against internalist accounts of science involves contesting how to portray the "individual geniuses" of scientific folklore. The move has been away from considering genius to be ineffably located in certain individuals, revealed in psychobiographic precursors to the intellectual feats later manifested, and turning instead to the question of how any individual ever accretes enough resources and authority to get authorial credit for breakthroughs. Where individual scientists have been studied in great detail, for example, in the study of Barbara McClintock by Evelyn Fox Keller, of Robert Boyle by Steven Shapin, or of Copernicus by Robert Westman, the tendency has been to show how cultural factors—here, gender, class, and humanist rhetoric, respectively—work to constitute the individual as a site of knowledge (see also Keller, this volume). Externalism has been eschewed for suggesting that the relations between science and culture/politics are relations of reflection, with the science simply piggybacking on politics. Inspired by Kuhn's attention to the technical norms governing the content of science, people in science studies have wanted to tell the history of science in ways that show interconnections between science and culture, without one presupposing the other. This attempt to give a fuller and more accurate, if more heteroclite and iconoclastic, portrayal of the history of science has much in common with the attempts of others in science studies to understand science as practice and culture (Pickering 1992).

Not all historians of science who consider their work to be "post-Kuhnian,"—that is to say, those who take scientific practice and scientific content seriously, and are historicist rather than Whiggish in their historiography—are practitioners of science studies. Many historians of science embrace some of these theoretical orientations, and yet remain primarily affiliated to history, rather than to science studies. In addition, the characteristic analytic interest of practitioners of science studies in knowledge, truth, and objectivity may be absent in historians of science who otherwise share many science studies tendencies (Biagioli 1990; Findlen 1991).

PHILOSOPHY OF SCIENCE

I will not comment in depth on the philosophy of science, but it is worth noting first of all that most practitioners of science studies, by virtue of their interest in questions of knowledge, occupy the philosophical wing of their home disciplines. Likewise, as noted above for the sociology of scientific knowledge, philosophical work has been important in the foundations of other branches of science studies for providing insights into such things as underdetermination, the theory-ladenness of science, and antifoundationalism.

Philosophy of science shares with the rest of science studies the view that science is a good place to look if one wishes to raise and answer questions about knowledge, truth, and objectivity. Naturalist philosophers of science in the science studies community, such as Philip Kitcher, differ somewhat from most other strands of science studies. Oddly enough, this means that the philosophy of science often sits somewhat awkwardly with the rest of science studies.[12] Instead of problematizing truth and objectivity, naturalist philosophers take the notions themselves for granted, and instead problematize *access* to them. This is a move to put truth itself out of the reach of epistemology (how we know what we know) and experience. This is in contradistinction to many in science studies who have wanted to blur the boundary between what kinds of things make up the world (metaphysics) and the ways in which we can know the world (epistemology), and related binaries: between experience and reality; between how one knows, and the objects of one's knowledge; between the objective and the subjective; between the social and the natural; between content and context. This difference means that naturalist philosophers of science tend to ask about things which pose prima facie tensions for a position that presupposes objective truth, such as how to account for progress in science. A typical question might be: If some scientific fact is transcendentally true, how do we account for the fact that the scientific truths of one generation are constantly being superseded by later discoveries? Other questions are things such as: What does it mean to get closer to the truth, if truth is not relative? When properties were ascribed to phlogiston, what, if anything, was really known, and how does it relate to our current understanding of oxygen?

Some philosophers of science have resisted with ingenuity the idea of multiple or contingent ways of describing the world as suggested by other contingents of science studies. The very things that might be taken to point to a constructed and contingent (but not random or "purely social") understanding of scientific knowledge are made commensurable with a unique and true way of describing the world. Other philosophers, including several feminist epistemologists, have embraced multiplicity. For example, in the aftermath of the wide dissemination of Kuhn's *The Structure of Scientific Revolutions,* the so-called "Kuhn wars" erupted among philosophers of science over Kuhn's thesis about periods of "normal science" punctuated by "paradigm shifts." Some philosophers maintained that Kuhn's thesis should be read as support for the idea that there are incommensurable worldviews from one paradigm to the next, while the philosophers associated with resisting multiple worlds attempted to patch together epistemological and metaphysical continuity between paradigms. It is an important dynamic of the division in science studies between those who argue for multiplicity, versus those who argue for the unity of the world, that each side tends to think itself the "realist" side. Those pushing multiplicity, sometimes called "constructionists," think their approaches account for the real world much better, and that their accounts identify adequate resources for constructing and maintaining the relevant normative notions (truth, objectivity, etc.). Other philosophers of science, sometimes called "realists," think that relativist tendencies such as an openness to incommensurability and multiple worlds disqualify the other side from the claim to being realist at all. These fundamental differences are augmented by lesser differences in style and orientation (see Hernstein-Smith 1997).[13]

Among philosophers who do not presuppose the fundamental unity of science, and who have been influential in the wider interdisciplinary science studies community, are philosophers such as Nelson Goodman (1955), who argues for underdetermination (that there is always more than one way adequately to account for empirical observations), and Ian Hacking, who has shown the constructed nature of naming terms in science (e.g., Cartwright 1983; Dupre 1993; Galison and Stump 1995; Hacking 1983; Goodman 1955; Rorty 1979). Similarly, feminist philosophers of science like Helen Longino, Alison Wylie, and Sandra Harding, and feminist philosophers of mind and language like Elizabeth Lloyd, Patricia Hill Collins, and Lorraine Code, have argued that knowledge is social, or that one's subject position or the context of one's life and work matters to what is known and how (e.g., Baier 1985; Butler 1990, 1993; Code 1991; Collins 1990; Harding 1986, 1991; Longino 1989; Wylie 1992; see also Wylie, this volume). There are also several philosophers trained in the Continental philosophical tradition who have begun to combine insights of that tradition, including antimodernist critiques of technology and transcendental idealist metaphysics, with findings in empirical science studies (see Feenberg and Hannay 1995). And finally, there are a number of science studies–type philosophers from a variety of backgrounds who have begun to produce empirical and theoretical studies of formal devices and languages for organizing or representing knowledge. These latter include studies of a variety of aspects of information technologies, and studies of such devices as bureaucratic forms (see Berg 1997; Cussins 1992; Smith 1996; Star 1990).

II. The Politics of Science and Technology

THE POLITICS OF SCIENCE, AND SCIENCE AND THE LAW

The study of the politics of science within science studies is gaining ground. Sheila Jasanoff, a lawyer and a contributor to national and international science policy, has pioneered this trend. She has studied the interactions between science and the law (two of the most important institutions in contemporary liberal democracies) in an international comparative framework (major books: Jasanoff 1990, 1994, 1995; see also Jasanoff 1996a, 1996b). Her work shares many sensibilities with SSK and with ANT (see above), such as a constructionist framework that refuses to take truth for granted and to separate out politics and knowledge, and a belief that stabilization of facts and standardization of scientific practice are achievements that need to be explained. While most of the studies in SSK and ANT focus either on the lab or on relatively discrete episodes of science, Jasanoff has taken science studies into the broader political realm, looking at different national styles of accommodating, promoting, and using science for governance. Characteristic topics of inquiry of others working in science studies on the politics of science include the relations between science and the military; lay participation in the production of scientific knowledge and the public understanding of science; the interactions between indigenous and scientific knowledge; and the affect of global regimes on national science institutions (Dennis 1994; Irwin and Wynne 1996; Epstein 1996; Watson-Verran and Turnbull 1995; Jasanoff 1996a, 1996c).

Political theorists like Yaron Ezrahi have equated the theatrics of liberal democracy with science, arguing that democracies need accountability and transparency, and that the apparently universal certainties promised by the modern sciences perform this function in those polities (see, e.g., Ezrahi 1990; Foucault 1979). In social theory this view is counterbalanced by the view, associated with Michel Foucault, that the rise of modern science has meant the increasing disciplining and surveillance of citizens. Whether democracy and science produce transparency or surveillance, or, as many think, some combination of the two, the fact that science and democracy are so intertwined underwrites Jasanoff's program of inquiry into the intersection of politics and science. Unlike the tradition of political science, however, where science is not problematized (if it is considered at all), Jasanoff started with a constructivist understanding of scientific practice, and then asked how science manages nonetheless to be so important in the functioning of the modern state. Jasanoff has argued that in modern liberal democracies science and politics "co-produce" each other, in ways that are specific to and characteristic of the political culture of the place in question.

Jasanoff's work is responding to historically specific aspects of late twentieth-century liberal democracies. One can think of the circumstances to which she is responsive as a dual crisis. First, there are reasons to think that the common good is no longer aggregable, and so no longer knowable by political representatives, and thus that a single centralized vision of the common good can no longer form the basis of social policy. Instead society seems to be splintered into a multitude of special interest groups each with some kind of voice, but with incommensurable visions of how society should be organized. Relatedly, a lot of recent science, especially in the fields of biomedicine (think of "mad cow" disease) and the environment (think of global warming), seems to be uncertain, incomplete, and complex. Social movements focused around health issues such as Gulf War Syndrome and Alzheimer's, and scores of other conditions, put pressure on the idea that the state should have sole responsibility for setting biomedical policy. Likewise, local environmental movements, and global environmental regimes, challenge the ability of a nation-states to recognize and sponsor "good" environmental science. Jasanoff's work has taken this double crisis seriously and begun to chart out and suggest responses to the new connections being forged between science and competing forms of legitimacy.

ANTHROPOLOGY OF SCIENCE

The cultural and social anthropology of science connects closely with much recent social and cultural anthropology which has questioned the epistemology, ethics, and power relations implicit in the distinctions between the "Us" and "Other" of classical anthropology. A logical element of that breakdown has been a focus of anthropological attention by scholars from "developed" nations on their own institutions of truth and culture production, and a focus of attention by scholars from "developing" nations on truth and culture production in their own "traditional" or "indigenous" cultures. The repatriation (in both directions) of anthropology has been felt in part as a challenge to anthropologists to take seriously the anthropology of Western science, treating it analytically no differently from other cultural belief systems. In this regard, the anthropological branch of science studies has both expressed a trend in the field of anthropology and inherited a way of looking at systems of beliefs about the natural world. In a sense, then, anthropology can be seen as coming to science studies very much from within its own tradition, and yet it has much in common with other branches of science studies. Certain prominent institutional sites of science and technology studies such as the program at MIT, under the direction of Michael Fischer, exemplify this. The anthropology of science part of science studies shares with several other strands a general injunction to treat science as practice and culture, and to examine the genesis of beliefs in science without knowing ahead of time, or transcendently appealing to, their truth or falsity. By "writing against culture," anthropologists are beginning to expose those things that are routinely naturalized (accepted by people as "just the way the world is" even though there are other ways things could be conceptualized—see below) or considered to be inevitable in a scientific culture.

The areas in which anthropologists have so far had the greatest impact in science studies have been the areas of biomedical anthropology, and the introduction of a more thoroughgoing attention to transnational science and technology studies (see Layne 1998; Franklin 1995; Hess 1997; Traweek 1988, 1993). Anthropological studies of biomedical sciences are particularly fruitful sites for the exploration of naturalization. Work on the biological distinctions between men and women, for example, has shown that there are no definitive biological grounds for sorting everyone into either male or female, yet those categories are so naturalized in our culture that we greatly exaggerate sexual dimorphism. Notoriously, we label sex hormones "male" and "female" hormones, although everybody needs both. Other prominent anthropological contributions to sciences studies include explorations of the theme of Western kinship around human reproductive technologies, which has been broached by Marilyn Strathern, Sarah Franklin, and others (e.g., Strathern 1992; Franklin and Ragoné 1998; Franklin 1997; Edwards et al. 1993; Yanagisako and Delaney 1995). Sharon Traweek (1993) has carried out comparative ethnographies of Japanese and U.S. high-energy physics cultures, and Pamela Asquith (1994, 1995, this volume) has compared the Japanese and North American traditions of primatology.

WOMEN'S STUDIES AND FEMINIST STUDIES OF SCIENCE

Feminist historians, sociologists, and epistemologists have long raised questions about who has access to knowledge production and the other goods of science, such as credibility, authority, and economic independence. Recent scholarship has asked what it is about the roles and stereotypes of women that they have only sometimes been considered to be reliable spokespersons for the natural world. This has given these inquiries a distinctly "science studies" flavor. For example, feminist science studies scholars have shown the role that certain women played in witnessing truth produced by male scientists, or ways in which women's exclusion from science helped to define what the characteristics of a true scientist were (e.g., Oreskes 1997; Terrall 1995). Other scholars have documented when and under what conditions women have entered scientific fields (e.g., Rossiter 1997).

If one considers science to be governed by meritocracy, science ought to be more accessible to women and minorities than other high-status institutions and jobs (see Bielby 1991). Feminist epistemologists have suggested, on the contrary, that science has historically been especially hostile to women, and they have hypothesized that science itself and the scientific method are gendered. Evelyn Fox Keller, for example, has argued that modern science developed at a time when metaphors of conquering scientists gendered male, and conquered earth and nature gendered female, were pervasive and reflected the general cultural valuations of women and men (e.g., Keller 1985, 1992b, 1997). She has argued for a successor science that is gender-free, rather than inherently gendered male. A gender-free science, Keller maintains, is more likely to be achievable by women scientists because they are socialized relationally and so do not make the masculinist scientific "error" of assuming the rupture of subject and object, and of scientist and nature. Other feminist epistemologists have also developed the idea that scientific knowledge is inherently gendered. One strand starts with the view that what counts as valuable knowledge in a society is always mirrored in that society's predominant configurations of sexuality; and that the rise of modern science was associated with a heterosexual model of sex difference and even with compulsory heterosexuality. This led to the view that the best way to change the valuing and distribution of knowledge and access to institutions of knowledge-making was to insist on other models of sexuality (e.g., Irigaray 1974).

The idea that there might be a feminine way of doing science has been resoundingly criticized for its assumption of a commonalty between women; for its disregard of the structural privilege of white women over women of color, of rich women over poor, and women from developed over developing countries; and for the circularity of arguing for a group identity for women so as to bring about the change or demise of that very identity. But some have argued that although there is not a women's way of doing science that is true for all times, all places, and all women, "women" at a given time in a given discipline are themselves historically configured so that they often have a particular orientation to certain issues and questions and methods. For example, in primatology, this would mean that in fact a number of women were attending to similar issues at an important point in the development of the field, and that in fact these issues co-varied with the gender of the primatologist, even though those issues and methods could not be said to be feminine per se.

Other feminist epistemologists such as Helen Longino have argued that the gender of a knower matters epistemologically in determining what and how someone can know (Longino 1989; Code 1991). Still other theorists take an ethical orientation to epistemological characteristics that they claim happen to flourish among women and others associated with feminine stereotypes. For example, Lorraine Code has argued for a model of knowing based on positionality and friendship (different from the psychoanalytic object relations theory to which Keller appeals), rather than erotic domination, which she and many other feminist theorists feel characterizes most scientific epistemology. Feminist standpoint theorists have put the case that women are epistemologically privileged; that is, that because of their domestic oppression and relative lack of access to the commodities of academic and other success, they are privy to the *work* that separating subject from object takes. They have a superior view of reality not because of some intrinsic or essential feminine qualities, but because of their social position. Sandra Harding, adding a notion of multiple marginal identities borrowed from feminist postmodernism, is the best known proponent of standpoint epistemology vis-à-vis the sciences.

Donna Haraway is one of the most exciting feminist scholars of science studies. She is especially well known for her views on objectivity, for her work on primatology, and for her championing of "cyborg" hybridity. In one of her most famous essays, Haraway asked her readers to follow a quest for a reclaiming of objectivity—one that was neither totalizing nor relativist (see Haraway 1989, 1991a, 1997).[14] She acknowledged that feminists have been attracted to empiricism as a way of rejecting male bias in science and to antirealist constructivism as a way of bringing out the political aspects of the language games of truth and objectivity. She suggested instead that feminists should reclaim vision, despite the associations of the gaze with masculinist knowledge. She maintained that if you take vision seriously, you have to come to terms with the fact that every subject of knowledge is located somewhere and uses various bodily and technological apparatus with which to do the looking. Knowledge is both constructed, mediated, situated, and necessarily partial, and firmly anchored in the real world. The cyborg metaphor captures these dynamic and hybrid situated interactions that form the basis of scientific knowledge.

The science of gender is another area of inquiry in the field of gender and science. The categories of sexual preference, anatomical designations, medical and social codings of sex and gender, and so on, have all begun to be historicized, so that gender itself has increasingly been theorized as multiple and boundary defining (e.g., Martin 1996; Kessler 1996; Fausto-Sterling 1995; Jordanova 1989). Likewise, feminist studies of science and technology have become significant sites for looking at the connections between identity and technology, and for understanding technologies that are associated with one or the other gender, or which reveal gender relations.

COMMUNICATION, CULTURAL STUDIES, INFORMATION TECHNOLOGY STUDIES

A significant number of science studies scholars are to be found in communication departments, or in the emerging areas of cultural studies, or in information technology studies. These areas are all intensifying efforts to introduce a serious transnational component to science studies; communication and information technology studies because of the importance of theories of globalization to those disciplines, and cultural studies through an interest in the cultural specificity of different areas of science and technology. Scholars in communication departments are contributing studies on the active role of the media in scientific practice, and in public understandings of science, and thereby highlighting the importance of the interactions between lay and expert understandings of science (e.g., Lewenstein 1992). The area of the representation of science to its various publics has begun to be explored in all its richness. Work such Greg Mitman's on zoo exhibits are revealing the links between the goals and achievements of science, our evolving ideas of nature, scientific patronage, and problem choice in science (e.g., Mitman 1996; and for "nature" see also Mitman, this volume). In cultural studies, work on such topics as technologies of visualization used in the sciences has provided the means to begin to understand the cultural authority of science through the shared and readily recognizable iconography of science and technology (e.g., Hartouni 1991; Cartwright 1992; Grossberg et al. 1992). Information technology studies are posing in especially urgent terms classical questions about technology, such as whether or not it can be emancipatory, and are examining changes in knowledge and representational practices, as well as in access to information and democracy, through the spread of the information age (e.g., Suchman 1994; Agre 1995).

CONCLUSION: SHOULD SCIENTISTS BE SUSPICIOUS OF THOSE WHO STUDY SCIENCE?

In his famous 1938 paper "Science and the Social Order," Robert K. Merton observed that

> It is true that, *logically,* to establish the empirical genesis of beliefs and values is not to deny their validity, but this is often the psychological effect. (Merton 1973a, 264)

Merton was not taking part in the recent so-called "science wars" between those who practice and those who study the practices of modern science. At a critical moment in the history of the twentieth century, Merton was talking about reasons why some people were hostile to *science.* It was the scientists, not those who study science, who were being accused of attempting to discredit others' practice through their study of that practice. Yet, to some who study science today, this might seem to express scientists' reception of their work: scholarly attempts to understand the empirical genesis of truth, beliefs, and values in modern science, using "detached scrutiny," or "organized skepticism," as Merton variously called it, are being read by some scientists as an attack on the validity of science and the integrity of its practitioners. Merton was decrying the anti-intellectualism and generalized hostility exhibited by National Socialism toward the democratic variant of science, and he was trying systematically to uphold the values of objective science in the face of the political and cultural expediencies of Nazi science. It is normal for any community, including communities of natural scientists, to experience discomfort and suspicion if it is subjected to scrutiny from without. But the temptation to let this natural suspicion lead to a shutting down of inquiry, to anti-intellectualism, should be as strongly avoided now as then.

A theme among the Teresopolis conversations has been the value or otherwise of the media's and others' (mis)representations of primatology, and how to foster (more) fruitful collaborations. Not surprisingly, some journalists and some who study science do a poor job of representing the subtleties of research and its findings, and others do it much better. But there are a multitude of good reasons why primatology should seek, and be subject to, "outside" scrutiny and publicity.[15] The antics of nonhuman primates move and inform more people when primatology's findings are widely disseminated. But it is also more responsible to keep communication between primatology and its various publics flowing. Taxpayers contribute to a fair proportion of education and research that goes on around the world, and live governed by the truths and technologies science produces. And, at least in the West, the general public cedes the sciences tremendous cultural authority. While many people participate in school and even university education, not many inhabit the places or understand the knowledge produced at the cutting edge of scientific research.[16] Science is about the independent and disinterested search for truth, and the love of knowledge for its own sake. It would compromise science if scientists and their findings were forced to subordinate their research agendas to serving the public. Likewise, excessive openness to rational bureaucratic scrutiny is not always in the best interests of science (see discussions of the infamous "Baltimore Case" for this cautionary tale; Kevles 1998; Shapin 1999). Nonetheless, it seems reasonable to demand minimum standards of accountability and accessibility from science, just as we do from the other major knowledge-producing institutions. As many scientists themselves have advocated, this is necessary so as to detect fraud, and so as to avoid the calamities of sciences past. Nazi science paradigmatically showed us the potential onslaughts to both truth and humanity of politically instrumental and closed science. All of us should shun the anti-intellectualism of cutting off our fields of expertise from outside scrutiny.[17]

There is a deeper way of expressing this call for anti-anti-intellectualism, however. If lots of different kinds of people come to claim a stake in primates and primatology, more connections will arise between primates and their needs, on the one hand, and the rest of the world, on the other. If we are lucky, this might lead to a securer future for primates simply because more people will be taking them into account in more situations.[18] It is also plausible that it would lead to better primatology because theories would have had to withstand testing and application in more circumstances and from more points of view (philosopher of science Helen Longino, among others, has made this argument: refer to Longino 1989; Wylie 1997; Harding 1986; 1991; Code 1991; Baier 1985; Collins 1990; Butler 1990, 1993). The Wenner-Gren meeting at Teresopolis and the ensuing e-mail conversations increased the connections between nonhuman primates, primatologists, and other academics, and forced the theories and assumptions about each to alter somewhat, to complexify, and to enrich each other. In staging constructive encounters between a range of perspectives both internally and externally to primatology, the meeting was exemplary in the service of truth. By introducing nonhuman primates and their concerns in new arenas, their fate will perhaps also be more secure after this meeting. Certainly, mutual comprehension between primatologists and those who study primatology has progressed.

NOTES TO CHAPTER SIXTEEN

1. The kinds of reflection coming out of this meeting, e.g., "Comments on Wenner-Gren Symposium, Changing Images of Primate Societies," commentary by S. Strum, August 1996, have been remarkably rich. Alison Jolly, in the e-mail exchanges (May, 30, 1997), captured the importance of having had the Teresopolis workshop and subsequent conversations, rather than the empty name-calling associated with much of the science wars: "My initial reaction is that the Briefing article (on "Bruno's" Princeton chair) is much less interesting than Teresopolis, because it is so empty of content: it is almost "'Tis-'Tisn't"—as in science is or isn't true/searching for truth/making explained. Our stuff with the passion fruit mousse of primatology at least makes more intuitive sense of what is worth arguing about." Compare this to the rhetoric of both sides of the so-called "science wars": e.g., Gross and Levitt 1994, and more recently, see the May 22, 1997, issue of *Nature* 387: 331–336, 325.

2. I thank the Science Studies program at the University of California, San Diego, the Centre de Sociologie de l'Innovation at the Ecole des Mines de Paris, and the Department of Science and Technology Studies at Cornell University for my education in the field.

3. There are those who view science studies as a theoretically coherent field, and they tend to argue that science and technology studies is an emerging discipline. Similarly, there are those who feel that it is a mishmash of loosely related areas of inquiry. As suggested in the preceding paragraph, there are many themes that crosscut disciplinary roots. It is thus not surprising that some—but not all—

centers where science studies flourishes find that scholars of science and technology studies share more with each other because of their interest in knowledge than with the majority of their colleagues in their home disciplines, despite working on very different subject matter and eras.

4. Other ways I could have divided up the field include: (i) those who were initially trained as scientists versus those who were not; (ii) those who believe science studies is about science primarily versus those who believe it is about knowledge, with science happening currently to be (among) the most important sites for the production of knowledge; (iii) those who want to inform scientists and/or science policy versus those who don't care about this; (iv) those who think understanding science involves knowing the technical content of the scientific field being studied versus those who don't; and so on.

5. The Strong Programme had four theoretical/methodological principles for the examination of scientific practice: that the account should capture causality, that it should exhibit impartiality, that it should be symmetric in the sense described in the text, and that the analyst should be reflexive as regards the status of her own truth claims.

6. In the Teresopolis conversations and elsewhere, Bruno Latour has referred to it as the turn from science to research, and from certainty to uncertainty.

7. The normative order refers to the complex of technical, institutional, and moral prescriptions—or norms—that govern how to behave as a member of a given group of people (e.g., a family member, a faculty member, a student, a jury member, a voter, etc.). If something is a norm, it has the double meaning everyone is familiar with from the related concept "normal." That is, it describes things both as they usually are, and as we feel they ought to be. For example, the "normal" body weight charts we routinely encounter at medical checkups for ourselves or our children are average height to weight ratios derived from a healthy population (the original adult U.S. ratios were measured from young, fit army personnel). But these charts also provide weights against which to judge one's own body. They confirm whether one's weight deviates from the norm, and they provide a target weight range that is considered appropriate for one's height and gender. Insofar as one wishes or attempts to conform to this norm, these normal body weight charts also direct and justify action. Likewise, failure to conform to widely accepted norms has consequences. Consequences range widely depending on the sphere of action and the nature of the transgression, from such things as poor body image in the case of being overweight, through to failure to pass exams or get promoted in professional and academic fields, imprisonment for violation of legal norms, or being disliked for a violation of accepted ways of relating to other people. Norms are absolutely pervasive in everyday life, as well as in specialist forums, and we are all extraordinary experts at recognizing and following them. Norms describe things, then, but they also classify them in an evaluative way, and provide guides to action and justify retribution for failure to conform.

8. John Law, who is currently a professor at the University of Lancaster in the U.K., is a prominent ANT theorist outside the Ecole des Mines. Andrew Pickering's work shares many analytic sensibilities with ANT, although it is not completely assimilable to the ANT program.

9. Facts become unblackboxed when the facts themselves are contested, and people start to question the conditions and assumptions under which the data was obtained.

10. Latour uses the philosopher Whitehead to make the argument that scientists choreograph the critical ontological shift from eliciting properties or existences to discovering objects with essences.

11. For early modern science, see Mulkay 1976; Barnes 1971; and Bijker et al. 1987; also, e.g., Westman 1990. For nineteenth-century field sciences, see, e.g., Rudwick 1985. For history of modern biology, see, e.g., Young 1986.; Star and Griesemer 1989; and Clarke and Fujimura 1993. For history of modern physics, see, e.g., Galison 1997; Gooding Pinch, and Schaffer 1989. For history of technology, see Mackenzie 1990; Hughes 1983. The history of medicine is an enormous field unto itself, which intersects somewhat with science studies. There is much more on contemporary medicine within science studies (e.g., from a huge range, Epstein 1995; Rapp 1995) than on the history of medicine, but key historical texts such as Foucault ([1968] 1994) and Canguilhem ([1966] 1991) inform science studies practitioners working on both contemporary and historical medicine.

12. I say "oddly enough" because philosophy of science is one of the core disciplines of science studies programs.

13. The politics of identifying with one's subjects and objects of study is complicated, and in the current "science wars" context, taken by the various sides to have more importance than they perhaps warrant in terms of the intellectual endeavors at hand.

14. In the words of her exegete Baukje Prins, "Haraway indicates that she wishes to participate in the 'general' debates within science research and epistemology, but that she insists on doing this from her recalcitrant position *as a feminist* . . . instead of placing (her)self outside the scientific realm by embracing the idea of 'different' or 'feminine' ways of knowing" (Prins 1995).

15. The geometry of "inside" and "outside" is misleading, because it invokes hostility by its very boundary-drawing. In using this geometry I refer to the diffuse differences between professional primatologists and others.

16. Shareholders, for example, would not be happy with a state of affairs where they had no choice about what to invest in, and no knowledge of the activities of their companies.

17. Again, all of us seek many kinds of outside notice, and not only because we crave an audience and funding, but also to acknowledge our findings. It is always only some kinds of scrutiny that seem dangerous. My argument is that this fear should be resisted as far as possible, in the interests of society.

18. As Latour has argued, think of all the places where nonhuman primates are now important actors and so have to be taken into account.

17 A Well-Articulated Primatology: Reflections of a Fellow-Traveller

Bruno Latour

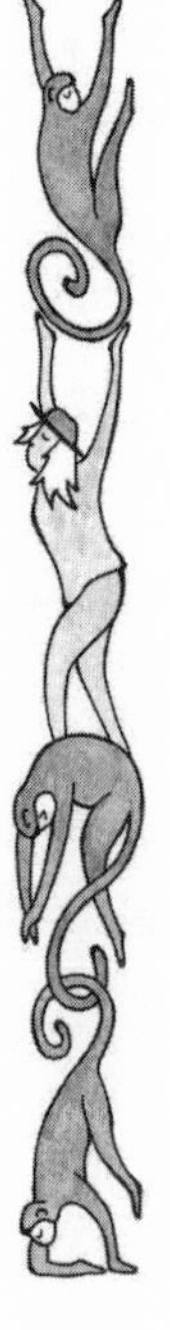

The lively interactions at the meeting in Teresopolis started badly and ended beautifully. At least three unbridgeable gaps threatened the discussion with an early dead end: Is science a social fabrication or an asymptotic access to reality? Do gender, theoretical biases, and methodological principles inevitably distort the quality of science, or is a good scientist the one who is able to escape from these shackles? Finally, is it culture or nature, sociology or biology that determines most of our (human and nonhuman) behavior? The first debate would pit primatologists against "science studies," the former scholars asking the latter, "Do you believe in reality?" while the science studies people would retort, "Do you really believe that monkeys are squashed flat inside the pages of your articles about them?" The second debate could degenerate into an endless purification rite, with everyone insisting on being protected or polluted by the "biases" of gender and paradigms ("I learned a lot from you as a woman," said one silverback, to which the other retorted pointedly, "You learned a lot from me because I am a good scientist, that's all"). The third debate would have subjected the participants to another reel of the nature-nurture fallacy, the "biologist" people defending the universality and constraints of their type of causality, while the "culture" people would have insisted ad nauseam on the variability, pliability, and historicity of human and animal behavior.

Fortunately for all of us, we did not get stuck in these three different, but interrelated, false debates. Instead, we slowly and painfully moved toward another agenda that I would like to outline in this chapter by following, in three different sections, how each of those old and tired discussions was reconfigured, thanks to the extraordinary setup devised by the organizers and for which I am infinitely grateful.

The Vascularization of Science and Society

The first originality of the meeting is to have brought together "science studies scholars" and scientists, who at first were presented to each other as two "camps"—which appeared to me as bizarre as presenting primates as one camp and primatologists as another. Yet it was only a matter of time before things became far more complicated: alliances began to shift without obeying party lines. It was soon impossible to consider that there were only two sides, the first made up of those who believed data were fabricated out of thin air, and the second being those who believed they possessed an unmediated and indisputable access to the reality "out there."

The reason for this high learning curve is easy to understand in retrospect: "science studies" is to scientific practice what primatology is to primates. Without primatology, in order to speak of apes and monkeys, we would have to rely on a few anecdotes brought back by missionaries and explorers; we would have no data, no comparative basis, no more than the shabby representations of wilderness and savagery with which Western culture has equipped us from the beginning. The change in the quality of our representation of apes and monkeys might not be terribly important to the primates—although it certainly does have an effect on the conservation of many animal troops (see Strum this volume)—but it is certainly of great import to us as a human community. To deprive ourselves of all the knowledge accumulated on primates in the last fifty years (see Fedigan and Strum this volume), would be an incredible loss.

The same is largely true for science studies, although the mass of knowledge is much smaller and the data softer (see Thompson Cussins this volume). Without the empirical studies of scientists at work, we would be limited to anecdotes and myths about a few stuffed "great scientists" hung on the walls of the University Hall of Fame. Science studies might make little difference to scientists at the bench, since, after all, they know what they do (as well as baboons know how they behave), but it would make a difference to all of us as a human community. To deprive ourselves of the knowledge on the practice of science that has been accumulated over the last twenty years in the history, sociology, and philosophy of science, in order to go back to a sort of mid-nineteenth-century positivism, would certainly be a pity—although I must admit I preach for my parish. Whatever the assessment of the two disciplines and the differences in quality and status, the discussion cannot be productive if one accepts the empirical revolution it brought about in one's own, while shrugging off the other as irrelevant. In my opinion, by the end of the meeting, there was some agreement that to return to a world made up of Dr. Livingstone's or Lord Zuckerman's baboons would be no more possible than to backpedal into an epistemology peopled by Auguste Comte's or Karl Popper's scientists.

This point being settled, the next problem to arise was that people at the conference were all well read in primatology, but very few had read much of the empirical work of the other discipline, science studies. To ask a sociologist of science, "Would you jump out of twenty-story building since you believe gravity to be socially constructed?" is like asking a primatologist, "Monkeys are disgusting and promiscuous wild beasts, aren't they?" No answer is to be expected and no answer should be given—except the one offered by Donna Haraway: "Push the heckler through the window!" To produce knowledge about scientific practice and to debunk the mythology of Science, capital S, is no more a denial of the reality of the scientific facts themselves than studying the social complexity of sexual competition and debunking the myth of the "wild beast" is a denial of the reality of the animals out there in the bush. Quite the contrary. To the question raised at one session, "What is responsible for the changing views of primates in the last fifty years?" the only answer I could find was, "But *the primates themselves* of course, it is they who forced us to modify our account of them," exactly as I would have answered, "We learn every bit of the new 'science studies' from the scientists themselves," if I had to explain the recent shifts in the definition of what science is and what makes it tick. "Please relax," I was tempted to say to some of my more anxious colleagues around the table. "Reality is not in question here." The debate does not oppose reality on one side and irreality on the other, but realities on both sides; or, more exactly, it opposes, on the one hand, a realistic version of what primates (and scientists) are and, on the other hand, a totally unrealistic or mythical vision of what primates (or scientists) do.

If we accept the comparison of two *empirical* disciplines and if we now leave aside the red herring of reality versus "pure social construction,"

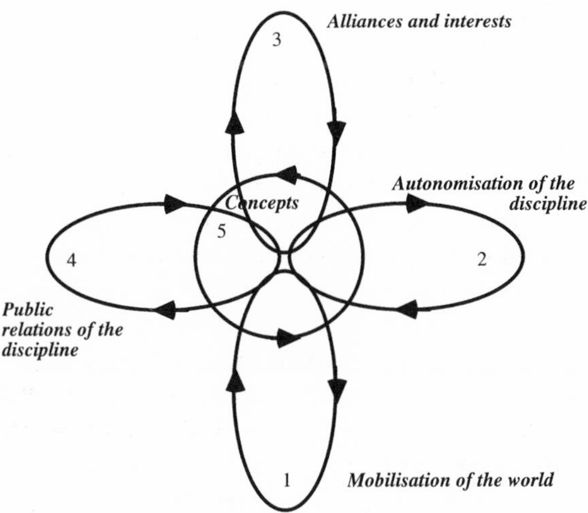

Fig. 1. A simplified view of the flows of knowledge (from Latour 1995a).

nothing is yet solved because the question becomes how did we *make the primates themselves relevant* to the questions we asked about them so that they could have a part in what we say of their behavior, while, before primatology started in earnest, they had so little to say in the representation Westerners had of them. A little summary of science studies is indispensable at this juncture if we are to continue. In order to be clear and, alas, sketchy, I will use a simple diagram to give an idea of the amount of work necessary to make the primates accountable for the facts produced by the discipline of primatology.

If we had to study primatology, we would be interested in five different horizons of practice, each of them being simultaneously necessary to make a science productive.

The first horizon—but one can start with any loop of figure 1 since it is a flow—is what can be called the "mobilization of the world," that is, all the efforts invested in creating a field site—or an enclosure or a laboratory—a data-producing unit. Every participant at the conference, including of course those in "science studies," knew only too well the immense effort that goes into obtaining a field site, maintaining it for any length of time, habituating the animal, mapping the territory, bringing the animals into the enclosures, caring for them, feeding them, equipping them with various devices, sampling them, etc. The beauty of primatology is the number of intermediary situations between field sites in unprotected areas all the way to brain surgery on animals held in laboratory conditions. The general point is that no matter how much natural historians like to get up early with binoculars and enjoy sunrise in the bush, and no matter how much neurobiologists like to watch single-neurons firing up, they would immediately stop being scientists if they ceased to *return* from their instruments—broadly conceived—with *data* that have the peculiarity of being highly *transformed* information. A practicing scientist is never presented simply with information, but always with a transformation that should nonetheless maintain, as intact as possible, the features she is most interested in—hence the circular form I gave the loop (Latour 1995b).

Although this work might account for 80 percent of the time spent, sweat and ingenuity is not enough to produce a discipline. Another vascularization is necessary. A scientist needs *colleagues* as much as data, and the former is no easier to obtain than the latter. The second loop we must take into account, is that which can be designated as "autonomization of the discipline," which is as much hard work as tending to the instruments. What is a primatologist? A zoologist first and foremost? An anthropologist interested in early man? A sociobiologist following selfish genes? A psychologist? We all know how difficult it is to solve these questions. Everyone at the conference was deeply aware of the fifty years of work that was necessary to produce professional associations, journals, institutions, conferences, and evaluation processes, so that the data painfully extracted from the instruments could be made relevant and the various benefits from different experimental or naturalistic setups would be available for comparison. Without colleagues, no quality control and thus no relevant data could be produced and made to circulate. A scientist who simply enjoyed fieldwork but who had no colleagues, would have no existence and no visibility. He might just as well have stayed in the bush fascinated by the beauty of the sunset.

This is not the end of his work, however. In order to have data and colleagues, another enormous amount of work has to be done, this time spent on the third horizon called "alliances." Field sites are expensive to keep up, zoos are huge organizations, journals cost a fortune if they have good referees, graduate programs to recruit future colleagues are expensive, and laboratory tests are time consuming. No matter how much a scientist is interested in her animals and no matter how many colleagues she enjoys having, she still has to interest *nonscientists* in her production system. A third vascularization is necessary that is in no way external or subsidiary, but internal and coextensive with the work to be done, and which can lead a primatologist very far away from his colleagues to the strangest people, even the military (Haraway 1989). Arguments for doing research must be provided, grant applications written, and relevant issues outlined. A scientist should approach his agencies and foundations with the same degree of enthusiasm he uses to convince his colleagues, or the same measure of concern he displays for his animals. No one said that being a scientist was an easy job!

Again, everyone in the room knew fairly well how many *nonprimatologists* were necessary to establish primatology as an autonomous discipline. A rough, but good, indicator would be to count the number of different institutions thanked in the acknowledgments of each of the papers produced by the people assembled in Teresopolis throughout their careers. One would quickly reach the hundreds. For each of them, much "networking" was necessary in order to persuade outsiders that their lives could not go on without first developing primatology. Without the translated interest of all these outsiders, the discipline as a whole would grind to a halt.

But there is a fourth loop that is as essential as the other three, especially in the case of primatology: the "public relation" or the "public appeal" of a discipline. In addition to the "science studies" people, the organizers had invited several scholars interested in the public representation of apes and monkeys (see Mitman this volume and Noble this volume). As with the others, the discussion started off badly, with scientists complaining about being either distorted and manipulated by the media or, worse, ignored. But for this issue as well, the learning curve was high. It was soon clear to all that for a discipline that claims to be relevant to everything from the origin of man and woman, to the genetics of violence, to the antiquity of emotion and sex roles, to the necessity of conservation, it was utterly impossible to exclude the public—all the more so since it was public opinions that historically generated the interest of those who had become the discipline's allies in the former loop. Long before Darwin's day, the impact of primatology and the question of the descent of man and its relation to the rest of the animal kingdom has made primatology an indispensable trading zone between ideologies and sciences. If we had forgotten this essential feature of the discipline, Donna Haraway's *Primate Visions* (1989) would have reminded us that films, museums, advertisements, and popular culture all play an enormous role in activating the whole of primatology, and providing much of its interest, passion, and energy. Here too, this vascularization is essential, and the comparison offered by the hyenas (see Glickman this volume) highlights how difficult it is to work on animals which have a "public relation problem" and, by contrast, how difficult it is to deal with animals which are, so to speak, "too much loved"!

A scientist, however, is not only sweating to produce good data, disputing with her colleagues to have papers evaluated and accepted, and convincing agencies and foundations to finance her field study and local authorities not to trap her animals. She is not only making popular films, preparing slide shows, and organizing conferences to raise money and redress the image given her argument. She also has to think about how the whole flow of information (assessment, argument, money, image, myth) holds together as one coherent whole. The fifth horizon is no more and no less important than the other four. It can be called "concept," "theory," or "paradigm." Contrary to many misconceptions, science studies are just as interested in this specific type of vascularization as in the four others. Concepts, however, are not colored lenses that would distort our view of things, to use the very unfortunate optic metaphor that will be criticized in the next section. Nor are they Platonic ideals floating far from the four other loops as if, in order to take seriously the "cognitive dimension of science," we had to escape to another world. Concepts are more like a beating heart that reoxygenates the blood, provided it is connected to the rest of the circulatory system. Yes, concepts are the heart of science, but one has rarely seen a functioning heart cut off from the rest of its body! "Male dominance," "kin selection," "proximate and ultimate causality," "bonobo scenario," and "selfish genes," to take a few examples, are highly complex integrations of masses of data, hunches, customs, and habits of thought, that cannot be easily discarded as irrelevant and cannot recapitulate what the whole discipline is about. When you isolate them from the rest you have nothing. When you have the rest without them, it is like holding a disheveled skein of wool. Theories are highly practical operators that do not constitute an "inner nucleus" of science which could be excised out of a protoplasm.

The point of quickly commenting on this five-horizon diagram is not to do the science study of primatology—although it would be a worthy task that Haraway and several others have already started—but to list the number of elements that should be taken into account to "make primates relevant to what they allow us to say about them." If a scientist were mad enough to brush aside all of these loops and scream in exasperation: "But let us get rid of all this sociology and history of science, of all these impedimenta—instruments, professions, journals, institutions, agencies, TV crews, exhibits, theories, concepts, paradigms—and let us go back to the animals themselves, let us seize them unfettered and unimpeded!" he would not produce a better knowledge. Instead, he would produce no knowledge at all and would be lost in the contemplation of a troop of fuzzy creatures even the names of which would escape him—since taxonomical labels reside in books, university training, databanks, and museums as well. Such a scientist might be enraptured by primates, but would be lost to science, and so would anyone who would have failed to fulfill at once the contradictory tasks requested by four different horizons. Yes, science is hard work, and each of these mediations is necessary to allow primates to have a say in our language. Our discourse can be accurate, but only on the condition that each of these transformations is carried out. Such is the great lesson of science studies: no one can jump outside of mediations and speak in truth about the outside world. To cut science off from its rich vascularization is equivalent to killing it.

It should be clear by now that the discovery of science studies is not that a science can be influenced or distorted by "outside" factors such as ideologies, politics, cultural biases, or psychological passions. The discovery—if this grand word can be used for such a humble discipline—is much more interesting yet, and slightly resembles that, if I dare say, of the great William Harvey himself! "Facts" are circulating entities. They are like a *fluid* flowing through a complex network, a rough sketch of which has been given in figure 1. What circulates is a certain type of transformation that allows the world some bearing on what we say about it. Thus, the triple notion of an outside world of nature "out there," an inner core of science "in there," and a political or social domain "down there," can no longer be sustained. When, during the conference, a silverback believes he is stating the obvious by saying: "We should not *confuse* our representation of the chimps and what the chimps are doing out there," he is in fact asking us to split in two the rich vascularization that "science studies" aims at describing without artificial interruption. The only goal of the primatology discipline is precisely to find *many ways* to mix, confuse, and intermingle what "primates are doing out there" and "what we say about them." But to understand this, a second false debate has to be pushed aside.[1]

From the Metaphor of Gaze to That of Proposition

The difficulty of integrating science studies and primatology was reinforced during the meeting by the organizers' original intention to probe "the role of theory, method, and gender" in "the changing images of primate societies." This earlier agenda, by its very formulation, could do nothing but paralyze the discussion since it imposed on each of us the fruitless task of purifying, in the sentences uttered about our animals, what depended on "them" and what depended on "us." If I have been right in the former section, this would have been tantamount to severing all the vascularizations that make up a discipline, and striving toward the impossible task of having animals, on the one hand, and statements

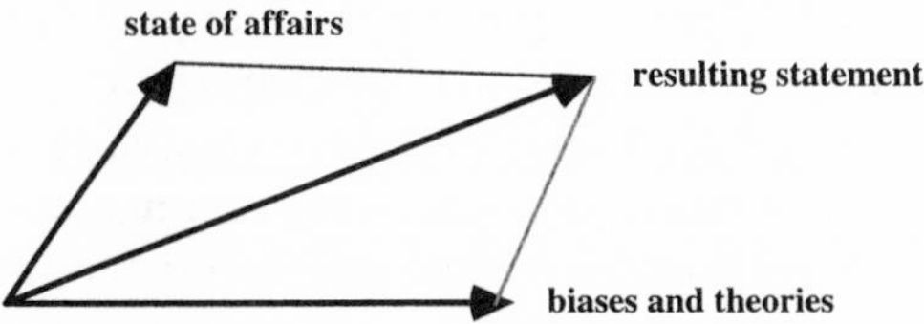

Fig. 2. In the dualist model, a statement is the resultant between two opposite forces, what the world is like and what we are equipped to say about it.

about them, on the other, with nothing in between. The dynamic of the meeting, however, slowly eroded this original intention, and began to nudge us toward a completely different set of metaphors. In order to be able to do the reflexive work required of us, we moved from an optical metaphor to a new one that I will call "proposition."

To be sure, the original intention relied on a perfectly sturdy and venerable intellectual resource. Like in the "parallelogram of forces" we all learned at school, any statement about a state of affairs can be considered as the "resultant" of two forces: what the world is like, and what we are equipped to say about it. If our biases are stronger, the resultant shifts toward one axis, while, if the world is somehow strong enough, the statement about it moves toward the opposite direction. With this classical model, we imagine our statements to be twice constrained, *not only* by the world *but also* by our mental and cultural equipment. It thus makes a lot of sense to try to weigh the different components and to measure, for each historical period, which one is stronger and which is weaker.

This model is obviously better than the naïve idea of science as an exact replica of the world, and it makes, I must confess, perfect common sense. Yet, it is utterly wrong, since common sense is rarely a trustworthy guide in scientific matters. A simple thought experiment demonstrates this point easily. What would happen if there were no counterforce coming from the axis that I have called "biases and theories"? According to the model, it means that the resultant would be entirely determined by the state of affairs at hand. Thus, if we had no theory, no preconception, no bias, and no standpoint whatsoever, we would benefit from an indisputable, unmediated, pristine access to things in themselves. No laboratory scientist would believe that for a minute. The same can be said of a natural historian—they know all too well the work needed to *make* a feature of the outside world *visible*. During the conference, Steve Glickman presented us with a simple and startling example. Hyena ethologists had waited for his paper published in 1990 to tell the difference between a male hyena penis and a female clitoris! Since Aristotle's day, this question had been, if I dare say, pending. After reading his paper, the difference could easily be discerned with a few minutes of training, as we all could witness for ourselves (Frank, Glickman, and Powch 1990). To "make something" visible is thus an entirely different task than to calculate the resultant of a parallelogram of forces.

The traditional model does not work any better if we push the couple of forces in the other direction and imagine a statement that is not counterbalanced by any constraint from the outside world. According to the diagram, it would mean that our statement about the world would be *solely* dependent on our repertoire of myths, prejudices, presuppositions, and biases, a pure story without any grounding in the world out there. No practicing scientist would believe this for a minute. How could such exotic notions as kin selection, Machiavellian intelligence, social complexity, matriarchy, social tools, and pulses of testosterone, be devised without a long and thorough intimacy with the animals themselves? Where in the preexisting culture would these notions reside so as to construct these stories from scratch? No imagination is fertile enough to produce even the most simple facts of primatology. To take up the hyena's example once more, what the millenaries of cultural biases have taught us is nil compared to what we have learned since the opening of the Berkeley enclosure: repetitive slurs, endless rumors—nothing that can even begin to articulate what makes hyenas' sexuality so specific. What the dualist model of a resultant between two opposite constraints does not explain is precisely what we want more dearly to understand: How do the animals out there contribute to how we imagine stories that no one without some level of intimate familiarity could dream up? How could the setup we ceaselessly devise elicit features in the lives of the animals that were invisible to all before we start making them up? Surely a tug of war between two contrary forces will not do the job.

One sentence by Thelma Rowell will clearly exemplify the alternative model that was slowly seeping into our discussions. Speaking about her new study on sheep, she stated one of her "biases" in the following way: "I tried to *give* my sheep the opportunity to behave *like* chimps, *not* that I believe that they would be like chimps, but because I am sure that if you *take sheep for boring sheep* by opposition to intelligent chimps they *would not have a chance*" (my emphasis). What on earth could this little clause mean: "give my sheep the opportunity to behave"?

A whole new philosophy of scientific practice resides in this extraordinary statement: "to give the opportunity to behave" is not the same thing as "imposing a bias onto" animals that cannot say a thing. Rowell states the difference between "a bias" and "an opportunity" very explicitly, since she insists that she does not believe sheep to be "like" chimps and since, left to their own devices, boring sheep will remain boring sheep forever. What does she mean, in my view? By importing the notion of intelligent behavior from a "charismatic animal"—another one of her treasurable expressions!—she might modify, subvert, or elicit, in the understanding of sheep behavior, features that were until then invisible because of the prejudices with which "boring sheep" have always been treated. She does not oppose, as in the dualist model criticized above, what sheep are really doing, with stories about them. On the contrary, it is *because* she artificially and willingly imposes on sheep another resource coming from elsewhere that "they could have a chance" to behave intelligently. But Thelma Rowell does not say that she is inventing sheep, socially constructing them, or making them up at her wishes. On the contrary, it is *because* of this very artificial collage between unrelated animals—charismatic chimps and boring sheep—that she can *best* reveal what sheep *really* are. Her sentence would make no sense in the dualist paradigm portrayed in figure 2, since she would have to choose features according to an absurd question: Are the sheep really intelligent or did you invent what they are? Or is it a combination, a resultant of both? "None of the above," she should answer. "By placing them, quite deliberately and quite artificially, into the paradigm of intelligent chimps, I gave them a chance to express features of behavior hitherto unknown. The more I work at it, the more autonomous my sheep may become."[2]

Thelma Rowell's sentence is in no way exceptional. It is, on the contrary, the common parlance of practicing scientists. For them, intensity of work and autonomy of what their object of study does, are *synonymous*. The better fabricated a fact, the more independent it is. Scientists behave as if they were "giving an opportunity" to phenomena that, in other settings, would not be "given a chance."[3] However, what makes this very common way of talking disappear from the scientists' own philosophy of science—not to mention philosophy of science itself, safely removed from all the empirical difficulties of benches, enclosures, and field sites—is the pervasive *optical metaphor* they have been made to use. If you transform all the actions that make the autonomy of scientific facts possible into "filters" that "color," "bias," or "distort" the view that a gaze should have of a phenomenon, then the very originality of scientific work becomes unaccountable. With the optical metaphor, the only reasonable outcome one can strive for is to *get rid* of all the filters in order "to see things as they are." Thus the work necessary to make things visible has itself been made invisible, and every reminder by sociologists, feminists, anthropologists, epistemologists, and psychologists that there are indeed "biases," "filters," "colored glasses," "prejudices," "standpoints," "paradigms," and "a prioris" will be considered as so many ways to *weaken* the quality of a science or to debunk its claim to truth. The only good gaze, according to the optical metaphor, is the one that is interrupted by nothing.

The dynamics of the meeting in Teresopolis were fascinating to watch—difficult not to use the metaphor even when criticizing it!—because the organizers rang a bell at every session trying to bring us back to a reflexive inquiry about "the role of theory, method, and gender in the changing images of primate society," while the dualist model under which we all operated to answer this question fell apart more and more completely as the days went on. Gender, for instance, did not appear as a filter that would make male and female scientists see things differently, blinding the males to some features while revealing others to the more perceptive female primatologists (see Keller this volume). At the conference, gender began to play the same role as "intelligent chimps" in the sentence analyzed above. Not that of a filter or of a bias, but that of a *trope,* to use Donna Haraway's favorite word. In the striking paper on the respective activity of eggs and sperm (see Tang-Martinez this volume), the importation to an unpredictable domain—reproductive physiology—of all the political debates in feminism over the dispatching of passivity and activity, allowed the ovum to "have the opportunity" of entering into a bewildering range of behavior instead of being considered as a "boring passive egg." The sessions on gender at the conference then shifted from a rather counterproductive soul-searching about whether or not a given primatologist was or was not "biased by gender," to a much more interesting research program: How much activity can be granted any given entity if we accept using the "indignation against passivity" gained by decades of feminist struggle as a resource to "give a new chance" to an animal or to one of its components?

The same shift quickly consumed the vague notion of theory and method. When it is said that the Japanese method was to stay in the field at least as long as an animal's life span (see Takasaki this volume), this obviously cannot be considered as a "bias" that would "limit" these scientists' vision of the animals. Quite the opposite. This simple decision entails, or reveals, different animals since it allows them to expand their interactions over a much longer period. This does not mean that those who go into the field for no more than a week with the sole purpose of collecting blood samples for their population genetics model are more "biased," but rather that they will give the animals a chance to behave differently. The same is true of important decisions like going to the field in Kenya instead of staying in an enclosure, or naming the animals individually, or following them on foot instead of watching them from the safe haven of a jeep. Each of these moves allows for new differences in the animal to be elicited or educed. The de facto abandonment of the optical metaphor was clear to all in one of the sessions devoted to the role of theory, when the following contradictory verbs were all used to describe what various concepts were making us do: "interact," "focus," "connect," "synthesize," "accelerate," "highlight," "raise a problem," "provide a solution," "polarize attention," "shift attention," "legitimate," "vindicate," "challenge," "stimulate," etc. Surely, all this very rich and active repertoire could not be squeezed under the label of "filter" or "standpoint" for an outside gaze looking at a thing out there. On the contrary, they made a lot of sense if theories are placed, like the fifth loop of figure 1, at the heart of several flows of data transformations.

What other metaphor would do justice to this practice and replace the old tired optical apparatus that limits the reflexivity of a scientific discipline to such an extent? In the paper prepared in advance for the conference, I had proposed, rather jokingly, to shift from the gaze metaphor to a gas metaphor! This had the advantage of keeping with the fluidity of facts introduced in the first section. When you put gas into the tank of your car, you are indeed connected with the oil fields of Saudi Arabia. These connections can certainly not be construed as so many "biases" which would have "distorted" the "real oil" out there. At the very least, if one wished away all these impedimenta, cracking, transformations, transportations, refineries, in order to gain access to the "oil itself," one would be left with no gas at all . . . The reality of oil in Saudi Arabia is proven by the number of transformations it undergoes before ending up as gas in your tank. So, with the gas metaphor, it is impossible to use the cracking and transformations of oil *against* the reality or against the quality of the final product. You have to choose either oil without transformation and thus no gas at all, or a lot of transformations but then you get gas instead of oil!

This gas metaphor, however, does not do justice to the originality of scientific transformation. To be sure, it outlines well the circulation and fluidity of the reference and it nicely emphasizes the impossibility of interrupting the flow. But the relation between what we say about animals and what animals are like, is not that of gas and oil. In the case of primatology, animals are much more than the raw material of our knowledge about them. The more knowledge we have of them, *the more visible they become*. It is as if the pipeline was *bidirectional,* providing more oil when we have more gas! It is because a scientist fabricates the fact that it becomes independent from his work—hence the puzzling double meaning of this little word "fact": what is made out, what is not made out. Clearly, this bizarre feature can't be sustained by the industrial metaphor.

Another way to better capture the practice of science is to consider preconceptions, biases, theories, methods, a prioris, and culture as so many roads that make it possible to gain access to the animals themselves. Surely, no researcher at Gombe, for instance, will call the tiny trails that lead to the field site a "bias." It does not stand "in between" the primatologist and his chimps. More exactly, it does *stand in between,* but instead of being what *hampers* the view of the chimp, as in the optical metaphor, it is, without any doubt, what *allows* the chimps to enter into view. The same is true of provisioning crates, clearings, sampling methods, statistical data reductions, binoculars, Jane Goodall's popular films, lectures given to the Leakey society, etc. All of these elements are in between, to be sure, but as so many indispensable *mediations* without which no knowledge would be produced at all. No one will call the tarmac on which planes are landing a "filter" that distorts what planes are supposed to be in themselves. The tarmac is, very commonly, what *allows* the plane to land . . .

The difference between the optical and the trail metaphor comes from the geometry and the position of the scientist. In the gaze paradigm, the observer is fixed and so is the thing to look at. It is a still life and probably comes from a mistaken interpretation of classical paintings. In such a metaphor, any addition of an intermediary is taken as detrimental to the quality of the view. In the trail metaphor, on the contrary, the observer is not fixed, but moves toward the thing to be seen, itself always in movement, and the more work that is done on the intermediary, the better the data will be. In the latter metaphor, it is as if the vertical position of the successive filters had been shifted 90 degrees to turn them into a platform allowing the spectator to move on it. It was soon clear to us, during the meeting in Teresopolis, that all of the possible effects of theory, method, and gender on our knowledge of primates would be evaluated differently if, instead of being what cut us off from the animals, it became what gave the animal an opportunity to be seen. The veils that until then had obscured the view of the animals, now became the red carpet allowing us an effortless walk towards them . . .

The trail metaphor is not without its defects, however, since it maintains the idea that knowledge is vision and that observer and observed are quite independent from the route they take. None of this captures the originality of Thelma Rowell's sentence. It is because she decided to treat sheep *as chimps* that they were lifted out of their condition of "boring sheep" and allowed the opportunity to demonstrate some intelligence.

How can we explain this action of making something else visible? Either it is made, or it is visible, but how can it be *made* visible? How can we replace the passive resultant of the model we have now discarded by an action that seems to have contradictory features? To be sure, we could use the traditional vocabulary of fabrication and construction, but this might entail artificiality, invention, and even deception. If we say that facts are fabricated or constructed, we clearly imply, in the common parlance at least, that they have some innate vice that makes them forever unable to "fly." It seems that we have no way—in modern Western language at least (Jullien 1995)—to entertain the possibility of saying at once, in the same breath, fabricated *thus* autonomous. We are always asked to choose "Is it real?" or, on the contrary, "Is it fabricated?" even though, in practice, we keep saying things like: "I tried to *give* my sheep the opportunity to behave *like* chimps."

One way out of this difficulty might be to talk about *propositions*.[4] A little bit of philosophy is necessary at this juncture, and I apologize to my colleagues for this little excursus, but it is crucial for allowing me, in the next section, to find another way of discriminating between good and bad science. As everyone will admit, the goal is worth a little pain!

Propositions should not be limited to statements made of words uttered by a human "about" a natural thing. As the name indicates, they are *offers* made by an entity to relate to another under a certain perspective. Propositions are not limited to the human domain of language and consciousness. For instance, Uexküll's canonical tick can be considered as a proposition and as a certain way of inhabiting the world by eliciting in its multiplicity a tiny number of relevant traits. To use another philosophical word, one could say that the tick "offers an interpretation of the world." But so does a field site with its research assistants on mopeds, its focus sample method, its archives, its portable computers, etc. The field site inhabits the world in a certain way and establishes certain types of connections that will modify the others. The passage of any hot-blooded animal will make the tick tick; the appearance of a new animal in the field site will make all the assistants suddenly attentive. A statement says in words what a thing is. A proposition designates a certain way of *loading* an entity into another by making the second attentive to the first, and by making both of them diverge from their usual path, their usual interpretation. A simple figure might help to grasp the abstract difference between a statement and a proposition.

A statement pertains to the human language and is utterly separated by an unbridgeable gap from the things it talks about. There is always an abyss between words and world, human and objects. This gap may be bridged, however, by the very mysterious act of establishing a correspon-

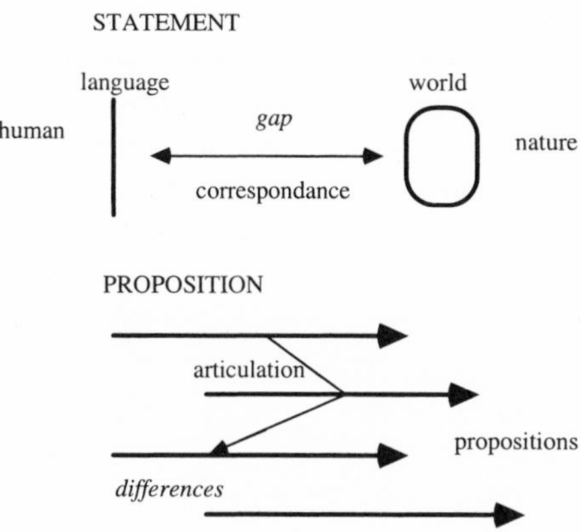

Fig. 3. The difference between a statement in language about a reference, on the one hand, and a proposition, on the other, lies in the situation of the two concepts: in the latter the difference between word and world is no longer pertinent.

dence between words and world so as to provide the statement with a truth value. If, and only if, the cat is on the mat will the sentence "the cat is on the mat" be verified. But since between the dimensionless sentence "the cat is on the mat" and a three-dimensional black furry cat on the mat there is no possible *resemblance,* the correspondence is always tentative and the gap between the two may never be filled, no matter how many hundreds of thick tomes the philosophers of language have thrown into it. The word "dog" does not bark any more than the word "cat" purrs. Because of their infinite distance with things—distance created artificially, for political reasons, by the erasure of all the intermediary steps of data construction[5]—statements are always running the risk of not corresponding to the world in an exact way, thus condemning the human locutor to life imprisonment, locked away in the cell of language. Skepticism directly descends from this implausible definition of truth as correspondence between words and the world.

Propositions, on the other hand, do not pertain to language but to the world. This world, however, does not resemble a nature made of things foreign to human consciousness that has been made to stand like an enemy camp opposite the human camp. It is made of interpretations, or propositions, sent *to* others so that they might behave differently. In between propositions there is thus not *one* gap but many *differences*. Meaning is not obtained by this very implausible correspondence between language and objects which have been made totally foreign to one another to begin with, but by connecting propositions which might or might not be slightly foreign to one another. While the notion of statements provides no possible avenue for the thing to be made relevant to what we say about it—except through the perilous footbridge of a correspondence between words and world—the notion of propositions allows things to be loaded into words. Whereas a statement implies the existence of a talkative human surrounded by mute things, a proposition implies that we are made to speak in this way *by what* is talked about. To understand this very tricky point in a less abstract way, we need to turn to the third debate of the conference—the most fascinating and also the most difficult to elaborate.

Articulated or Inarticulate Propositions

After having circumvented the false debate to decide if primatology is or is not "socially constructed," and after the rather sterile discussion on the role of method, theory, and gender in "shaping our views" of primates, the conference could have gotten stuck in the traditional opposition between "nature" and "nurture," and all the more so given the feelings for and against sociobiology, which were as intense as those triggered by science studies. This is where, from my point of view, the meeting was most rewarding. It seems that we avoided the usual pitfalls by exemplifying in practice a new set of criteria to *distinguish good from bad science* that completely cuts across the old tired distinction between biological and cultural determinism. It is this shibboleth that I want to render more explicit in the last section of this chapter.

I hope that Thelma Rowell will forgive me for analyzing her assertion about her sheep in more detail, but it provides the essential clue for my demonstration. As I said above, statements are different from propositions. This is especially clear if we consider how we make judgments about their quality. Statements are true or false depending on whether or not there is a state of affairs corresponding to the statement—with all the difficulties outlined by the philosophy of language. I will propose to say, however, that propositions are good or bad depending on whether they are *articulate* or *inarticulate*. "Boring sheep are boring sheep" is an inarticulate proposition since it repeats tautologically what a sheep is, as if refusing to enter into a connection with anything else. "Sheep are intelligent chimps" is an articulated proposition since it offers to establish a connection between two completely different entities that will give meaning to both: in the first sentence, sheep "are not given a chance," as Thelma said; in the second, they "will be given an opportunity to behave differently." The first sentence is a repetition—A is A. The second is, to use a philosophical term, a predication—A is B—that is, something else, on which it now depends to gain its meaning.

The difference between articulate and inarticulate is not the same as between true and false. It is more like the difference between a music made of one note that remains at the same pitch and another that modulates the tone by shifting from one note to the other. Contrary to popular beliefs about science, it is very difficult to capture with some precision what scientists mean when they say that one piece of science is "interesting" and another "rubbish." If dictionary science is made up of statements that are simply true or false, science on the run, that is, research, is made up of propositions that also have rhythm, color, temperature, and tempo. When a scientist says that a proposition is "interesting," she does not only mean that it is accurate, but that it has a lot of other qualities as well: it can be warm, hot, surprising, fecund, productive, exciting. . . . When we insist on the distinction between bad and good science, we not only distinguish between truth and falsity, but also between repetitive and articulate sentences. When we say that "data are meaningful," we thus designate another type of circulation than the one between a referent out there and a statement in the language (see top of figure 3). We seem to designate a very specific kind of movement among propositions that rearrange themselves into new and unexpected combinations (see bottom of figure 3).

We now understand why the debate about "social construction" or "biases" was so fruitless. It was still connected with the linguistic and optical metaphors. The more intermediaries there were between the eye of the observer and the object—or between the statements of the scientists and the thing in itself—the *less* direct, and thus, the less accurate was the correspondence and the quality of the information produced. Ideally, according to this view, if there were no intermediary at all, no complication whatsoever, the knowledge would be more complete. The situation is entirely different with propositions. The more activity there is, and the more intermediaries there are, the *better* the chance to articulate meaningful propositions. The difference between settings is not between those where the scientists are inactive, remote, detached, disinterested, and autonomous, and those where they are active, constructive, busy fabricating, or being biased, and full of a prioris and presuppositions. Instead, the distinction is between the settings where all the activity ends up producing repetitive data and those where the activity produces interesting articulations. Once again, Thelma offers an excellent example when she castigates the farmers for constantly separating sheep from one another as soon as they demonstrate behavior that is not sheepish and Panurgian.

Farmers do not give the sheep a chance, whereas Thelma's setups are unusual because she actively counteracts the farmer's customs, allowing the sheep to establish hierarchies and social relations wherein they stand a slightly better chance of being socially complex.

This case also nicely shows that propositions are in no way confined to language. The forms of life, in their entirety, in which farmers interact with sheep by selecting them for their docility and sheepishness, will not allow *them,* i.e. the farmers, to let *them,* i.e. the sheep, *be talked about* in any other way. Speech is not an exact property of the human phonic apparatus, of the human inner subjectivity, of the human consciousness. It is more a property of the whole setting in which very heterogeneous elements have been gathered and connected: barns, enclosures, grass fields, and wool in the farmers' form of life; barns, enclosures, grass fields, libraries, genes, papers, and ethological meetings in Thelma's form of life. Whoever enters the farmer's setting will be *made to say* that boring sheep are boring sheep; whoever enters Thelma's quasi-laboratory will be made to say that sheep too may be "charismatic animals," in spite of the distance between primates and ruminants!

In practice, it is *never* the case that we utter statements by merely using the resource of language and then only afterward check to see if there exists a corresponding thing that will verify or falsify our utterance. No one has ever *begun* by saying the "cat is on the mat" and *then* turned to the proverbial cat to see whether or not it is sitting grandly on the proverbial mat. Our involvement in what we say is at once much more *intimate* and much more *indirect* than that of the traditional picture: we are allowed to say new things when we enter well-articulated settings. Articulation between propositions is much deeper than speech. We speak because the propositions of the world are themselves articulated, not the other way around. More exactly, *we are allowed to speak interestingly by what we allow to speak interestingly* (Despret 1996). The notion of articulated propositions establishes entirely different relations between knower and known than the traditional view, but it captures much more precisely the rich repertoire of scientific practice and is much better adapted to the reflexive task requested by the organizers' brief.

A simple diagram might clarify the shift from one shibboleth to the other. The traditional touchstone tries to distinguish scientific statements about primates from nonscientific statements. The first type is said to correspond to a state of affairs "out there," while the other will be elaborated by using only the resources provided at one time by the available stock of presuppositions, clichés, myths, a prioris, or paradigms, without the benefit of anything "out there." To be sure, this touchstone provides a very useful dimension for sorting out dictionary science. But it does not cap-

STATEMENTS	PROPOSITIONS: articulate	PROPOSITIONS: inarticulate
scientific	A	B
non-scientific	C	D

Fig. 4. The classical distinction between scientific and nonscientific statements is not enough to capture the most important distinction between articulate and inarticulate propositions.

ture the constant evaluation going on among scientists engaged in doing research *before* the facts have been well established. The other touchstone, the one that is used in practice to separate out "interesting" from "boring" science, aims at distinguishing well-articulated from inarticulate propositions. The key difference between the two dimensions is that for a statement to be evaluated positively it is *no longer enough to be simply scientific, it should also be well articulated.* There is a lot of rubbish and non-scientific non-sense in primatology, everyone agrees on that, but there are also many impeccably scientific statements that are utterly useless because they are simply repetitive. Conversely, there are, to be sure, many repetitive clichés in the nonscientific literature about primates, but it is also entirely possible that practical situations that have none of the characteristics of science provide decisive insights since they allow for a new articulation between original propositions. Community-based conservation offers many examples of a sudden modification into the knowledge produced about animals by people who do not wear white coats (Cussins in press; Western, Wright, and Strum 1994). To be brushed off the sacred domain of science, it is no longer enough for a statement to be simply nonscientific, it must also be inarticulate and repetitive.

How, if at all, can this new shibboleth bear on discussions about sociobiology? In what way could a different distinction between good and bad science have prevented us from falling into the third false debate between biological and cultural determinism? My gloss on the interactions during the conference is not that we hesitated between zoology and anthropology, determinism and history, necessity and contingence, nature and nurture, science and critique of science, naturalization and historicization, naïveté and reflexivity—to use some of the common couple of oppositions—but that we were opposing, quite simply, the *smallest* number of active mediations to the greatest number of active entities. We were not trying to move from a naïve science to a critical science, nor from a non-science to an eventually scientific outlook, nor from a pure science to a polluted science, but from a science made from a certain number of active entities to another made of a greater number of them.[6]

Let me give a few examples to illustrate this point. The difference between the classical picture of active spermatozoa trying to penetrate a passive egg, and the new picture provided by Tang-Martinez, is not that the second is freed from gender biases while the former had been distorted by them; nor is it that the former was solely scientific while the second brought in external factors borrowed from the larger culture to "contaminate" the scientific facts of physiology. The difference is that the former leaves a large number of entities inactive, simply transporting necessity, while the second generates at every point active entities that, in part, modify the causality exerted onto them by the other. An egg that actively selects out spermatozoa differs from a passive egg not because it is more "feminine" or because it is studied by a radical feminist, but because it *does more things* and it is composed of more elements, more *articles,* and more mediations, none of which can be reduced to a simple input and output blackbox. The same is true of Glickman's rats before and after the demise of Skinnerian psychology. The postbehaviorist rats are not more scientific than Skinner's rats. They *do* more things. They are made up of more elements. They are more articulate and so are the psychologists who use more diverse elements to talk about them. Simplification becomes harder; transports of necessity less easy. The same is true of "smart baboons," "charismatic chimps," and "Machiavellian animals." No one claims that these represent more scientifically what the animals are like—although they obviously do; nor does anyone pretend that these are pleasing stories that are more in keeping with the prejudices of the age—although they obviously are. Smart animals are more active in their own behavior and thus allow for more awareness on the part of the scientists studying them, forcing them to take more precautions, obliging them to become in turn more intelligent, more respectful.[7]

The point of this very crude indicator—the number of active entities and the number of active scientists—is to point out that articulation was a much more important element in the discussion than the *type* of entity each of us tried to deal with. That they pertained to nature or to culture, to environment or to physiology, to science or to the history of science, to genetics or to feminist studies mattered *less* than their degree of activity and their ability to reconfigure their inputs and outputs. This shibboleth can be applied to various elements, genes, hormones, physiology, brain waves, and behaviors, without forcing us to resort to the image of "levels"—the inferior level being considered as more determinant than the one preceding it. Reductionism is no longer an interesting issue. Thus, for instance, dealing with smart baboons who live in socially complex groups does not mean that primatologists coming from anthropological departments will be unwilling to deal with genetic determinism. They will simply be wary of a certain type of genetics that connects elements so as to *decrease* the number of active mediators. Also, *inside* genetics the distinction between articulate and inarticulate propositions will sort out which parts of genetics are repetitive and simply scientific, and which parts are articulate and deal with a great many active "smart" genes and "smart" proteins, the pathways of which cannot be used for transporting an indisputable necessity.

Glickman's hyena enclosure at Berkeley offers a magnificent illustration of what it means to deal with articulation. Every possible discipline is brought in, from endocrinology to ethology, from genetics to psychology, from anatomy to natural history, from media studies to the history of science, but not one of them is introduced to decrease the number of active entities elicited by another. Quite the contrary: every time a new discipline appears, a new active entity is made visible that complicates the straight path of another. Glickman is not trying to integrate all the disciplines, each of them dealing with a certain "level" and defining an inflexible type of necessity. Instead, he is forcing all of the disciplines brought to the Berkeley enclosure to reconfigure their definition of action *at the occasion* of new and puzzling features offered them by those hyenas to whom he has offered "the opportunity" to behave intelligently and to be made up of entities, whereby each of them can be described as slightly *smarter* than before. "Off the shelf" endocrinology will not do the job any better than "off the shelf" population genetics or "off the shelf" history of popular misrepresentations of hyenas. Articulated propositions cannot be easily traversed by indisputable necessities. From this enclosure, where scientists are rendered smarter by smarter hyenas who at last escape the terrible fate of being despisable Disney-like hyenas, no simplification about determinism can escape.

The great advantage of this definition is that it also applies to the critical discourse bearing on primatology itself. When someone says that "women primatologists see things differently because they are women" or that "Japanese scientists see things differently because they are Japanese," it can mean two different things that are easy to distinguish if we use the new touchstone. This introduction of an outside element might play exactly the same repetitive inarticulate role as the one I've used thus far as an example: "boring sheep are boring sheep," "women are women," "Japanese are Japanese." The new entity is introduced—sex, culture, etc.—not in order to elicit a new feature from the other, but so as to *maintain* the essentialist character of the causality that has been brought in. It gives no more critical edge to the proposition than to say that "genes are genes" or that "Westerners always see things as Westerners." Tautology is always a tautology, no matter if it comes from the inside of science or from the outside, no matter if it deals with nature or culture, from ultimate or proximate mechanisms.

But these new features may also mean something entirely different. By bringing in women scientists or Japanese researchers, this new original standpoint will introduce a difference that will lead *away* from the standpoint. Once again, it is Thelma Rowell who provided the best example of this shift when, to ridicule the notion of standpoint, she said in passing, "If a female scientist studies female baboons and sticks to them she will end up studying males because female baboons are very much interested in males!" That's exactly the quality of a standpoint: it allows movement in a different way than what was intended. Standpoints never stand still! Because of the new attention given by female researchers to female baboons, a new attention will be given to males that differs entirely from the original focus on the domineering males *and that differs also* from the focus on the female that was originally intended. When we contend that primatology "betrays" many Western cultural biases about animals, monkeys, apes, or Dark Africa, that's exactly it: primatology *betrays* these original standpoints by turning them into something completely different. What is true for the gene, for hormone levels, and for aggression is also true for history and the sociology of science: the carrying over of indisputable necessity is always less interesting, and, in the end, less scientific than the revelation of active mediators all the way down.

If I am allowed some Gallic exaggeration, I would be tempted to say that the meeting in Teresopolis had an historical significance. Too often, scientists believe that their science will be better served if they ignore as much as possible all the untidy connections that make it work, with which they deal on a daily basis. They might be gathered to reflexively evaluate their discipline, but this evaluation, in their eyes, can only be carried out by using an off-the-shelf philosophy of science that dates back sixty years—to be charitable. The aim of such reflexive gathering could only be to purify the discipline ever more from the last remnants of adherence to subjectivity, politics, mythology, ideology, or biases. Shirley and Linda led us along an entirely different trail, which I have tried to map in dotted lines using my own system of projection. What would happen to the collective understanding of a discipline, if scientists were no longer trying to extirpate themselves from the sin of being connected, but accepted the vascularization as so many positive features that would turn their science into a well-articulated one? Primatology would not only be crucially important as a trading zone between anthropology, zoology, evolutionary theory, ethics, conservation, and ecology, but also an exemplary site for the renewal of philosophy of science.

NOTES TO CHAPTER SEVENTEEN

English in this chapter kindly corrected by Duana Fullwiley.

1. I am well aware that this distinction between representation and things, or, to speak more philosophically, epistemological questions and ontological ones, is built in the culture for much stronger political reasons that have nothing to do with primates or Teresopolis. I have traced elsewhere part of this genealogy (Latour 1997).

2. The sentence is all the more interesting since it deals with a purely observational ethology which has none of the usual features of laboratory experiments where it is always easier to show the artificiality of the setup (Hacking 1992). For a treatment of a similar sentence by Louis Pasteur, see Latour 1996a.

3. See the beautiful case studies by Despert (1996) on the theories devised by A. Zahavi about Arabian babble. See Glickman (this volume) on what happened to laboratory rats in the cages of the behaviorists.

4. I have tried to work out this limit of the philosophy of action by devising the concept of "factishes" (Latour 1996b). For one possible use of this notion in epistemology, see Stenger 1996. The notion of proposition is a central concept in Whiteheadian metaphysics (Whitehead 1929 [1978]). It has close connection with the debates between "saltationist" and "deambulatory" conceptions of truth-making (James 1907).

5. For a more complete demonstration, see Latour 1999.

6. The question of why is it that the "greater number of active entities the better" cannot be tackled within the confines of this chapter since it depends on a further redefinition of the difference between science and politics. For a first-effort go at it, see Latour 1999.

7. This extends, as Despret (1996) has so elegantly shown, to those people who watch primatologists or ethologists at work: intelligence, so to speak, is infectious—stupidity too . . .

18

Women, Gender, and Science: Some Parallels between Primatology and Developmental Biology

Evelyn Fox Keller

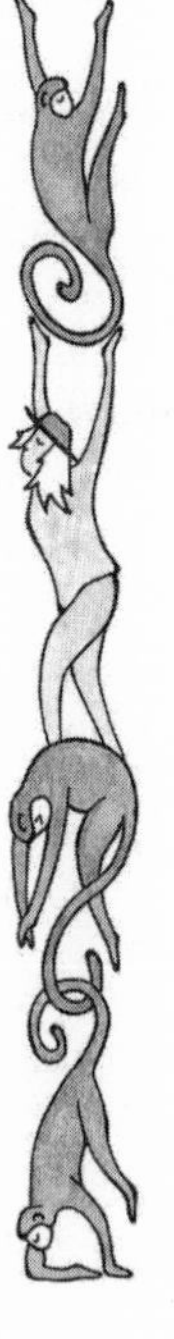

Both primatology and developmental biology have claimed considerable interest among feminist historians of science: both are fields in which women are prominent; both fields hold out more promise than most other scientific disciplines of the realization of full gender equity; in both fields, traces of implicit and explicit gender coding can be found in the historical structuring of the subject; and in both fields, the entrance of significant numbers of women has been associated (at least in the minds of some) with the introduction of new paradigms. In short, both fields can be seen as sites of victory for contemporary feminists. As such they are especially useful sites in which to explore particular issues of women in science, of the symbolic work of gender in the natural sciences, and of the relation between these two quite different kinds of issues (one participatory, and the other epistemological). In the present paper, I want to use the history of developmental biology to examine certain continuities, tensions, and even oppositions between the participatory and epistemological goals of contemporary feminism, to ask how they have operated, and continue to operate, in this particular disciplinary history, and to raise questions about possible parallels with the history of primatology.

I take my responsibility in this conference on "Changing Images of Primate Societies: The Role of Theory, Method, and Gender" to be that of addressing the relevance of the third factor listed in the title, namely, gender, and my specific task that of responding to the questions Shirley Strum and Linda Fedigan raised in their initial "position paper" (1996). Here they wrote,

> One of the initial impetuses for our joint discussions, and for this paper, was the attention given to women in primatology, by both the media and by feminist scholars. In particular, a claim has been made that women have had a major influence on . . . our changing images of primate society. (61)

And a few pages later, somewhat more specifically,

> What are we to make of the increase in the number of female practitioners in primatology over the years, and their specific concern, both empirically and theoretically, with female primates? (66).

Elsewhere, the organizers distinguish between the different terms "sex," "gender," and "feminism"—allowing them at least potentially separable kinds of influence; here, the clear implication in the wording of these initial questions (as it indeed seemed to be throughout the conference) is that the term "gender" can be understood as interchangeable with sex, and particularly with the female sex—be it of the primatologist or of the primate. While I have no authority to speak about primatology itself, I may be able to help distinguish between the different kinds of roles that "gender" can be expected to play in scientific change—above and beyond, i.e., the presence or absence of either female investigators or female objects of study. Furthermore, by drawing some obvious parallels between issues of women, gender, and science in primatology with these same issues in the history of developmental biology (a subject with which I am rather more familiar), I will suggest some ways in which the questions raised by Fedigan and Strum might be more usefully recast.

Meanings of "Gender"

In good part, the sophistication that feminist scholarship of the last twenty-five years has added to our understanding of the roles of "gender" has depended on a methodological resistance to two commonplace conflations: first, between "women" and "gender," and second, between biological and social meanings of the term "woman."[1] With regard to the first (the interchangeability of "women" and "gender"), we are cautioned to note the obvious asymmetry of the two terms (women are not the only members of the human race who are marked by "gender") and to explore the implications of that asymmetry (e.g., are we to suppose, in this case, that gender would not have been an influence on primatology before women entered the field?). But it is probably the second point, the importance of distinguishing biological from social meanings, that has provided the clearest hallmark of early feminist scholarship. In what at first seemed like a simple and unproblematic separation, feminists claimed the category of "gender" as the social category, employing the term to refer to cultural definitions of "masculinity" and "femininity," leaving biological definitions of "male" and "female" to the category of "sex" (assumed to be both fixed and universal). To quote Donna Haraway (1991a, 131), "gender is a concept developed to contest the naturalization of sex"; by counterposing gender to "sex," it served to underscore the manifest variability of attributes of "femininity" and "masculinity" observed across both culture and time. Among feminist scholars, this distinction between sex and gender is now old, and it has been greatly complicated by more recent scholarship. But even so, it has proved to be of considerable value in thinking about the different kinds of social (and hence, also of scientific) roles that gender may play and may accordingly prove useful in the present context.

Finally, a third distinction, one whose importance for feminist theory has grown steadily over the last two decades, may also prove useful. This distinction is between "gender" as a normalizing force in the socialization of individual men and women, and "gender" as an organizing force in the world of objects and attributes beyond the mere classification of human bodies (as, for example, between mind and nature, competition and cooperation, power and love, or active and passive). Elsewhere, I have distinguished between these two kinds of work by referring to the former as the "socializing" and the latter as the "symbolic" work of gender (see, e.g., Keller 1992a, 17).

To a considerable degree, these three distinctions—between women and gender, between sex and gender, and between the socializing and the symbolic work of gender—have also helped to organize much of the work that has been done in "gender and science." Indeed, they provide the logic of the tripartite structure that is commonly employed for courses on the subject, where material is roughly grouped into three subcategories: (a) women in science; (b) the science of gender; and (c) gender *in* science. Under "women in science" falls work on the numbers and experiences of women scientists; under "the science of gender" falls work on scientific constructions of gender and the ways in which these constructions have historically influenced and been influenced by conventional cultural understandings of "masculinity" and "femininity"; and under "gender *in* science" fall analyses of the symbolic work of gender, i.e., of the ways in which metaphors of gender have been employed (now by both male and female scientists) in structuring the various questions, goals, and conclusions that have historically been deemed scientifically legitimate.

Meanings of "Feminism"

While these various kinds of analyses may be (and often are) referred to as "feminist," such a label is confounded by the many meanings of "feminism." Again, there are some obvious caveats. Some feminists may be "essentialists" (in the sense that they embrace a particular set of cultural attributes of "feminine" as naturally belonging to women), but this position has long been shunned by the majority of those who have contributed to what is commonly called "the feminist critique of science." Perhaps the only point on which these scholars agree is the necessity of including gender (in all of the senses of that term) as a category of analysis in the consideration of any cultural product, including, of course, the natural sciences. It may also need to be said, obvious as it is, that not all women are feminists, nor are all feminists women. Finally, and perhaps the source of greatest difficulty, is the fact that the many political and intellectual differences among those who call themselves "feminists" have now become so overt and so contentious that even I am no longer sure whether the label still has a function. In the face of all these difficulties, is it possible to usefully respond to Fedigan and Strum's query about the impact of "women," "gender," and "feminism," either separately or collectively, on the field of primatology? I think it is, but not quite in the terms in which they pose their questions. The key, I believe, is to return to the problematic of the relation "women" and "gender."

It is abundantly obvious that, for all of their pains, those scholars who have, at least until recently, felt they could unproblematically call themselves feminists (and I count myself among them) have failed miserably (sometimes even in their own writings) in their efforts to implement the distinctions I have tried to enumerate. A far more powerful force seems to be at work (at least in the U.S., and perhaps especially in the last few years) luring a generic "us" into the regressive conflation of "gender" with "femininity," and of "femininity" with "female." And to this lure, even Fedigan and Strum, for all their evident sophistication, at times fall prey.

Thus, my proposed reformulation begins not by assuming either a direct or a transparent relation between women and gender, but by posing a question—indeed, the first question that we need to ask: What does the presence of women in science have to do with gender and science? And, as a second and closely related question, asking not about the direct influences of second-wave feminism on scientific values, but about its indirect impact—not only through improving the access of women to science, but by way of its contribution to changes in our collective assumptions about gender.

To illustrate a possible approach to these questions, I want to draw on some of my own research on the history of developmental biology, a field that exhibits some striking parallels to primatology. Both developmental biology and primatology have claimed great interest among feminist historians of science; both fields are distinguished by an unusually large number of successful women; in both fields, traces of implicit and explicit gender coding are evident in the historical cognitive structuring of the subject; and in both fields, the arrival of a number of prominent women has been associated (at least in the minds of some) with the introduction of new and more interactionist paradigms. In short, both fields can be seen as sites of victory for contemporary feminists. In order to more concretely situate my particular question about the relation between "women" and "gender," I want to look at the particular role of an especially pivotal figure in the recent history of developmental biology, namely, Christiane Nüsslein-Volhard, who was recently honored for her contributions with the award of a Nobel Prize. First, however, I need to provide some background.

Women in Developmental Biology

Despite the undeniable gains for women in American science over the past two decades, for most women scientists, real equity is a long way off. The gains have been conspicuously uneven, more in some fields than in others, more in some parts of the country than others, and more in some institutions than others. To those women who continue to feel beleaguered, it is a matter of considerable interest that there are some areas in which the promise of full gender equity actually appears as a realizable goal. Developmental biology (like primatology) is a prominent example. The marks of success are not simply to be found in numbers, although numbers do tell part of the story, but in the status and visibility of women and, perhaps above all, in the public display of a new kind of confidence. In some ways, the actual numbers are the hardest part of the story to tell.

The history of women in developmental biology is to be found mostly in individual stories, in impressionistic accounts, and in a few woefully incomplete cumulative records. The inadequacy of cumulative records derives in part from the fact that counting women in particular disciplines is a relatively recent activity, stemming partly from the fact that "developmental biology" as a disciplinary label is itself fairly recent (dating only from the 1960s) and partly from the failure of the records that do exist to tell us much, if anything, about the actual position of women in the discipline. Even if the data were available, a story that is in many ways more interesting is one gleaned from a finer-grained analysis of particular laboratories, publication records, and conference proceedings. It is here that the recent success of women in developmental biology is most clearly in evidence. Today, women head a significant number of the major laboratories in the field, they comprise over one-third of the membership of professional societies, and almost as high a proportion of the papers presented at major conferences. Indeed, it is the intellectual space occupied by women in developmental biology today that, while still well under 50 percent by any objective count, has led to the subjective impression among some biologists that developmental biology is a field now "dominated" by women. In my reading, what is responsible for such impressions is not simply the novelty of so many women in positions of authority, but, perhaps more importantly, the ability of some of these women to publicly inhabit that authority and to present themselves with a kind of assurance and self-confidence that I at least have in the past seen only in male scientists. It is the newness of this confidence that, to me, puts into stark relief—far better than statistics can do—the toll that past (and in most fields, enduring) obstacles have taken on those women scientists who have managed to survive.

The question of how developmental biology (like primatology) came to be such an exemplary model is of course an important one and one that has yet to be adequately explored. But, any attempt to answer this question must begin by situating its history in the larger context of the history of twentieth-century biology and, simultaneously, by situating the history of women in that field in the larger context of the history of women in American culture. Let me therefore turn to the history of developmental biology and, especially, to its history of gender coding.

A Brief History of Developmental Biology

One hundred years ago, biologists defined inheritance as subsuming concerns about transmission along with those about development. The central question of what was then called embryology—how does an egg cell develop into an organism—was also the central problem of biology. But with the rise of the American school of (Morganian) genetics in the 1920s, what had earlier been a single subject split into two rival fields: genetics and embryology. Throughout the 1930s, the two disciplines ran neck and neck, but by the advent of World War II, embryology began a decline from which it did not recover. Only in the last fifteen to twenty years has that subject, and its question, returned to center stage.

Relations between the two disciplines in the prewar period are neatly captured by a drawing by the Swiss embryologist Oscar Schotté, depicting two views of the cell. As seen by the geneticist, the cell is almost all nucleus, but, as seen by the embryologist, the nucleus is barely visible (from Sander 1986). In this drawing, nucleus and cytoplasm are employed as tropes for the two disciplines—each lends to its object of study a size in direct proportion to its perceived self-importance.

But to speak of the rivalry between two disciplines, troped here by two separate domains of the cell, suggests the possibility of (and need for) coexistence. Geneticists, however, had a program for colonizing the cytoplasm, and with it, the discipline of embryology. That program is captured in a metaphoric field that I have elsewhere described as the "Discourse of Gene Action" (Keller 1995).[2]

By the discourse of gene action, I mean a way of talking about the role of genes in development, introduced in the 1920s and 1930s by the first generation of geneticists, that attributes to the gene a kind of omnipotence—not only causal primacy, but autonomy and, perhaps especially, agency. Development is controlled by the action of genes. Everything else in the cell is mere surplus. As H. J. Muller put it in 1926,

> the great bulk . . . of the protoplasm was, after all, only a by-product of the action of the gene material; its "function" (its survival-value) lies only in its fostering the genes, and the primary secrets common to all life lie further back, in the gene material itself. (Muller 1929)

The discourse of gene action actually evokes a Janus-faced picture of the gene in relation to the rest of the organism—part physicist's atom, and part Platonic soul, at once fundamental building block and animating force. This way of talking not only enabled geneticists to get on with their work without worrying about what they did not know, but it also framed their questions and guided their choices, both of experiments worth doing and of organisms worth studying.

Nowhere is this more striking than in the reframing of the problems of embryology. Alfred H. Sturtevant, for example, was explicit in 1932 when he wrote:

One of the central problems of biology is that of differentiation—how does an egg develop into a complex many-celled organism? That is, of course, the traditional major problem of embryology; but it also appears in genetics in the form of the question, "How do genes produce their effects?" (304)

Between "the direct activity of a gene and the end product," he went on to argue, "is a chain of reaction." The task of the geneticist is to analyze these "chains of reaction into their individual links."

This rephrasing of the embryologist's question guided research in developmental genetics for the next forty years. It encouraged the view that research on "gene action" was primary, that developmental "chains of reaction" could as well if not better be studied in single-celled organisms than in higher organisms, and that cytoplasmic effects were at best of secondary interest or, in Morgan's terms, "indifferent." The phenomenal success of this research program, first in classical and later in molecular genetics, goes without saying. But it also had its costs—not simply in the long eclipse of the discipline of embryology and its original problem, but also of a genre of experiments and even of organisms (the *Drosophila* embryo, for example).

Over the last fifteen to twenty years, this problem has returned to center stage. And with its return has come a change in discourse. As we have learned more about how genes actually work in complex organisms, talk about "gene action" has subtly transmuted into talk about "gene activation," with the locus of control shifting from genes themselves to the complex biochemical dynamics (protein-protein and protein-nucleic acid interactions) of cells that are in constant communication with each other. *Scientific American* glosses this shift as the "news" that "organisms control most of their genes" (Beardsley 1991).

New metaphors abound. Fred Nijhout has even suggested that it would be better to think of genes as "suppliers of the material needs of development," as "passive sources of materials upon which a cell can draw" (1990, 441). Nijhout's proposal may be extreme, but there is no question that a new way of talking is in the air, in keeping with the emergence of a new biology: molecular biologists seem to have "discovered the organism."

How did this happen? Certainly, the introduction of labeled antibodies and of new technologies for cloning and manipulating genes has been immensely significant, but they do not tell the whole story. Consider, for example, the work on maternal effect genes and cytoplasmic rescue in *Drosophila* begun in the early 1970s, and later carried to such remarkable fruition by Christiane Nüsslein-Volhard and her colleagues. This work, establishing the critical role played by the cytoplasmic structure of the egg prior to fertilization, is widely regarded as pivotal in the recent renaissance of developmental biology. But it did not depend on new techniques. Indeed, as Ashburner writes, "it could have been done 40 years ago, had anyone had the idea. . . . All [it] required was some standard genetics, a mutagen, and a dissecting microscope, all available in the 1930s." So why had it not been done earlier? Ashburner says no one had had the idea, but this is not quite right. Rather, I suggest, it was the motivation that had been missing. These experiments are immensely difficult and time-consuming and one would have needed the confidence that they were worth the effort. Or, put another way, there was no field in which the "idea" could have taken root. Earlier, the discourse of gene action had established a spatial map that lent the cytoplasm effective invisibility and a temporal map that defined the moment of fertilization as origin, with no meaningful time before fertilization. In this schema, there was neither time nor place in which to conceive of the egg's cytoplasm exerting *its* effects. Indeed, the preferred term for "maternal effects" was "delayed effects." As long as one believed that the genetic message of the zygote "produces" the organism and that the cytoplasm is merely a passive substrate, why would one go to all that trouble? By the 1970s, however, the discourse of gene action had already begun to lose its hold. A number of different kinds of changes, above and beyond the obvious technical progress of molecular biology, contributed to its decline; here, I will mention only three.

I have already referred to Oscar Schotté's invocation of the nucleus and cytoplasm as tropes for the disciplines of genetics and embryology. In his sketch, each discipline lends to its object of study a size reflecting not only its self-importance, but also its own self-attributes of agency, autonomy, and power. In addition, however, the nucleus and cytoplasm also came to stand as tropes for national importance, agency, and power, with the former as the domain in which American genetics had come to stake its unique strengths, associated with American interests (and prowess), and the latter associated with European, and especially German, interests and prowess. German biologists were often explicit about what they saw as the attempt by American geneticists to appropriate the entire field. In 1926, for example, Haecker described the field between genetics and development as the "no-man's land" of somatogenesis—"a border field which by us has been tilled for quite some time. . . . The Americans have taken no notice of this." This tension persisted throughout the interwar years and was resolved only with the resounding defeat of Germany (and the virtual destruction of German biology) in World War II.

But the most conspicuous metaphoric reference of nucleus and cytoplasm is surely to be found in sexual reproduction. By tradition as well as by biological experience, at least until World War II, nucleus and cytoplasm are also tropes for male and female. Until the emergence of bacterial genetics in the mid 1940s, all research in genetics and embryology, both in Europe and the U.S., focussed on organisms that pass through embryonic stages of development. For these organisms, a persistent asymmetry is evident in male and female contributions to fertilization: the female gamete, the egg, is vastly larger than the male gamete, the sperm. The difference is the cytoplasm, deriving from the maternal parent (a no-man's land indeed); by contrast, the sperm cell is almost pure nucleus. It is thus hardly surprising to find that, in the conventional discourse about nucleus and cytoplasm, cytoplasm is routinely taken to be synonymous with egg. Furthermore—by an all-too-familiar twist of logic—the nucleus was often taken as a stand-in for sperm. Boveri, for example, argued for the need to recognize at least some function for the cytoplasm on the grounds of "the absurdity of the idea that it would be possible to bring a sperm to develop by means of an artificial culture medium" (published posthumously in 1918, 466, and translated in Baltzer 1967, 83–84). Thus, many of the debates about the relative importance of nucleus and cytoplasm in inheritance inevitably reflect older debates about the relative importance (or activity) of maternal and paternal contributions to reproduction, where activity and motive force were routinely attributed to the male contribution, relegating the female contribution to the role of passive, facilitating environment. The egg is the body and the nucleus, the activating soul.

But with the conclusion of World War II, the world began its irreversible slide out of modernity. The aftershocks of Hiroshima spread far and wide, transforming cultural landscapes along with political landscapes, and the many spin-offs of the war (including, for example, the computer) soon destabilized the grounds of all the categories that had seemed so secure to modernism.

Change, of course, did not come overnight to biology. While embryology was no longer a thriving research enterprise after the war, the memory of that disciplinary struggle took time to abate. It also took time (roughly two decades) for German biology to be rebuilt. Lastly, it took the women's movement to change our ideas about gender, and perhaps the hiatus of bacterial genetics (where no one had to think about male and female contributions) for these changes to creep into biology. By the 1970s, the entire world had changed and so did the ways that seemed natural to talk. Embryology was no longer a rival, Germany had become a friend, and gender equity was all the rage. And overlain on and interwoven with these "external" changes were, of course, the extraordinary developments internal to molecular biology, especially the techniques of recombinant DNA.

From this, many questions arise. For one: Did the change in discourse lead or follow our understanding of the complexity of regulatory dynamics? The answer is neither and both—they piggybacked off each other. The technical developments of molecular biology, working alongside and interacting with changes in the way we talked and thought, soon effected dramatic changes in what we could know. The language and the science worked, as they always do, in concert and in mutual reinforcement.

But for us here, the central question is not about the relation between language and science, but about the relation between women and gender. What about the actual women in developmental biology? What, for example, about the fact that Nüsslein-Volhard, a major protagonist in this story, is a woman? And not only Nüsslein-Volhard. If one looks at the maternal effect mutants identified over the years, especially in the early years, one sees that more than half of them were identified by women. Why should this be so? Could it be that women are, after all, *naturally* drawn to the study of embryology and that they are *natural* allies of the egg? I don't think so. For the record, let me clearly say that I believe that women have no more *natural* affinity for embryology, qua women, than they do for nature. Some women may be so drawn, but probably more as a consequence of the cultural insistence on equating women with reproduction, or on equating women with nature, and less because of their actual sex. But others will be propelled in just the opposite direction, in, as it were, natural rebellion against the coercion of cultural stereotypes. My guess is that the large number of women in developmental biology we see today has a lot to do with timing—with the fact that the increase in the number of women entering biology coincided in time with the rise of developmental biology as a field. Some of it has to do with the presence of successful "role models" and even with patterns of direct tutelage, such as, for example, the success of Nüsslein-Volhard and the women she trained. As to all the women who labored over the identification of maternal effect mutants in the earlier part of the century—well, as I've said, that was hard and often backbreaking work and widely assumed to be unrewarding. What more natural job to assign to women?

Situating Gender[3]

By the time Nüsslein-Volhard entered the story in the 1970s, many changes had already begun. Elsewhere, I have argued that a significant part of Nüsslein-Volhard's scientific acumen (and her success) derived from the ambiguities of her location—first, with regard to discipline (she is a molecular biologist working on problems in classical embryology and genetics); second, with regard to nationality (she is a German in a post–

World War II scientific world), and third, with regard to gender (she is a woman in a profession that, at least in Germany, is entirely controlled by men). Nüsslein-Volhard found herself at the intersection of multiple crossroads, able to make productive use of the ambiguities of her location, in large part because of the timing of her intervention. The time was right for a molecular biologist to take on developmental problems, for a German biologist to win international acclaim, and for a woman scientist to make it big. From her identity as a molecular biologist, she drew confidence, a kind of arrogance, and a cultural style. As a German, she had imbibed a tradition of finding particular interest in complexity (she cites Driesch as an example). As a woman, she was able to benefit from (and perhaps contribute to) many of the achievements of postwar feminism.

To many American women scientists, Nüsslein-Volhard is a heroine, both for the example she has set and for the large number of women scientists she has trained and helped launch into successful careers. She might even qualify as a heroine to American feminists for the role her work has played in legitimating a more interactive conceptual framework. But she is no heroine to many German feminists. To those who are aware of her existence, she is more likely to appear as an enemy than an ally. Her ambition, her phenomenal drive, and her all-consuming investment in her research are anathema to a generation of feminist scientists in Germany known for their advocacy of a kinder, gentler, and more "relevant"—in a word, "greener"—science. Fiercely opposed to genetic engineering, they see her as a member of the biotechnology establishment, doing her science "like a man," and they complain bitterly of her impatience with the interference of family obligations with the work in the lab.[4]

In turn, Nüsslein-Volhard has her own complaints about women scientists in Germany: "They have such a hard time taking themselves seriously, being professional"; "They so often give up because they don't want to work hard, they aren't really devoted to science. . . . They can't understand that one would just want to understand nature. It is not accepted here." Observing that in the U.S. there is now a real culture for women who "take themselves seriously," she notes that "here, it is still very lonely." And she thinks back to a time when it may have been different in Germany, the time of her grandmother, when women could "take themselves seriously."

She even has some complaints about American-style feminism. Although she acknowledges her fierce ambition, she does not see herself as doing science "like a man," but rather takes pride in "a distinct and perhaps more 'motherly' style in running a lab," teaching even the male graduate students "how to make Christmas cookies." And she criticizes her American postdocs for their insistence on "never making coffee" and, by contrast, recognizes similarities between her own ways of reacting to students and the ways her mother had dealt with her as a child.

Finally, there is the matter of the consonance between Nüsslein-Volhard's intervention and the interactionist agenda of much of recent feminist scholarship. Here we are dealing with gender neither as a biological nor even as a social category,[5] but as what might be described as a cultural category—i.e., with the culturally symbolic work of gender. In my discussion of the discourse of gene action (Keller 1995), I argue that part of the strength of that discourse derived from the tacit association of nucleus (or gene) with sperm, and of cytoplasm with egg, and of the declining strength of that discourse (at least in the U.S.) with (1) the changing valences of gender that were part of the cultural transformation wrought by the contemporary women's movement, and (2) the weakening of these gender associations enabled by two decades of biological focus on single-celled organisms. Nüsslein-Volhard's work on the role of maternal effect mutants in disrupting the anlagen of the developing embryo was both facilitated by this shift and itself instrumental in restoring prominence to the importance of the informational content of the egg's cytoplasm in initiating gene activation. But once again, her ambivalence is evident. Although a focus on the informational content of the egg cell was explicit in her work from the start, much of her language conforms to this day with the framework of "gene action." Some of the women she trained may speak unhesitatingly of their pleasure in finding evidence of "maternal control," but she herself is more likely to adhere to the traditional discourse of genetics, referring unhesitatingly to the "genetic control of development." Yet, at other times, she speculates that perhaps there is a special attraction for women in the very subject of development and embryos.

Perhaps it no longer matters. Once the linear narrative, beginning with fertilization and ending with maturity, is disrupted, as it has now been, it matters little where in the generational cycle one begins. The significant point is that the linear progression has now been replaced by a circular one in which neither chicken nor egg can any longer be prioritized. And despite her extensive and multifaceted ambivalence, Nüsslein-Volhard was so placed in time and in both personal and cultural space that she was able to play a significant role in bringing this about.

What does this say about the relation between "women" and "gender"? Or about the influence of women, gender, or feminism on developmental biology? Because of her fierce ambition, her aggressivity, and her intolerance of those either unable or unwilling to survive on the "cutting edge" of what she regards as "good science," some have argued—notwithstanding either her fondness for serving tea and for baking Christmas cookies or her focus on the developmental importance of "maternal effects"—that Nüsslein-Volhard has betrayed a "feminist vision of science" and that she practices her science "just like a man." But in the final analysis, it needs to be recognized that neither of these notions has much meaning apart from its context. The particular meanings of gender, of feminism, and of science that are relevant to Nüsslein-Volhard are functions of the particular subject positions that have been available to her, and these are not only specific to her historical and cultural location, they are also multiple and contradictory. She identifies herself simultaneously as "scientist" and as "woman," where, as Wendy Hollway has emphasized, the meanings of each of these terms are themselves multiple and variable. The images evoked by "scientist" oscillate among images of an American-style molecular biologist, a new breed of modern German biologist, and a more traditional German developmental biologist, while the images of "woman" oscillate between "American-style woman scientist," "path-breaking feminist," adversary of German feminism, caring "mother," independent "grandmother," and "lonely" victim of German sexism. And the form of this oscillation—her frequent shifts in subject position—depend, as became abundantly clear in my interviews with her, on the particular context and the particular struggles (or power relations), in which she finds herself engaged.

Perhaps the real moral of Nüsslein-Volhard's story is to be found in her very ambivalence. She has not needed to be an unequivocal supporter of either feminism or of women in order to make an intervention of immense value to women in science, just as she does not need to be an explicit proponent of a new discourse for the work she has done to play a pivotal role in dislodging the discourse of gene action. In turn, she has not needed to embrace feminist concerns in order to benefit from what such concerns might lead us to regard as victories. Gender matters to this story not because of her intent, but because of her situatedness, as a woman, in a field in which gender (now biologically, socially, *and* culturally) has mattered for a very long time—both for its practitioners and in the culture at large. Once again, gender matters for women in science not because of what they bring with their bodies, and often not even for what they may bring with their socialization, but for what the cultures of science bring to community perceptions of both women *and* gender, and in turn, because of what such perceptions bring to the communal values of particular scientific disciplines.

The days when one might have expected the needs and goals of women and feminists to naturally, as it were, cohere, or even that one might speak of the needs and goals of either "women" or "feminists" in a single breath, are long since gone. The great strength of feminist scholarship over the last decade has been the deepening of its understanding of what I might call the "situatedness" of gender. We have become exceedingly wary of sentences that begin with "Women are . . . ," realizing that just about the only way one can complete such a sentence is to say that women are people, situated by many social variables, and both adaptive and resourceful in the face of the pressures and opportunities they encounter. For women working in developmental biology or in primatology over the last couple of decades, some of those opportunities arose out of the social movement we call second-wave feminism and others out of different kinds of political, methodological, or conceptual/linguistic shifts.

As to Nüsslein-Volhard, her success may finally owe less to her genius and her phenomenal energy than to the fit between the singular bricolage out of which she has constructed her persona, her alliances, her career, and the scientific and cultural crossroads at which she found herself. She was an effective bricoleur in part because of the extent to which she was able to make opportune use of local alliances—with those strains in feminism that accord with, and those women who share, her goals. I suspect that a similar use of local alliances would also be found to hold for the most successful women in primatology.

Clearly, not all women primatologists think of themselves as feminists, and even those who do, hold very different notions about the meaning and the goals of feminism. Nor (as both the conference discussions and the subsequent e-mail exchanges abundantly indicate) do they necessarily agree on the best ways to study primates, or on the conclusions that can safely be drawn from these studies, no more so than do male primatologists. Indeed, one of the greatest values of this conference was the opportunity it provided for open and friendly discussion of the many different perspectives represented. Some common ground could of course be identified—provided, for example, by shared beliefs (at least among the conference participants) about equity for women scientists and in the productivity of recent shifts of focus in primatology. But on the question of the particular contribution of women primatologists, qua women, to these shifts in focus, widespread disagreement was manifest. Equally evident was the formation of constantly shifting local alliances around particular positions in particular arguments. I would argue that it was precisely the formation of such local alliances that enabled a clear articulation of what was at issue in the disagreements and, accordingly, what was responsible for the productivity of the discussions. Just as it was the formation of local alliances that had enabled new perspectives in primatology to take hold in the first place.

To be sure, important differences separate the history of primatology from that of developmental biology. But, in addition to the similarities I have already mentioned, the two disciplines also share a more general characteristic. All scientific change needs to be understood in the context of a complex network of interacting social, technical, and cognitive factors. In their initial position paper, Fedigan and Strum performed a useful service in identifying the many factors relevant to primatology over the last couple of decades. My problem—both with the way in which their questions were posed, and with the methods they proposed for answering these questions—is that neither the questions nor the methods leave space for the local and situated interaction of the factors (or variables) that contribute to determining the course of scientific change. Without attending to the locality of these interactions, we can neither understand the meanings of "woman," "gender," and "feminism" in any particular context nor the ways in which they might contribute to changes in paradigm. This, I believe, was one of the take-home messages of the conference—demonstrated, as it were, in action.

NOTES TO CHAPTER EIGHTEEN

An earlier version of parts of this chapter appeared in *Osiris,* volume 12 (1997).

1. A demand implicit in de Beauvoir's famous dictum, "Women are not born, rather they are made."

2. The notion of "gene action" is by now so much a part of our language that it must be considered a dead metaphor, and like many other conspicuously dead metaphors—e.g., "the first three minutes," "genetic program"—accumulating in power as it declines in vitality (or visibility).

3. Adapted from Keller 1996.

4. Even the day-care center she worked so hard to get established at the Max Planck Institute for Developmental Biology in 1990 (one of the few such examples in western Germany) meets with their disapproval; only partially subsidized by the Max Planck Society, the costs seem prohibitive to them.

5. In the sense, i.e., of the social roles of men and women.

19

Morphing in the Order: Flexible Strategies, Feminist Science Studies, and Primate Revisions

Donna Haraway

Scientific Practice and Tough Love

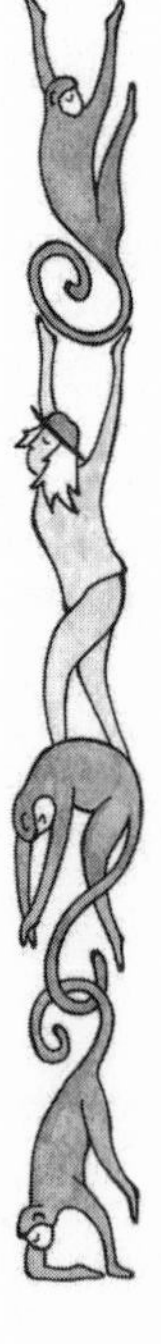

"For thus all things must begin, with an act of love." In a 1980 broadcast on South Africa Radio, "The Soul of the White Ant," Eugene Marais, a naturalist who published his observations on baboons in 1926, enunciated a central ethical and epistemological point about where to start a scientific account. Marais's comment applies to more than the sexual and reproductive capers of the animals he observed. I understand Marais's statement, which also opened my book *Primate Visions* (1989), to state the relationship between scientists and their subjects and between science studies scholars and their subjects. The greatest origin stories are about love and knowledge. Of course, love is never innocent, often disturbing, given to betrayal, occasionally aggressive, no stranger to domination, and regularly not reciprocated in the ways the lovers desire. In the practices I inherited, the love and knowledge of nature—embedded inextricably in the histories of colonialism, racism, and sexism—are no Edenic legacies. A white South African does not accidentally grace my text. Finally, love is relentlessly particular, specific, contingent, historically various, and resistant to anyone having the last word. Colonialism, racism, and sexism are not the last words even on a Westerner's love of nature; but they are an enduring part of the lexicon.

The major ethical and epistemological issue for me, in trying to understand what kinds of undertakings the biological and anthropological sciences are, is that knowledge is *always* an engaged material practice and *never* a disembodied set of ideas. Knowledge is embedded in projects; knowledge is always *for,* in many senses, some things and not others, and knowers are always formed by their projects, just as they shape what they can know. Such shapings never occur in some unearthly realm; they are always about the material and meaningful interactions of located humans and nonhumans—machines, organisms, people, land, institutions, money, and many other things. Because scientific knowledge is not "transcendent," it can make solid claims about material beings that are neither reducible to opinion nor exempt from interpretation. Those solid claims and material beings are *irreducibly* engaged in cultural practice and practical culture; i.e., in the traffic in meanings and bodies, or acts of love, with which all things begin. Semiosis is about the physiology of meaning-making; science studies is about the behavioral ecology and optimal foraging strategies of scientists and their subjects; and primatology seems to me to be about the historically dynamic, material-semiotic webs where important kinds of knowledge are at stake.

In this introduction I want to offer a slightly confessional account of my own noninnocent act of love in writing about primatology and primatologists over the last two decades. I care about primatology for many reasons, not the least of which is the pleasure of *knowledge* about these animals, who include us, *Homo sapiens,* but who exceed us in their varied ways of life. The other primates who are different from us are at least as interesting and consequential as those beasts taken to be like us. But mainly, it is a personal fact that I *identify* as a member of a species and a zoological order. My view of myself is shaped by bioscientific accounts, and that is a source of intense interest and pleasure. My life has been shaped at its heart and soul by material-semiotic practices through which I know and relate to myself and others as organisms. To identify, internally and subjectively, as a member of a zoological species and order is an odd thing to do, historically speaking. I am intensely interested in how such a practice came to be possible for many millions of different kinds of people over a couple hundred years. So, my act of love with primatology is more like sisterly incest than alien surveillance of another family's doings.

I am in love with words themselves, as thick, living, physical objects that do unexpected things. My paragraphs are peppered with words like "semiosis" because I am in love with the barnacles that crust such seedy, generative, seemingly merely "technical" terms. Words are weeds—pioneers, opportunists, and survivors. Words are irreducibly "tropes" or figures. For many commonly used words, we forget the figural, metaphoric qualities; these words are silent or dead, metaphorically speaking. But the tropic quality of any word can erupt to enliven things for even the most literal minded. In Greek, *tropos* means a turning; and the verb *trepein* means to swerve, not to get directly somewhere. Words trip us, make us swerve, turn us around; we have no other options. Semiosis is the *process* of meaning-making in the discipline called semiotics. Primatologists, beginning with C. R. Carpenter, have drawn richly from the human science of semiotics, and I have a playful and serious relationship with the ways communications sciences, linguistics, information sciences, and their motley offspring have infused primatology since the 1930s.

Science and science studies depend constitutively upon troping. Unless we swerve, we cannot communicate; there is no direct route to the relationship we call knowledge, scientific or otherwise. *Technically,* we cannot know, say, or write exactly what we mean. We *cannot* mean literally; that negative gift is a condition of being an animal and doing science. No alternative exists to going through the medium of thinking and communicating, no alternative to swerving. Mathematical symbolisms and experimental protocols do not escape from the troping quality of any communicative medium. Facts are tropic; otherwise they would not matter. Material-semiotic is oneword for me. I also know that there is a fine line between an exuberant love affair with words and a pornographic fascination with jargon. Tropes are tools, and, female or not, endowed with only the minor instrument of the *mentula mulieribus,*[1] I am a practicing member of *Homo faber.*

Embedded in narrative practices, stories are thick, physical entities. If storytelling is intrinsic to the practice of the life sciences, that is no insult or dismissal. Stories are not "merely" anything. Rather, narrative practice is a compelling part of the semiosis of making primatology. Some sciences reduce narrative to the barest minimum, but primate studies have never had the questionable privilege of an antiseptic narrative sterilization. Many other practices make up primatology, but not to attend lovingly to stories seems worse than abstemious to me; it seems a kind of epistemological contraception. "For thus all things must begin, with an act of love."

There is a doubled quality to what I love in primatology, or in any science. *First,* I am physically hypersensitive to the historically specific, materially-semiotically dense practices that constitute science made, as well as science in the making (Latour 1987). Science is practice and culture (Pickering 1992) at every level of the onion. There is no core, only layers.

It is "elephants all the way down," in my purloined origin story about science. Everything is supported, but there is no final foundation, only the infinite series of carrying all there is. Or, in another tropic effort to say what I mean, nothing insoluble precipitates out of the solution of science as cultural practice (Hess 1997). Remaining in "solution" is the permanent condition of knowledge-making.

Linked to my meaning of a solution, science and science studies are both generative mixings of natures and cultures, where the lively and heterogeneous actors blast the categories used to contain them. Conventionally in Western discussions, nature is both outside of culture and posited as a resource for culture's transforming power. Culture is tropically layered onto nature in a quasi-geological sedimentation. Simultaneously, culture is figured as the force that transforms natural resource into social product. Nature is needed for foundations; culture—in its manifestation as science, for example—is indispensable for direction and progress. Elaborate linked binaries are like stem cells in the marrow of this conventional discourse, in which primatology initially differentiated. Animal-human, body-mind, individual-society, resource-product, nature-culture: the monotonous litany is interrupted by science studies, which both foregrounds the tropic quality of these tools to think with and also suggests the possibilities of other tropes, other tools. As the British social anthropologist Marilyn Strathern insists, "it matters what ideas one uses to think other ideas (with)" (Strathern 1992, 10). This approach to primatology strengthens the perception of diverse, complex, nonanthropomorphic actors.

Second, my primate loves demand an appreciation of the solidity and nonoptional, if also open and revisable, quality of scientific projects. How else could I continue to argue that teaching Christian creationism in biology classes in the public schools is serious child abuse? The articulations that constitute knowledge are fragile, precious, historical achievements. If no god intervenes to grant certainty to the representations we call knowledge, no devil reduces knowledge to illusion either. The strength of scientific articulations is a practical matter, involving the development of analytical tools, narrative possibilities, representational technologies, patterns of training, institutional ecologies, structures of power, money, and, not least, the ability to craft diverse connections with nonhumans of many kinds. Primatology is made of such things.

My first job after graduate school in Yale's Department of Biology was in a general science department in a large state university from 1970 to 1974. My task was to teach biology and the history of science to "non–science majors," a wonderful ontological category, to make them better citizens. I was part of a team of young faculty led by a senior teacher, who had designed a course to fill an undergraduate science requirement for hundreds of students each year. In the middle of the Pacific Ocean, home of the Pacific Strategic Command, so critical to the Vietnam War with its electronic battlefield and chemical herbicides, this University of Hawaii biology course aimed to persuade students that natural science, not politics or religion, offered hope, promising secular progress not infected by ideology. I and the other younger members of the course staff could not teach the subject that way. Our post-Enlightenment epistemological confidence was messier than that.

For us, science and history had a more contradictory and interesting texture than the allegory of purity and prophylactic separation we were supposed to teach. Many of my graduate-school biology faculty and fellow graduate students were activists against the war partly because we knew how intimately science, including biology, was woven into that conflict—and into every aspect of our lives and beliefs. Without ever giving up our commitments to biology as knowledge, many of us left that period of activism and teaching committed to understanding the historical specificity and conditions of solidity of what counts as nature, for whom, where, and at what cost. It was the epistemological, semiotic, technical, political, and material connection—not the separation—of science and cultural-historical specificity that riveted our attention. Biology was not interesting because it transcended historical practice in some positivist epistemological liftoff from earth, but because natural science was part of the lively action on the ground.

I still use biology, animated by heterodox organisms burrowing into the nooks and crannies of the New World Order's digestive systems, to persuade readers and students about ways of life that I believe might be more sustainable and just. I have no intention of stopping and no expectation that this rich resource will or should be abandoned by others. Biology *is* a political discourse, which we should engage at every level—technically, semiotically, morally, economically, institutionally. Besides all that, biology is a source of intense intellectual, emotional, and physical pleasure. Nothing like that should be given up lightly—or approached only in a scolding or celebratory mode. From the establishment of biology as part of the curriculum in urban high schools early in the twentieth century to the training of environmental managers and molecular geneticists in the 1990s, the teaching of biology in the United States has been part of civics. From complex systems to flexible bodies, imploded natural-cultural worlds are modeled and produced. As biologist Scott Gilbert argues, in the post–Cold War United States, biology is the functional equivalent of the once required course in Western Civilization as the obligatory passage point through which educated citizens must pass on their way to careers from law to finance to medicine, as well as to a liberal arts degree (Gilbert 1997, 48–52). By the late 1990s, biology has become the foundation for the educated citizen in complexly global and local worlds.

What I want to understand in science and science studies—for which I use primatology metonymically as part for whole—is the *simultaneity* of both the facts and explanatory theoretical power and also the relentlessly tropic, historically contingent, and practical materiality of science. The world we know is made, in committed projects, to be in the shapes our knowledge shows us; humans are not the only actors in such projects; and the world thrown up by knowledge-making projects *might* (still) be otherwise, but *is* not otherwise. Science is revisable from what its practitioners think of as "outside" as well as from "inside." What counts "semiotically" as inside and outside is the result of ongoing work *inflected by* and *constitutive of* power of all sorts.

One instance of the simultaneity of power-laden historical contingency and material facticity that characterizes scientific objects of knowledge has riveted my attention since that first job in the militarized tourist fields of Hawaii. I came to know what a command-control-communications-intelligence (C^3I) system was. As a late twentieth-century American organism, I was such a system, both literally and tropically. Teaching biology as civics to non–science majors was a revelation. I began to get it that discourse is practice, and participation in the materialized world, including one's own naturalcultural (one word) body, is not a choice. Practitioners of immunology, genetics, social theory, insurance analysis, cognitive science, military discourse, and behavioral and evolutionary sciences all invoked the same eminently material, theoretically potent stories to do real work in the world, epistemologically and ontologically. That is, I learned that I was a cyborg, in culturalnatural fact. Like other beings that both scientists and laypeople were coming to know, I too, in the fabric of my flesh and soul, was a hybrid of information-based organic and machinic systems.

The term "cyborg" was coined by Manfred Clynes and Nathan Kline to refer to the enhanced man—a cyb(ernetic) org(anism)—who could survive in extraterrestrial environments. They imagined the cyborgian man-machine hybrid would be needed in the next great technohumanist challenge—space flight. A designer of physiological instrumentation and electronic data-processing systems, Clynes was the chief research scientist in the Dynamic Simulation Laboratory at Rockland State Hospital in New York. Director of research at Rockland State, Kline was a clinical psychiatrist. Enraptured with cybernetics, Clynes and Kline (1960, 27) thought of cyborgs as "self-regulating man-machine systems." Their first cyborg was the "standard" laboratory rat implanted with an osmotic pump designed to inject chemicals continuously to modify and regulate homeostatic states.

Brain-implanted lab animals showed up regularly in psychiatric research projects in the U.S. Following up ideas he had in 1938 for behavioral experiments using primates with brain ablations, C. R. Carpenter, despite misgivings recorded in his research notes, took advantage of the convergence of Cold War and psychiatric research agendas for a brief study of experimentally brain-damaged, free-living gibbons on Hall's Island in Bermuda in 1971. The gibbons had been operated on before Carpenter entered the story; he addressed questions of "anti-social behavior" among the one adult and five juvenile gibbons. Carpenter was brought into the work by José Delgado, then at Yale's psychiatry department, who collaborated with the Rockland State researchers on behavioral-control technologies and psychopharmacology in the context, stated in their research applications, of "stress" and "alienation" in U.S. cities. Like Delgado, Carpenter lectured widely in the 1960s on the relations of war, aggression, stress, and territoriality; he believed the other primates could teach "modern man" important lessons about these subjects (Haraway 1989, 108–10). Evident everywhere by the late twentieth century, the implosion of informatics and biologics has a cyborg pedigree.

This account of myself and other organisms as communications systems is a representation; but it is more than that. This kind of representation shapes lived worlds, even as the account is shaped by all the naturally/culturally situated human and nonhuman collaborators that have to be articulated to do such representations. In the late twentieth century, in a globally distributed pattern affecting billions of people, we really do know and relate to the biological world, in material-semiotic-practical fact, as energetic, economic, and informational processes. Similar formulations can and do show up interchangeably in economics textbooks, immunology journals, evolutionary discourses, family policy documents, and military strategy conferences. What is going on? How does the "mangle of practice" (Pickering 1995) that is science, including primatology, produce the zoological order to which I am committed? How do commitment, anger, hope, pleasure, and work all come together in the practice of love we call science? And science studies?

Primate Revisions

Near a drawer with a chimp skull implanted with a box-like telemetric monitoring device dating from the Cold War's space race, the bones of

Gombe's old Flo—the first chimpanzee to receive an obituary in the *New York Times*—rest in the laboratory of my colleague at the University of California at Santa Cruz, Adrienne Zihlman. By 1980, baboons at Amboseli were understood to be dual-career mothers (Altmann 1980), juggling the demands of making a living and raising kids. At the same time, many of these monkeys' trans-specific human sisters in the U.S. represented their lives in similar terms. In the 1930s, howler monkeys in Panama and gibbons in Siam lived in different sorts of societies; but both types seemed to be socially managed by a mechanism called a "socionomic sex ratio" (Carpenter 1964b). In the 1980s, a sign language–using, middle-aged, middle-class gorilla living in California while awaiting permission to move to Hawaii sought IVF in a desperate effort to conceive a child (Haraway 1989, 143–46).

If primates in past decades were sometimes represented by monomaniacal models of species-typical behavior, impoverished minds, sexual rigidity, and ecological stereotypes, by the 1990s human and nonhuman primates alike appear to be flexible strategists, with multifactorial cost-benefit analyses guiding the order's behavioral and evolutionary investments. Diversity is everything.[2] If communications dominated professional and popular scientific discourse in the 1970s and 1980s, diversity and flexibility name the high-stakes game at the turn of the millennium. Indeed, in their position paper for the conference on "Changing Images of Primate Societies," the organizers argued—with a touch of irony—that, even in a fiercely dangerous world, "[b]y 1995, baboons everywhere have more options than ever before" (Strum and Fedigan 1996, 45). The biological world these days acts like a flexible accumulation system. Appropriately, Strum and Fedigan emphasized multifactorial flexibility: "[P]rimates are smart actors. Individuals, regardless of sex or age, are strategists in an intricate evolutionary game. Their options, choices and successes depend on a variety of factors including environment, demography, age, sex, development, personality, biology, and historical accident" (1996, 48).

It is too easy reading the last two paragraphs to mumble about bias, cultural relativism, ideology, popularization, storytelling as opposed to proper conceptual models and testable hypotheses, and all the other things that are supposed to act as obstacles to the hard-won prize of real scientific knowledge. I don't believe it. In particular, I don't believe bias is a very interesting idea for thinking about primate studies. Bias exists, and goddess knows primate studies (as well as feminist theory and science studies) provide truckloads of examples. It is edifying for an historian of science to watch a notion—say the "man-the-hunter" hypothesis or the competitive sexual access model of macaque social organization—move from state-of-the-art theory to surpassed science to pseudoscience and sometimes back into fashion again among folks with doctoral science credentials, no math anxiety, and solid fieldwork in the right subfields. But "bias" tells the scientist or historian little about how a field practice, story, or theory travels and the work that gets done with the miscreant tools' aid. Scrubbing away bias is like cleaning one's toilet—it's got to be done, but more has to be said about how life gets lived in different sorts of houses.[3]

Rather, I want the intersecting worlds of primate lives and scientific practice—in which "[b]y 1995, baboons everywhere have more options than ever before"—to signal a different kind of approach to the historical contingency, tropic and narrative thickness, situated interactions of subjects and objects (like simians and people), and explanatory power. I see primatology (like feminist theory and science studies) to be a zone of implosion, where the technical, mythic, organic, cultural, textual, oneiric (dream-like), political, economic, and formal lines of force converge and tangle, bending and warping both our attention and the objects that enter the gravity well. Imploded zones interest me because that is where knowledge-making projects are emerging, at stake, and alive. It is possible to discuss mythic and textual axes separately from technical and organic ones, but the (often hidden) work it takes to keep the lines separate is stupendous and counterproductive. Getting important things done in the world—like building a creditable scientific account of primate lives—requires forcefully converging threads. My mode of attention causes me to mix things up that sometimes others have high stakes in keeping separate, and I might often be wrong-headed. But my way of working will also, sometimes, usefully avoid reductive notions of what is "inside" and "outside" scientific primatology, what is popular and professional, and what is "cultural" or "political" and what is "scientific" about our notions of primates.

I will enter a zone of implosion by concentrating on a persistent question in primate studies: What is the correct social unit of analysis for behavioral and evolutionary understanding of one species, the common chimpanzee, *Pan troglodytes?* Most of my account will draw from and revise bits of two chapters in *Primate Visions,* "A Pilot Plant for Human Engineering: Robert Yerkes and the Yale Laboratories of Primate Biology, 1924–1942" and "Apes in Eden, Apes in Space: Mothering as a Scientist for National Geographic" (Haraway 1989). I am also instructed by *The Chimpanzees of the Mahale Mountains* (Nishida 1990b) and a spate of reports in 1997. I will highlight analytical moves taught me by science studies scholars, feminist and antiracist writers and activists, and primate scientists. Sometimes the same people inhabit all of those categories, and the boundaries among them are permeable. The point is to learn how to navigate in a gravity well.

Primate Visions was often reviewed as if it were about gender and science. I read the book to be about race, gender, nature, generation, simian doings, and primate sciences, as well as about many other things as they co-constitute each other, not as they are retrospectively narrated as already formed variables. Neither gender nor science—or race, field, and nation—preexist the heterogeneous encounters we call practice. "Gender" does not refer to preconstituted classes of males and females. Rather, "gender" (or "race," "national culture," etc.) is an asymmetrical, power-saturated, symbolic, material, and social relationship that is constituted and sustained—or not—in heterogeneous naturalcultural practice, such as primate studies. Doing science studies, my eye is as much on "gender-in-the-making" or "race-in-the-making," as on "science-in-the-making." Category names like "gender" or "science" are crude indicators for a mixed traffic.

In this light, "Apes in Eden, Apes in Space" cannot make sense outside of its dramatic setting in the theaters of Cold War; nation-making; oil multinationals sponsoring natural history television specials in the age of ecology; changing field practices in primate studies; expatriate practices situated in decolonizing white settler colonies; relations between foreign scientists and primate habitat–nation populations, field staff, and officials; publication conventions; "first-world" feminism; racialized gender narratives in both of the mythic-material spaces called "Africa" and "the West"; the histories of academic disciplines, institutions, and cohorts in several nations (U.S., U.K., Japan, the Netherlands, Tanzania); and pedagogical and popular magazines, film, and TV.

From the point of view of many practicing scientists, perhaps the part of "Apes in Eden, Apes in Space" that seems to be about what they do—as opposed to what affects scientists from the outside—occupies two subsections, "Intermission at Gombe: History of a Research Site" and "Crafting Data." There, at last, we hear about such things as changes in data collection and analysis, from field diaries to computerized databases and efforts to make variously collected data comparable; field site development; theoretical alternatives with varying kinds of empirical support; and career patterns and contending cohorts of primate scientists.[4] The wording of the subtitle "Intermission at Gombe" was an impolite troping device meant to urge the reader to swerve in order to pay attention to the traffic between professional and popular practices, to the unpredictable direction of arrows of influence, and to the *constitutive,* and not merely *contextual,* doings of many communities of practice in shaping our knowledge of other primates.[5] My analysis is full of promiscuously presented material that perhaps ought to be hygienically sorted into science, on the one hand, and contexts for and influences on science on the other. No such luck, or so my mind, shaped on Hawaiian beaches and mountains as much as in labs at Woods Hole and Yale, insists.

"Apes in Eden, Apes in Space" opens with a section on the fetish that ruled popular and technical practice in post–World War II transnational contexts: "communication." A serious joke, "fetish" is a trope that stands in for a missing organ (guess which) and represents a disavowal of the dangerous, castrated condition of the matrix of our origin, or "mother." In my story about post–World War II science, including primatology, "communication" stands in for a disavowed "history." "Communication" was everywhere in 1960s and 1970s scientific and popular discourse, not least in the intercourse that humans engaged in with simians. In most representational practices in professional and popular primate studies, "nature" and "science" substitute for the trauma-inducing traffic of naturecultures called history, which I think is what science is really all about.

I pay attention to "communication" as a fetish in a web of practices: (1) National Geographic television advertising for the Goodall, Fossey, and Galdikas specials ("Understanding Is Everything," one Gulf Oil ad proclaimed); (2) Goodall's ethograms; (3) streams of data pouring into the space race's computers from the bodies of orbiting captured chimpanzee children acting as "surrogates for man"; and (4) representations of the Ameslan-using gorilla, Koko, who showed the (dehistoricized) universal signs of being "man" in her naming pets, referring to herself, knowing what is naughty and nice, and taking her picture in a mirror with a Polaroid camera. In the face of such thick histories and dense collections of human and nonhuman actors, I am fascinated by the technologies that accomplished magical things like making "Jane Goodall" appear to be "alone in nature"—especially in 1960, the year fifteen primate-habitat nations in Africa achieved independence. With their own social, technical, and rhetorical practices, Japanese primatologists were not much taken by the device of representing themselves to be "alone in nature."

The chapter continues its investigation of ways of "reading out history" by exploring scientific/cultural productions from the 1960s through the 1980s. Beginning with Frans de Waal's and Dian Fossey's hybrid technical/popular books, the chapter turns its attention to a synergistic triple code—gender, race, and science—needed to read *National Geographic*'s accounts of monkeys and apes. That task required paying attention to the details of how and why U.S. and U.K. "white" women filled the narrative function they did in those stories, how "black" women and men got the kinds of scientific credentials they did in the same years in the U.S., and how writers for *National Geographic,* like Shirley Strum, struggled with modest success to control visual and prose narration of their scientific work. Missing throughout these accounts was the contemporary ape fieldwork being done near Gombe in the Mahale Mountains by the Japanese.

Just before the Intermission, "Apes in Eden, Apes in Space" heads for the movies. The bill of fare includes both *King Kong* and his ongoing mutants and the cascade of sober pedagogical films on primate behavior. These edifying celluloid records began with Julian Huxley and Solly Zuckerman's 1938 *Monkey into Man*—which predictably linked family, race, and technology in a functionalist and evolutionary great chain of being—and then moved to C. R. Carpenter's "positivistic" films about free-ranging primate species, made in the 1940s from his prewar field footage, and lastly settled on Sherwood Washburn and Irven DeVore's (1966) "objective" baboon behavior and society films. The visual and verbal rhetoric of the films produced the epistemological and aesthetic *effect* of objective vision. *How* such important effects get produced commands rapt attention among science studies scholars, who investigate the relation of the filmic effect of direct, objective observation to the messy doings of human and nonhuman primates. It would be hard to overestimate the influence of Carpenter's and DeVore's films, which for generations of novice viewers of nonhuman primates warranted belief in species-typical behavior and grouping patterns.

By the time we get to Gombe as a research site, the reader of "Apes in Eden, Apes in Space" is saturated with the messy cross-traffic in the midst of which knowledge of primates is crafted. I mean materially solid knowledge, not biased opinion or ideological illusion. Like any good science, primate studies produces revisable and complexly progressive knowledge that travels beyond its sites of emergence. Scientific practice never yields knowledge that precipitates out of the solution of situated histories and material-semiotic apparatuses. If communication was the fetish that "read out history," then my tool for learning to inhabit natureculture will be "memory"—i.e., a trained practice of retelling scientific accounts to situate them as thickly as I can.

As promised, my focus will be on one kind of contested object, namely the unit of chimpanzee social life. I will trace a few threads in a complex fabric; but perhaps enough can be said to show what I mean by a material-semiotic object of knowledge located in apparatuses of knowledge production, for which the concepts of bias, ideology, and cultural relativism are weak tools. Not a neutral observer, I, like the primate scientists, am less "biased" than "engaged."

I can't start my story in East Africa, at Gombe and in the Mahale Mountains. Instead, I have to go to caged pairs of adult chimpanzees engaged in a test of motivation for taking food treats at Robert Yerkes' Florida breeding station that was part of the Laboratories of Comparative Psychobiology at Yale University in the late 1930s (Yerkes 1939). Yerkes firmly believed that "the family" was the organic unit of primate social life, and that "dominance" organized "cooperation" and "integration." None of these words between quotation marks was transparent; all were "boundary objects," grounded in action, which traveled among many communities of practice with just enough continuity in reference to sustain projects and debates (Star and Griesemer 1989). Believing that chimpanzees approximated the state of monogamy in nature, Yerkes caged his animals in male-female pairs when he could. Chimpanzees were a model for man; their natural family life, occurring just on the other side of the border from "culture," was a mirror and testing ground for theories and policies. In Yerkes' framework of functionalist associationism, the family was its members, which could be analyzed into constituent organic drives functionally integrated by the nervous system. The family economy, like the mental one, involved division of labor, (re)productive efficiency, and unity resulting from an integrating hierarchical principle of higher functional adaptation in an evolutionary (but not Darwinian) chain of being.

Yerkes was committed to the intelligent interaction of apes and people in the cooperative enterprise called the laboratory. "Dominance" did not mean exploitative domination, but rather assured natural positioning in organic hierarchies that maximized group efficiency and harmony. That was true among animals and between animals and people. Yerkes' secular, New England, Protestant love for his science and for the animals he studied was intimately intertwined with his beliefs in both himself and the chimps as servants of science for a better world (Yerkes 1943, 11).

Organic drives, such as the "hunger for social status" (Yerkes 1943, 46), shaped role differentiation. Drives varied in strength and effectiveness of expression, so it was important to measure them, just as it was important to measure cognitive capacities in the plethora of mental tests that Yerkes excelled at designing for both human and nonhuman primates. Neither males nor females were inherently dominant; position in a hierarchy was a question of relative strength of organic drives. Status motivation was conditioned by sexual hungers and opportunities. The food-chute test for caged pairs of "mates" measured the interaction of drives for dominance and sex, as observers registered who grabbed bits of banana against stages of sexual swelling of the female and the personality of the animals. "Personality" was "the product of integration of all the psychobiological traits and capacities of the organism" (Yerkes 1939, 130). Individuality mattered, but functional integration of organic systems was a higher level of organization. If females were seen to trade sex for tasty favors, that was simply a view made possible by the research apparatus.[6]

For Yerkes, dominance was a physiological, psychological, and social principle linked to the processes of competition and cooperation, both of which were central to his overall project, which he called human engineering. The chimpanzee lab was a pilot plant for human engineering. Yerkes' work in the Personnel Research Federation, the Committee for Research in Problems of Sex, the Yale Laboratories of Psychobiology, the Boston Psychopathic Hospital, the Surgeon General's Office in the Army in World War I, the Rockefeller Foundation, and many other locations was geared toward fulfilling his vocation of shaping man for more efficient organic modern social life. Such scientific projects were intrinsically part of building democracy in the contest with "authoritarianism," especially fascism. Racial hierarchies, sex-role relations, and democratic cooperation were part of the great evolutionary, non-Darwinian organic scheme that chimpanzees were asked to clarify. Classifying individuals, ape or human, in accordance with their organic capacities—whether through intelligence tests or scales of motivation—was a fundamental scientific practice. Cage design, building architecture, experimental protocols, and data collection practices in Yerkes' laboratory only make sense within these frameworks.

So, for Yerkes, the monogamous heterosexual pair was the natural chimpanzee social unit, and the food-chute test yielded important data about role differentiation in the family. Concepts of bias and ideology get us almost nowhere in understanding this woeful situation. Yerkes was practicing good science, and he got good data, by the standards of his community of practice (which, it must be said, was a bit short on statisticians). That does not mean he was right about chimpanzees or immune from criticism in 1939 or now; but he was not doing science "influenced" by subjective and cultural "biases." Rather, his science as naturalcultural practice *built* an apparatus of knowledge production that crafted the world in a particular semiotic-material way. Strip the "biases" and not much is left of the scientific apparatus. "The family," "personality," and "intelligence" were solid material-semiotic entities that Yerkes' apparatus helped put together in the world. Their real materiality was an effect of their constructedness (by humans and nonhumans). In humans as well as animals, sexual "role differentiation" (the word "gender" would not have made sense to any of the communities of practice in 1939) was as much a product as a preexisting variable in the Laboratories of Comparative Psychobiology. The *projects* and *commitments* were what Yerkes engaged in as a scientist. The whole messy web of articulations was Yerkes' science. It deserved critical engagement over the practical material-semiotic work that produces knowledge, not ideology critique or celebratory hagiography in the history of primatology.

Jane Goodall's (1967a) early descriptions of chimpanzee social organization at Gombe identified only one stable social grouping, the mother and her dependent offspring. Otherwise, chimpanzees were described to associate fluidly and mostly peacefully in nomadic bands, without defended social or territorial boundaries among bands or parties. The first phase of research, 1960 to 1966, when Goodall got her doctorate and more observers began arriving at Gombe, seemed to reveal a primate utopia—mother-centered, but with outstanding male personalities engaged in status competitions, which did not seem to be the organizing axis of chimpanzee society. The material-semiotic unit of the mother-infant pair remained deep in Goodall's naturalcultural practice (Goodall 1984). The unit was crucial to many scientific constructions of objects of knowledge, including Robert Hinde's. Goodall's later collaboration with David Hamburg of Stanford's psychiatry department, in the context of work on "stress" in modern society, is another part of a wide-ranging process of constructing a natural-technical unit of observation in the field.

Symbolically, in Goodall's writing the chimpanzee mother and infant, especially Flo and her newborn, constituted a perfect model after which she portrayed her own relationship with her infant son, Grub. Her personal motivations are unknowable, but the textual narrative of the personal emphasizes the congruence of her own mothering, the utopian model, and the scientific inquiry. The forest's peaceable, open chimpanzee society, full of strong personalities, was a counterpoint to the dominance-organized and closed baboon unit on the dangerous dry savanna. Narrative mattered. Culturally, politically, and technically, the early Gombe accounts participated in contemporary European and Euro-American concerns. The accounts offered a peaceable kingdom, one part of the dual code of a culture obsessed with psychological explanations and therapies for all kinds of historical conflict and pain. Male aggression concerned Goodall, but it did not define what counted as chimpanzee society. These matters were part of the practices that shaped research at Gombe, not some suspect "outside" to the real action "inside" science.

In Japanese accounts of their chimpanzee study population in the Mahale Mountains, observed from 1965, the concept of a "unit-group"—a multimale, bisexual group of 20–100 animals—was emphasized (Itani and Suzuki 1967; Nishida 1990b). The group was described as fluid, breaking into different subgroups with exchange of members among neighboring unit-groups. Resulting from their search to identify the social unit as the first task of a proper study, the Japanese emphasis on a unit-group was consistent with their general methods. For the Japanese, the rational starting point of an explanation was not the autonomous individual; they did not begin by seeking to explain the slightly scandalous (to a Westerner) fact that many animals live in groups whose members, beyond the mother-infant primal One, seem mostly to like being with each other.

After the Japanese reports, Gombe workers began to describe chimpanzee groupings in terms of the concept of a "community." The community concept at Gombe was constructed from observations of male associations and interactions (Bygott 1979; Wrangham 1979a, 1979b). Females were assigned to communities as a function of the frequency of their interaction with males, whose own interactions were the independent variables. Chimpanzee males engaged in more overt violent and affiliative behaviors with each other than the females did, and the patterns established the core and boundaries of a social unit. Bygott described females as living in the male community, more or less as valuables within the shared male ranges (Bygott 1979, 407). This meaning of a bisexual community was not what Japanese workers in the Mahale Mountains meant by a bisexual unit group. The focus among the European and American men who followed male chimpanzees at Gombe seemed to be the problem of "human" aggression, as that essentialized attribute of "human nature" was materially-semiotically constructed in psychological, evolutionary, and mental health practices, including primate studies. Goodall shared this framework with the students, and the issue was basic to Hamburg's interest in the chimps. The chimpanzee community was the ahistorical natural-technical object for examining "male" violence and cooperation. These kinds of studies were tools for constituting what it meant to be male in Western scientific societies.

In a situation that would later be seen as a logical scandal, female behavior was not at the center of early sociobiological formulations of natural selection and inclusive fitness, as they began to seep into the increasingly Darwinian Gombe accounts. The gender-stereotypic (and gender-constituting) interest of male observers in chimpanzee male behavior of certain types, leading to a natural-technical object of knowledge called a community defined in terms of male associations, was initially unchallenged by the emerging "new" explanatory frameworks.

Meanwhile, at Gombe women quite different from Goodall were producing accounts of female lives, like Anne Pusey's study of female transfers between the male-defined communities. She noted the similarity of her picture to Japanese descriptions of female movements. The absence of data on female-female interactions and female behavioral ecology began to be remarked in the literature, and graduate students planned field studies to explore the topics (Pusey 1979, 479; Smuts, interview, 18 March 1982). Primate workers began to understand that sociobiological explanatory strategies destabilized the centrality of male behavior for defining social organization. Female reproductive strategies came to look critical, unknown, and complicated, rather than like dependent (or silent and unformulated) variables in a male drama.

Human female observers at Gombe pressed their arguments with their male associates in the field and in informal transnational networks. In general, since the non-Tanzanian men were not then taking many data on females, they were not in a position to see the new possibilities first. I think the Western women generally had higher motivation to reconsider what it meant to be female. Several of the women in my interviews in the mid-1980s reported personal and cultural affirmation and legitimation for focusing scientifically on females from the atmosphere of feminism in their own societies. The men I interviewed also reported a growing sense of legitimation in the 1970s for taking females more seriously, coming from the emerging sociobiological framework, from the data and arguments of women scientific peers, from the prominence of feminist ideas in their culture, and from their experience of friendships with women influenced by feminism. It is not possible in principle to build a causal argument from these reports, even if unanimous, but the construction of scientific knowledge is implausible without these dimensions, where "inside" and "outside" are unstable rhetorical emphases. Implosion is more evident than separation.

In that context, especially in light of the scientific-personal friendship with Barbara Smuts in the key period of rethinking, I read Richard Wrangham's use of sociobiological resources to formulate his papers on chimpanzee behavioral ecology. In interviews, Smuts and Wrangham both recalled a rich brew of conversation about females, selection theory, Robert Trivers's ideas about females as limiting resources for males, and missing data on female behavioral ecology. Published during this period of intense interaction, Wrangham's papers developed the theoretical perspective of behavioral ecology to redraw ape society. His explanations centered female foraging and social strategies as independent variables, in relation to which male patterns would have to be explained. Simultaneously, similar ideas were developed for evolutionary theory of vertebrate society generally (Wrangham 1979a, 1979b, interview on 13 August 1982; Wrangham and Smuts 1980).

Sociobiological theory really must be "female centered" in ways not true for previous paradigms, where the "mother-infant" unit substituted for females. The "mother-infant" unit had not been theorized as a rational autonomous individual; its material-semiotic functions were different, located in the space called "personal" or "private" in Western narrative practice. The sociobiological kind of female-centering remains firmly within

Western economic and liberal theoretical frames and succeeds in reconstructing what it means to be female by a complex elimination of this older special female sphere. In sociobiological narrative, the female becomes the calculating, maximizing machine that males had long been. In locally relevant gender symbolism, the "private" collapses into the "public" (Keller 1992a, 148). The female is no longer assigned to male-defined "community" when she is restructured ontologically as a fully "rational" creature. The female ceases to be a dependent variable when males and females both are defined as liberal man, i.e., "rational" calculators. The practical effect of constructing this "female male" was to legitimate data-collection practices that made both men and women watch females more and differently. The picture that emerged of female lives has been full of rich contradictions for the logical model of stripped-down individualism that legitimated the investigation.

It is impossible to account for these developments without appealing to personal friendship and conflict, webs of people planning books and conferences, disciplinary developments in several fields (including practices of narration, theoretical modeling, and hypothesis testing with quantitative data), the history of economics and political theory, and recent feminism among particular national, racial, and class groups. The concept of situatedness, not bias, is crucial.

Female-centered behavioral ecology, however, is not the "good" ending to a story that began with Yerkes' caged mates bumping each other aside for food while modeling heterosexual family life for rapt scientists with data sheets. The boundary object called the unit of chimp society remains in the hot trading zones of scientific practice, where data systems, personal and cohort friendship and enmity, theoretical narratives, national and institutional inheritances, local chimp doings, gender-in-the-making, and more are the machine tools for crafting scientific knowledge.

The chimpanzees of Gombe structure my program for "Morphing in the Order." And so, appropriately, on the front cover of *Science* magazine for August 8, 1997, a touching portrait of old Flo's adult daughter Fifi (now thirty-eight years old with seven surviving children of her own) and baby grandson Fred highlights updated accounts for my primate revisions. Several threads come together. The lineage of Gombe workers reproduces itself, even as the scientists focus on the differential reproductive success of the chimpanzees. A graduate student at the University of Minnesota (Jennifer Williams) publishes with her senior mentor Anne Pusey, from the generation of sociobiologically influenced researchers that followed and in many ways challenged Jane Goodall, who is the third author of the 1997 *Science* article (Pusey, Williams, and Goodall 1997).

The central achievement of the publication is a statistically significant demonstration that differential female reproductive success—measured as infant survival, rate of maturation of daughters, and the rate of annual production of babies—can be correlated to female dominance rank. At least at Gombe, for these fifteen adult female chimpanzees, the correlation holds and is pregnant with testable behavioral ecological questions gestating in the pages of *Science*. That is, the correlation holds if the highest ranking female is excluded because she remained sterile for the twenty-eight years the scientists could account for. Appropriately, Flo's lineage shines with reproductive achievement at Gombe. The achievement of the scientists rested on thirty-five years of collective work by Tanzanian and foreign observers embedded in diverse institutional, cultural, and individual matrices. The central artifact that allowed the important new knot to be tied in the web of collective knowledge was the record of the exchange of pant-grunts between chimp females from 1970 to 1992. It is on just such homely stuff that the credibility of Darwinian understandings of life depends. That labor-intensive examination was only possible because of the initial and subsequent systems of record-keeping at Gombe and the transcription of those data into a computer-based data retrieval system beginning in the late 1970s at Stanford. Those data systems are materializations of mostly invisible conflictual and collaborative work to produce "good enough" categories, practices of collection, and mobility and comparability of records.

Examination of the noises made infrequently by female chimps to each other over twenty-two years made sense because of prior narrative and theoretical transformations. In particular, the drama of evolution had to feature the idea that *females* evolved—that they differ *from each other* in ways consequential for natural selection, that is, for differential reproductive success of individuals (or some other bounded unit in the story). Females had to be "strategists" in the great games of productivity and efficiency. Females had to be inventive in Darwinian terms. The ability to state such a thing explicitly, in testable formulations, took the same kind of conflictual and collaborative work by scientists and nonscientists as did sustaining a field site with its transnational and multimedia data tendrils (Brody 1996). In Yerkes' world, such a thing was unthinkable, literally.

Such a thing is also unthinkable in naturalcultural worlds that do not *think* action in terms of bounded possessive individuals. Again, I recall Strathern's admonition that it matters what ideas are used to think other ideas with. The people she worked with in Papua New Guinea do property, reproduction, gender, and dominance differently (Strathern 1988, 1994). If these Melanesians did primate studies *with their own categories for thinking person, action and interaction,* the Japanese, Europeans, Americans, and other primate science producers would have to reimagine and retheorize the history of life in order to do good science. Tropes matter, literally.

Comparing three commentaries on the Pusey, Williams, and Goodall (1997) paper collects up remaining lines of force imploding in the gravity well of Gombe. The first is by the authors themselves, who speculate about the consequences for genetic diversity in this endangered species if the kind of reproductive skew at Gombe over the last thirty years prevails in other populations. The authors' speculation highlights the consequences of habitat destruction and fragmentation that intensify genetic depletion for endangered species worldwide. In "Perspectives" in the same issue of *Science,* besides commenting on the importance of support for the idea that female chimps vary in fitness, Wrangham (1997) translates the rare, low-key expressions of dominance among females (the pant-grunts) into the idea of "covert rivalry," which he analogizes to "cuckoldry" by chimp females in a study in the Ivory Coast; these females got pregnant at high rates from copulations with males outside their "community." Females seem mighty secretive in Wrangham's story. Speculating about costs and benefits for these extra-group matings, he did not suggest that genetic diversity, rather than his inference of "choosing genes," might be the payoff for females. He emphasizes that "until this year no one suspected that female chimpanzees were so active in pursuit of their reproductive interests, yet they are probably doing still more than we appreciate" (1997, 775).

Despite his provocative tropes, Wrangham abstemiously cautions against analogies with humans, but my third commentator is not so severe. The prize-winning *New York Times* science writer Natalie Angier could never be accused of deemphasizing sex, competition, and violence in her riveting accounts of life's ways for the Science Times. True to form, Angier finds grounds for reading the cunning and power of Federal Reserve Chair Alan Greenspan in the doings of chimp supermom Fifi. Her message, however, is close to Wrangham's and Pusey et al.'s, and her language is no more ripe than Wrangham's. "Beneath the females' apparently distracted exteriors skulked true political animals" (Angier 1997, B11). Pusey et al. and the commentators agree in the speculation that the females' subtle, consequential dominance might be exercised through conferring "better access to food, both by enabling a female to acquire and maintain a core area of high quality and by affording her priority of access to food in overlap areas"—a testable socioecological idea favored in 1990s evolutionary biology (Pusey, Williams, and Goodall 1997, 830).

Several things imploded in the 1997 Gombe report. First, the bidirectional traffic between professional and popular science remains thick. The way of troping individuals and action by an Angier shapes a Pusey or Wrangham just as much as the reverse, and all are shaped in the cauldron of naturalcultural life in the New World Order, Inc., where flexible accumulation and diversity management are the high-stakes transnational games. Second, science is crafted in this world not as ideology, but as materialized action in thick histories of work, where tools for thinking and doing are relentlessly tropic. That means the ways of doing science are contingent on levels that make many people, scientists and not, nervous. Next, in the 1990s the power and agency of human and nonhuman females are still produced as front-page news in science journals and the daily press, at least in the United States. Finally, studying primate science shows the precious achievement of such knowledge even while emphasizing the situated character of the achievement.

I want to give the last word in this unfinishable essay to workers with experience watching chimpanzees in the Mahale Mountains. In his conference paper for this volume, Takasaki Hiroyuki[7] noted that members of the Kyoto school of primate studies do not generally find Wrangham's approaches to describing or explaining chimpanzee social grouping very fruitful (e.g., Hasegawa 1990). Part of the explanation Takasaki offered is a Japanese "culture-language complex" that, from the angle of common Western perspectives, reverses the relations of part and whole, individual and society, and other organizing polarities for explanations in biosocial sciences. Happy as I was with his account of difference, I had trouble sleeping on the soft bed of cultural relativism. Like the seed under the mattress in the fairy tale about the princess and the pea, there was something unsettling to the soothing surface of contrasts between East and West.

I found the pea in Hiraiwa-Hasegawa Mariko's (1990) paper on "Maternal Investment before Weaning" in *The Chimpanzees of the Mahale Mountains*. There, in a study rigorously focused on the mother-infant pair, but not in Goodall's 1960s social-functionalist frameworks, Hiraiwa-Hasegawa fluently deployed sociobiological explanatory narratives and associated quantitative methods to examine

> aspects of chimpanzee maternal care before weaning: nursing, infant transport, and grooming. The first two aspects were selected because they apparently inflict costs on the mother. The third is a typical primate social activity on which a considerable amount of time is spent during the day. . . . Because an individual's time for social behavior is limited, the time a mother allocates for grooming her infant is regarded as a form of maternal investment. (1990, 257–58)

The languages of cost-benefit investment strategies flow freely in this paper. Her acknowledgments hint at the transnational webs of primate studies, where interactions among the bisexual and multinational groupings of scientists weaned at Gombe and Mahale are a microcosm of the disciplinary, institutional, narrative, personal, and other trading zones in primate studies.[8]

This seems a good place to close my own far-from-innocent account. Remembering that in the world of human and nonhuman primates, "all things must begin with an act of love," I hope that my shaggy-dog story of intercourse among science studies, primate studies, and feminist studies can participate a little bit in making it true someday that "[primates] everywhere have more options than ever before." For that hope to be realized, the old naturalcultural issues of survival, justice, diversity, agency, and knowledge in science and politics are as sharply relevant as ever in the primate order. If, as Strum and Fedigan put it, the lines between science and advocacy and between basic and applied science are increasingly blurred for field biologists—and, I would add, for science studies and feminist scholars—perhaps primatology is, after all, mission science.

NOTES TO CHAPTER NINETEEN

1. *Mentula mulieribus,* the "little mind of women," is an Early Modern term for the clitoris.

2. For "diversity" as an object-in-the-making, see Wilson 1992, Shiva 1993, and World Resources Institute et al. 1993. On the biological world in terms of flexible strategies and the traffic between political and biological economies, see Martin 1992, 1994, Harvey 1989, 147–97, and Haraway 1991a, 203–30. This traffic dates from circulations of the concept of division of labor among political economists and biologists from the late 1700s (Limoges 1994).

3. Who cleans up after whom can tell much about how the world is built, including the scientific world, even in its squeaky-clean theoretical game rooms. Studying phenomena from the angle of those who do the cleaning up—from the position of those who must live in relation to standards that they cannot fit (Star 1991)—can be the most powerful scientific (and moral) approach. What happens in 2000 to the humans and nonhumans who cannot be flexible strategists?

4. Early doctoral dissertations by Gombe researchers, with graduate and/or undergraduate degrees mainly from Cambridge and Stanford, showed men largely writing about males, and women about the females and kids (Haraway 1989, 174, 404). The significance is neither self-evident nor the pattern necessarily typical. Tanzanian male field staff, without Ph.D.s, have also shaped and been shaped by primate studies (Goodall 1986, 597–608).

5. Useful for primatology, Clarke and Montini (1993) use social arenas analysis to show how communities of practice constitute and contest for the abortifacient RU486.

6. One female ape dissented from Yerkes' scoring practices (Herschberger 1948, 7, 11).

7. I keep the Japanese convention for ordering names as a reminder of the ways Japanese practices have to be translated into Western formats to be known by Westerners; the reverse is not true.

8. Nishida was Hiraiwa-Hasegawa's advisor; Kelly Stewart, Sandy Harcourt, and Timothy Clutton-Brock got thanks; English-language translation and Japanese science-funding systems were noted.

20 Life in the Field: The Sensuous Body as Popular Naturalist's Guide

Gregg Mitman

The lasting pleasures of contact with the natural world are not reserved for scientists but are available to anyone who will place himself under the influence of earth, sea and sky and their amazing life.

Rachel Carson, *The Sense of Wonder* (1965)

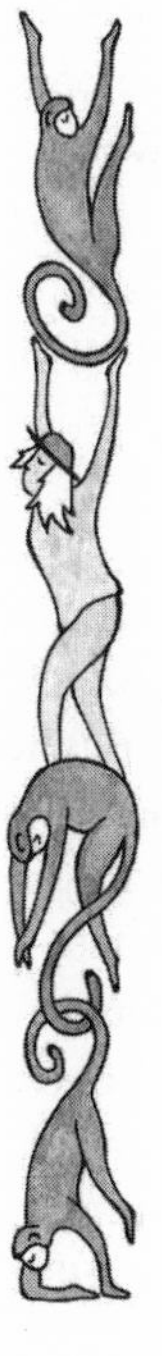

In the last ten years, anthropologists, historians, and sociologists working in the field of science studies have largely undermined a view prevalent within the scientific community that equates the popularization of science with dissemination from an elite group of knowledge producers to a passive lay culture of consumers (Cooter and Pumfrey 1994; Hilgartner 1990). Frustrations inevitably arise among scientists when they find that such a diffusionist model fails to account for what happens. To understand science within popular culture, it is necessary, as Stephen Pumfrey and Roger Cooter have recently argued, to "begin to comprehend the ways in which science was actually practised in popular culture, and recover a deeper politics and ethnography" (Cooter and Pumfrey 1994, 243). The term popularization is itself an obstacle because it highlights modes of transference and the distorting influences contained within dissemination processes rather than focusing on what scholars in popular culture studies have long known—meanings are not given, but made: through acts of resistance, struggle, and appropriation.

In this paper, I wish to explore the subject of science in popular culture, as distinct from the popularization of science, because popular science has itself been pivotal in shaping our understanding of primates and claims made about ways of knowing animals. I focus on the ways in which popular culture in 1950s and 1960s America framed nature and life in the field. During that time, a postwar generation of animal behavior researchers developed defined procedures and instrumentation technologies to transform nature into a laboratory. They adopted a remote analytic view that they hoped would introduce greater distance between subject and object and guarantee a high degree of scientific reliability and control (Mitman 1996). Yet within public culture, a number of these same biologists found and contributed to a popular natural history that emphasized the experiential knowledge of the body, a knowledge grounded in the sensory and emotional experience that individuals could possess regardless of professional training. As ethologists like Konrad Lorenz and Niko Tinbergen realized, amateur bird watchers, photographers, and hunters often displayed a patience and passion for observation that led to a craft knowledge of wildlife habits which often surpassed the understanding of trained biologists (Lorenz 1990).[1] In exploring the contours of an alternative epistemology found within the practice of popular natural history, I suggest that a methodological approach centered on individual experience, patience, and emotion was not confined to the practices of "women in the field" as some have argued, particularly in our understanding of primates. It was, instead, part of a "public" science of nature that came to intersect with the revival of naturalistic field studies in animal behavior.

Disney's Nature

The postwar revival of naturalistic field studies, not only in primatology, but in other areas of animal behavior, coincided with a surge of public interest in nature as a commodity for consumption. Rachel Carson's *The Sea Around Us,* published in 1950, sold over 200,000 copies and stayed on best-seller lists for 86 weeks (Carson 1950). It marked the first of a preponderance of natural history books that became popular during the fifties. An emerging interest in the environment, coupled with increased leisure time and disposable income, prompted Americans to travel to places like Florida's Marineland, national parks, and distant continents such as Africa where glimpses of wildlife could easily be had (Hays 1987;

Mitman 1993; Nash 1977). The expansion of nature subjects on television shows such as Marlin Perkins's *Zoo Parade* and the financial success of Walt Disney's *True-Life Adventures* made nature accessible for mass consumption. *The Living Desert,* for example, the first full-length feature in Disney's *True-Life Adventures,* grossed four million dollars at the box office and is largely credited with saving the studio from financial bankruptcy. In this nature of popular culture, a distinct popular science of natural history emerged, one that needs to be considered in its own right and not as a vulgarized form of professional science (Mitman 1999).

Disney's *True-Life Adventures,* which began with the short-subject *Seal Island* in 1948 and expanded to full-length features with *The Living Desert* in 1953, presented an image of nature in which the public came to know wildlife on intimate terms. Skillful editing and spectacular photography, in addition to the democratic ideology of the camera and an epistemology that underscored experiential knowledge, combined to create an image of intimate contact and a sense of the universality of aesthetic experience in nature. To make the audience feel a part of nature, Disney had to eliminate any sign of human presence or any hint of artificiality. In a directive sent to nature photographers interested in selling footage, the Disney studio emphasized that there was to be "no evidence of civilization or man's work in the picture" (Walt Disney Studios 1952). Even though scenes in Disney's *True-Life Adventures* were staged, and footage shot a thousand miles apart by different photographers was pieced together to create dramatic sequences, Disney took great pains to emphasize that in these films nature wrote the screenplay; the eloquence, the emotion, and the drama were nature's own. Unlike later *National Geographic* films, where the scientist often served as mediator between nature and popular culture, Disney's *True-Life Adventures* offered the public a direct experience of nature—they were meant to be perceived as a faithful rendition of nature in the raw.

Disney's portrayal of a timeless, pristine nature and his choice of settings such as the Pacific Northwest, and particularly Alaska, for many *True-Life Adventures* coincided with the postwar public campaign efforts for the preservation of wilderness areas in the far north by leading conservation organizations such as the Conservation Foundation and the Wilderness Society. The completion of the Alaska Highway in 1948 threatened what many conservationists like Robert Marshall, founder of the Wilderness Society, had hoped in 1938 would become a permanent place to relive "pioneer conditions" and the "emotional value of the frontier" (Nash 1982, 288). In 1956, under the sponsorship of the Conservation Foundation, the New York Zoological Society, and the Wilderness Society, the prominent conservationist and naturalist Olaus Murie, together with his wife Mardy, the young naturalist George Schaller, wildlife photographer Bob Krear, and University of Alaska zoologist Brina Kessel, spent three months conducting an extensive survey in the Sheenjek Valley of the Brooks Range in northern Alaska. The information collected and the film narrated by Olaus and Mardy Murie, *Letter from the Brooks Range,* proved influential in the 1960 establishment of the nine-million-acre Arctic Wildlife Range (M. Murie 1962; O. Murie 1973).

Throughout the fifties, Disney's naturalist-photographers and conservationists alike often appealed to the frontier conditions of the wilderness as the place where the values of American individualism could still be found. In the Alaskan wilderness, where people were "farther removed from the central hive," Murie and his associates found "individuality and what [they] believed was a promising outlook on the world stemming from the strong life in the wilderness" (Murie 1953, 12). To Murie, the freedom of wilderness and the individualism engendered through the hardships of a strenuous life served as important counters to mass society and the organizational man. His beliefs in the values of wilderness were echoed by his friend Supreme Court Justice William O. Douglas. In 1954, at the age of 65, Murie was one of 37 people to accompany Douglas on his famous 175-mile hike along the Chesapeake and Ohio Canal to fight the federal government's plan to turn the canal into an interstate highway from Cumberland to Washington. Two years later, Douglas visited the Brooks Range expedition for a few weeks at the end of July. In a series of essays published in 1960, Douglas spoke of the therapeutic values of wilderness for those uneasy with the conformist trends found in an affluent, consumer culture. "The struggle of our time," wrote Douglas, "is to maintain an economy of plenty and yet keep man's freedom intact. Roadless areas are one pledge to freedom. With them intact, man need not become an automaton. There he can escape the machine and become once more a vital individual" (Douglas 1960, 101). The image of Mardy Murie in *Letters from the Brooks Range* washing clothes by hand in frigid, Arctic lake water affirmed the values of individualism forged by the hardships of frontier life.

The experiences recounted by Disney's naturalist-photographers gave overwhelming support to the image of wilderness as a symbol of freedom and a place where individual identity could be restored. Nowhere is this more evident than in the accounts written by and about Herb and Lois Crisler, pioneer photographers traversing Disney's nature frontiers, whose film credits include *The Olympic Elk, The Vanishing Prairie,* and *White Wilderness.* The Crislers' careers as wildlife photographers began when Herb Crisler, an avid sportsman, waged a 500-dollar bet with the *Seattle Times* in 1931 that he could travel across the wilderness of the Olympic Peninsula with nothing but a pocketknife and come out, thirty days later, fat. Although he won the bet, Crisler came out of the wilderness thirty pounds leaner and transformed by the experience. Abandoning the gun, he picked up the camera in the hopes of defending wildlife and the last remnants of wilderness. His wife, Lois Crisler, left her position as instructor at the University of Washington, and the couple ventured into the Olympic Mountains and the Colorado Rockies to photograph wildlife (Muir 1954). Upon seeing their first film, *The Living Wilderness,* Olaus Murie wrote to Disney describing it "as the most beautiful picture on an outdoor subject" he had seen. "Here is a real life, real wilderness, film," he told Disney, "produced by a rare combination of wilderness understanding and imagination, that seemed to me to have many of the qualities of *Fantasia.*" Murie encouraged Disney to pursue this theme in future *True-Life Adventures,* and the Disney studio contracted the Crislers as wildlife photographers.[2]

In her 1958 book *Arctic Wild,* Lois Crisler describes the couple's experiences during eighteen months spent in the remote regions of the Brooks Range, with a winter interlude in Point Barrow, Alaska, to photograph wildlife, in particular, caribou and wolves, for Disney's feature film *White Wilderness.* The book is both a story of wilderness survival and an ethological study of wolves. In fact, Crisler portrays her gradual understanding of wolves as dependent upon her own discovery of freedom and independence.

Before departing for the Brooks Range, Herb Crisler made arrangements with a bush pilot to have a litter of wolf pups captured that the couple would raise and film during their stay in the remote wilderness. In July, two pups were transported to their summer camp, and Lois Crisler's journey into the world of another species had begun. At first, Lois was unable to free herself from the "filaments dragging [her] heart back toward civilization" (Crisler 1958, 11). She herself was imprisoned in the tameness of civilization; her understanding unable to break from the "myth-wolf pattern" that portrayed wolves as ferocious animals, an understanding that reflected more the self-righteous fury, anger, and nervousness of the human species than an insight into the nature of wolves. Immersed in wilderness, she gradually realized a freedom previously unknown to her. "People are free and of equality in a wild sunlit wood," she wrote. "No man is slave, no man is master, facing the sunlight on wild wood and wild fur and eyes. Liberty seeps like health into your heart . . . these past months, I had for the first time been exposed to genuine freedom" (Crisler 1958, 118). In the wilderness, where each creature was free to pursue its own destiny, Crisler discovered that wildness was not ferocity, but "independence—a life commitment to shouldering up one's own self" (Crisler 1958, 158). Only in the complete freedom of wilderness, Crisler argued, could one come to know the true nature of one's individuality and that of another being. Lois Crisler experienced the wolf's "selfness" as a flash of creative awareness realized through an understanding love. It was an exchange of pure emotion, made possible by the abandonment of security for freedom, an act of abandon which Herb Crisler understood to be as vital a part of human nature as of wild nature.

If Disney's naturalist-photographers emphasized the values of individualism and freedom in coming to know wildlife on intimate terms, Disney himself emphasized their craft knowledge and the ways in which nature was equally accessible to the public. The cultural appropriation of Disney by white, middle-class Americans was in part the result of Disney's open disdain for elitist views of mass society (Finch 1973). This egalitarian stance spilled over into the making of the *True-Life Adventures* and helps explain the enthusiastic reception of his nature films by the public and scientists alike. In filming nature, the Disney studio drew upon a wide range of talents, social backgrounds, and educational training for its photographers in the field. Disney emphasized that these photographers were "naturalists first." Under this category, he included "scientists, teachers, Park Rangers, and reformed hunters" (McEvoy 1955, 20–21). What united these individuals was not a common theoretical framework—they were sent into the field with no script in hand—but a patience and passion for watching nature. In refusing to distinguish between folk authority and scientific expertise, Disney privileged an experiential knowledge acquired not through academic training but through labor in the field (White 1995). "We find more and more that there's an awful lot of nature lore that gets preserved even in books and in print that scholars themselves—or supposed experts—tend to preserve even though it's false," remarked James Algar, director for many of the *True-Life Adventures*. Such errors occurred, Algar reasoned, because these experts spent more time digesting each other's articles than living with animals in the wild.[3] In the Disney version, nature was not a space accessible to only a select few of scientifically trained experts, but was open to anyone who had the yearning and diligence for understanding. The camera itself reinforced this democratic image, since photography homogenized class, ethnic, and social differences by "binding and unifying," in the words of art historian Jonathan Crary, "all subjects within a global network of valuation and desire"—in this case, the desire for contact with nature (Crary 1994, 13).

Disney boasted of the many discoveries his naturalist-photographers made while observing nature. His exultations were not completely unfounded. When Julian Huxley organized a Royal Society symposium on

"Ritualization of Behaviour in Animals and Man," for example, he wrote to Disney to obtain rare footage of the vertical "racing" of the western grebe and the courtship display of scorpions.[4] It was the contingent, however, not the repeatable and predictable, that Disney emphasized most. "We hope [the naturalist-photographer] will capture those unexpected and unpredictable happenings that cannot possibly be written into such a story ahead of time," remarked James Algar (Walt Disney Studios 1952). These "nuggets," as Lois Crisler described them, were chance episodes that occurred perhaps two or three times in an entire field season and revealed the individuality of animals in the wild (Crisler 1958, 34).

Through the use of techniques similar to the ones adopted in his animated cartoons, Disney sought to capture the "seeming personality of an animal," in order to help the audience "sympathize with it and understand its problems better" (Walt Disney Studios 1952). Disney expressed his own sentiments clearly when he told Olaus Murie how he loved to observe squirrels for hours at his desert home—"to me they have personalities just as distinct and varied as humans."[5] And Murie agreed. "All Nature," he wrote Disney, "has much in common among its various forms, certain general laws, certain general reactions, and much that can be predicted under many circumstances. But, and I hope this is not too paradoxical, there are many distinct facets that have individuality."[6]

It was this individuality in nature that naturalist-photographers looked for and which Disney sought to project on screen. Jack Couffer, a freelance wildlife photographer for many *True-Life Adventures,* believed that every animal possessed "some quirk of personality which sets that individual aside as one unique." The task of the photographer was "to find each animal's eccentricity and to somehow exploit it and incorporate the individualism into the story" (Couffer 1963, 152). The nature films of Disney had strong links to the artist-naturalist tradition of the nineteenth century which endeavored to bring an expressive, emotional quality to wildlife painting that elevated the "uncommon or unique event over the repeatable" (Blum 1993, 113). This was the contingency of nature, the "nuggets," that Disney photographers sought to capture. Not surprisingly, James Algar, the principal director and writer for *True-Life Adventures,* often found inspiration in the writings of the early twentieth-century nature writer Ernest Thompson Seton, who emphasized the "personality of the individual . . . rather than the ways of the race" in crafting animal stories (Seton 1917, 9–10).

Since Disney believed animal behavior revealed the "instinctive beginnings of the deepest, most basic human emotions," eliciting those emotions became an important means for getting the audience to identify with animals on screen (Disney 1953, 106). Musical accompaniment was a critical component of the *True-Life Adventures,* because the emotional motifs of theme music provided continuity to the story, while variations on the major themes, synchronized to the actions on screen—a technique borrowed from animated cartoons—added personality to the individual characters. Although music added a human emotional dimension to the personalities of individual animals, those involved with the making of *True-Life Adventures* objected to criticisms that they had anthropomorphized nature. James Algar defended the technique, since it permitted the "audience to identify with the creatures."[7] The photographer Jack Couffer also saw anthropomorphism as a helpful means for understanding. "Since no one *knows* what an animal thinks," wrote Couffer, "what an animal does must be interpreted—put into human terms—for us to understand" (Couffer 1963, 15).

Disney's stance on anthropomorphism was a position that ethologists and primatologists such as Julian Huxley, George Schaller, and Jane Goodall would similarly come to defend, although largely within the confines of popular culture. The naming of individual animals and the family portraits that appear in field chronicles, from the sketches of The Dandy, Granpa, and Robber Mask in Adolph Murie's *The Wolves of Mount McKinley* to the photographs of Flo and her family in Jane Goodall's *My Friends the Wild Chimpanzees,* convey a sense of intimacy, as though something is revealed to the viewer about the personalities and relationships of these individual lives (Murie 1944; Lawick-Goodall 1967).

While professional biologists occasionally objected to the cute commentary, or criticized scenes such as the woodcock choreographed to samba music in *Nature's Half-Acre,* in general, the scientific community either overlooked or chose not to object to Disney's humanizing of animals. Factual errors that appeared in films were also casually dismissed in reviews. Disney's *True-Life Adventures* won the respect and admiration of the professional biological community, not because of their factual content, but because of their aesthetic appeal. Anthropomorphism was just one means by which Disney captured the emotional elements of nature, and it was his ability to touch the emotional plane of viewers that biologists appreciated most. Viewed from the perspective of art, biologists found much in Disney's nature films to praise. Robert Cushman Murphy of the American Museum of Natural History summarized the opinion of many biologists in his review of *Water Birds* when he wrote: "A foremost aim in our branch of education is to instill a love of nature that will redound to its appreciation and protection. There is no better way to accomplish this than by taking advantage of aesthetic opportunities. This Walt Disney has done supremely well" (Murphy 1952, 330).

Popular Primates

Having established the prevalence of an epistemological space for the practice of popular natural history separate from the professional sphere, I now want to consider two popular accounts of primates that appeared in the early sixties: George Shaller's *The Year of the Gorilla* and National Geographic's documentary *Miss Goodall and the Wild Chimpanzees.* A diffusionist model of popularization would consider these works as derivative of scientific research in zoology and ethology, the respective disciplines in which Schaller and Goodall received their doctorates. I suggest that the primate studies of Schaller and Goodall achieved widespread popular acclaim, not by virtue of their elite scientific status, but by the ways in which they appropriated from and participated in a tradition of popular natural history that offered an alternative to elite professional science. This tradition, with its emphasis on individualism, sensory bodily experience, and emotional engagement, was familiar to a certain white, middle-class segment of the American public during the fifties and early sixties. Although Schaller dedicated his 1963 scientific monograph on the mountain gorilla to Olaus Murie (Schaller 1963), his 1964 book *The Year of the Gorilla* was more in keeping with the tradition that Murie symbolized.

In *Arctic Wild,* Lois Crisler believed that only in the freedom of wilderness could humans come to understand something about the nature of another species. George Schaller echoed these sentiments in *The Year of the Gorilla.* Upon his first sight of a gorilla, Schaller expressed the "desire to communicate with him, to let him know by some small gesture that I intended no harm, that I wished only to near him" (Schaller 1964, 35). Because the human senses are akin to those of gorillas, Schaller believed establishing a relationship would be relatively easy. But the senses dulled by the drone of human civilization would not suffice; one needed the senses of a "man attuned to the wilderness" (Schaller 1964, 113). Despite his wish for intimate contact with gorillas, Schaller had to first immerse himself in wilderness before he would come to know anything about their personalities and individual lives. "It takes time," wrote Schaller,

> to cease to be an outsider, an intruder, and be accepted once again by the creatures of the forest. . . . Once the senses have been relieved of the incessant noise and other irrelevant stimuli that are part of our civilization, cleaned, so to speak, by the tranquility of the mountains, the sights, sounds, and smells of the environment become meaningful again. Slowly the courage and confidence of man, previously nurtured by his belief in the safety of his civilized surroundings, slips away. Finally he stands there, a rather weak and humble creature who has come not to disturb and subdue but to nod to the forest in fellowship and to claim kinship to the gorilla and the sunbird. (Schaller 1964, 107)

Having crossed the threshold between culture and nature, Schaller found himself in another world.

The intimacy Schaller achieved is revealed by the introduction of individual group members and their personalities, such as The Outsider, Big Daddy, and Junior in group IV or Mrs. Blacktop, Mrs. Patch, and Mr. Shorthair in group VII. Different captions for the same photograph in Schaller's scientific monograph and popular field narrative illustrate the significance of individuality in popular natural history. In Schaller's scientific monograph, Junior is "a blackbacked male [that] displays the strutting walk" (Schaller 1963, plate 30). In *The Year of the Gorilla,* "Junior struts along a log" (Schaller 1964). While the caption in the scientific monograph conveys a sense of distance, the description in the popular narrative evokes a sense of sympathy and draws the reader in. Schaller defends anthropomorphism as a vehicle for understanding the behavior of wildlife. With support from Julian Huxley, one of ethology's self-proclaimed "fathers," Schaller argued that the resemblance of human emotional and instinctive behavior to that found in other animals made it both legitimate and operationally necessary "to ascribe mind, in the sense of subjective awareness, to higher animals." In the case of gorillas, Schaller believed that only by looking at them as "living, feeling beings" was he "able to enter into the life of the group with comprehension, instead of remaining an ignorant spectator" (Schaller 1964, 176). After months spent in the surrounding forests of Kabara, where he and his wife Kay "found a freedom unattainable in more civilized surroundings," Schaller found a common understanding between human and gorilla in which the gorillas "talked" to him at times "with their expressive eyes" (Schaller 1964, 137, 162).

Alone in the wilderness, surrounded by the Virunga Volcanoes of the Congo, Schaller felt Kabara as his own, shared only with his wife Kay, the gorillas, and other creatures of the African forest. His journey into wilderness and the life of the gorilla had taken him on a journey of personal freedom and independence, a quest ended abruptly by a different journey of freedom and independence, one undertaken by native Africans who sought to reclaim Schaller's personal paradise as their own. Schaller ends his chronicle of life in the field with a chapter titled *Uhuru,* Swahili for freedom. When decolonization of the Belgian Congo turned violent in the summer of 1960, the couple fled their idyllic Kabara home in Albert National Park for Uganda. Like the conservation networks that supported his study, Schaller viewed the freedom of native Africans as a threat to the freedom of wildlife and the personal liberation found in watching wildlife on intimate terms. His account of the mountain gorilla lent support to international conservation efforts that upheld tourism as the cornerstone to the continued survival of Africa's wildlife and the economic future of independent African states. Sponsored by the New York Zoological Society and the National Science Foundation, Schaller's two-year study of the distribution and life history of the mountain gorilla in the vicinity of the Belgian Congo, Uganda, and Rwanda was one of many ecological reconnaissance missions funded by the Conservation Foundation and the New York Zoological Society's African Wild Life Fund in the late fifties to develop a general, systematic plan of game management and land use in eastern and central Africa in which tourism played a central role. Undertaken in a national park, aided only by binoculars and a camera, Schaller's study conveyed an impression that anyone with a certain stamina and passion for wildlife could discover something of these creatures for themselves. One needed no scientific training; immersed in the wilderness, the senses and emotions, the experiential knowledge of the individual body, would serve as guide.

While Schaller implicitly legitimated a space for the wildlife observations conducted by outdoor enthusiasts, Jane Goodall and her study of the wild chimpanzees at Gombe similarly endorsed the life of the amateur in the field. In *Primate Visions,* Donna Haraway provides a compelling cultural analysis of the National Geographic specials about nonhuman primates that began with the 1965 documentary *Miss Goodall and the Wild Chimpanzees.* Within these dramas of touch, Haraway argues, white woman fulfills a mediating function, enabling nature to approach man through a triple code of gender, science, and race. "Gender in the western narrative works simply here," writes Haraway. "Woman is closer to nature than Man and so mediates more readily" (Haraway 1989, 149). The camera, however, "rests firmly in the hands of men," an act which Haraway reads as a sign of distance that is permanently maintained in "Man's mediation of the touch with nature" and why women are constructed in these films to play emissarial roles (Haraway 1989, 150). Haraway's perceptive reading of these National Geographic documentaries is pursued from an angle that emphasizes the scientific origins of cinema; the camera is an investigative and reproductive technology that enabled natural history disciplines to mirror more closely the distanced analytical view integral to the highly mediated world of twentieth-century life sciences (Cartwright 1995). But the mythic origins of cinema are rooted in both art and science. As an artistic form, nature film could capture and reproduce the aesthetic qualities of wilderness; the camera's ability to trade in emotion was the quality that individuals like Olaus Murie and Walt Disney prized. In the following analysis, I wish to approach *Miss Goodall and the Wild Chimpanzees* from a different vantage point than Haraway's, one that considers the film in relation to well-established conventions in nature documentary that foreground the formative tendencies of cinema to capture elements of the human heart (Kracauer 1960).

The film's opening sequence of a chimpanzee performing tricks at a circus and cut to chimpanzees in the wild reinforces the distinction between knowledge of a species in captivity, where the gaze of spectator and scientist is veiled in the trappings of civilization, and understanding of a species in its natural habitat, where chimps "play unguarded, unobserved, wild and free." Arriving by boat on the shores of Lake Tanganyika at the site of the Gombe Stream Chimpanzee Reserve in Tanzania, Jane Goodall is identified by the narrator Orson Welles as a "twenty-six-year-old English girl," who apart from an African cook "is on her own." While "her discoveries will startle the scientific world and lead to the possible redefinition of man," she is "a girl with no special training, but with natural aptitude." Goodall's amateur status in the film is significant. As someone with no scientific training, Goodall is just one among the many viewers of *National Geographic* television specials who, like her, "simply" want "to be amongst animals in the wild." Her powers of observation and understanding are honed not by theory but by her love of animals. The film evokes a sense of desire among viewers that they too might become part of nature and discover something of wildlife, either for science or for themselves. The rite of passage into this animal world is not science, but one's gradual immersion in wilderness (*Miss Goodall and the Wild Chimpanzees* 1965).

The narrative structure in the film closely follows conventions familiar to audiences well versed in popular natural history lore and films of the fifties. Although Goodall was accompanied by her mother, the film screens out any signs of human companionship, except for that of an African cook, his family, and an African aide. Editing has rendered Goodall alone in the African wilderness, an effect mirrored by the darkness and silence of the theater, which accentuates the viewer's isolation in what is admittedly a very public space: both heighten the sense of intimacy with the animals on screen. "Armed only with binoculars," Jane Goodall sets out in search of contact. Two months pass, but Goodall still has no direct sight of wild chimpanzees. After two months, an opening appears. On top of a mountain peak, Goodall sights a group of chimpanzees, but she cannot get closer than 500 yards.

The audience awaits with eager anticipation of closer contact. Footage of the rainy season, of Goodall crossing a torrid stream barefoot in the rain, and of the chimpanzees getting soaked elicits the sensuous feelings experienced by Goodall in the wilderness. Only with Goodall's senses alive to nature can intimacy be achieved. In her book, *In the Shadow of Man,* Goodall wrote of her ability to get closer to chimps in the rain, a closeness which came after she became attuned to the pungent "smell of rotten wood and wet vegetation," the cold and slippery tree trunks, alive under her hands, the "water trickling through [her] hair" and "running warmly into [her] neck" (Lawick-Goodall 1971, 57). In the film, she is rewarded with the observation of tool-use among the chimpanzees, as she observes an individual making a leaf into a sponge to draw water out of a hollow in a log. But her attempts to move closer are thwarted. Only after almost two years "alone" in the wild is Goodall able to make contact with the chimpanzees, observing them from a distance of thirty feet. And only when Goodall crosses the threshold from culture to nature is the audience introduced one by one to the individuals and personalities of this chimpanzee group, beginning with Flo's family. The events leading to Goodall's moment of contact follow the same sequence found in almost every popular field narrative of the period: isolation from civilization, a gradual immersion in wilderness through an awakening to the senses of the body and nature, and understanding.

In a letter Jane Goodall wrote to Julian Huxley, which accompanied a copy of her general-interest book *In the Shadow of Man,* Goodall confessed that *In the Shadow of Man* "is a book I have wanted very much to write and, though written for the general public, it says I think, a good deal more than my scientific monograph. I can't help feeling it is a more meaningful statement."[8] Goodall was not alone in her belief that professional science lacked the means to fully capture and express the meanings of life in the field. In *The Year of the Gorilla,* Schaller similarly remarked on the inadequacy of his scientific monograph on the mountain gorilla. "My previous work," wrote Schaller, "is a compendium of facts, discussing the apes as subjects to be studied, not as acquaintances whose activities my wife and I discussed at the end of each day. I had no space to reveal the enjoyment I derived from roaming across grassy plains and uninhabited forests and climbing mist-shrouded mountains" (Schaller 1964, 12). The field accounts of Goodall and Schaller should not be conceived as transmissions of knowledge from science to the lay public; they were instead part of a vernacular natural history tradition that valued the sensuous body over disembodied observation in the quest for intimate contact with animals in the wild.

A historical analysis of popular naturalists challenges the diffusionist model of science popularization. It also calls into question explanations that tend to equate empathy, relationality, and concern for other individuals with the practices of women in the field. For example, in *Made from this Earth,* Vera Norwood argues that narratives of intimate contact and communication with wildlife written by women such as Lois Crisler and Dian Fossey were a product of the high value placed on empathetic skills in the gender socialization of women (Norwood 1993). Both Crisler and Fossey, however, were part of a tradition in popular natural history which, while seemingly more in keeping with feminine gender constructions, was not confined to the practice of women in the field. This tradition of popular natural history in postwar America had strong roots in the artist-naturalist tradition of the nineteenth century, in which an aesthetic vision defined the art of seeing. Through an immersion in nature's language—the language of the senses—empathy, a truth of feeling, a knowing of the individual animal, took place. Certainly, an emphasis on emotional understanding within popular natural history could help code it as a feminine practice, particularly when contrasted with professional animal behavior studies of the period. And women, as a result of their exclusion from professional science, certainly contributed to this popular tradition. But their participation alone is not a sufficient explanation to account for the distinctive epistemological framework of nature observation within popular culture. Furthermore, we have as yet no comparative historical studies that consider how the gender coding of popular natural history shaped the construction of gender identities for both male and female practitioners like Schaller and Goodall.

Conclusion

Treating popular natural history as a watered-down version of scientific study found within the professional fields of animal behavior, ethology, and primatology misses the point. As Roger Cooter and Stephen Pumfrey argue, "popular culture can generate its own natural knowledge which differs from and may even oppose elite science" (Cooter and Pumfrey 1994, 249). In this case, postwar American popular natural history centered on an alternative epistemology based on the experiential knowledge of the body. As John Fiske suggests, popular culture "often centers on the body and its sensations rather than on the mind and its sense, for the bodily pleasures offer carnivalesque, evasive, liberating practices" (Fiske 1989, 6). The intimate contact with wildlife achieved through an immersion in wilderness and bodily sensation was itself portrayed by naturalist-photographers of the fifties as a liberating experience, one that freed them from the constraining forces of mass society and a conformist culture.

Some historians have suggested that a tradition which highlighted individual experience, patience, and emotion was confined to the practices of women primatologists in the field. I have argued instead that it represented a popular naturalist tradition in which the body served as a guide. However, this understanding of popular natural history exposes the complexity of the interaction of science and popular culture, a complexity that certainly affects the questions about "primate encounters" and their answers.

I have dealt almost exclusively with the meanings that naturalists made of their experiences of nature and life in the field. But just as popular natural history should not be conflated with the sciences of animal behavior, so the meanings embedded within the practice of popular natural history should not be equated with or reduced to the meanings that a heterogeneous public made of the final version of nature that appeared on screen. The naturalist-photographer encounters with wildlife experienced in the field was quite distinct from the encounter witnessed on the television screen by the typical middle-class American family within the comforts of their suburban home. Understanding the heterogeneity of the public and the meanings made of nature, not just in the field, but in social settings such as the movie theater or home, is a topic in need of further study if we are to begin to comprehend the ways in which primates and the nature they inhabit become meaningful within and are mediated by diverse cultural experience and practice.

NOTES TO CHAPTER TWENTY

This chapter was adapted from *Reel Nature: America's Romance with Wildlife on Film* by Gregg Mitman for the Teresopolis workshop, and has been revised based on discussions that took place there.

1. Niko Tinbergen to John Sparks, 28 November 1982, MS.Eng. C3157, E36, Niko Tinbergen Papers, Bodleian Library, Oxford University.

2. Olaus Murie to Walt Disney, 29 July 1949, Box 264, Correspondence 1965–1949 File, Olaus Murie Collection, Western History Department, Denver Public Library (hereafter OMC).

3. Algar interview, 1968, Walt Disney Archives (hereafter WDA).

4. Julian Huxley to Walt Disney, 5 November 1964, Box 37, Folder 5, Julian Huxley Papers, Rice University (hereafter JHP).

5. Walt Disney to Olaus Murie, 4 December 1953, OMC, Box 264, "Correspondence, Jan. 8, 1963– " File.

6. Murie to Disney, OMC, Box 264, "Correspondence, Jan. 8, 1963– " File.

7. Algar interview by Richard Hubler, 7 May 1968, WDA.

8. Jane Goodall to Julian Huxley, 16 September 1971, JHP, Box 44, Folder 1.

21

Politics, Gender, and Worldly Primatology: The Goodall-Fossey Nexus

Brian E. Noble

Among the modern natural sciences, primatology has distinguished itself over the last four decades as an exceptionally public science, especially in and among those nations where science has been adopted as both an everyday and an institutionalized dimension of social life. Of all primatologists the most excessively public have been Jane Goodall and Dian Fossey,[1] and of all nonhuman primates, the most public are chimpanzees and gorillas.[2] Several matters are interwoven here: the lives of a number of African apes; fixations on ape-human commonalities; gendering of science; agendas of the National Geographic Society; the centrality of English-language and post-Darwinian science in the cultural production of "nature"; and changing social dynamics within the practice of primatology. Underwriting all this is the persistent matter of how certain technical and public communities of Euro-American folks work to *make* sense of themselves by selective, specialized contact with some nonhuman cohabitants of a common planet.

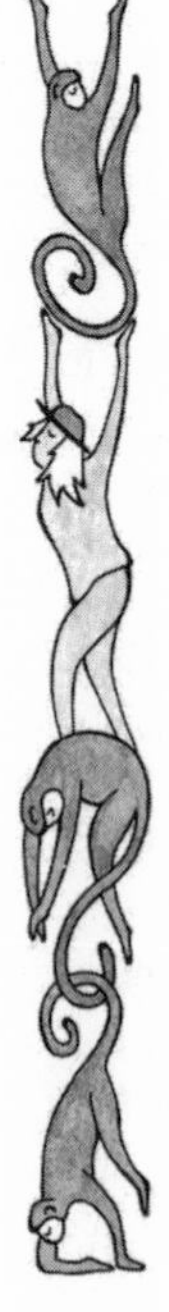

While rich in its own right, a consideration of how this complex of simian science as public spectacle came into being would present longer and denser histories than those explicitly bracketed by this symposium.[3] Rather, my discussion launches midstream into that meandering set of histories, following selected aspects up to the current moment. My aim is to consider the *nexus of mediations* of Jane Goodall and Dian Fossey in the relations of simian science and simian spectacle.[4] Let me first lay out what I mean by "nexus of mediations." "Nexus" refers to the collectivity of interacting agents and includes the primatologists, the apes, the instruments of scientific practice (e.g., vehicles, binoculars, notebooks, money, etc.), the local environmental elements, field assistants, camps, film crews, and so on. "Mediation" is what takes place in each engagement between two or more of these agents. At the most obvious level, through their technical and public writings, Goodall and Fossey have mediated their ape encounters. Before that, however, the field techniques in which they engaged have mediated the way encounters take place: gazing and note-taking, photographing and filming, producing charts, writing field narratives. Furthermore, the apes mediate their actions with one another, and in turn with the primatologists, in response to the local situation and moment.

The more familiar notion of mediation, as in the term "the media"—e.g., the mediating of the entire action via a National Geographic film on the Gombe or Karisimbe field camps—is also part of the nexus. It impinges not just on how the primatologist acts vis-à-vis the apes, but often has more distributed consequences, such as providing an additional source of funding for the primatologist's field program. Money changes the way research takes place. Journalists, local officials, scientific colleagues (indeed, science studies scholars!), field permit applications, availability of supplies and communications, corporate sponsorships, and so forth, all mediate in varying degrees what is produced in knowledge, visions, and outcomes of the science and spectacle activity. "Mediation" is used in both discrete and correlated senses, in that each individual moment of mediation counts, but so does the collective set of mediations. Therefore, to discuss a "nexus of mediations" is to try and sample the complex and trace its contours, knowing that any sampling can only be partial but will attempt, all the same, to register the heterogeneity of impinging actions and mediations.

In this sense, "mediation" is not singular but multiple, not isolated but integrated, not bounded but fluid, not determined solely by the primatologist nor by the primates, but as a consequence of all the mediating actors moving, trading, and changing through a complex set of actions in space and in time, from that which is recognized as the very local to the very global and vice versa.[5] Mediation is human and nonhuman, technical and public, momentary and historical, contingent and regulated. It is composite and multidirectional: apes change the humans and vice versa, public mediations change technical mediations, national and international policies change and are changed by all the mediations, as are environments and ideas about how people engage in environments. Primatologists, while exercising great agency in the local terms of the privilege they are accorded as knowing mediators of primate societies, are acting simultaneously in conscious and nonconscious response to this complex. I refer to this larger, dynamic set of relations both as the "mediation nexus" or, alternately, as "worldly primatology."[6] Primatologists and primates are anything but insulated from the world—they are locally and globally mediated, which is to say, they are *of the world.* I propose that it is in this extended complex that the shifting matters of gender, theory, method, and primate societies come to be forged. To discuss how these focal matters precipitate and become transformed, I want to turn attention to the workings of the <Goodall + Fossey + Ape> mediation nexus in this "worldly" manner. I will represent the nexus through various versions of this bracketed shorthand formula: <x + y + z>.[7]

In the past thirty-five years, this nexus of mediations of and by Goodall and Fossey and their ape allies—which includes the mediations of them by other agents of institutional and public culture—has expanded the terms of feminized primatological practice. Early on in this time span, women primatologists were often cast on the side of nature, as having a special, near empathic, relationship with the apes they studied. This took place within a larger network of masculinist mediation agents, such as National Geographic, objectivist science, male-dominated mass media. Today women primatologists are also positioned as political activists working as autonomous, self-conscious agents. Indeed, Euro-American women primatologists appear to be at the forefront of this activism, mobilizing the power of the mediation nexus itself. As I will argue, this signals a shift from dichotomous "woman-is-to-nature as man-is-to-culture" propositions (MacCormack and Strathern 1980; and cf. Keller 1989a) to heterogeneous gender propositions. It also signals shifts from passive to active politics in scientific engagements with primates, and an apparent decentering of theory by politics. A "political primatology" has been effected in the face of a dense, shifting mix of conditions. These conditions are complicated, encompassing transnational, environmental, biocultural, and political economic forces. Yet, Euro-American interests remain in a predominant position.[8]

In the Goodall-Fossey nexus of mediations, *primate societies* tend to mean the social lives of great apes. *Theory,* such as it is, attends to sex, reproduction, evolutionary transitions to humanness, and reports of "human-like" behaviors. *Practice* tends to be associated with white English-speaking women and *gender-coded* with the approaches—or *methods* one might say—of Fossey and Goodall. Ape protection, welfare, and conservation are highlighted. The dominant narrative frame is that of adventuring into spaces of pure, "African" nature, though arguably less so recently as politics enters the stories. In all, the nexus attracts and generates media coverage and is very well sponsored.

The exposés of Jane Goodall in Gombe by National Geographic in the 1960s mark out a pivotal period for the surfacing of the nexus of <primatology + spectacle + ape + woman researcher> in this tumbling, mixing stream of actions. The next section of this paper considers the Goodall nexus principally from the standpoint of how her highly participatory orientation in engaging chimpanzees came to have enormous cultural force. The following section, focussing on Dian Fossey, extends on these matters of participatory engagement with the apes, but samples more fully the way in which the mediation nexus has operated in sociopolitical terms. The 1988 film *Gorillas in the Mist* marks a resurgent, reconfigured public primatology, brought into high relief by Fossey's tragic death at Karisoke. Adding local and recent emphasis, in a 1995 series of interviews with eighteen-to-twenty-two-year-olds in a college coffee shop in Edmonton, Canada, I learned that Dian Fossey and Jane Goodall still stood out, by far and largely alone, as identifiable primatologists. National Geographic films, books, and articles along with the 1988 gorilla film were reported as the key resources for these formulations, so it is to them that I pay critical attention in this paper.[9]

Jane Goodall and the Gombe Chimps: Naturalizing Woman-Ape Communion

This study throws important light on general anthropoid behavior in animals that have almost human brains but lack the power of speech or the ability to transmit from generation to generation social discoveries by the use of written symbols. (Goodall 1963a, 274, inset text by Leonard Carmichael, then secretary of the Smithsonian Institution and chairman of the National Geographic Society's Committee for Research and Exploration)

[Figan] approached Flo in his typical cocky fashion and brushed her face with his lips. How similar to the peck on the cheek that is all a human mother can expect from a growing son! (Goodall 1965a, 825)

In these two statements in *National Geographic* magazine, there are some marked contrasts in the degree of liveliness, sophistication, and agency accorded to the same animals—chimpanzees. Carmichael's patronal comment about Goodall's research displays his investment in a distancing, dominating *view* of chimpanzees (and of Goodall). It suggests productive knowing from a superior position to chimpanzees, in terms of the lesser adequacy of their brains. The implicit contrast is with the grander mental instruments of another group of beings—humans. Goodall's confessional comment betrays a personal engagement with chimpanzees utterly opposite to that of Carmichael, yet one that pits a masculinized formulation of impetuous cocky son against feminized formulation of patiently caring mother.

Carmichael speaks from a historically particular scientism, Goodall from a historically particular humanism; Carmichael from above and beyond, Goodall from within and between. However, Goodall's positioning of chimps as human mother-son dyad suggests that this logic of self-assertive, dismissive son (i.e., male) against caring, perseverant mother (i.e., female) is an expression of universal human-chimp nature—maleness begets distance, femaleness connection. And, while Carmichael's endorsement of Goodall's research positions her efforts within a globalist theoretic frame, he also benefits from Goodall's affirming within the localist frame of her fieldwork the sex/gender logics which undoubtedly aided his climb to his position aloft. He speaks with paternal assurance both as the National Geographic Research Committee chair with insider knowledge and considerable say over who does and does not receive Society funding and as Smithsonian director wielding special authority over what stories do and do not get presented in the national museum of the world's most assertively powerful nation. In this logical if problematic way, the Carmichael and Goodall statements are oddly complementary, displaying an asymmetric and hierarchic order of power, position, and gender.

The two passages also encode gendered vocality of 1960s (and later) public science writing practices—Carmichael's mode associated with masculinized vocality, Goodall's mode with feminized vocality. The sense of separation in the Carmichael text and that of continuity in Goodall's is striking. In the Jane Goodall National Geographic productions, her commentaries are predominantly in a personal mode, and much less commonly in a remote scientistic mode. A sampling of her language and descriptions in her first *National Geographic* article is particularly revealing of the experiential approach she enjoyed, which in turn allowed her to make her observations in a particular manner. The personalized descriptive narratives are pervaded by anthropomorphic and sentimental characterizations, as when Goodall extends a description of the specific case of a chimpanzee constructing a leaf nest (Goodall 1963a):

Finally he bends in all the little leafy twigs that project around the nest, and the bed is ready. But the chimpanzee likes his comfort, and often, after lying down for a moment, he sits up and reaches out for a handful of leafy twigs which he pops under his head or some other part of his body. Then he settles down again with obvious satisfaction.

The projective association in Goodall's participation with the chimpanzees is well illustrated by an anecdote related to naming and physical appearances:

One chimpanzee had a pale, flesh-colored face instead of the dark color common in adults. It gave me a slightly eerie feeling when I first saw him close to, and ever after that he was "Count Dracula."

Goodall's unquestioning sense of communion with the chimps is expressed in descriptions of her provisioning interactions with the male David Greybeard:

Gradually he became tamer and tamer, but it was not until the last five months that David showed complete confidence in his human friend. Two of the palms in camp were ripe so I got in a great supply of bananas and devoted myself to David for a whole week.

Such passages reveal the spontaneous and unrestrained responsive orientation of Goodall, her sense of personal participation. Even such affective responses as "eerie feelings" were valued enough to be fixed symbolically in the naming of animals.

As those of a woman scientist, Goodall's modes, both of engagement and vocality, came to be encoded as "feminine." Her practice as such became emblematic as that which counts as a natural "women's science." In spite of its nonscientistic tenor, this approach did allow something very important, however, in that Goodall was positioned to report some very unexpected observations about chimpanzee behavior—most notably the more nuanced aspects of maternal-offspring relations. From *National Geographic* and other texts written by her or others on the approach she took, she made these accomplishments as a result of her highly personal, participatory practice—that is, by her capacity for communion with wild chimps. Biologist Charles Birch summed up this altered practice:

The non-mechanistic student of animal behavior tries to study animals in their complex relations with a complex world, as Jane Goodall has done with chimpanzees in Gombe Reserve. Her success was dependent upon her establishing a rapport with the chimpanzees and, presumably, vice versa. She tried to think like a chimpanzee and to imagine what it was like to be one. She was, in fact, taking into account what she perceived to be critically important internal relations in their lives. (1988, 73)

What Birch's point leaves unquestioned however, is whether Goodall's practice is peculiar to Goodall, or whether it may be taken as something inherently or contingently gendered.

Essences and Contingencies

On the cusp of the 1980s and 1990s two highly contrasting positions on the life, research, and public presentations of Jane Goodall were published as components of the more wide-ranging books *Walking with the Great Apes* by Sy Montgomery (1991) and *Primate Visions* by Donna Haraway (1989). For Montgomery, Jane Goodall, along with Dian Fossey and Biruté Galdikas, were described as having "shamanic" characteristics and special "ways of knowing," communicating, and interacting with their primate "subjects."[10] Haraway argued that Goodall's communion practices were historically contingent outcomes which, when packaged as documentary media, served the corporate interests of such organizations as Gulf Oil and the National Geographic Society. They reinstantiated a capitalist, androcentric ordering of nature. This order drew upon racializing and gendering stories to extend the position of Western white "woman as [Western white] man's emissary to nature." Haraway's position could be seen to entail Montgomery's.

The represented and enacted practices of Goodall, Fossey, and Galdikas in "communing" with their ape subjects gave the impression of this communing capacity as being "essential" or "natural" to these women, and so by association to women generally. The effect was forcefully produced by the interlaying of pictorial, textual, and mediation practices of the women on their own, in the context of concurrent primatological discourse, and of National Geographic stylization at the particular historical moments of their respective field activities. The structural logics required to effect these mediated senses of what counted as essential feminine practice are recognizably "modern,"[11] first of all requiring the separation of culture from nature, and the assigning of gender to these domains. The next move is to "naturalize" or "biologize" the connection of women to nature, and men to culture, a point effectively contested by feminist cultural theorists over the last two decades (cf. Keller 1989a, 1989b; MacCormack and Strathern 1980; Yanigisako and Delaney 1995). The complementing move is to circulate and sediment the naturalized position in a wide public domain—enter National Geographic as major player in the mediated production of "nature" in the "modern world system" (cf. Lutz and Collins 1993; Haraway 1989; Marcus 1995).

Louis Leakey's patronage of Goodall, Fossey, and Galdikas credited them with an alternate feminine "way of seeing" or "quality of perception" (Montgomery 1991, 78). He reinforced this recognition by lending them the privilege of his influence as lauded scientist and National Geographic figure and by mustering his funding connections to support their fieldwork (Goodall 1986, 2–3, 649). National Geographic had funded and popularized Leakey as man-scientist in the 1950s, and would do similarly now for three woman-scientists through the 1960s and '70s.

This gendering of scientific modalities within the mediation nexus also resonates with certain anthropological positions of the 1980s and early 1990s, which posited different modes of epistemic engagement with the world, and tended—overly so—to dichotomize these into such oppositions as "participation/causality" (Tambiah 1990), "continuity/separation" (Willis 1994 [1989]), "orality/literacy" (Goody 1982). These parallel such equally crafted dichotomies as Lévy-Bruhl's (1985) "pre-scientific/scientific" or "pre-logical/logical" orderings of reality, or Honigmann's (1963) "personal/impersonal" worldviews, which aligned with notions of "primitive" and "modern" thought. Mimicking the Goodall-Carmichael contrast that I noted, all of these contestable positions contrast a rich knowing of the world through supposedly immediate engagement against a more constrained, analytic, scientific, theory-informed and therefore highly mediated engagement. But the very work of circumscribing the dichotomy serves as well to reinforce it, setting it out as a performative as well as observational expectation. Writing on the scope of rationality recognized throughout the history of anthropological inquiry, anthropologist Stanley Tambiah echoes the binary (1990, 105):

> . . . it is possible to separate analytically at least two orientations to our cosmos, two orderings of reality that woman or man everywhere are capable of experiencing, though the specific mix, weighting, and complementarity between the two may vary between individuals and between groups within a culture and between cultures taken as collective entities.

We may read of the dichotomy, witness it in others, and enact it in our own lives and produce it as a practiced universal, a normative ordering of action. In the enacting of either modality, either of which may be privileged or gendered differently in different settings, real and meaningful knowledge is produced all the same. A similar mapping of binary, gendered modalities via 1980s feminist considerations was realized in Evelyn Fox Keller's position on Barbara McClintock's "feeling for the organism" (Keller 1983) and Carol Gilligan's consideration of psychologically embedded male and female vocalities (Gilligan 1982). In all these instances, "ways of knowing" were aligned in binarist terms, a practice easily accommodated to gender logics. What Goodall's work at Gombe over several decades provides, with its repeated translations and reinforcements in public media and scientific fora, is a longitudinal extension of the mediation nexus over a long period, building up the gendered ape-communion expectation through this history of citation and recitation. Her working between technical culture (i.e., primatology supported by National Geographic grants) and public culture (i.e., stories produced and distributed by National Geographic's media agencies) over this period served to legitimate without ever needing to prove an essential difference between male and female sorts of orientations to the animals being studied.

Bringing all these matters now into a larger coordination—that is, the <Goodall + National Geographic> nexus plus the anthropological and feminist positions on modes of orienting to "nature"—it becomes possible to imagine a culturally forceful continuum of modalities ranging from *participatory* through to *causal* orientations to the animals. Each binary cited above becomes a gradient: from personal to impersonal orientation, from a sense of continuity with things to a sense of separation, from "prelogical" to "logical" orderings. Social, historical, and privilege vectors then work to align differently embodied and gendered folks along the continuum. In modern Western science, certainly into the 1980s, the gender-privileging of causality, separation, impersonalism has been historically and *habitually* assigned to masculine formations. Those oriented in this epistemic direction have stood to garner outright power and privilege—and this has applied as much to entrepreneurs, workers, technicians, and masculinist politicians, from American presidents to British prime ministers (including Margaret Thatcher), as it has to scientists. Those oriented elsewhere on the continuum struggled, or advanced only by enlisting allies in relation to this historically contingent, epistemic benchmark, as with Goodall supported by Leakey and Carmichael for instance. In effect, the delimiting of Goodall's productivity within the binary terms of a naturalized feminine modality served to bracket out and so produce a naturalized masculine modality. Notwithstanding its usefulness, the feminine would simultaneously be reckoned as subordinate to the masculine.

The Binary Logic of the Goodall National Geographic Productions

Returning to the *National Geographic* magazine productions, Goodall can be seen to alternate between causal (i.e., explanatory, impersonal, analytic) and participatory (i.e., emotive, personal, anthropomorphic) presentation in her *National Geographic* writing practices from the earliest

Gombe article in 1963, becoming increasingly participatory in the later articles of 1965 and 1979.[12] In the 1963 article, reinforcing this causality/participation split, there is a two-level narrative apparatus: one narrative follows the photo-visual sequence of the article, the second narrative follows the textual flow of the article. Individual photos display Goodall making contact with apes, while the total photo-narrative cunningly guides the viewer/reader in orderly, causally progressing fashion on an adventure narrative to Africa. This, then leads into the treetops following the chimps, eventually bringing them to the ground to engage in behaviors indicative of a sought-for "proto-human" status—tool use, toolmaking with termite sticks, hunting, even a purported proto-ritual behavior in response to a rainstorm.

The working narrative and visual logics of the 1960s articles consistently reproduced the above-noted "down from the trees" moment as the evolutionary modus operandi for Goodall's scientific engagements. In effect, Goodall is posed as witnessing, in a microcosm, that evolutionary transition from a naturalisitic, animalian life in the trees to a human-like life on the ground, ready now for the emergence of culture—i.e., upright walking, toolmaking, etc. This narrative places apes both phylogenetically *and* ethologically as the closest beings in nature to humans, and by putting women at the interface with apes in the scientific-media-public practice, the naturalized woman-nature communion is effected. This narrative does nothing to prove that woman has a shamanic connection to nature, or alternately that woman is man's emissary to nature, but it does show how the netted practices of <science + media + visual-narration> could have produced a faith in these effects, to the point they could be transformed into matters of natural fact, remaining largely uncontested models of the relation of woman-to-nature, man-to-culture.

The replaying of such evolutionary "culture emerges from nature" stories in the technoscientific material world is crucial in producing those formulae of gendered practice which are to become dominant. Critical contestation, however, brings such formulations to the fore, exposing them as contingent historical productions of a set of "natural facts." A smaller result is to show that neither Carmichael's discontinuity comment nor Goodall's continuity comment are right or wrong. Rather, both are inadequate not only because they imply a radically disproportionate privilege of Euro-American meaning-making, but also because they imply a singular master story of ape/human separateness, in Carmichael's statement, and a singular, essentialized mother/son relation in Goodall's statement. There are too many always unreliable essences, unquestioned universals, and uncontested power relations here.

Similarly, the limitations of Sy Montgomery's position, while a valid effort to "empower" women in natural science, are in essentializing the participatory as a female capacity, just as Goodall's moves tended to essentialize the maternal in female biology. Where Tambiah and Haraway come together is in recognizing worldly practices such as social mediation (i.e., admittedly more "cultural" for Tambiah, more "natural cultural" for Haraway) which could, for example, produce a maternalist-participatory primatological practice for Goodall, a paternalist-causal practice for Carmichael and Leakey, a lively and powerful iterative (and directing) contribution by National Geographic in bourgeois Euro-American mediations, and indeed the possibility of differently active ape behaviors and ape societies throughout this nexus of mediations. In this way, essences, epistemic orientations, gender and gendered practices, theories and methods, natures, and ape behaviors are all forcefully constituted and reconstituted by a multiplicity of practices and actors (cf. Haraway 1989, 133–85).

Goodall's Political Turn

I am able to draw three key points from the Goodall-nexus considered. *The first* is that the notion of feminine empathic orientation to engagements with primates is recognizable as historically contingent and ephemeral but, at the same time, so forceful through incessant repetition and reperformance as to produce a very thoroughgoing practiced effect of being essential, commonsensical, natural.[13] It is evident that the recitative force of her Gombe National Geographic exposés came to shore up the feminine-natural field observer capacity—reinforcing a woman-nature empathy frame in literate, "mass-mediated," predominantly white, middle-class capitalist public and technical domains. At the same time, however, the public overdetermination of this orientation in National Geographic Goodall productions has played very well in setting the opposition of woman-scientist with "feminized" mode of participatory science-practice against man-scientist with "masculinized" and distanced science-practice.

The second point is that while Goodall's approach to her research may be an effect rather than a "nature," it certainly permitted a richer, more particular, and revised configuring of chimpanzee society. Goodall's "mediation" practice effectively if not intentionally brought other impulses into field techniques aside from otherwise influential Anglo and European theory and method as being developed, say, by Robert Hinde or Hans Kummer (Goodall 1986, 649 ff.). These other mediating matters were "exuberant" storytelling and visual practices, National Geographic storytelling, Leakey patronage and financial connections, maternalist concerns, chimp personal names, etc. In this sense, the typical "cleansing effect" of concurrent scientistic technique—which has tended to refuse and demean such added descriptions as "subjective," "biased," or "nonempirical"—was at least partly bypassed by her practice, bringing to bear not lesser but different mediation resources.

What is possibly most interesting in terms of "on-the-ground" practices of primatology is that Goodall's work—via provisioning, via personal contact, via long-range study, via uninhibited "projective" engagement, via National Geographic photo-op demands, via hegemonic expectations of her feminine empathic and observation capacity, via naming and identity-building of individual chimps—did effect a rich range of different options for chimpanzee social behavior. Indeed, these actions of "engaging" the chimps activated them in her life without any significant resistance—in fact, with conspicuous willingness on the chimps' parts, as they gradually entered the camp in search of bananas, or initiated grooming of Goodall. Goodall's social world and theirs melded in a way not instanced or recorded before.

That, as it turns out, was immeasurably productive for primatology, especially once it was compiled in her culminating scientific monograph *The Chimpanzees of Gombe: Patterns of Behavior* (1986). That volume translated her cumulative narratives into systematized descriptions, graphics, and ethological rhetoric, under the tutelage of Robert Hinde, who, she states, "guided my exuberant writing style into a mold more suitable for scientific expression" (1986, 649). In effect, the same stories were repackaged, charted, and—here is the irony—"biased" in a scientifically "useful" direction. It is more than apparent that her research had always been of importance, but the language was not in the "mold" of canonical practice. What finally confirmed the importance of her findings was the highly positive reception given the monograph by most in the professional primatological and ethological community (Morrell 1993).

The third point relates to the recognition that Goodall's 1986 lifework monograph also marks the crucial turn in her primate-articulating activities from field study to chimp and ape welfare—perhaps not surprisingly the welfare of orphaned chimps in addition to lab and zoo captives. This mid-'80s turn, corresponding with her achieving of scientific credence, also seems to have gradually effected a tempering of her less complicated, passive nature-communion associations by the infusion of an active politics-engagement. Her way into these politics, following on the ideological interests of others, from Leakey to the National Geographic Society, still carries much of the baggage of that history—but she arrived all the same. The mixing of Goodall as "chimp-empath" (my term) with Goodall as "chimp-crusader" (Montgomery's term 1991, 193) is increasingly evident through the later *National Geographic* articles profiling her activities since the 1980s. The 1995 article "Jane Goodall—Crusading for Chimps and Humans" (Miller 1995) provides the striking contrast of the 1960s wonder-struck field researcher at Gombe, with a 1990s weary-looking activist engaging her human "publics" and chimp "victims" in diverse "cultural" sites around the world.

Goodall's turn came on the heels of Dian Fossey's murder in 1985, and also coincided with a newly peaking involvement of women in Euro-American primatology, the latter being documented in complementary but different ways by Linda Fedigan (Fedigan 1986b, 1994) and Donna Haraway (1989). In 1995, I returned to these questions of women in primatology, wanting to track further the flows of gender-making and nature-making knowledges and practices through public, educational, and technical domains. I began by interviewing nonexperts—college-aged (i.e., 18–23-year old) women and men—and found to my surprise that here, at 53° north latitude, far from Africa, America, and the U.K., Dian Fossey had equaled and possibly surpassed Goodall in recognition as the foremost iconic figure of field primatologist. Their main source was the 1988 film *Gorillas in the Mist.*

Dian Fossey: Reckoning with Woman-Man-Gorilla in the Mist

In a scene from Michael Apted's *Gorillas in the Mist: The Adventure of Dian Fossey,* the 1988 Hollywood feature film partly based on Dian Fossey's autobiographical book bearing the same title—the actor-portrayed figures of Dian Fossey, her hired Rwandan tracker Sembagare, and National Geographic photographer Bob Campbell are seen cutting out wild animal traps. The traps have been set by the local Batwa people in the forests of the Virungas volcanic mountains, also home to the mountain gorillas studied by Fossey from her Karisoke field camp. Sembagare assists the work, silently watching and listening to the following exchange:

FOSSEY: Goddamn Batwa!

CAMPBELL: You can't put all the blame on the Batwa.

FOSSEY: No?

CAMPBELL: They've been feeding their families like this for generations. If you're going to blame anyone, blame the doctor in Miami. He's the one that hires the bloke that hires the Batwa. The

Batwa get to feed their kids, the middleman gets a silk shirt, and the doctor gets a gorilla-hand ashtray for his coffee table and a great big gorilla head for his walls.

FOSSEY: Well, I can't get to the damn doctor in Miami!

CAMPBELL: Ever been to a doctor's office that didn't have a copy of *National Geographic?*

Fossey, the woman primate scientist, silently reflects on this last point of Campbell, photographer and man of the world, as if in this instant she is seeing a great net of associations that had hitherto escaped her. It is like a political awakening.

A later scene follows the slaughter by Batwa hunters of five gorillas in a local government-approved capture of a baby gorilla for a high-paying Belgian zoo supplier. Following this, the white American primatologist, portrayed by *Aliens*-famed Sigourney Weaver, cagily negotiates for the funding and assignment of game wardens for the Virungas National Park to help protect the local gorilla population from "poachers." She offers effectively to "nurse" the captured baby gorilla back to health, the baby now cradled lovingly in her arms after she has "stolen" it back from the zoo supplier's van. Negotiating with her is the black Rwandan government official Mukara, played by Waigwa Wachira, who remarks: "Ms. Fossey, your problem is decreasing gorillas, mine is increasing people. We're on opposite sides of the same problem!"

This scripted remark of Mukara replays a longstanding Western conceit that emphasizes population growth, rather than unequal distribution of capital and of knowledge as power, as the salient source of global human suffering. Nevertheless, it still signals the fraught local-global political conditions in which the contemporary forging of postcolonial primatological field practice regularly takes place today. Fossey has personally chosen to engage the political economy, now using her savvy to trade nurturance of one gorilla for government protection of many gorillas in their habitat.

Much more is signaled here about the history of commodification and spectaclization of great apes, the competing value for and destruction of wildlands, species extinction, human and animal welfare, presumptions about feminized nurturance and mothering, racialized political encounters, late twentieth-century British film sensibility within the American film industry, the history of colonization and nascent attentiveness to postcolonial conditions, the marketability of conservation ethics, the exoticizing of field science, class politics, the financing of science, etc. The otherwise simple adventure of Fossey's gorilla encounters and conservation struggles also contains buried implications of theory, method, gender, and "changing images of primate societies" which get played into this rather thick broil of politics, science, media, and economics. Surely, this imbroglio located in a murder mystery indicates the worldliness of primatology, and may account for some of the box office success of the Apted film.[14] These Fossey-located mediations contrast significantly with the Goodall mediations in this very complexity: whereas the 1960s and 1970s Goodall accounts speak little about the complex local relations and even erase the presence of "Africans" as people with agency, this Fossey account suggestively restores everyday, albeit dramatized, action and interaction of political and economic consequences to the "work" and outcomes of primatology.

To bring some further intelligibility to the political dimensions of Fossey's transformations, I want to consider how "space" is used to align the stories, semiotics, and depictions in the film. In the discussion that follows, "space" refers to the sites of action: on the trail, in camp, with the gorilla groups in the bush, in the cabin, in the town marketplace, etc. These spaces are the multiple locales occupied by Fossey and other, mostly male, figures in the film, and which come variously to effect, limit, or mobilize her agency. As with Goodall in association with vocality and narrative, the gendered logics of nature and culture are ever-present as, more particularly, are the logics of "culture out of nature" articulated within this play of space. The key relations are of how domestic cultured space (i.e., signified "home" locales such as Fossey's field cabin and the U.S.) produced a gender-constrained, feminized, maternalist Fossey, while the foreign nature spaces (Virunga forests) produced a gender-mobile feminine, masculine, androgynous Fossey. Notably as well, the story turns far more on matters of overtly political action than any of the Goodall/National Geographic presentations to that time. That, in itself, is significant.

From "Culture" to "Nature" via Politics

Where Jane Goodall's lifework trajectory through National Geographic productions (i.e., 1963–1995) appears as a movement from nature (in the field) to culture via political activism (in the ape welfare campaigns), Fossey's lifework trajectory in *Gorillas in the Mist* (i.e., 1966–1985) runs in the opposite direction—but politics are the crucial fulcrum in either case. The narrative genre for the film is the "journey to the dark continent," or in other terms, the masculinist, colonial adventure narrative. This narrative is more overdetermined than in Fossey's original book of the same title (which I found was not the principal source for this narrative form—see below). In her journey Fossey is beset by a host of male interlocutors, starting in America at a public lecture given by Louis Leakey, who sustains a paternalist relation with Fossey (via personal meetings and tape-recorded correspondence) throughout the film. The next important character is her chief tracker, Sembagare, who acts as a locally savvy natural and cultural mediator and guide for the expressly civilized, uptown woman Fossey. And then there are the very intimate eroticized relationships Fossey forges: an explicit one with Australian National Geographic photographer Campbell, and an implicit one with the silverback gorilla Digit. The only other woman figure whose character is developed in the film is Ros Carr, an American expatriate who provides an Eden-like refuge for Fossey after she has been ejected as a result of a coup d'état from her first camp, Kabara.

The film plays out a transformation of Dian Fossey from a naturally destined "mother-to-be" in cultured America, to an ambiguously feminine/masculine figure in the natured setting of Karisoke. This odd inversion—of being at first "naturally" fitted to maternity in a civil setting and later losing that "nature" in a natural setting—sets up some curious implications about gender mutability, the instability of any "normal" gender, the related instability of nature and culture along normalized gender lines. Indeed, it is Fossey's encounters with the apes and her propensity for "aping" or becoming ape that proves the gender expectation to be contingent. The film spins this "aping" and gender-play as both an empathic communion as well as a fall into madness. The maternalist orientation is set as the norm against which Fossey's politics, gender, and species moves can be reckoned as deviations. Though akin to the Goodall maternalist inflections in National Geographic, Fossey's moves are much more radical and even anarchic in relation to the norms set. She is permitted to be a silverback through chest pounding and other "male" displays in encounters with other silverbacks, and in adoptive paternity of Digit's group after his brutal murder in the film. Her forays into gender and species slippage and, with those, into political activism as a member of her gorilla kin group, are preempted by her murder—signaling the film's moral sanction against her transgression in so many directions across gender, political, cultural, and species lines.

Yet another and strikingly different exchange occurs in the film: that between Fossey and Sembagare, who also resides in the politically charged spaces of Karisoke at the nature-culture intersection along with Dian and the gorillas. In Fossey's book *Gorillas in the Mist,* there is no person quite equivalent to Sembagare—rather, he is an excessively developed character in the film. While starting out as bush savvy, Sembagare increasingly takes on a Westernized domestic-savvy and culture-savvy position, coming to attend more and more to Fossey's physical and localized political needs while she moves from American civility to Virungan gorilla nature. Jane

Goodall of the National Geographic nexus had no such local transcultural guide. As others have pointed out (Haraway 1989; LeBihan 1992; Mittman this volume), Goodall was continually noted in the texts to have been "alone" in the wilderness, even though her camp included several local African workers and was not far from several villages. As such, the making visible of a local African as an active agent marks a contrast to the Goodall stories. Nonetheless, this visibility turns on the logics of colonial exchange in that Sembagare is predisposed to "improvement" by his mission-school upbringing, and follows his ongoing "civilizing" as he spends increasing time in Fossey's Karisoke domestic-space cabin. This enhances further the politics of transgression, including lurking possibilities of yet another "love" story, between Fossey and Sembagare.

In many reviews of the film, in Apted's comments in the film press package, and indeed in critical analyses (cf. Sippi 1989), the Sembagare character and his portrayal by John Omirah Miluwi is noted as pivotal to the film's workings. My point is that Sembagare marks the elided, forgotten other, and is accorded new if ambivalent agency along with the "madwoman" Fossey and the Virunga gorillas. This agency is mobilized at the oft-recited zone of contact between nature and culture, between the domestic and the alien, between the empowered and the oppressed.[15] His presence signals a turn which now acknowledges that folks other than the "Western" primatologist and their ape subjects are implicated in the widening political nexus of postcolonial practice. In 1988, this was effective filmmaking and "based-on-a-true-story." In 1999, the postcolonial politics of local community engagement is appropriate and necessary practice for primatologists.

Connecting the Goodall nexus now with the Fossey nexus, and to summarize somewhat, in both cases, we see a personal practice forged at first in passive response to institutional conditions, in which both participated in an androcentric prescription of how women do primatology. Both personally engaged their ape subjects as more than anthropomorphized animals, as Marianne Torgovnick points out, as "containers of Beingness . . . rendered identical in value to the 'human'" (Torgovnick 1996, 24). Eventually, Goodall diverted from the passive path to attain a more forceful agency in chimp welfare projects. Fossey, on the other hand, was a figure who has been reckoned as working against a prescribed passive feminine practice, but this work was undermined by the outcome of the story, that is by her death. Her resistance story, an ambivalent one, is fraught with failure against forceful masculinist directives, and instead glorified by masculinist logics which prescribe a return of Fossey to maternity in domestic home-space, or if resisted, to psychopathological collapse in nature-space which, in this tale, is fatal. Both stories—however you look at them—put the women primatologists back into nature's embrace. Both position them as guardians of nature. Yet, superficially at least, it appears that both struggled against cultural prescription to forge their own political actions in primatology, and so to animate primate lives, societies, and behaviors with a participatory practice, supported by the visualizing institutions of mass nature mediation.

The Fossey story, however, has added poignancy given the (at least) dual effects of her being reckoned (a) as nature bound, or (b) as a martyr for nature and apes in the material and semiotic "terrain" where Western culture meets Western-recognized nature. This latter effect leaves the door open to newly reformulated insurgencies by women, "indigenous" players, and others into such transgressive terrains, promising an even more thorough reconfiguring of the nature/culture apparatus via political action across ever more sweeping nexuses.

Recirculations, Reformulations

Hollywood and National Geographic do not just circulate limited ideas about theory, method, and gender. They circulate and so *extend* an entire complex of mediations, borrowing from and translating those seemingly partial nexuses evident in the very practice and lived actions of primatologists and primates in their various locales. These "packed" nexuses stand as a complex trading project, where stories, practices, assumptions, and "ways of being" circulate complexly throughout the entire apparatus. The quirky re-routing of these forms and matters is notable when productions such as those of National Geographic or Apted's film are aimed at the public "imaginary" through commercial television and cinematic distribution, but then also come to affect teaching regimes of college-level primatology, anthropology, and ethology courses: National Geographic films are often used as teaching tools in such settings, and many who take such courses will have seen the Apted film and productions by National Geographic at some point in their lives. This manifold remediation in multiple viewing situations helps, in part at least, to make sense of the persistence of "Fossey's Mist" observed in my 1995 interviews, ten years after her death and seven years since the release of the film.

Attending to this longer span of recirculations and reworkings of the nexus, I was able to note a trajectory starting provisionally with Louis Leakey in the '60s, moving through the National Geographic mediations of the '60s through the '80s, Fossey's own book in 1985, and then, crucially, a 1986 article in *Life* magazine on Fossey which proved to be the linchpin with its hyper-expressed adventure narrative and gender logics.

This article by Harold T. P. Hayes, an old-school Africa "hand," was made in the colonialist mold of Joseph Conrad and Arthur Conan Doyle, and is the base document that provided the initiating logics animating Anna Phelan's script for the Apted film.[16] In this article are explicit statements from an unnamed California primatologist that pose Fossey as gender-, profession-, and species-deranged:

> The very fact that the animal is so intelligent in human terms makes it difficult to withhold human response. This is particularly true with women researchers, and especially those who are childless. There comes an overwhelming need to protect. When this happens, look out, objectivity goes. Then scientific credibility goes. (Hayes 1986)

Comments like this clearly work to abject Fossey as a woman, and to cast a normative behavior of women as child-bearers. This also came to brace up the drama of her portrayed descent into madness in the film which, as I have pointed out, is premised on a binary logic of her going astray of a presumed biological imperative to mother, her drift from a natural order in culture to an unnatural order in nature. Stirring up the neatness of Hayes's African colonialist, normative sex/gender plan, however, is the screenplay's addition of postcolonial working conditions in the story. This permitted some appropriate though limited foregrounding of what was going on all along in the Fossey lifework by acknowledging local political-economic dynamics associated with the presence and agency of local workers, indigenous people, struggles in state formation, and postcolonial administration. So, while the Apted film is no postcolonial tour de force, the figure of Sembagare becomes crucial, signaling the "worldly" turn from the colonially imagined local Africans in the early Goodall articles to the postcolonial conditions of contemporary primatology.[17] Sembagare's increased agency as displaced, postcolonial interlocutor in the film coincides with the increasing agency of apes and of women primatologists in the making of an increasingly worldly primatology since 1988.

The conflict and counterposing of a masculinist, binary-gender, colonialist spatial strategy (from Hayes, National Geographic, etc.) against a postcolonial spatial strategy (from Apted and Phelan) appears to have produced a much more articulated and so publicly interesting being in the murdered Fossey (and also in the murdered Digit) than that of National Geographic's stellar figure Goodall, even though a normative gender prescription is sustained in the film. That enhanced articulatedness of the Fossey story with the public imaginary surfaced as recently as 1995 with the young adults I interviewed, most of whom recognized Fossey and her actions as much or more than Goodall. The agency of students and instructors permits the setting up of one more trading site in the <primatol-

ogy + public culture> nexus—that being the encounter in the classroom, where nonexpert (such as National Geographic and Hollywood) and expert (professional primatologist) knowledges engage yet again. Whereas the colonial, chimp-empath woman researcher of Jane Goodall was a highly recirculated learning resource available up to 1988, the colonial/postcolonial gorilla-mime woman researcher and political activist has been added since 1988.

Empaths, Activists, Animateurs: Political Mediations in Worldly Primatology

Finally, I'd like to draw together the mediations I have discussed from the outset and to describe some of their practical consequences. I should also point out that this nexus is necessarily limited and partial, and could be extended further yet, most notably in regard to the specific technical and field practices and mediations of Goodall and Fossey, and in regard to the mediations of the apes themselves.[18] In following the organizers' suggestion that I connect my two earlier papers on Goodall and Fossey respectively (cf. note 4), I've come to trace an extensive network along an historical axis as well as a technical-public axis. In the process, a very important effect has emerged: the gender-associated politicizing of primatology in the rise of "worldly primatology" (that is again, primatology practiced by primatologists who consciously and intensively engage the natural-cultural worlds in which they work rather than more narrowly following technical prescriptions of what counts as primatological practice). This effect may also bear upon how the organizers identified matters of theory and method (as rather disarticulated actions) were so diffusely discussed, while the identified matters of gender and the constitution of primate societies came to be more intently discussed.[19]

Primatology's indebtedness to the <primatology + women + apes + National Geographic + Hollywood + spectacle> nexus should not be underestimated. That nexus offers other primatologists a set of resources which they may secretly or unselfconsciously borrow from, reject, set up as an opposition, etc. The Goodall and Fossey actions known through personally, technically, and publicly mediated accounts have never been disavowed outright by primatologists. Instead, they are embraced as part of the domain of primatology. Presumably, this is because for most in primatological circles, too much has been gained from the Goodall and Fossey work, not just in the limited sense of their technical activities, but, importantly, as it is situated in a technical-public complex. Gains include, for example, ongoing rich funding of great ape research, higher public profile for other, nonpongid primatology, effecting gendered "role models" etc.[20]

While National Geographic, Hollywood, and other "public" mediations are inadequate on their own and should insistently be problematized, much as the lived actions and technical work of Goodall and Fossey or any primatologist on their own are inadequate, these large-scale public mediations add in aspects of practice which technical discussion regularly leave out. These include guiding narratives, local social conditions, the force of sentiment and desire, marketing of primatology, conservation and postcoloniality as political frames for research activity, the agency of the animals and of locales to change the researcher, gendering of the authors and animals in mutual relation, space, the force of funding agencies, mentorship, race, class, power, voice, history, etc. All are part of the trading system, and all could positively be subjected to articulate study and critique. More to the point, however, it appears that shifting gender participation, gender critique, and public mediation have combined to shift the terms of primatology from a limited technical practice to a worldly, political, technical practice.

In general terms, from the Goodall-Fossey productions I've reviewed, it is possible to recognize a forceful spectrum from natural associations to cultural associations in the various accounts of the actions of Goodall and Fossey. This is paralleled by a long-practiced set of socially reinforced gender logics which correlates scientism with masculinized nature-description, and humanism with feminized nature-mediation. Goodall and Fossey—and so "women-in-primatology" as hyper-presented to the public via mainline middle-class presentation like National Geographic and *Gorillas in the Mist*—are still widely understood as humanist nature-empaths. However, the spectrum is increasingly destabilized by engagements with the politics of ape lives and ape social and environmental conditions in contact with non-ape and human social and environmental conditions, especially since the mid-1980s (e.g., Strier 1992; Strum 1994b). Certainly in public mediations (and in technical mediations also as Fedigan and Haraway indicate) women in primatology have increasingly populated the political terrain set up by happenstance at the Western-made intersection and dividing point of nature and culture.

That said, the multiplex mediations of <Great Apes + Primatology + Goodall + Fossey + Apted + National Geographic>, while motioning toward an increasing collapse of nature and culture via politics at that arbitrary junction, are still more or less circumscribed as a middle-class discourse of nature communion. This nexus is supported by and based in educational, research, and entertainment institutions of wealthy northern nations—the United States, Great Britain, Canada, continental Europe, Japan. The capitalist productivity of this nexus is one based on the excess commodification and spectaclization of panhuman connection with (and/or disconnection from) a nature reckoned in evolutionary discourses—women and apes are still among the most highly replayed mediators at the modern boundary between culture and nature. That is something which appeals (and is made to appeal) to a more or less modern range of scientistic/humanistic consumer sensibilities. Nature/Culture logics are produced and productive (if problematic) and are not easily dismissed, but rather need to be engaged in such a manner as to expose their constraining effects. Gender transformations in primatological action has triggered such engagements. Indeed, it is the very engaging against the grain of these logics which has been so very productive for primatology itself and in the politicizing and testing of the regulated ordering of nature/culture and sex/gender.

The reconfigurations are apparent within the practice of several women contributors to this meeting who have opened up the bounds of primatological practice. Examples include the conservation-engaging, and therefore necessarily worldly, political work of Karen Strier, Emília Yamamoto, and Shirley Strum, and other moves highlighted in the comments by Alison Jolly. What is also interesting about these worldly and political turns in primatology—incipiently signaled in the lifework of Fossey and *Gorillas in the Mist,* and now practically, in technical work—is the presence of other actors in the mediation nexus, which is now increasingly postcolonial. Most notably among these are human members of local communities whose interests intersect, conflict, or contrast with those of scientists or of the primates themselves. These are people who have otherwise been portrayed, beyond any "reason," as deserving to be radically backgrounded or composed in the Western narrations, depictions, discourses, technical publications, scientific meetings, acknowledgment pages, keynote addresses, and on-the-ground in-the-field practices. More primatologists—many of them women—recognize and actively engage the worldly dimensions of their practice, while others still maintain a well-secured, limited practice adopting a "worldview" which sees science as detached, detachable, and in a special supposedly apolitical position outside of *both* culture *and* nature. At the same time, transnational conditions even in technical practices of field science also make it clear that primatologists from, for example, Japan, Brazil, Colombia, and many "non-Western" nations are working with resources both alternate to and exchanged with "Western" practitioners.

This movement into local politics of primate lives is exposing still more players and sites in which primatologists turn up as a matter of everyday practice[21]—in local communities coextensive with the lands in which the primate communities abide, in zoos, in primate confinement facilities of biomedical laboratories, in the public media, in public advocacy lectures, in corporate boardrooms, in international and governmental policy meetings, etc. Worldly practice now moves primatologists, and mobilizes resources, through a complex array of sites, beyond a misconstructed site at an imaginary boundary between an uncultured nature and an unnatural culture. However, the making of that site is intimately tied to an evolutionary telos which sees culture as emergent from nature. This is mutually reinforced in the trading among technical and public mediations such that the impulse to place "emissaries" at the imagined/materialized boundary seems likely to continue.

Returning finally to the thematics of the organizers, it certainly appears that theory and method are being subsumed by matters of overt, active, self-conscious politics. Indeed, the diffuse discussion of these matters during the meetings themselves at least in part bears this out. Discussion of gender is necessarily political discussion, and this along with other politics continue to expose and destabilize the Western logics of nature and culture which seem in worldly primatology to operate according to an often unproblematized evolutionary telos. These meetings point to how science studies, feminist, and cultural scholars may trade critically and productively with scientists, policymakers, environmentalists, as well as with other primates, environments, political-economic systems in accordance with or against the grain of the evolutionary, nature/culture telos.

Perhaps even more importantly, they also suggest the ongoing but less critically engaged trading with those standing for local communities with tangible economic and life and death stakes in the issues. An Associated Press story in June 1997 underscored this point with the headline "Four mountain gorillas died in crossfire between Congolese soldiers and Rwandan insurgents in eastern Congo, a conservation group said yesterday." In addition to gender, the politicization of primatology takes in wire services like Reuters and AP, natural and cultural territorial contests, dreams of better lives for people, animals, and environments alike, and endangered habitat markers or "flagship species" including mountain gorillas and muriquis (Strier 1992, 101, 123). Taken together, such multiple engagements signal an even *more* worldly practice than that which I have been able to contour here. The effects of the complex are tangible changes in the lives of primates, primatologists, local people, gender, theory, and method. The nexus of Goodall, Fossey, Flo, and Digit allow primatologists to act variously as empaths, activists, and animateurs. Such a nexus can extend to include yet-to-be-named people whose lives associate with yet-to-be-named primates everywhere. This distributed nexus of mediations is already acting to configure a newly critical, political and, therefore, worldly primatology. As Alison Jolly points out in her contribution, ". . . whatever and whoever is responsible, the results are action in the world."

Acknowledgments

I would like to acknowledge the support of the Wenner-Gren Foundation in enabling my participation in this symposium as meeting monitor and contributor. I also wish to thank Linda Fedigan and Shirley Strum for their facilitating of my participation, and for their critical comments on drafts of this paper. I'm also grateful to Hans Dieter Sues, Constance MacIntosh, Trish Salah, Gwen Burrows, and Charis Thompson Cussins, whose very useful comments on drafts of the paper have helped it along considerably.

NOTES TO CHAPTER TWENTY-ONE

1. Note the litany of popular, quasi-popular, or scientific books, films in several languages, magazine articles, and websites, the Jane Goodall Institute, lecture series, children's books, etc. associated with the two of them.

2. This is not to discount orangutans, baboons, or any others as massively public, but rather to note the apparent. It is also apparent that the orangutan researcher/activist Biruté Galdikas is a prominent public figure, but Goodall and Fossey have been presented even more widely. As such, the latter provide well-instanced examples of the historical stream being considered here, which could be further textured by the addition of others.

3. Certainly Donna Haraway's (1989, 1991a) multilayered discussions of North American primatology course through some of these histories. Londa Schiebinger's work (1993) on eighteenth- and nineteenth-century ape-human discourses and practices help to contour a lengthier modern Western tradition.

4. My discussion draws on two earlier papers I prepared on the place of women in public primatology. The first paper, "Learning and Unlearning Separation: From Jane Goodall to Epistemic Possibility" (Noble 1993), undertook a critical consideration of the four-way intersections of middle-class public media in National Geographic, Goodall's Gombe studies, feminist engagements with Goodall in practice and in visual-textual exposition, and the cultural production of multisubjective epistemological orientations in gendered scientific practice. My second paper, "Leaky Visions of Nature, Women, and Apes: The Persistence of Fossey's Mist" (1996), constituted the original paper submitted for the Teresopolis symposium, and continued in like manner to trace intersecting networks, but with less attention on epistemology and more on the transformation and movement of colonial

and postcolonial race-gender-spatial logics in the equally mobile narratives leading to and following the film *Gorillas in the Mist.* I try here to hook these two discussions together and extend them somewhat, in order to bring attention to issues raised in the Teresopolis meetings.

5. Please see Marilyn Strathern (1995, 177–85) for a more detailed discussion of local-global knowledge transformations as "relocated relations."

6. This counters the unsatisfactory notion of mediation as reducible to a factor or a bias which is seen more or less to distort translations of the "reality" of ape social behavior by scientists, by the mass media, by the nonexpert, etc., and presents a much more extendible proposition of how primate societies or primatology comes into being. I take the media presentations as coextensive features of worldly primatological practice.

7. Throughout this paper, I use variations of this notation practice (e.g., x + y + z) to indicate the nexus I am addressing. This is a rather shorter version of the form, which at other points adds in or replaces some of these terms. All the same, I am referring throughout more or less to the same set of connections. Combining all the terms I use, the longest notation (with some evident redundancies) is <Fossey + Goodall + apes + great apes + women + woman researcher + primatology + spectacle + NG + Apted + gorillas + chimps + Hollywood>. Clearly, even that notation, and any notation, is still only suggestive, incomplete, partial.

8. I note here "Euro-American" interests. Some important work is called for to redress interests outside of this high-powered "mainstream" in primatology. Here, I am thinking particularly of primatology of southern nations (e.g., Yamamoto this volume), and also of Japan (e.g., Takasaki, Asquith this volume).

9. In all, there have been seven articles on Goodall and Gombe in *National Geographic,* and three on Fossey and the Virungas. Subscription figures for the magazine ranged between six million per issue in the early 1960s to over ten million in the 1990s (Ogburn 1991). A rough calculation puts the total number of copies produced with articles on these two at around sixty million—and those figures do not indicate multiple readers of copies, which presumably would extend the "audience" to an even larger number. *National Geographic*'s shelf life is remarkably long, often "kept" by subscribers for years and even decades, possibly as a consequence of its high production values and visuality. As such, the articles have had an enormous circulation effect. These indications do not even indulge the plethora of additional National Geographic productions including books for adults and children, several television specials, educational publications, etc. The audience effect for National Geographic, Goodall, Fossey, and for the apes (and others activated in the nexus) as well has been very big, as has the revenue effect, though it would be interesting to see the relative scaling of the revenues accruing to productive benefit of each of these agents and others not so visibly implicated in the complex—that political economy itself is a worthy scholarly matter.

10. This position was restated in different terms by Barbara Noske (1997) in her call for "ethnography" of animals, including primates.

11. Bruno Latour has discussed the binarist relation of nature and culture as illusory yet constitutive of "modern" sensibilities (1993).

12. The sequence of major Goodall National Geographic productions to 1992 is as follows: 1963a, 1965a—Goodall chimpanzee articles; 1965—the film *Miss Goodall and the Wild Chimpanzees;* 1967—Goodall book *My Friends the Wild Chimpanzees;* 1979—Goodall chimpanzee article; 1984—the film *Among the Wild Chimpanzees.* Additional, non-Goodall, primate productions include: a 1995 article, "Crusading for Chimps and Humans"; a 1970 Fossey gorilla article; a 1971 film, *Monkeys, Apes, and Man;* a 1975 Strum baboon article and an orangutan article by Galdikas; a 1980 Galdikas orangutan article; a 1981 Fossey gorilla article; a 1987 Strum baboon article; a 1992 general article, "A Curious Kinship: Apes and Humans"; a 1995 article on the threat of human war on mountain gorillas. In addition, several issues of the magazine dealt with women involved in primate research, including topics such as ape language training and "temple monkeys" in India.

13. This is akin to the reiterative, discursive, performative force of lived-out, embodied sex and gender discussed and theorized by Judith Butler (1993) and Rosalind Morris (1995).

14. As a film maker, the Cambridge-educated Apted has moved back and forth between "documentary" and "docudrama" genres, obviously finding public currency in both forms. Moreover, his films show he is no stranger to decolonization issues. I note here his complimentary drama "Thunderheart" (1992), which plays off his documentary "Incident at Oglala" (1992), dealing with the events and struggles associated with the American Indian Movement of the Sioux Nations in the 1970s.

15. For a discussion of colonial and postcolonial "zones of contact" see Pratt (1992) and Clifford (1997).

16. The credits to the film read "Screenplay by Anna Phelan" and "Based on the book by Dian Fossey and the article by Harold T. P. Hayes."

17. On this point I am quibbling with Diane Sippi (1989) who reads Sembagare as a colonial figure only, caught up in the politics of desire associated with fantasies of miscegenation. Augmenting this reading, I add that there is an incipient postcoloniality in the way in which Apted accords agency to characters like Sembagare and the local administrator Mukara to redirect Fossey's action. In one sense, Fossey—and indeed the Virunga gorillas—can be seen to be caught in this strange collision of the figurative, practiced colonial adventure story and the historical postcolonial conditions of 1970s and 80s Central Africa.

18. Such a project might entail an analysis of technical publications, interviews with Goodall or with coworkers of the two researchers, tracking of the adoption by other primatologists of field techniques which were first used by Goodall and Fossey. To consider the agency of the apes as actors that change the world including the primatologist, one might take guidance from Barabara Noske's proposals for an ethnography of animals (1997). Goodall's role as a model researcher for Japanese primatologists is also challenging to notions of essentialized gender-prescription, where cultural relocation changes the game all over again. See Takasaki, this volume.

19. I note this point as the meeting monitor whose task it was to keep notes on the entirety of the meeting roundtable discussions. Indeed, as an anthropologist

of contemporary culture and nature, I take the meetings themselves as a case of the mutations taking place in primatology.

20. I should say this has also been true in the case of spectaclized, masculinized dinosaur research in relation to the wider taxonomic interests of nondinosaurian vertebrate paleontology (Noble, in press).

21. One newly emerging area in which the inextricability of biologists in the real-world politics of transcultural, postcolonial economics and ecology is taken seriously is that of "political ecology," an area in which an increasing number of anthropologists are becoming involved (cf. Brosius 1996; Brosius, Tsing, and Zerner 1996).

E-Mail Exchanges

At the end of the workshop in Brazil, Alison Jolly said she understood and was more sympathetic to Donna Haraway's goals and approach in *Primate Visions* than she was in her original reading (and in her *New Scientist* review) of the book. This was because we had made new contacts, built connections between people holding different views. Despite this work, uncertainties remained just below the surface, and the tension between those who do science and those who study science never disappeared. The e-mail exchanges below illustrate, perhaps more clearly than do the chapters, the participants' different perspectives on science and how, although the within-group diversity was high for both scientists and analysts, each group shared a greater intellectual affinity among themselves than with the other. The first part of the e-mails concerns science, scientists, and science analysis; the second part focuses on the media as a broker for scientific ideas and popular culture. Together they give a flavor of the actual exchanges during the week in Brazil and the unresolved controversies.

Part 1: Science, Scientists, And Science Analysts

Discussions about science surfaced in many places in the e-mail exchanges, with the same tensions between positions and participants as witnessed in Teresopolis. If anything, these were more obvious, since by the time electronic messages were flying around the globe, we all had a better understanding of what was at stake. As a result, the material in the e-mails is more explicit and incisive than what was said at the workshop.

Question 8: Why Has There Been So Much Resistance by Scientists to Science Studies and the Critique of Science? Is This Because of an Association with Postmodern Deconstructionism, Because Scientists Don't Like to Be Subjects, Because the Two Groups See Science Very Differently (Making Communication and Discussion Difficult), or Is It Because Rejecting or Critiquing "Scientific" Socialization Means That Scientists Lose Their Credibility (and Risk Having Their Scientific Ideas Devalued or Dismissed)?

To: <teresopolis@majordomo.srv.ualberta.ca>
From: "Naomi Quinn" <nquinn@duke.edu>
. . . One of the things that captivated me about the Wenner-Gren workshop was the dance between the primatologists and the science studies folks. It was all very civilized, but the primatologists were really poking and probing—doing science on the science studies representatives to see what they were all about. I noted the animal watchers' fascination with this new kind of animal. Attention was deflected onto the science analysts by the assembled animal researchers' willingness to let them take center stage. This is because as a group, the animal researchers are "watchers."

To: <teresopolis@majordomo.srv.ualberta.ca>
From: "Dick Byrne" <rwb@st-andrews.ac>
. . . it seems to me that some people have been conspicuous by their absence from the discussions. I'm particularly thinking of those more towards the "sociology of science" end of the spectrum, though some plain ol' scientists have also been silent.

Why is this? Perhaps these people are shy, retiring, or have little to say. If so, I must have been at some different Teresopolis. I hope the words that the more talkative ones keep typing in aren't just somebody else's "data"!

To: <teresopolis@majordomo.srv.ualberta.ca>
From: "Charis Thompson Cussins" <cmc34@cornell.edu>
. . . There are all sorts of science studies folks. Of these, probably the majority are not interested in a "critique of science," at least not in a wholesale one . . .

Science studiers are more likely to be interested in normativity and knowledge. And in the amazing power of the modern Western sciences to garner the resources of objectivity to order the world and the people in it. . . . We are pointing to and then trying to understand the historic significance of this system of knowledge, taking the good and the bad as (and for whom) they fall. Mostly individual scientists are only incidentally an object of study. . . . It seems to me a deeply intellectual question to want to understand powerful cultural systems of normativity, whether it be politics, religion, ethics, or scientific knowledge making.

. . . Having been trained in the sciences (like many in science studies) and teaching many science students, I care a lot that scientists and science studies calibrate with each other. The thing that I would most like to see vanish is the idea that studying the production of knowledge is a way of saying that science isn't real, or doesn't capture reality . . .

To: <teresopolis@majordomo.srv.ualberta.ca>
From: "Thelma Rowell" <thelma@ingleton.demon.co.uk>
What is normativity and how do you calibrate with someone? Or maybe, as a mere biologist, I don't want to know?

To: <teresopolis@majordomo.srv.ualberta.ca>
From: "Charis Thompson Cussins" <cmc34@cornell.edu>
What is a population and how do you establish inter-observer reliability, or maybe as a mere science studies person I don't want to know?

To: <teresopolis@majordomo.srv.ualberta.ca>
From: "Shirley Strum" <sstrum@weber.ucsd.edu>
What happened in Brazil was both confrontation and bridge building. I think we went further towards understanding and connection as we danced around the topic of primates and primatologists than in the general domain of science or in the domain of popular culture. I would like to know why—what are the obstacles and how can they be overcome?

To: <teresopolis@majordomo.srv.ualberta.ca>
From: "Brian Noble" <brian.noble@utoronto.ca>
Taking up this amazing quite American artifact called "Science Wars," which rings out like a "call to arms," to me seems fruitless and bound only to generate multiplicities of defensiveness and repetition of outmoded discussions.

. . . so I want to attend to usefully productive work. One of the most interesting comments comes from Zuleyma who suggested that scientists who work on the "critical edge" may end up moving out of science into science studies. The engagement of science and science studies has done something very important in the sense that those who count themselves as scientists "read" the texts of those who count themselves as "science studiers." As a result we now have a very savvy intellectual trading zone, populated with those who adamantly defend their positions as scientists or science studiers vis-à-vis one another, those who meet and exchange views, those who more or less just carry on in their fields quietly ignoring or incorporating the actions of the other, and I'm sure more sorts of possibilities than these. But [there] . . . is the idea that one is either "in" science or "in" science studies, even if they choose to move. Donna, Charis, Evelyn, Alison Wylie have all "been trained in the sciences," but are now recognized as "out" of science (correct me if I'm wrong). Linda and Shirley, Zuleyma, Alison Jolly (and others of you among the practicing scientist) so far as I can tell, are "in" science, but reaching out into science studies . . . they didn't have to leave altogether.

These are interesting dynamics partially associated with disciplinary boundary maintenance, but you will notice that the direction is from science either to science studies, or from science as an internally-regulated practice to science as a situated (the word I use is "worldly") practice . . .

To: <teresopolis@majordomo.srv.ualberta.ca>
From: "Naomi Quinn" <nquinn@duke.edu>
I agree wholeheartedly with the contributions of Charis, Brian . . . about the role of science studies. Once again, though, I find myself wanting to distinguish the intellectual issues from the dynamics of academic politics, and to not lose sight of the latter. I believe that, however, inadvertently, science studies has tapped into, and fueled, a powerful "antiscience" sensibility in the humanities and neighboring disciplines like my own. Thus, just as there is a vulgar Marxism and a vulgar sociobiology, there is a vulgar science studies out there today. It plays out in departments and professional associations and other local venues . . .

I have also wondered whether we are not dealing with a generation (the one after mine) of U.S. academics who went to college at a time when the pendulum had swung to no distribution [general education] requirements. Having no direct knowledge or understanding of how science is done, these humanists find it very easy to caricature and stereotype. Both science studies folks and primatologists, being scientists yourselves, may find the depth of antiscience feeling out there difficult to fathom. But it's a real problem in disciplines like mine, where the so-called "science wars" are being fought.

To: <teresopolis@majordomo.srv.ualberta.ca>
From: "Thelma Rowell" <thelma@ingleton.demon.co.uk>
Isn't it interesting that nobody has answered my questions? I did not understand Charis and I picked out a couple of words that, just possibly, if I had understood might have provided a key. I wasn't trying to be clever or cheeky or whatever.

To: <teresopolis@majordomo.srv.ualberta.ca>
From: <brian.noble@utoronto.edu>
This exchange has helped me to recognize a couple of matters about manners (as in "techniques," and as in "etiquette") of exchange . . .

Whereas

—postmodernism is not equal to science studies.

—some science studies looks to be postmodern, but so does some science.

—also, see Bruno's book *We Have Never Been Modern* which logically also means there's nothing to be "post" of.

. . . The problem with postmoderns is that they parody really well and generate lots of that "wondering what it means" stuff. People who feel they are implicated by the parody often get really offended, they turn defensive, sometimes they get ferocious. But it depends on how people take the joke and moreover, if the joke is simply in such bad taste and funded by the same agencies that fund science, it really is nothing but insulting and hurtful.

There are, however, many other possibilities which I've witnessed in our discussions at least, all of which involve a really important peaceable common ground: play.

I underscore how Donna has always suggested that we think about how certain words, "tropes," figures of speech, can make us "turn." A good metaphor can change the world, as we all know in the case of "GENESIS" and "EVOLUTION" and "CODE."

In a post-science-wars economy in which postmodern is now pretty much an artifact, the manners are changing, languages are getting exchanged; I think we've made an important turn from "wars" (with words and more) to "games" (with words and more) among those who work as "scientist" or as "science studyist" or as both. The examples are everywhere in our exchange . . .

[from the next day's e-mail] . . . I think a really powerful mode of exchange (is) "irony, parody" which can work in a lot of different ways, sometimes producing anxiety and confrontation, sometimes really changing things in productive new ways, and more ways. In short, I think the way it has been used has hurt alliances between science studies and science folks, at the same time garnering huge attention in scholarly AND public circles. . . . Others have suggested that the matter is differences between the various disciplinary lexicons drawn upon by science and science studies. . . . I think the use of different manners compounds this . . . whether used consciously to disrupt things or simply as a default. But I think in the late 1990s, manners are changing to forge reconfigured alliances . . . (that's something a bit different than building or mending bridges).

To: <teresopolis@majordomo.srv.ualberta.ca>
From: "Thelma Rowell" <thelma@ingleton.demon.co.uk>
. . . I fervently believe that science is about communicating what one has observed in the briefest, clearest way possible using new words only when there is a new entity to be defined. Speculation should be limited to a last paragraph carefully separated from data and logical conclusions. Obfuscation is blasphemy in science. Irony and parody and punning etc. are for fun, not work. Science as a puritanical endeavor?

To: <teresopolis@majordomo.srv.ualberta.ca>
From: "Alison Jolly" <ajolly@princeton.edu>
. . . I still cannot take the science wars seriously. In the mild form, which involves people taking a critical interest in science and scientists, I find it flattering—as authors must feel about even critical critics. The extreme "strong program" I find ludicrous, to the extent that all my information is second-hand, and I have things to do which I find more interesting than reading the originals or even such rebuttals as *Higher Superstition.* So I haven't answered your questions because to me the core of the science wars is someone else's sandbox . . .

To: <teresopolis@majordomo.srv.ualberta.ca>

From: "Bruno Latour" <latour@paris.ensmp.fr>

I don't want to give a retrospective shudder of horror to the participants, but I am a strong programmer! I was thus greatly surprised by the reaction of Alison Jolly; let me try to explain so that it could be clearer to imagine how the "other" feels, which is always, after all, a good exercise.

First, strong program means something perfectly respectable: science can be thoroughly subjected to a scientific enquiry that uses normal types of causality to explain its qualities and certainties. It is the equivalent of the rejection of vitalism in the nineteenth century when biologists used to say: life is entirely subjectable to a thorough scientific enquiry; some people said "No. There is something you will never be able to study: the vital movement or such things." Well, it is very much the same thing in the science wars. Don't be mistaken; the vitalists are those who say that science cannot be studied as a matter of principle (citing *Higher Superstitions* in this discussion is, for me, exactly like bringing a creationist pamphlet to an evolutionary biology meeting and requesting equal time for the presentation of both! I don't believe in equal time for those who reject the "strong program" and those who have worked on it for twenty-five years!).

Strong program also means a "methodological principle." In order to study how science works it is important that an equal balance be struck between different competing research programs so that the victory of one or the other can be causally explained (and not just assumed to be inevitable). Now this produces a swarm of difficulties. But so does, for instance, the idea of comparing within primatology vastly different primates like humans and chimps. At the level of methodology, this procedure is a useful caveat against simplistic assumption about the history of science (Darwin was right and Spencer wrong, for instance, because one was doing science and the other ideology). After that, of course, we run into all sorts of difficulties!! But that's quite usual in any scholarly field. So please don't use strong programmer as a curse!! We are pretty proud of it!

Hope this helps.

To: <teresopolis@majordomo.srv.ualberta.ca>

From: "Gregg Mitman" <gmitman@ou.edu>

. . . I am also resistant to [an] "us-versus-them" mentality. . . . Each of us has different entry points to these debates. Yet we are so quick to lose sight of that individual situatedness and categorize into binary oppositions. I was struck at how much productive work at Teresopolis got done during individual conversations, on walks, at meals, in the sauna, later at night. But when we got around that big table together, the investment in professional identity and turf boundaries became forces too powerful to resist.

. . . the core of what is at stake for me . . . really centers on finding meaningful ways to live and work productively in a world of process. And that is bound to raise deep-seated anxieties, both personal and professional, because we then find ourselves trying to move through a world where the categories are no longer familiar. In my experience, this is why interdisciplinary work is so difficult, because we all fall naturally back to the forms that are most comfortable and what we know best. But when you get a group of people together like we had in Teresopolis and lock them in a room for eight days, about the third day, egos begin to soften, the defense of professional identity and territory seems less important, and productive work gets done. . . . What is missing in the science wars debates is that kind of shared labor . . .

Part 2: The Media

Workshop discussions about the media, science, and society began a conversation that was never finished. Analysts suggested that the media was not an instrument of "representation" but a vehicle through which knowledge is appropriated (and changed) by diverse constituencies. They also proposed that despite the strength of science within its own particular "setups," its power is limited when it comes to the much larger networks producing and conditioning "what counts as nature." During our e-mail exchange, a science/media event occurred that gave us the chance to delve further into these issues. The journal *Nature* published a short report on sexual behavior among the Gombe chimpanzees as revealed by DNA paternity data. The title of the report was "Cheating by Chimpanzees." The report made headlines in newspapers across the United States and prompted TV and radio interviews by national and international media. The earthshaking news was that females had infants sired by males living outside their social unit. While these results are important and interesting, they are not entirely new; a growing body of data (going back more than ten years) shows that females (primate and otherwise) successfully mate with nongroup males. Scientists were upset by the words used (adultery, cheating) and the media response. They began to wonder why they get upset, why they care, and what other options there might be.

Question 9: Why Do Those Who Study Primates Agonize about the Media and Ideas of Popular Culture? In Contrast, Astronomers, for Example, Do Not Worry about the "Nonsense Talk over UFOs," or "Life on Mars." Who Has Control? Who Is Responsible? What Can and Should We Do?

To: <teresopolis@majordomo.srv.ualberta.ca>
From: "Gregg Mitman" <gmitman@ou.edu>
It seems to me that this episode shows precisely why diffusionist models don't provide a very good account of the interaction between science and the media. You [scientists] seem upset that the media once again got it wrong and have sensationalized science. But implicit in such a view is that scientists are the sole guardians of information and that the press is

there to interpret their findings to the public. My point about meaning being made, not given, I think explains much here. Maybe it would be fruitful to suggest (for example) that chimps in popular culture are different animals from those within primates studies, just as was suggested in previous e-mail exchanges that chimps in Gombe are quite different animals [now] than when Goodall went to study them in the early sixties. So whatever findings primatologists come up with about chimps, that information is going to be appropriated and transformed in ways that reinforce the symbolic meanings of chimps within popular culture. And, whatever may motivate primatologists for studying chimps, let's face it, within popular culture, one prime fascination with chimps is as surrogate humans.

[from an e-mail the next day] . . . I think we need to come to terms with the issue that primates have different meanings for different constituencies and scientists aren't going to be able to control all those meanings and how their information gets appropriated by others. I am reminded of Dick Byrne's comment at Teresopolis that the media more accurately represent his ideas than many of his peers. But in thinking about science and the media we seem to be locked into representational models: the media is there to represent science. Bruno was really trying to push us in other directions, and this was how my statement about chimps being different animals in popular culture was intended. Why is it that this not-so-new information about chimps (that they mate outside their "group") gets transformed into a story about adultery despite the cautions of the scientists involved? To begin to appreciate this, we have to begin to unravel the deeply entrenched narratives and cultural and scientific practices that constrain the ways in which chimps are, in Bruno's language, allowed to speak. There is a very good book by Roger Silverstone entitled *Framing Science: The Making of a BBC Documentary* that follows the making of a BBC documentary on the new green revolution and beautifully shows how the film is not a representation but a product of a negotiation with a set of political, aesthetic, technical, and bureaucratic constraints. So it is in this sense that my provocation was intended: that the negotiations operating within the sphere of media and popular culture may entail quite different conversations than any one of us speaking to a reporter or filmmaker could imagine. And those negotiations are going to transform chimps in unexpected and not so unexpected ways . . .

To: <teresopolis@majordomo.srv.ualberta.ca>
From: "Hiroyuki Takasaki" <takasaki@big.ous.ac.jp>
The media releases as news what their subscribers would like to hear as news. We cannot expect much more than that. Primatologists constitute only a minority within the whole human population, just as Kyoto school field primatologists constitute a tiny minority among all the primatologists. What can we do? *Nature* attracts much news media attention, while *Primates* does not. Sugiyama et al. (1993, *Primates*) reported a similar result [to the Gombe chimp extra group matings], although only a single case from the Bossou chimpanzees. Also Ohsawa et al. (1993, *Primates*) did so for patas monkeys. These are also notable but didn't get noted by the media. So if a primatologist wants to live a quiet life, *Nature* is not a good choice as a journal to submit his/her papers, particularly on topics vulnerable to

distorted amplification in news media. *Nature,* which must "sell," has to publish such articles preferentially in order to survive as a commercial magazine. The amount of contribution a primatologist makes, whether he/she lives a quiet life or a noisy one, won't differ much in the long run, I think. Therefore, the choice is a matter of taste. The outside world, which is made up by the majority who will hear only what they want, won't change much overnight anyway . . .

To: <**teresopolis@majordomo.srv.ualberta.ca**>
From: "Alison Jolly" <**ajolly@princeton.edu**>
I am now back from Madagascar and the lovely limbo of a forest without e-mail. I have been chuckling over the correspondence. . . . But I still don't know why everyone is so righteous about television. It seems to me an art-form, not a science-form, subject to rules of visual composition and communication, which are fascinating in themselves. It's frustrating when it goes wrong, but that is true of any attempt to communicate. I confess to thoroughly enjoying playing at TV, just as much of a thrill as research, only different . . .

To: <**teresopolis@majordomo.srv.ualberta.ca**>
From: <**Pamela.Asquith@ualberta.ca**>
. . . with regard to why (many) primatologists "agonize" about media and pop culture while astronomers do not, I submit the following thoughts.

Ursula Franklin's little gem "The Real World of Technology," along with several others, note that media, especially TV or documentary film, creates a pseudoreality or a reconstructed reality based upon edited "bytes" of information about an event. Yet the media claim to represent "reality," or in this case, "what happened". . .

Besides the impossibility of conveying all aspects of any event, the "contamination" of the "reality" occurs in part with the media's effort to "sell" something to the audience. It is packaged to be marketable and/or to reflect a certain viewpoint. The "good news" newspaper did not last many months. Consider primatology in North America. People have been primed to expect primate studies to shed light on our behavior. What will be chosen by the media to "sell" will be something considered of interest to and a reflection of human society: "adultery," "harems," etc. in our society, "boss monkeys" in Japan's explicitly hierarchical society.

Because primatologists are indeed trying to say something about human society (among many other things), one would expect them to be sensitive to misrepresentation of their findings. Astronomers perhaps don't have their current pet hypotheses unduly misrepresented by what the media chooses to report. It is a fairly level playing field in talk about aliens as the scientists do not know much more about them than anyone else. On the other hand, anything to do with understanding human behavior will always get a rise out of people—both the scientists and the public, and primatologists do know far more than the public about primate behavior.

To: <teresopolis@majordomo.srv.ualberta.ca>
From: "Thelma Rowell" <thelma@ingleton.demon.co.uk>
Ah, that's the point—"primatologists are trying to say something about human nature," says Pam, and therefore care about the media talking nonsense. So those of us who study monkeys because they are interesting ANIMALS, and not because they are little furry people, are also free to ignore the media nonsense, to treat it as irrelevant and not part of some joint effort as others have suggested? Should we split the grouping "primatology," now found to be a highly convergent bit of radiation, from two quite different foundation stocks?

To: <terespolis@majordomo.srv.ualberta.ca>
From: <Pamela.Asquith@ualberta.ca>
No, not all primatologists are trying to say something about human nature. However, the media usually is. It becomes a double whammy I suppose when popularizers not only recreate the world of little furry people despite what the scientists actually say, but when/if they also do it with material that was not meant to say anything about human nature at all.

To: <teresopolis@majordomo.srv.ualberta.ca>
From: <brian.noble@utoronto.ca>
. . . What's the honest motivation (with high stress on the term "honest") of primatologists to be more concerned with "human nature" questions, and those who tend to be more concerned with "animal nature." And is there an alignment between those motivations and the individual scientist's effectiveness in engaging well with the media and getting what they feel are "good" translations of their science?

To: <teresopolis@majordomo.srv.ualberta.ca>
From: "Naomi Quinn" <nquinn@duke.edu>
Just to . . . show how perverse the world is (and to prove I've been here listening all the time): in the context of my giving a talk at her institution that would be "relevant," a cultural anthropologist colleague commented to me a couple days ago, "I'm tired of seeing the archeologists and the physical anthropologists get all the headlines!"

SECTION 6

Reformulating the Questions

22 Science Encounters

Shirley C. Strum

Ideas about science, like ideas about primate society, are in flux. At Teresopolis, the debate about science came to be as important as (and perhaps more heated than) the discussions about the nature of primate society. Of course the two were inextricably linked, but not exactly in the way I had expected. This chapter is about our "science encounters." These highlighted the hidden "science" questions embedded in our project (and offered alternative formulations for our questions). As well, we played out the current controversy over science known as the "science wars." What happened makes the Teresopolis "science encounters" useful in understanding how science wars originate and what might be done to end them.

This is not an academic chapter. I will not review or synthesize the large literature on science, the burgeoning literature on science studies (see Thompson Cussins this volume), and the growing material on the "science wars." Instead I try to represent the nature of the discussions we had at Teresopolis and their consequences. The real significance of the science issues emerged after the workshop when I tried to evaluate what had happened. This chapter is based on a commentary I wrote (using the discussions, the papers, and my notes) and circulated to all the participants (Strum 1996). Many replied with their own thoughts about my interpretation of the issues. I have tried to take these into consideration but, ultimately, what you read below is a personal "take" on what happened and what it means.

In many respects this personal perspective which talks about process and gives details of context fits very nicely with the direction that our science experts championed (as I will discuss). But while I am positioned in the trading zone between those who study science and those who do science (and often the former have been the latter)—I am still firmly on the science side, a location that colors my perspective. My goal—to use "science encounters" to build a better science—may also not be shared by either set of colleagues (scientists or science analysts), although for different reasons.

Outrageous Ravings

"I haven't changed my ideas about primates since the 1960s." "My thoughts about what science is began at age six." "I have never changed my mind. I have done work that is a cumulative process. What is later is different from what is earlier because of this accumulation of evidence." "Our way of studying primates produces both gems and pebbles." "Insisting on theory is like extracting the nutrients from food as 'pills' rather than eating it as whole food." "Scientists are incredibly vulnerable, not about paradigm issues but about control and dissemination of their knowledge." "My colleagues get me wrong more often than the media does." "The only thing that is more abused than sociobiology in this meeting is science." "There is a difference between what scientists do and what science is." "Science is not about making things up but about making them real." "It is striking how scientists flee from being studied like their subject animals flee from being watched." "That is only when they have had a 'bad' experience as their first contact. Naive scientists are as approachable as naive baboons!!" "The philosophy of science ignores the practice of science and the practice of science uses bad metaphors about what science really does." "We do better science if we connect not isolate ourselves from the world." "Journalists and science studies share the same relationship to scientists: probing personal questions are part of an opaque agenda." "A conference is like a beast which has a life of its own. This was an interesting event for the conference beast because it had many other 'beasts' to play with, chimpanzees, baboons, bonobos, marmosets, hyenas, even ground squirrels."

The Approach and Framework

The workshop was stimulated by two provocative claims: that there have been significant shifts in the interpretation of the nature of primate society and that women scientists have played a major role in these revisions. We approached this controversial topic as committed scientists wanting to take the argument out of the realm of opinion and into the domain of "evidence." But to do so required several steps and additional resources. The first was "history"—before we could say why ideas changed we needed to demonstrate that they did. Our potted history (this volume) was the result.

We were specifically interested in the relative contribution of a set of "factors" to shifts in scientific perceptions about primate society. These were: changes in theory, changes in methods, increasing proportions of women in primatology, changing societal concerns over the past several decades, and the interaction of science and society through popular culture and the media. The aim was to broaden the scope of previous single-factor analyses and expand even Haraway's multifaceted sociocultural analysis (1989) by going across cultures and by including elements intrinsic to science itself.

What was needed next, after a history and an identified set of factors, was an analytic framework. The most important element of the proposed framework was its comparative and historical approach, which could offer evidence about how ideas have changed, as well as provide a larger context for specific interpretations of important events. The disciplines most closely related to primatology (anthropology, psychology, and animal behavior) seemed a good place to start. Other national/cultural traditions of primatology provided an additional comparative angle. The notion of a "larger context" also included looking at the interaction of science and society.

Setting up the historical and comparative framework rested on certain assumptions and posed some challenges. First we assumed that we could get "data," that is, we could gather empirical evidence both about which ideas changed and about how these changes took place (that is, the role of our factors in these changes). Second, we needed a way to be able to zoom in on specific factors despite their obviously intricate interaction. We took our inspiration from a particular experimental design used in biology. Our version involved multiple paired comparisons which when put together might be able to highlight, although at a gross level, the importance of only one factor at time (theory, methods, gender, or culture) in changing ideas about primate society. Our "experiment" had to use preexisting data which, of course, reduced its power. Despite these shortcomings, the logic seemed sound: in comparing different disciplines which shared theory but not methods, which shared methods but not theory, which shared methods and theory but not with the same representation of women, we might learn more about the role of "theory," of "methods," of "gender." The cultural/national comparison of primatology was intended to work in the same way relative to "culture."

We were also self-conscious about issues of "science," at least we thought so. This was the reason for placing changing ideas in primatology within their *larger* social and cultural context. What counts for knowledge is often hotly contested. Science and society intersect and interact. Our topic was particularly problematic because of its "location." Several popular writers, feminist historians, and philosophers of science have singled out primatology as a discipline where the presence of women practitioners has made a major difference, even labeling primatology as a "feminine science" or a "feminist science" (see Fedigan, chapter 23 this volume: "Gender Encounters"). More broadly, the study of primates has a special interest in modern American society as our closest relatives are used to reconstruct a behavioral genealogy of what was and therefore what should be (Latour and Strum 1986). This rationalization of human behavior takes place within the context of powerful images of nature, of science, of women in nature, and of women doing science. Therefore, it also seemed important to consider how these images were created and their impact on science. We put a special emphasis on the role of the media as the interface between science and popular culture.

Once the framework was designed, the actual analysis was obviously beyond our limited expertise. We needed new resources and looked to science studies to help with understanding how knowledge is constructed and how scientific ideas change, to feminist studies to help examine the impact of feminist concerns on science and the role of women scientists, and to studies of popular culture to help us investigate the role of the media in changing images of primate society within the American cultural context.

The framework informed our choice of participants and the topics they were asked to address. It even influenced the organization of discussion sessions. In the end, the workshop design revolved around our assumptions, our questions, our history, and our framework. The participants were remarkably willing to help move this agenda forward. Yet as time progressed, it became clear that others did not share our "point of view."

The Science Issues in the Project

The workshop was about primate studies and the role of specific factors in changing scientific ideas about primate societies. Soon it was obvious that the essential questions were really about science: what it is, how it works, and the importance of who does it. In the process of the week's discussion, small disagreements led to other disagreements that exposed what turned out to be very different views of science. Equally important, the science issues that emerged from the discussions were fundamentally different from the science issues that had been self-consciously embedded. The resulting "science encounters" were a surprise and had major repercussions for the questions and the framework we had developed.

THE HISTORY FALLS APART

I had been astonished and delighted while writing the potted history that Linda and I were in such agreement about the stages, the factors, and the important ideas. This I took as evidence that we were correct. It was therefore a shock to find strong disagreements from other primatologists. Robert Hinde and Sarah Hrdy, to name the two who actually took us to task in their original workshop papers (Hinde 1996; Hrdy 1996), didn't exactly agree with each other on "the history," but concurred that our version was wrong.

What had happened? Although we had tried hard to be specific, we did not go far enough. First, we didn't appreciate the impact of "location" on our point of view. Our history was just that, "our" history. Linda and I agreed because we were from the same academic lineage and the same generation. We both descended, me directly and Linda indirectly, from a North American tradition of anthropological primatology founded by Sherwood Washburn. Even within American anthropology, the history of ideas about primate society is seen differently from Harvard, from Chicago, in different descent lines and at different historical periods within specific institutions. No wonder Robert Hinde disagreed. He was the founder of another lineage at a different institution in a different country addressing ethological rather than anthropological questions.

I now recognized a new issue: How, when, and why can we generalize (in this case concerning the history of ideas about primate society)? The week's discussions indicated that the first step must be to "situate" ideas, to locate them in time and space. Culture, gender, institution, discipline—perhaps even mentor (see new title for our potted history, this volume)—must be specified. In this way, our history was no less correct than Hinde's or Hrdy's, because there were multiple histories. In fact, the multiplicity of histories spawned new questions about how ideas diverge and how divergent perspectives interact with each other. For now, it seems incorrect and inappropriate to have one history "stand in" for another or for all others, since even within what looked like a homogenous tradition of primatology, there was great diversity. This added another element to the larger cross-cultural perspective that was already part of the analytic framework.

Bloodied, the history suffered another blow because of what historians of science call the "presentist" bias. The past seems different when viewed from the present than the same events appeared at the time. Although we didn't do "Whig history," focusing on the great men/women of science to the exclusion of all else, our interpretations used the present as the window into the past. This process could have imposed rather than exposed trends, making the trajectory of ideas appear more progressive and more deterministic than they actually were (see Thompson Cussins, this volume, for a fuller discussion of the presentist bias). Ironically, we sensed this without understanding its full import during the analysis of Stage 4 (see chapter 1), the period of field research from 1985 to the present. We had more difficulty with Stage 4 than any other period perhaps because the outcome could not be identified in advance. The hidden presentist stance might actually have sabotaged our project by masking the very contingencies of "ideas" and "factors" that we sought to elucidate. Jolly's chapter (this volume) is a good illustration of the tension between the two types of perspectives and the discrepant interpretation that this produces.

THE FRAMEWORK FALLS APART

My expectations about the comparative framework were less grand than about the history. Although rudimentary, I thought it would offer provocative hints about whether having more women made a difference, whether sociobiological theory made a difference, whether naturalistic observations made a difference. Sadly, the framework, like the history, turned out to be flawed. For example, the innovative comparison with closely related disciplines was jeopardized by the selection of participants. Could we now assume that one person might represent an entire field? Worse yet, what seemed an irrelevant piece of information before became critical. Since Glickman was Tang-Martinez's mentor, could they be counted as independent voices? If not, the psychology/animal behavior comparison would collapse, becoming interesting "cases" that might be part of a *future* analysis.

The national/cultural comparison was similarly at risk. Being sensitive to who could speak for whom we had invited Asquith, who studied Japanese primatology, and Takasaki, who was a Japanese primatologist; Strier, an American who does research in Brazil with Brazilian primatologists, and Yamamoto (who despite her last name is not Japanese), a practicing Brazilian primatologist. Were two people enough? Asquith cautioned that essentializing "Japanese" is dangerous since it is not a stable category and Takasaki noted that he was discussing only one of several schools of Japanese primatology. The same reservations and difficulties emerged in the Brazilian case although primatology in Brazil has a shorter history. Even the European view, as represented by "indigenous" practitioners (Hinde, Rowell, Jolly, Byrne for "British" and lonely; Kummer for "Continental"—Kummer couldn't come, in the end, but sent a paper that stimulated great discussion), seemed inadequate. Hinde claimed that there were no distinct characteristics of "British" primatology but rather a set of lineages housed at specific institutions; Byrne disagreed. If Hinde was correct then our British participants disproportionately represented a particular "lineage," further reducing the comparative potential. American primatologists had already dissolved into "schools" and factions with specific disciplinary roots, and although better sampled than the British, the representation now seemed inadequate. Obviously a new starting point was needed, here as well as in the history, one in which people and ideas were contrasted taking into consideration institution, discipline, mentor, and even geography, not just national tradition.

If the framework's cross-disciplinary and cross-cultural heuristics were weakened, perhaps the "factors" could rescue the analysis. But what were these "factors"? By the end of our discussion on the "role" of theory, we had used the term "theory" in at least thirteen different ways. The Japanese view of theory puzzled Western scientists (see Takasaki, this volume), particularly their insistence that theory was ephemeral while good "description" would last forever. Theory, it seems, was a much more heterogeneous "actor" than the simple "factor" of the framework's imaginings. Things didn't get better when we went from abstract to concrete. Sociobiology was the only theory explored in any detail. But there was no agreement about what sociobiology is, where it came from, or its impact. The brief foray into sociobiology suggested that the history of ideas, in this case of specific theories, is a lot messier than we reconstruct it to be, and that theories are not free floating but have disciplinary and institutional constraints.

Methods didn't fare much better. First, there was no consensus on what was a "method." What criteria should be applied? Air travel and antimalarial drugs together made it possible to study animals in the tropics. Should they be included? We even debated the terms: Were we talking about *methods* or *methodology?* Whatever the label, they were more connected and intertwined with theory and with techniques than the framework's list of factors implied.

If the more solid factors like theory and methods ran into difficulty, it is no surprise that "gender," too, became problematic. We agreed that gender is a cultural construct not identical to "woman" or "female." Beyond that, gender seemed remarkably unstable in time and space and between different disciplines (see Fedigan, chapter 23 this volume). There were disagreements about what questions to ask (e.g., about differences between women and men and their impact on science, or about the hegemony of androcentric ideas and the possibility of their replacement) and uncertainty about what counts as evidence.

Culture, our final factor, was identified as "messy" from the start; the framework offered multiple entry points. One was through cross-cultural comparison using the hybrid category "cultural/national traditions." Another was by making the distinction between popular culture and scientific culture as explored through the media. The third was through society as the larger context for ideas about primate society. But these entry points rested on simplifying assumptions and, in light of the workshop papers and discussions, on perhaps false dichotomies such as that between science and society, between popular and professional science, between Western and "other" cultural perspectives. In this way, the framework had inadvertently tidied up "culture"!

All this would have been depressing except that the underlying reasons for the unraveling of our carefully designed approach were both fascinating and important.

Science and Science Studies, Scientists and Science Studiers

What is science? Many, including some scientists, felt it didn't make sense to ask this question because there are many different meanings and boundaries—science is not a unitary enterprise. Nonetheless, we do use the term frequently. Some notion of science informs our behavior as scientists as well as society's expectations and different understandings about science fueled many of the workshop discussions. Yet the fact that participants subscribed to different models of science was not explicit until the end of our time together, which unknowingly hampered many exchanges and produced controversy.

Two distinct "models" of science emerged, each with its own sense of history and its own focus. The first was primarily held by scientists who believed in Science with a capital "S" or what Latour calls "science-already-made" (I use Latour's terminology [1987, 1998, this volume] because it seems a useful way to highlight the contrasts). In this model, science is isolated from society and consists of the norms of scientific method and its products, solid facts. The history of scientific discovery looks straight and narrow; credibility is based on "truth value." This is what Shapin calls the "legend of science" (Shapin 1994—legend, not myth!), where the qualities of science have reached legendary proportions. By contrast, those who study science and scientists subscribe to a different model that focuses on the practice of science or "science-in-the-making" (or "practice" or "research" [Latour 1998]) where the outside world becomes central to and inside scientific practice. History gets messy, facts don't and can't speak for themselves. Instead, humans have to work hard to make scientific discoveries travel in time and space. This model doesn't question "reality" but attempts to make the understanding of the reality of science more realistic. Gender and culture, for example, get transformed from "biases" polluting science into part of the cultural practice of science which, at a minimum, influences the idiosyncratic paths of individual practitioners. Although the two models are not mutually exclusive, the different emphases (one on outcome and norms, the other on process and practice) have major implications for how scientists and science analysts argue and for their ability to talk to each other (see the section "Science Wars," below). And for our framework.

Perhaps things had "fallen apart"—the history, framework, and analysis—because I had approached science in the wrong way. Certainly, some of the tenets of science-in-the-making seem particularly suited to the project's needs and goals. For example, since the study of primates is a new and hotly contested field, what counts as "true" is not yet stable—we should expect controversies about who gets to speak and who is considered a credible voice. It becomes necessary to study rather than presuppose "the stabilization of facts and the standardization of scientific practice" (see Thompson Cussins this volume). Both were goals of our project.

I first began to be convinced of the relevance of the science-as-practice model when participants gave individual introductions (How did you begin your work?) and then individual research histories (Did you ever change your ideas about primates and why?). These were as diverse and multifactorial as the science analysts had claimed. Interesting patterns emerged in the research histories. Generally, our images of primates change because of the animals themselves rather than as the result of innovations in theory, for example, although we might reconstruct the events differently afterwards. I was surprised by what a small role theory plays in molding scientific practice when compared with the impact of institutional setting and mentors, for example. It also seemed that the assembled group seldom changed their minds about scientific ideas (like images of primate society) late in careers or in the heat of ideological battles. There were even confessions of what could only be called "contrariness" as being important in all aspects of scientific practice. Several primatologists acknowledged the role of the "Thelma effect" (after Thelma Rowell) in motivating their research: thinking thoughts opposite to the currently accepted ones. Others found unexplained variation more interesting than the mainstream paradigmatic questions and shifted their research accordingly.

As the analysts predicted, these rich messy histories did lose their richness (and their reality) when forced into a discussion of abstract factors like "theory," "methods," "gender," and "culture." Zigzag routes mysteriously became straight lines and the vast work that had to be done to get from *a* to *b,* not to mention all the way to *z,* went missing. Science-in-the-making really did look different than science-already-made.

Reconfiguring the Questions and the Framework

If science-made is the wrong model in the context of our project, the switch to science-in-the-making requires a number of shifts in approach. These can be summarized as: *from norms to practice, from global to situated,* and *from isolated to embedded science.* Implementing the alternative view requires both a change in perspective and in methods. For example, instead of starting at the top, with broad generalizations about global categories like theory, method, gender, culture, the process should start at the bottom, utilizing case studies of scientific practice located in their proper context.

What is gained by looking at "science" in this way? For a start, the diversity of practices, of primatologies, of lineages—what was missed before—are transformed from repudiations of each other into new data capturing a fresh reality of science. Armed with these data we can elucidate larger patterns and build out to broader but more robust generalizations.

The new orientation changes the questions as well. Below I discuss how to reformulate the old questions using a few rules as guides: situate and unpack the question, aim for more richness and recognize the complexity of science, make sense of the diversity and complexity. One participant described primatology as a "rich, thick passion fruit mousse." The challenge with this new approach was both how to make the mousse and then how to eat it. I have used gender and our discussions about gender at Teresopolis as the template both for the form of the new questions and for the type of data that we will need (for a substantive discussion of gender and science at Teresopolis, see Fedigan, chapter 23 this volume).

STEP 1: SITUATE AND UNPACK THE QUESTION

The first step is to situate and unpack the question. Keller's work on gender (this volume and in discussions) suggests that large, all-encompassing questions about the role of some factor, in this case the role of gender in science, are more tractable when divided into their constituent parts. These are not necessarily identical or commensurate with each other and often require different types of evidence. According to Keller, there are at least three issues embedded in the gender and science question. The first is about *women in science*—historical projects tracing individual careers. This kind of research uses case studies of women scientists in place of often ungrounded generalizations about how "women" behave "in science." The second issue is about the *science of gender.* Here the question is how science contributes to constructing our ideas about gender, ideas that limit or open up opportunities for women both inside and outside science. This requires very different types of evidence than the first question, although the two will logically and practically connect in many ways. The third concerns the *symbolic work that gender does in science.* Gender is an organizing principle in human interactions. Specifically for science, this is an investigation of the way images of gender come to shape scientific practices such as experimental procedures or our expectations of what constitutes scientific data or acceptable scientific explanations.

It is easy to see how Keller's gender "model" might break down monolithic questions about the other factors, theory, methods, and even culture. In each case, the "role of x" is decomposed into constituent parts, minimally three questions: about *x in science,* about *the science of x,* and about *the symbolic work that x does in science.* This would mean that to understand the "role of theory," there are (1) historical projects tracing the careers of individual theories, situated in time and space (and institution and by individual scientists); (2) studies of how science contributes to creating and maintaining ideas about what theory is that open up or constrain opportunities in the real world; and (3) investigations of the symbolic work that theory does in structuring scientific practice and expectations.

Let us consider sociobiology and its potential impact as an example of how this reformulation of questions might work. In the Keller framework, sociobiology could have (1) raised sensitivity and awareness about certain key ideas, (2) changed the configuration of practitioners in a way that

(3) had an impact on science by creating new possibilities for science and for theory in science, and (4) marginalized some scientific questions while highlighting or eliciting others.

There are other gender models which might also be useful. Wylie's discussion of gender at Teresopolis suggests, by analogy, the steps needed to understand the impact of sociobiology. First, essentializing and stereotyping "theory" needs to be rejected and replaced by a focus on specific practices. Next, attention should be paid to institutional and disciplinary settings and to situated histories of individuals. This maintains the richness and diversity of routes by which sociobiology may have influenced practice. Finally, we should expect a variety of connections and disconnections between sociobiology theory and scientific practice rather than just one. "Methods" and even "culture" could be treated similarly.

STEP 2: AIM FOR MORE RICHNESS AND RECOGNIZE THE COMPLEXITY

The analysts were unanimous that practice, whether a specific method or theory or by an individual scientist, was the key to understanding how science works. Their point was that we needed to recognize the real labor that is done in science and those who do the work rather than gloss these.

The analysis of gender and science gives us some examples of the fruitfulness of this approach in offering a different sense of "science." Keller's study of Nüsslein-Volhard (this volume) demonstrates the multiple and diverse connections and disconnections she had to feminism, to science, and to the history of developmental biology. Keller argues that this complexity cannot be simplified into essentialist statements about gender and science.

Analysts supported this position during discussions: personal histories of individual women may show very different trajectories and different links between gender and science because timing, institutional setting, and other variables are important. The examples that were offered included the influence of a nonfeminist woman scientist in the right place at the right time (see Keller this volume), feminist women scientists in the right place and right time (see Quinn this volume), and feminist women scientists in the wrong place/time (Quinn's description of feminist anthropology displaced by postmodernist anthropology, this volume).

The comparison of gender issues in the subfields of anthropology also illustrates the importance of the details in understanding the outcomes. For example, politically aware feminists helped to create and popularize feminist anthropology among cultural anthropologists in the 1970s (see Quinn this volume) but were displaced by the rise of postmodernism within anthropology. The result was that both women and science were marginalized in mainstream cultural anthropology. By contrast, gender issues entered archeology later, during what could be called the 1980s "postfeminist" era (see Wylie, this volume) when a postmodernist processual reassessment was already under way. Gender research became both a critique of standpoint and a way to make empirical statements about the deficiencies of current models despite female practitioners distancing themselves from feminism. The outcome for archeology was better science and a better position for women in the field. By comparison with other subfields, primate studies has had a slow and small influx of politicized feminists, later than in cultural anthropology but earlier than in archeology. What their impact has been is one of the questions motivating the current project. Silverman, drawing on her familiarity with all six subdisciplines, succinctly described how the impact of feminism and of women varied within anthropology. In her scheme, paleoanthropology could be placed at one extreme where there has been little impact and cultural anthropology at the other end with the greatest interaction. But Silverman also dispelled the common view that cultural anthropology was dominated by women from the start. While there were many early women anthropologists, including the "greats," Ruth Benedict and Margaret Mead, she argued that these women did not have high status or power, at least at the time. Today, by contrast, women have risen to positions both of visibility and influence in cultural anthropology.

The gender and science issues change when examined more closely from a cultural/national traditions perspective. Recent claims about the special virtues of women scientists, particularly women primatologists, allude to a different way of "knowing" and of doing science: empathy, holism, attention to detail, long-term commitment, and interest in certain topics such as cooperation instead of conflict. Japanese and Brazilian primatology provided a different take on what "gendered" science might mean. The Kyoto tradition of primatology has methods that can only be characterized as "feminine" if we use the American idiom, although the founders and practitioners are mostly men (see Takasaki this volume). It is clear that Japanese and U.S. science are not gendered in the same way. In Brazil, according to Yamamoto, the methods used in primate studies are not gendered at all. Instead, to use Keller's formulation, the symbolic work that gender does in science in the U.S. is coded as "power" in Brazil. The concern is with struggles between the science of "North vs. South" and between the credibility of the "periphery vs. center." The politics of gender are replaced by the politics of international science.

Even in the North American context, the feminization of methodology is quite recent. As Mitman reports (this volume), early field biology, primarily done by men, explored an alternative way of knowing which centered on individual experience, emotion, patience, and empathy. Only later was the science of animal behavior transformed in the United States into something more remote and analytical. Sympathetic, empathetic, and connected methodology did not begin with "women in the field" but was part of a long tradition dating back to the artist-naturalists of the nineteenth century.

STEP 3: NEW SCHEMAS AND MODELS OF INTERPRETATION

Once we have new questions and new data, how do we put it all together? Some science analysts offered schemas and models that might help.

Latour (this volume) suggests a way to track the multiplicity of elements and their interaction during the discovery and survival of a scientific fact (figure 1, page 361). The figure sketches five horizons of practice: the mobilization of the world, the autonomization of the discipline, the creation of alliances and interests, the mounting of public relations, and the application of central scientific concepts. These are *simultaneously* necessary to productive science. According to Latour, we can intercept the circulation of the entities we call "facts" anywhere in these five loops, but we cannot understand what happens unless we see facts as the product of them all. Certainly, Latour's picture helps to organize the complexity and perhaps even explain how it is generated.

Haraway takes a similar view although she uses different metaphors in her "primate revisions" (this volume). Primatology is a zone of implosion where primates and scientific practice intersect in multiple ways. A variety of actors, "technical, mythic, organic, cultural, textual, oneiric (dream-like), political, economic . . ." (p. 406) converge and tangle. The convergences and entanglements make primatology both interesting and complex. Haraway rejects reductionism, suggesting that to distill the elements and treat them in isolation is not only inappropriate but impossible. On the contrary, she shows how to enter the "zone of implosion" by investigating a specific question (in this case, what is the correct social unit of the common chimpanzee). Her rendition of the contestations, connections, and entanglements of the answers to this question demonstrates how to collect and interpret complex data.

Wylie and Quinn suggest other approaches to the interpretation of complexity. Wylie (this volume) proposes that "standpoint" is important. Standpoint allows us to see how contingencies get organized into more stable and critical perspectives. Standpoint theory may also explain the manner in which complex interactions come to make a specific difference in the practice of science. In her discussion of the fate of feminist anthropology, Quinn (this volume and discussions) draws attention to unconscious "cultural schemas." Since schemas influence our everyday behavior, we should expect that they would also affect our scientific behavior. Quinn discusses how this would work in science. Science is gendered male according to Quinn. Before the second-wave of feminism in the U.S., both men and women shared the same male-gendered science "behavior" schemas. Then feminist "discourse" consciously tackled these schemas. Women were probably affected earlier than men but eventually the awareness initiated by feminism reached everyone, even if indirectly. In this way, feminism created a contestation that shifted the cultural schemas that influence our scientific behavior.

In addition to new "models," analysts also felt that we needed new "metaphors." The notion of "bias" came under frequent attack. Bias may have done useful work in the past but analysts felt that the metaphor of bias divides the world into two simple and unrealistic categories: the biased world and unbiased science. Instead, Latour suggested an "access" metaphor, images of runways that allow science and scientists to "take off," or roads that bring scientists and animals closer together making it possible to see new things and to be surprised. Thus when Rowell said she had decided to treat her new subjects, sheep, like chimpanzees, she didn't believe they would act like chimps. But she felt this orientation freed them from past constraints and gave them the chance to demonstrate (to scientists) whatever new "sheep" capabilities they had. In Latour's terms, Rowell created a new access, not a bias. This switch in metaphor may make a difference; access enhances, bias distorts—one is considered good, the other bad.

Other metaphors were criticized and replaced. Mitman called into question the way scientists think about the interaction of science, the media, and popular culture. The current "diffusionist" assumption claims that the media is a conduit for scientific facts which should travel into popular culture without distortion. But this implies that the public is a passive consumer of scientific information. Instead, popular culture is always actively constructing new meanings, often in ways that authors never intended, whether it is *Star Trek* or science. According to Mitman, diffusionism should be replaced by an interactionist metaphor. Latour went even further. If we take seriously his five-horizon model, scientific facts get transformed at each step as they change audiences. There cannot be any "undistorted" communication of science. It is not just the media and popular culture that change the meaning of scientific facts, but every move from field to university to journals to films to corporate boardrooms, even within science, creates changes. Analysts argued that ideas of diffusion and of "distortion" should be discarded and replaced with the metaphors of "circulation" and "transformation."

The Teresopolis discussions were critical in helping me to understand why analysts want to concentrate on practice and how this focus changes so many things. Switching to a new model and different metaphors explodes my past way of seeing and of talking about science. The world is no longer divided into simple dichotomies: biased or unbiased science, isolated or connected science, real facts and fabricated facts. And I can now say why. The shift is not a politically correct leap of faith (although some can make it into that), but is based on a growing body of evidence about what scientists do and how their work gets accepted and transmitted. I also came to believe the claim (made by many at Teresopolis) that studying how scientists behave doesn't deny "reality" but instead makes "the reality of science more realistic."

THE BUMPY ROAD AHEAD

Taking an alternative model of science for this project, one which emphasizes practice, richness of detail, situated complexity, and embedded science, transforms the old framework, reformulates the old questions, and suggests new types of evidence (and new questions) that will be necessary for understanding how and why our ideas about primate society have changed.

Not everyone in the "science" camp shares my enthusiasm for switching approaches. Even those who might be interested will face some serious problems. These will be evident in the nitty-gritty of data collection, the selection of questions, the type of analytic framework, and the ultimate goals of the inquiry.

Science studies employs a variety of approaches (see Thompson Cussins, this volume, for examples). Most are descriptive and qualitative, often ethnographic, making them vulnerable to the same criticism as leveled at cultural anthropology and processual archeology. Although I value descriptive data, I am concerned about how to use and go beyond description. The study of primates faced a similar dilemma in the 1960s (see section 1, this volume), which it solved, in part, by the use of a comparative framework and the shift to quantification.

I was uneasy with the position taken by some analysts during our discussions. They felt that thick description of scientific practice was not a means to an end but an end in itself—that there was no way to build back up after shifting down, by situating. And no need to do so! I have championed the value of case studies during my twenty-seven-year study of wild baboons, but my goal has always been to find patterns, make generalizations, and ultimately derive principles from these details that can be used more broadly. For baboons, for primatology, for ideas about primate society, and for science, I want to be able to talk about more than the local. "Science" (capital S) offers some tools.

The focus on practice transforms the questions about how and why ideas about primate society have changed. If situated case histories are to be our new data, as the science analysts suggest, then these should be embedded in a well-designed and systematic comparative framework. Juxtaposing a set of carefully selected cases can be a powerful way to reveal their similarities and differences, to find patterns in the complexity and diversity, to build out from the local. The lasting value of our original framework was its comparison of different disciplines, different cultural traditions, and different generations.

Two other matters may prove troublesome. The first is whether certain questions are off-limits. This was clearest in the case of gender and science: some analysts felt that particular avenues of inquiry were prohibited because they were dangerous, while others contended that collecting empirical evidence on gender was impossible. The scientists strongly disagreed with the analysts on both of these.

The second area of contention may be the role of norms in science. Sussman echoed the silent thoughts of many others when he proclaimed that "there is a difference between what scientists do and what science is." Hinde addressed the same point by suggesting that there are better and worse ways to do practice, supporting Rowell's position that replicability and testability are central tenets of science. We never discussed the part that "scientific method" plays in differentiating science from other types of cultural practices. Partly this was because the different views of science did not clearly emerge until the end. It was also the result of different implicit assumptions. The analysts' focus on descriptive rather than normative approaches to science skirted around the issue while the scientists' believed that the outcome of scientific practice was predetermined by the efficacy of the "scientific method."

How Does the New Approach Resolve Some Old Issues?

I have often found myself defending the study of primates as "science" to practitioners in other disciplines. They select certain aspects that appear to undermine its scientific credibility, for example the controversy over the use of "anthropomorphism," or the broad public interest in the field which has led to its extensive "popularization," or even the new conservation "mission." I now see that these can be threats to scientific legitimacy only if one adheres to the science-made model where any connection between science and society diminishes the science. Shifting from norms to practice, from global to situated, and from isolated to embedded science reworks the logic of the argument and produces a different conclusion.

THE SPECIAL POSITION OF NONHUMAN PRIMATES

The study of nonhuman primates has a special position in both science and society as part of "embedded" science. As our closest living relatives, other primates look and act a lot like us. Scientific discoveries about primate behavior add to knowledge about ourselves in this "trading zone" between science and society. Primates' special location, biologically and socially, motivates scientific interest, specific scientists, and public attention. The more "articulate" nonhuman primates become through this growing body of work, the more they can talk to everyone, specialist and nonspecialist, scientist and nonscientist. But the location also creates difficulties. Language is one. Primate studies imports language about human behavior to describe nonhuman primates and then re-exports nonhuman primate models to interpret human action. The process is heavily laden with extra, sometimes hidden, meanings. From the perspective of "science-in-the-making," these should be open to investigation rather than become grounds for dismissing the scientific value of primate studies.

THE QUESTION OF BIAS AND EMPATHY

When the scientists and their subjects are so similar it is hard to decide on proper methods and appropriate types of analyses. A broader view of scientific practice also argues against applying rigid, inflexible rules. This is relevant to the current controversy about the place of "empathetic understanding" in science. Empathy is part of standard practice in the Kyoto tradition of Japanese primatology, while for North American traditions it is considered "bias." However, the "status" of empathy is unstable even among North American primatologists. It is not unusual for a scientist to accept empathy and anthropomorphism in one context like the study of nonhuman primate cognition while rejecting it as "bias" in others (such as the claim that female scientists have special insights into female primates). Rather than prejudging empathy, it may be more productive to test its scientific worth using the variety of innovative criteria that have been proposed by those who study scientific practice.

OPEN VERSUS CLOSED SCIENCE

Viewing science as "already-made" carries with it other assumptions. For example, that science is separated and insulated from society; that any crossing over from societal concerns into science creates distortion and "bad" science. Latour argues (1998) that during the last 150 years, ideas about science have not caught up with the historical transformations of science and society. The opening up of science to make information more available and more relevant to nonspecialists and to society at large is a fact of the twentieth century. Primatology has been at the forefront of this trend. The media created and enthusiastically responded to public curiosity about primates. Today, information, interpretations, and implications leap from the wilds into living rooms—and back again—almost instantaneously. Media images feed back into science in a myriad of ways, ranging from recruitment of future practitioners to implicit expectations about methodology to explicit notions about the nature of primate society. This broad involvement and interest in primate studies has, in the past, yielded rich sources of funding and, more recently, created a growing constituency for biodiversity conservation.

Primatology has had no option but to be an open science. The benefits are obvious. What is currently unknown are the costs: what it means to make the public (for now simplistically homogeneous and ill-defined) and the media (currently totally unaccountable) part of a more broadly distributed scientific enterprise. Of immediate concern is what happens to "facts" and to the scientific process in "open" science. Who has the right to speak? What should they be allowed to say? What happens when the number of steps required in the scientific transformation of information is dramatically shortened by the early intervention of the media (which sidelines the other horizons in Latour's model of the circulation of entities on their way to becoming scientific facts)? Making the public part of a distributed science also puts new obligations and responsibilities on scientists, for example, to actively engage nonspecialists, perhaps even creating "corrective" models that take account of what the public already believes in order to present new data most effectively.

Rather than rejecting "open" science as "bad" science, it may be more fruitful to concentrate on developing appropriate rules about rights and responsibilities, trying to capitalize on the good (better funding, more caring, more conservation) and minimizing the bad (gutter press and sensationalism).

MISSION SCIENCE

If society is inside science and science is inside society, it is impossible to think of scientists unconnected to societal concerns. The growing involvement of primatologists in both conservation and animal welfare/rights suggests that having a "mission," which has often been labeled "bad" science, is instead a pragmatic recognition of a changed reality. Fortunately there are now criteria to assess primatological practice that don't rely on isolating science from society. It is also interesting to think about the differential value assigned to various "missions." For example, primate conservation (science in the service of saving biodiversity) has not only acquired legitimacy and status, but interest in conservation is about to eclipse more academic pursuits among graduate students and young professionals. Science in the service of animal rights is more controversial, yet as the *Great Ape Project* (1993) demonstrates, it has already enlisted many respectable scientists. A much more contested area is science in the service of feminist rights.

These missions are increasingly relevant to daily decisions in primatology about what data to collect, what to do with differing interpretations, how to apportion professional time and which activities are appropriate to science. The future of ideas about primate society cannot help but be affected. Conservation foregrounds populations, demography, and ecology and gives questions of "society" low priority. The fight for animal rights focuses on the basic nature of socialness and the need for social groupings rather than the evolution and diversity of primate societies. By contrast, the nature of primate society is a critical issue to feminist concerns.

SCIENCE WARS: THE USUAL SUSPECTS

The Teresopolis "science encounters" taught me about more than the science questions embedded in our project. It illustrated just how wars over science come about and how they might be avoided or resolved.

We were an unusual gathering, to begin with. At least in the American context, this was the first time that the different "parties" were face to face to discuss science with enough participants to create two critical masses. Both "sides" were deeply suspicious of each other, mostly about hidden agendas like "is there a reality," "are you for or against science." This was partly because we weren't equally familiar with each other's work. The analysts had read a lot of "science" and some were acquainted with the primate literature. By contrast, the scientists at Teresopolis were mostly unfamiliar with the new field of science studies. Worse yet, to them it was about "cultural criticism" and the postmodern "critique" of modernity and of science.

This asymmetry of knowledge made discussion difficult. Other barriers included the contrasts in technical jargon, the use of common words in very specific and uncommon ways, and the fact that both scientists and analysts represented diverse schools of thought (and often disagreed among themselves). It was only at the end of the week that the real source of the problem became apparent—differing views of science. It is easy, in retrospect, to sort the misconceptions from the misunderstandings from the substantive differences. At the time they mingled together to heighten the sense of threat posed by the "other."

Scientists also didn't like being the object of study (although these scientists made the study of other primates their life's work!). There was a deep suspicion about the process of studying science and its motives. Scientists and analysts disagreed as well on who had the right to speak about science (and whether analysts had to corroborate their claims with scientists).

The misunderstanding about the meaning of practice and about the complexity of science that fueled our mini "science wars" provides insights into the larger battle over science in society. Perhaps as Latour (1998) suggests, ideas about science have lagged behind changes in science, in society, and in the interaction of the two during the last two centuries. To make matters worse, we lack "a way to talk about" a process which is messy, includes controversy, and whose outcome is connected to the rest of the world. Exposing or emphasizing this aspect of science (practice/research) actually threatens the "ideal" of Science, since scientific controversy and uncertainty, in our current popular idiom for science, gets "mis"interpreted as evidence that science doesn't work. The practice of science becomes a weapon to be used against Science. The media catalyzes the "science wars" because they tend to focus on the uncertainty and highlight the controversy, undermining the public's view of the solidity of the scientific enterprise.

Analysts at Teresopolis were very concerned not just about our little disagreements but about the larger *Science Wars*. Latour even predicted that if "we" lose the fight, in twenty years primatologists will have to change their motto from "save the primates" to "save the scientists." The solution proposed was that scientists, science analysts, and the media team up to engage the public with new ideas about scientific practice and make a compelling "research"-based argument for science. Otherwise the public (and their politicians) will continue to believe that science requires only the "reason" of isolated "brains" and does not need laboratories or other essential (and expensive) resources.

But how could we team up when we were so deeply divided? The first step was to minimize conflict. The next was to build bridges. During the week we partially diffused our mini "science wars" through hard work—long walks and even longer talks. Slowly suspicion faded. Everyone made a greater effort to use common language or to try and explain the meaning of unfamiliar word usage. Good will counted for a lot (so did not being able to escape from each other for seven days!). A better understanding of what was at stake emerged, even if there was no consensus about the exact next step. The scientists also realized that the analysts weren't interested in debunking science but in understanding how science worked. That was not incompatible with the goal of the scientists—to do better science.

Furthermore, the different views of science that we hold need not necessarily negate each other. The workshop conversations suggest why, although we never discussed this point directly. Scientists and science analysts, by virtue of their divergent "practice," focus on different aspects of science. Since they do different work, we should expect that the two sets of practitioners might develop different "standpoints." As we know, different standpoints often lead to different conclusions. Another way to think about the disagreements is to consider scientists as the "natives" that cultural anthropologists/analysts study. When "natives" are asked to describe how their culture works, they often answer in terms of norms, ideals, and outcomes. However the description that emerges when the anthropologist/analyst watches what the natives do (rather than what they say they do) is often messier, more complex, and more idiosyncratic. In this way the analyst captures a different sense of the "culture" than the practitioners of that culture. These reconfigurations do not resolve all the tensions. Some analysts and some scientists will not be so flexible as to concede multiple views. For the others, those not forced into choosing one model over the other, interesting issues remain. How do the two versions fit together? How do norms and practice, process and outcome relate? Perhaps different views of science are more appropriate to different contexts, projects, and as means to different ends. And perhaps as "science" circulates to different audiences, like the facts in Latour's five-horizon model, its reality gets transformed. At least these moves bring scientists and analysts closer together rather than pushing them further apart, perhaps smoothing the way towards the new teams needed to educate the public about the realities of science.

Conclusion

During the week, Linda and I frequently repeated our very modest goals for the workshop, often to the disbelief of the participants. We wanted to learn how to ask better questions and to get better data in order to resolve the controversy about how and why ideas about primate society have changed. I think we succeeded on both counts because of the framework we developed, the structure of the workshop, and the contributions of the participants. A much larger set of issues was also addressed by broadening the perspective and enlisting resources outside of primatology. We were forced to reconsider the very nature of science, of scientific practice, of scientists, and of connections of science to the rest of the world. The project and the workshop have opened up new opportunities and new constraints for research about "changing images of primate society" which take seriously the title of our last session: "Understanding the Production of Knowledge about Primate Societies."

23 Gender Encounters

Linda Marie Fedigan

Gender has collected a history of both uses and abuses, of political purposes and deviations, of slippages and confusions; and it brings this history along with it wherever it goes. (Oakley 1997, 53)

Introduction

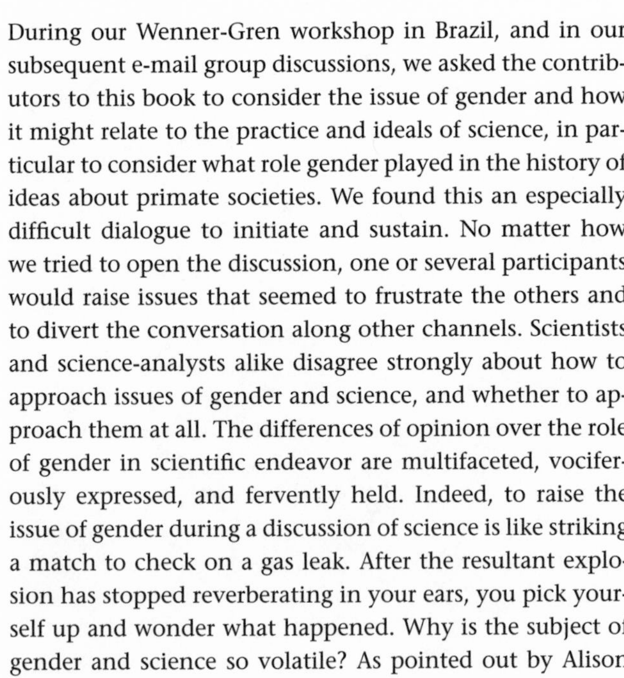

During our Wenner-Gren workshop in Brazil, and in our subsequent e-mail group discussions, we asked the contributors to this book to consider the issue of gender and how it might relate to the practice and ideals of science, in particular to consider what role gender played in the history of ideas about primate societies. We found this an especially difficult dialogue to initiate and sustain. No matter how we tried to open the discussion, one or several participants would raise issues that seemed to frustrate the others and to divert the conversation along other channels. Scientists and science-analysts alike disagree strongly about how to approach issues of gender and science, and whether to approach them at all. The differences of opinion over the role of gender in scientific endeavor are multifaceted, vociferously expressed, and fervently held. Indeed, to raise the issue of gender during a discussion of science is like striking a match to check on a gas leak. After the resultant explosion has stopped reverberating in your ears, you pick yourself up and wonder what happened. Why is the subject of gender and science so volatile? As pointed out by Alison

Wylie during our workshop, and in later conversations, many listeners make two immediate transpositions when they hear the phrase "gender and science"—the first is that by gender, the speaker must mean woman, and the second is that by woman, the speaker must mean biological female. Whenever we attempted to discuss gender and science, someone took the speaker to be making these two reductionist moves—from gender to woman and from woman to biology. In spite of the fact that I think all of the participants were aware of the pitfalls of such reductionist assumptions, the suspicions about essentialism were pervasive, in part because the members of our workshop come from such different backgrounds and hold such divergent models of gender and of science (see Strum, chapter 22 this volume: "Science Encounters").

It became obvious that there were some preliminary questions that needed to be addressed before we could talk together about the interaction of gender and science: What is gender? Is it relevant? Can it be measured? Can gender be considered separately from other factors? Are there good and bad questions about gender and science? Once we had aired these issues, even though we cannot claim to have come to universal agreement on the answers, we found that they had been productive questions to raise and had helped us to better understand each other's positions. And we were able to move on to a beneficial exchange of ideas about how to conduct research on the interaction of gender and science. The participants in our discussions made many suggestions about what would constitute good and useful approaches to the study of gender and science—general questions that were in turn useful to my own thinking about how to approach the more specific study of gender and primatology.

Thus, the structure of this chapter will follow the route charted by the interactions of our contributors. We started out by asking how gender is related to science. We found that the answer depends on how we define gender. Discussion can only begin if we can first sort out the various meanings of gender, even if we cannot expect to agree on one definition. To provide a context for the approaches taken by our contributors, I provide a short history of how the term gender has been used by scholars from different disciplines. I also briefly review some of the other difficulties (for example, disagreements over relevance and measurement) that arise in any discussion of gender and science—controversies whose resolution depend on our understandings of gender and of science. Then I argue that even given the complexities of how we understand these two concepts, and also given the fact that both gender and science are ever-changing and interrelated with other factors, it is still fruitful to study how gender makes a difference to science. We must take account of the controversies and the complexities to help us ask more nuanced questions, but it is still appropriate and important to ask the questions about how gender relates to science. Therefore, I turn to the heart of the paper, which is a consideration of suggestions from our contributors about how to best study the relationship of gender and science. I then use these suggestions to generate specific questions about gender and primatology.

A Brief History of the Definitions and Uses of the Term "Gender"[1]

Somewhat surprisingly, the controversies over gender extend to the very basic level of what the term means—how should we define it? In my training as an anthropologist, I learned that all the different human cultures take the basic clay of biological sex differences and construct this clay into quite variable models of how men and women should behave. Thus, I think of gender as the cultural transformation of biological sex differences into stereotyped and dichotomous expectations of the attributes of men and women. But there are many definitions and several disciplines (e.g., linguistics, psychology, anthropology, feminist studies) that hold claim to the concept, and the term has metamorphosed more than once during its history of usage.

The grammarians may have the original claim. Those of us who have struggled to learn Indo-European languages other than English have had to contend with the classification of nouns, pronouns, and accompanying modifiers into masculine, feminine, and sometimes neuter categories that require specific articles and variable spelling. This feature of grammar is known as gender. Not only is the idea of a feminine or masculine noun an alien concept for English speakers, the gender of most nouns also fails to correspond to any obvious or consistent principle, biological or otherwise. The reason why a particular noun—such as "el arbol" (the tree) in Spanish, or "la plage" (the beach) in French, or "das Madschen" (the girl) in German—should be respectively masculine ("el"), feminine ("la"), or neuter ("das") is lost in the mists of Indo-European language history. But, perhaps it was this subdivision of nouns by gender with little correspondence to biological facts that first attracted English-speaking social scientists to the use of the term "gender" to refer to the sociocultural constructions of sex differences outside of the realm of language. According to the dictionary (Oxford English Dictionary 1979; Webster's New World Dictionary 1980), the word "gender" is also related to the term "genus" in the sense of a class of species related by descent. The basic morpheme of both gender and genus—"gen"—means to beget. Thus, it may have also been the taxonomic and procreative connotations of gender that attracted social scientists to the term. As Donna Haraway (1991b) has described, there have been, and still are, strong differences in the way German, English, French, and Spanish (to name a few) speakers understand and use the term gender.

In any case, the borrowing and transformation of this word from a grammatical concept to a psychological term that refers to the cognitive and emotional attributes of masculinity and femininity was initiated by psychologists in the 1930s (see Oakley 1997), and the study of gender has continued to be an important branch of social psychological research since that time (Moghaddam 1998). John Money and Anke Ehrhardt's classic 1972 book *Man and Woman, Boy and Girl: The Differentiation and Dimorphism of Gender Identity from Conception to Maturity* was a pivotal exploration and development of the concept of gender in psychology and psychiatry, and was also influential outside the discipline. Money and Ehrhardt divided gender into two parts, which they defined as follows:

> Gender identity: the sameness, unity and persistence of one's individuality as male, female, or ambivalent, in greater or lesser degree, especially as it is experienced in self-awareness and behavior; gender identity is the private experience of gender role. . . . Gender role: everything that a person says and does to indicate to others and to the self the degree that one is either male, or female, or ambivalent. (Money and Ehrhardt 1972, 4)

Robert Stoller, another psychologist/psychiatrist who was interested in the unchanging core of personality formation, and who helped to develop the concept of gender for use in the social sciences, referred to it as "areas of behavior, feelings, thoughts and fantasies that are related to the sexes and yet do not have primary biological connotations" (Stoller 1968, viii–ix).

The first (or the most influential and early) introduction of the term gender into the feminist literature has been variously attributed to Kate Millet's 1969 book *Sexual Politics* (Oakley 1997), Ann Oakley's 1972 book *Sex, Gender and Society* (Segal 1987; Komorovsky 1988), and Gayle Rubin's chapter "The Traffic in Women: Notes on the 'Political Economy' of Sex" in Rayna Reiter's 1975 book *Toward an Anthropology of Women* (Haraway 1991; Melosh 1993). Soon after these three publications, other feminist writers (many of them anthropologists) began to employ the concept of gender, and by the late 1970s, it became obvious that this term had once again been adopted and reformulated by a new set of scholars. Taking as their starting point Simone de Beauvoir's assertion in *The Second Sex* (1952) that women are made rather than born, feminist scholars originally employed the concept of gender to reinforce the distinction of cultural from biological factors. They conceptualized gender as the "meaning of masculinity and femininity that a given culture attaches to the categories of male and female" (Keller 1992a, 16). Later, according to Evelyn Fox Keller (1992a, and see her chapter in this volume), feminist scholars also made an important distinction between gender as a socializing force in the psychosocial development of men and women, and gender as an organizing force in the world of objects and attributes beyond human bodies (the symbolic work of gender). One example of gender symbolism, pointed out by the biologist Zuleyma Tang-Martinez (this volume and during our e-mail exchanges), is that cultures (and scientists) often ascribe gender to objects that are not biologically male or female (e.g., nature, hurricanes, ships, the nucleus and protoplasm of cells). There are many discussions in the literature on feminist understandings of gender (see overviews in Haraway 1991b; Keller 1989a, 1992a; and this volume), but perhaps the most cogent and comprehensive description is Sandra Harding's (1986) presentation of the three aspects of gender in feminist studies:

gender identity (an aspect of self-concept, this is how social psychologists tend to see gender and probably the way most nonspecialists think of it);

assigned or ascribed gender role (an aspect of social organization);

gender symbolization of objects, people, and attributes (the way that cultures and individuals ascribe gendered meaning to both sexually differentiated and non–sexually differentiated phenomena). The view of Harding and many other feminists is that gender symbolism is about power relations and the devaluation of whatever is associated with women, females, and feminine.

As an analytical tool for feminist analysis, the concept of gender has succeeded almost too well. Indeed, Londa Schiebinger (1999) noted that the fashionability of the term gender has resulted in its appropriation to almost any description of feminism, femininity, women, females, or sex differences. For example, many people are wary of the powerful connotations of the word "feminist," and therefore gender is often used as a substitute for feminist or feminism (e.g., gender studies rather than feminist studies), when these terms are not in fact the same. There are as many definitions of feminism as there are kinds of feminists (for overviews, see Rosser 1997; Schiebinger 1999; Harding 1986; Wylie 1997). But Lancaster and di Leonardo's definition may be general enough to cover a large part of the spectrum of feminisms: "the critical examination of gender relations from the position of protest against women's unequal status." (1997, 2). Thus gender refers to cultural constructions of masculinity and femininity, whereas feminism refers to a critical examination of, or perspective on, these cultural constructions. Furthermore, gender has come to be used synonymously with "women" and "femininity." This is an inappropriate conflation of women and gender because it implies that only women have gender attributes, whereas gender also refers to cultural constructions of masculinity—what it means to be a man in a given time and place (see Keller this volume)—and, for some cultures and some individuals, gender also refers to third and multiple genders or androgynous identity and roles.

Along with what some feminists see as the overgeneralization of its use, the concept of gender has also experienced "slippage" back into the realm of sex differences—it never has been that easy to keep gender differences separate from sex differences in any sustained discussion, just as it is so difficult to distinguish more generally where sociocultural interpretations begin and biological facts leave off. Some people now talk about the cultural construction of sexual identity (maleness and femaleness) as well as the cultural construction of gender identity (masculinity and femininity), which further contributes to the confusion by collapsing the distinction between constructed genders and essentialist sex differences. Indeed, gender has became so prevalent a term that it is now common to see references in the animal behavior literature to the "gender" of animals (Pearson 1996; Walker and Cook 1998), a usage that definitely obscures the distinction between cultural genders and biological sex differences.

Thus, Oakley argues that "gender slips uneasily between being merely another word for sex and being a contested political term" (1997, 30). Similarly, Keller (1989a) has discussed what she refers to as the "dynamic instability" of the term gender. In her opinion, the uses of gender are tilted by popular and scholastic forces toward one of two poles—the pole of biological determinism (where gender is the same thing as sex differences) or toward the pole of infinite plasticity and relativity (a postmodernist world where any reference to biological sex differences or natural grounding has effectively disappeared altogether). Indeed, some feminist scholars have placed more and more emphasis on the constantly shifting and infinitely variable nature of gender (see discussions in Hawkesworth 1997, Heckman 1997). They do so for several reasons: as part of the recognition of complexity of the issues, as part of a larger postmodernist trend (Bordo 1990), and as part of a response to the allegation that early second-wave feminism was developed exclusively by privileged, white, Euro-American women who share few gender attributes with women less privileged (e.g., chapters in Oakley and Mitchell 1997 describe ways that feminists have responded to this "backlash"). Although it is certainly the case that the ideology of gender attributes (what it means to be a woman or a man) varies across races, classes, ethnic groups, cultures, and historic periods, feminists who bend over backwards to avoid *any* generalizations that might imply gender stereotypes run the risk of qualifying the concept right out of existence. Or certainly out of usefulness. This point was made at our Wenner-Gren workshop in Teresopolis by Emília Yamamoto, a Brazilian primatologist, who noted that if gender is conceptualized as altogether unstable and completely situated, it becomes meaningless and little more than a reference to individual characteristics that themselves shift over a lifetime—case studies and no more. Keller (1989) has written of her struggle to occupy a "middle ground"—an attempt that must contend with the difficulties of formulating a medial position, and with the constant pressure to move toward one of the two poles of biological determinism or infinite plasticity. She concludes that gender is neither determined by sex differences nor entirely independent of them. Along somewhat similar lines, Wylie (this volume) describes how feminists have worked towards increasingly nuanced and nonessentialist understandings of the forms and effects of gender roles, relations, and norms, while at the same time ensuring that gender is never preemptively "disappeared" as an important factor in science and society.

Let us also briefly consider some of the related controversies that arise in almost any attempted discussion of gender and science. Is gender relevant to science? From an idealistic perspective on science, gender—and any other sociocultural baggage—is irrelevant to science and to scientists. Such an idealistic view has been called "the legend of science" by Steven Shapin (1994), and "Science with a capital S" or "science-already-made," as opposed to science-in-the-making, by Bruno Latour (this volume). According to such an idealistic perspective, the objective scientific method eliminates, or severely constrains, the effects of sociocultural influences. Many people who are not scientists believe that science operates outside the boundaries of normal human behavior, and some scientists encourage this belief. As an anthropologist, I find it unlikely that we perform any actions outside the context of our sociocultural selves/beliefs; and as a practicing scientist I am painfully aware of the limitations on objectivity, even though objectivity (and reflexivity) are certainly ideals toward which we can strive. For those who seek to maintain the legend of pure science, free of sociocultural effects, it is heresy to suggest that gender plays any role in science. But even those who hold a more pragmatic (as compared to idealistic) view of science may see gender as irrelevant or unimportant, and may appear to dismiss it as only one of an interrelated constellation of factors that can affect scientific practice.

Can gender be considered or even discussed separately from other factors? When we ask about the role of gender in science, a very common response is for someone to say: But what about socioeconomic class and other aspects of social position, ethnic background, sexual orientation, institutional training—don't these affect scientific practice as well? My own answer is that yes, these other factors probably influence scientific practice as well, and are likely to interact with gender, but anthropologists have shown that gender dichotomies (cultural constructions of sex differences) are fundamental and pervasive categories of human thought and practice in all cultures, and as such, they organize our lives on all kinds of levels. As Naomi Quinn noted in our e-mail discussions, gender dichotomies are such psychologically powerful models for dividing up the world that they find their way into category systems everywhere. And these gender dichotomies are learned so early and are so integral to the developing self that they color our perceptions of all other categories.

Some feminists who have examined the interaction of gender and science have simultaneously analyzed factors such as race and class in a multiplicative or mutually constituted rather than additive manner (e.g., Haraway 1989; Harding 1993, 1997a; Rothenberg 1998; Stepan 1996). Others see the request to consider other factors as a diversionary tactic, particularly if it is suggested that it is impossible to talk about gender without simultaneous consideration of all the other sociocultural phenomena that may affect science. I believe that gender, however we conceive of it, is ubiquitous, and while it cannot be studied in isolation from factors such as race and class, neither can the latter be studied in isolation from gender. To recognize that other factors are interactive with gender does not lead to the conclusion that gender cannot be studied.

Even if we could agree on a definition of gender and agree that it is relevant to science, could we find a way to measure gender and evaluate its effects? Many feminists consider that gender is not the property of an individual, rather that it refers to the power relations between the sexes. Thus, they say that one cannot *have* a gender, and that gender cannot be said to adhere in individuals, because gender-appropriate behavior is constantly changing at both the societal and individual levels (Schiebinger 1999; Keller and Haraway during our workshop in Brazil).[2] Those feminist scholars who consider that gender-appropriate behavior is constantly changing and is not a lifetime characteristic of an individual would argue that gender is too variable, too socially contextualized, and in too much flux to be measured.

However, remember that the use of the term gender as a set of human attributes first appeared in the 1930s in the work of psychologists and psychoanalysts who wanted to describe what they considered to be lifelong and widespread mental features of women and men that were not determined by physiological aspects of sex differences. Subsequently, social psychologists have published scores of papers that measure gender in individuals and use it as a predictor of other behaviors. In contrast to the conclusion of some feminist scholars that a person does not *have* a gender, psychologists such as Robert Hinde argue that gender *is* a property of individuals and can be measured for a given culture and time period (Hinde this volume). Standard and widely used questionnaires, such as the "Bem sex-role inventory" (sex and gender are sometimes used synonymously in this literature), were developed by psychologists to measure individuals for degrees of masculinity or femininity. On the Bem inventory, individuals receive independent scores on both a femininity scale and a masculinity scale, and may be classified as "feminine," "masculine," "androgynous," or "undifferentiated," according to the pattern of their scores on the two scales (Bem 1981; Hinde this volume; and Moghaddam 1998). The characteristic gender patterns of individuals are then used to predict their behavior and attitudes in other situations. According to Moghaddam (1998), social psychologists recognize the importance of culture in constructing masculine and feminine roles, and they realize that these roles are malleable over time and across cultures. Furthermore, some psychologists have criticized sex-/gender-role inventories, particularly the use of a bipolar scale. Nonetheless, social psychologists clearly conceptualize gender as adhering in individuals over their lifetimes as a result of socialization experiences, and they see gender as a measurable trait of individuals which is a demonstrably useful predictor of other patterns. Thus, although there are feminists who are also psychologists, it is clear that these two groups have a fundamentally different understanding of where gender is located and therefore a disagreement about whether it can be measured and used as a predictor.

In sum, the concept of gender has a complex and winding history. As Alison Wylie noted during our Wenner-Gren workshop in Brazil, it "seems to be an idea that doesn't stand still." Ann Oakley's 1998 paper on the history of gender refers to the "rise and fall" of gender as an analytical tool for understanding the position of women. Not long after gender entered the public's and the general academic's lexicons as a popular code word for sex, women, and feminism, as well as for cultural constructions of masculinity and femininity, it seems to have begun to fall out of favor with at least some feminist scholars because of its unstable, contested, and heterogeneous nature. However, what this history suggests to me is that although no one group can lay exclusive claim to the concept, nor can we control its evolving linguistic uses, the idea of "gender" does perform a number of important functions, and to discard it now would definitely be a case of discarding the grain with the chaff. If we can agree that gender is complex, unstable, situated, and multifactorial, can we move on from there? For the purposes of the rest of this paper, I will adhere to Sandra Harding's comprehensive understanding of gender as composed of three components (identity, roles, and symbols), which allows for much plasticity and diversity.

Fruitful Avenues for Research on Gender and Science—More Controversy

To begin this discussion of fruitful avenues for research, I will describe what some of our contributors consider to be a very *non*productive question about the relation of gender to science, and why. This is the question of whether men and women "do science differently," an issue that is, in my experience, commonly raised by scientists and the scientifically literate public during any discussion of gender and science. Once people become convinced that gender identity, roles, and symbolization strongly influence the behavior of men and women, they also become interested in whether the gender identity of the scientist affects the practice of science. Indeed, some feminists and some sociologists of science have addressed this question. However, at least some social analysts, three of whom I will mention, consider the question to be inappropriate, a blind alley, or nonproductive.

Londa Schiebinger, a feminist historian of science, devotes a lengthy section in the introduction to her new book *Has Feminism Changed Science?* (1999) to the question: "Will women do science differently?" She argues that this question falsely equates "women" with feminism and political action. In her view, it is feminism and not women qua women who make the difference. Thus she sees it as inappropriate to map the successes of feminism onto women, or to collapse "political outlook onto sex, oversimplifying the process of democratizing science by making women alone the agents of that change" (1999, 15). According to Schiebinger, such an approach would overlook the role of feminist men, of women's studies programs, of feminist theory and practice, of institutional changes, etc. in changing science. Also, Schiebinger notes that the question as formulated is too monocausal and too easily implies a "universal" woman, whereas it is obvious that change in science, like change in any complex social phenomenon, results from an array of factors.

During our Wenner-Gren workshop, Evelyn Fox Keller, another feminist historian of science, responded to a question about the impact of being a woman on one's practice of science by saying the question is meaningless. It presupposes a commonality among women and among men. Which women? Which men? When? Where? Thus, like Schiebinger, Keller objects to what she sees as the implication of the question that women everywhere are alike and will take a similar approach to science.

Elsewhere, Keller (1989a) has argued that historically any acknowledgement of gender-based differences has been employed as a justification for exclusion. Thus, she feels that a question like this one can be harmful to women, and that women scientists are justifiably wary of it—for many women the best strategy for success in science has been to deny all differences between men and women. Furthermore, Oakley (1997) argues that second-wave feminism was built on the disavowal rather than the celebration of differences between men and women. Seen in this light, the question of whether "women do science differently" is dangerous because it undermines women's chances of achieving equality through avowed similarity with men.

During e-mail discussions between the contributors to this book, Bruno Latour, an anthropologist of science who is primarily associated with actor-network theory (see Thompson Cussins this volume), argued that the question of whether women have made a difference to primate studies is unrewarding in this formulation. In his opinion, it is obvious that women scientists have made a difference, but then so have many other factors. He would reformulate the question as: Do women make a difference that provides (or provided) a useful, powerful contrast, one which renders (or rendered) visible important new features? Latour (this volume) contends that most people view the issue of gender as a type of "bias" that shackles and distorts the perspective and practices of scientists. And he does not think that the characterization of science as "biased/unbiased" or "good/bad" is a useful way to analyze science. A more productive approach, according to Latour, is to ask whether scientists grant agency/activity to the entities that they study. For example, do scientists "make the primates themselves relevant to the questions we ask about them so that they can have a part in what we say of their behavior?" (Latour, this volume). To the extent that women primatologists have granted agency to their primate subjects because of their own "indignation against passivity" gained through decades of feminist struggle, he sees women as having made important, original contributions to the science. Thus, for somewhat different reasons than Schiebinger and Keller, Latour also rejects the original question as too simplistic.

Not all of the contributors to this book agreed with the rejection of the question "do women do science differently?" For example, the psychologist Robert Hinde argued that women have mean values that differ from those of men on any factor, including how they do science. Thus, in his view, it is legitimate to ask, and feasible to determine, if women do science differently. According to Hinde, human cultures may experience shifting gender roles over time, but at any one time and place there are fundamental differences in masculine and feminine gender identities. Furthermore,

Naomi Quinn, a cognitive anthropologist who participated in our workshop and subsequent discussions, argued that women and men may have partially different cognitive schemas, or unconscious maps, that influence how they practice their science because they are attuned to slightly different aspects of their environment. The feminist critique of science, according to Quinn, helped to make these unconscious cognitive schemas more explicit.

Although I wouldn't presume to label Naomi Quinn, she does echo the views (at least with respect to cognitive schemas) of what have sometimes been called "difference feminists" (Schiebinger 1999), who emphasize psychological dissimilarities between men and women. Many "difference feminists" have worked to revalue the qualities that have been traditionally devalued because of their association with the "feminine"—attributes such as cooperation, empathy, and holism. One feminist psychologist whose work has been very influential is Carol Gilligan (1982), who argued that women "speak in a different voice" when it comes to moral reasoning. Prior research on gender differences in moral development by psychologists such as Lawrence Kohlberg (1981) had concluded that the ontogeny of moral reasoning unfolds in universal stages and that girls/women are less developed in moral judgements than boys/men—women only achieving the third stage of a six-stage sequence. However, Kohlberg developed his theory of moral development from a sample of boys only, and those groups not included in his original sample seldom reached his higher stages. Gilligan's research led her to conclude that girls use different criteria than boys (e.g., context, responsibility, and community rather than abstract principles of justice and rights) to make moral decisions, and do not proceed through the universal stages postulated by Kohlberg. Another influential book was *Women's Ways of Knowing* (1986), by the psychologists Mary Belenky, Blythe Clinchy, Nancy Goldberger, and Jill Tarule. These researchers argued that the women they studied used connected, contextual, and "constructed knowledge" rather than "separate knowledge" (i.e., abstract, dispassionate principles) as tools to understand truth and reality. Thus, in contrast to the views of some leading feminist theorists today, Gilligan and Belenkey et al. suggested that women as a group bring a distinctive perspective (a "voice" or "way of knowing") to their intellectual practices. Although the ideas of Gilligan and Belenky et al. drew criticism from feminists and nonfeminists alike (see Schiebinger 1999; Wylie this volume), others found that these two books resonated with their own experiences.

At least one feminist scholar has addressed at length the question of whether women do science differently from men. Sue Rosser in her 1990 book *Female-Friendly Science* devotes one chapter to women scientists who have worked differently from men in their discipline through the use of connected, constructed knowledge. Using examples from many branches of science, Rosser argues that at least some women scientists have differed from men in many aspects of scientific practice—for example, in methods of data collection and observational practices, theories employed, conclusions drawn from data, application of the findings, and the use of competitive models in carrying out the science.

Sociologists of science such as Gerhard Sonnert have also addressed the question of whether men and women scientists "act differently." His study (1995a, 1995b) tracked the careers of 699 high-achieving scientists (former NSF and NRC postdoctoral fellows) through the use of questionnaires. Two hundred of these scientists were also interviewed and asked whether they thought "men and women do science differently": 60.8 percent of the female scientists and 49.4 percent of the male scientists answered yes, leading Sonnert to conclude that a sizeable proportion of the scientists in his study considered gender a relevant variable for interpreting the behavior of working scientists. Sonnert also evaluated other aspects of their responses to see if there were differences in the socialization and styles of men and women scientists. From this analysis, he drew attention to different levels of self-confidence, ambition, independence, competitiveness, and perfectionism in women and men scientists, and concluded that these differences lie less in the domains of epistemology and methodology and more in the social realms of science: for instance, collaboration with peers and professional conduct (see Morse 1995 for a similar argument).

During our Wenner-Gren workshop, Shirley Strum was asked whether her work as a baboon researcher had been affected by her being a woman, and she answered that it had not, rather her work had been affected by the long-term perspective of her many years spent studying baboons. Evelyn Fox Keller then interjected: "Why would anyone even want to ask that question?" and Shirley answered: "It is asked over and over again." This little exchange perfectly captures the conundrum: we have a question that is repeatedly asked, and that is even researched by some scholars, and yet others raise strenuous objections to it and clearly feel that it is an oversimplifying, vexatious, and even potentially harmful question.

What can we conclude about whether "women doing science differently" is a fruitful or nonfruitful way to conceptualize the issue? I suggest that the controversy itself highlights the way different schools of thought encounter each other at the intersection of gender and science. Some see the question as an essentializing move, as dangerous and unproductive, others see it as a testable proposition that will shed some light on the relationship of gender and science. After our discussions, I understand how the question can lend itself to various distortions of gender, of women, of feminism, and of science and how science changes. At the same time, I think that if this is a question in which scientists themselves and the public are interested, then it is important to find ways to modify and adapt the question to meet some of the objections described earlier, to break it down into more specific questions that are more contextualized and multifactorial, and then to address these reformulated questions.

How to Study Gender and Science—Suggestions from Our Contributors

Several of the participants in our discussions proposed what they consider to be fruitful questions that can be addressed in relation to issues of women, gender, and science, and I will briefly describe some of these here. Although I will often attribute a particular question to one individual, the interactive nature of our discussions in Brazil, and later through e-mail, means that more than one person may have helped to articulate a particular question.

Naomi Quinn suggested that we consider the role of men in "gendering" academia. Her argument is that in order to understand the role of women and the role of gender, we need to realize that academia is a man's world to begin with. In her chapter for this volume, Quinn portrayed the field and history of sociocultural anthropology as an arena where feminist approaches briefly held the "center ring" in the late 1960s and 1970s, only to be displaced by the resurrection of postmodernism in anthropology. Furthermore, during our e-mail discussions, she argued that men designed academia, their psychology maintains it, and their preoccupations dominate and bias academic theory. In her view, men designed academia to be an idealized man's world by rewarding an uninterrupted career trajectory and by rewarding those qualities associated in our culture with masculinity—assertiveness, competitiveness, individual glory-seeking. She says the question should be not how women could be more successful in this system, but how to change the gendered reward structure. We will return to this question.

Following on his view that "do women do science differently" is an unrewarding question, Bruno Latour argued that we should instead ask this question: Do interesting women studying interesting primates in interesting ways make an interesting difference? Obviously, "interest," like beauty, is in the eye of the beholder, but what I take Latour to be saying here is that it is not women, but creative new approaches that are important in science. Similarly, he asked if women were part of a new "setup" that lets primates speak out more about their lives and become more articulated and articulate. Latour was particularly taken by a comment from Thelma Rowell that in her recent studies of sheep she allowed the sheep to be as intelligent and socially astute as chimpanzees and then sheep turn out not to be boring animals, but just as complex and interesting as primates. Thus, what Latour suggests we research is whether women practitioners are associated with inventive and productive ways of conceiving and expressing their science, with novel avenues of approach that render aspects of our subject matter visible in a new light.

Evelyn Fox Keller suggested in her chapter for this volume that the first question we need to ask and answer is: What does the presence of women in science have to do with gender and science? I assume by this she means that we need to determine how the numbers or proportions of women in the sciences both affect and are affected by gender identity in individuals, gender roles in society, and gender symbolism in culture. This would help us to avoid the unwarranted assumptions that the term "women" is synonymous with gender and that the relationship between women and gender is straightforward or homogenous. As suggested by Robert Sussman in our Brazil workshop, we could begin by studying the history of diverse sciences that have included quite variable proportions of women across disciplines and over time, and try to trace the role that gender plays when a scientific discipline includes few, none, or many women as practitioners. Indeed, this is what we were trying to do in the original comparative framework for our workshop in Brazil. Naomi Quinn suggested more specifically that we examine what has been the positive impact of women practitioners on the history of ideas in primatology, particularly those ideas related to the behavior of males and females.

Keller further suggested that we ask how the second wave of the women's movement has changed primatology, not only directly, but indirectly, through improving the access of women to science and by changing our collective assumptions about gender. During our workshop in Brazil, Keller outlined her view that the influx of women into science and the shift in the cultural ideology of gender occurred simultaneously as a result of the women's movement, rather than in a causal sequence such that more women came into science and these women then brought about the shift in visions of gender.

During our subsequent e-mail discussions I asked why "gender consciousness" (an awareness of, and interest in, gender issues) arose in closely related disciplines at such different time periods. If the women's movement did affect science by changing the cultural ideology of gender, then why (to take the example of the subdisciplines of anthropology) did feminist thinking blossom in sociocultural anthropology in the 1960s and

1970s (Quinn this volume), in primate studies in the 1970s and 1980s (Fedigan 1994, 1997; Hrdy 1984b), in archaeology in the late 1980s and 1990s (Wylie this volume), and in some areas such as human paleontology hardly at all (for exceptions see Fedigan 1986b; Hager 1997a; and Zihlman 1978, 1987, 1997). In answer to this question, Naomi Quinn suggested that a decade is not really a big time difference, and that the subdisciplines of anthropology could be in a dynamic relation to each other, such that changes in sociocultural anthropology may have had a domino effect on primate studies, and so forth. Similarly, Brian Noble pointed out that one field science (e.g., social anthropology, primatology) can have effects on others (e.g., archaeology, paleontology). Alison Wylie suggested that archaeology may be the most recent subdiscipline of anthropology to take up gender issues because of a combination of factors—a more machismo disciplinary culture, a slower influx of women, a differently structured workplace (more teamwork), and more recent changes in dominant theory and methodological conventions. Hrdy (1986) has argued that feminist revisions of science occur after a motivational change which is itself subsequent to an increase in the proportion of women in the discipline. Note that this is a different reconstruction than Keller's sequence of how the factors—the women's movement, the influx of women into science, and the changing ideology of gender in society and in science—interrelated and occurred in time. These admittedly preliminary suggestions imply that we could shed some light on the general question of how gender and science have interacted by attempting to trace the specific histories of how different disciplines have developed "gender awareness" over time: Does a change in gender ideology (or the publishing of internal feminist critiques) seem to coincide with, or follow from, the reaching of some sort of "critical mass" (proportion of women in the discipline)? Or with the rise of particularly influential individual women scientists? Or subsequent/prior to major theoretical and methodological shifts in the discipline? Or some combination of these factors? When and why do "backlashes" occur? Do the specific case histories of various disciplines build up into any more general pattern or patterns?

Focusing on the feminist argument that gender is about power relationships, and arguing that gender issues are a subset of larger power issues in science, Shirley Strum asked whether women might influence a discipline such as primatology because of their gender, or more generally because of the historic link between gender and marginality in science. It is not difficult to document the marginal position of women in science. For example, in our e-mail discussions Alison Wylie presented many examples of how women have been marginalized in North American archaeology—from being trained into secondary laboratory and technical roles (rather than primary fieldwork and directorship roles), to being less well funded, employed by less prestigious institutions, and less widely cited than males. She also argued that the defining preoccupations (i.e., which questions, explanatory models, factors, data, and methods are considered important) of a discipline such as archaeology are shaped by the conventions and constituents of a scientific community into core and marginal issues. As one example, Paleo-Indian big-game hunting (an activity that archaeologists have associated with male tool assemblages) has been widely studied and interpreted, whereas other Paleo-Indian subsistence activities (that archaeologists associate with women's activities) have been treated as marginal, not because there is no evidence for them, but because the evidence comes largely from women archaeologists working on tools assumed to be associated with women's activities.

Keller (1983, 1985, 1989a) has discussed at length the case of the cytogeneticist Barbara McClintock, who won the Nobel Prize late in life for her work on corn chromosomes, after a lifetime of having her work largely ignored by the other scientists in her field. McClintock had taken a very unusual approach to her science, which Keller captured in the title of her 1983 book on McClintock as "a feeling for the organism." Many feminists embraced McClintock's story as an example of a "woman's way of knowing," but McClintock herself believes that gender is irrelevant to science. Furthermore, what Keller emphasized was the contradiction of McClintock's successful but marginal position in science (McClintock has never held a regular tenure-track position and has always felt isolated). Keller also framed McClintock's story as one of deviance and drew attention to the repudiation by traditional, established science of those very attributes on which McClintock's prize-winning science relied—empathy, respect for individual difference, connectedness, and relatedness. In a later discussion of another Nobel Prize–winning woman scientist (the developmental biologist Christiane Nüsslein-Volhard), Keller argues that

> . . . gender matters for women in science not because of what they bring with their bodies, and often not even for what they may bring with their socialization, but for what cultures of science bring to community perceptions of both women and gender, and in turn, because of what such perceptions bring to the communal values of particular scientific disciplines. (Keller this volume)

In other words, the main reason that gender is relevant to science is because one set of gender attributes (as well as the women to whom these traits are rightly or wrongly attributed) has been devalued, subordinated, and marginalized by science.

There is a now a sizable body of literature documenting that women even today do not fare as well in scientific careers as do men—they obtain relatively fewer tenure-track positions and those they do obtain are at less prestigious institutions, and they do not go as far or as fast in their career trajectories, etc. (e.g., Rossiter 1995; Sonnert 1995). Much of this literature deals with the probable *causes* of the lesser success and marginal positions of women, but our conference participants suggested we also investigate the probable *outcomes* of this association of women with marginality. Might women be influenced to approach their subject matter in distinctive ways at least partly because of the marginal position in which they "stand" in science? And might we expect similar stances from other groups marginalized by science? For excellent descriptions of how the standpoint of a scientist might affect the way they practice their science, as well as discussion of the history of feminist standpoint theory, see Wylie (this volume), Harding (1997b), and the debate recently published by the journal *Signs* (Heckmann 1997).

Some participants, such as Strum, suggested that it is all types of marginality and power relations in science rather than just the marginality of women that we should investigate. One reason for doing so is that it is possible that since all marginalized or underrepresented groups in science share powerless positions, they may also share some similarities in their approaches to science (although this is a controversial proposition) or similarities in their critiques of science. A second reason is more ideological—if we think it right and proper to rectify the historic marginality of women in science, then it would behoove us to also redress the marginality of other groups. Yet another reason is the argument that we cannot fully understand one type of exclusion without understanding other types—without understanding the nature and pattern of exclusion.

In our e-mail exchanges, the anthropologist and Japanologist Pamela Asquith pointed out the similarities—in her view probably superficial similarities—between the way that a "feminine approach" to science has sometimes been described and the way that the predominantly male Japanese primatologists have been described as approaching their subject (e.g., holistic, connected, contextual, empathic). In her chapter for this book, and elsewhere, Asquith has documented how the characteristic "empathic" and descriptive Japanese approach to primate studies originated and evolved simultaneously and independently of Western primatology. In Japan, primatology developed from the thinking and ideas of influential founding fathers of the science such as Kinji Imanishi and Junichiro Itani. Although Asquith argues that it is the history of Japanese ideas about nature and science (and not the powerless position of Japanese versus Western scientists) that has determined their approach to primatology, she does develop the case that Japanese primatologists feel constrained on the international stage of science by the requirements that they publish and discuss their ideas in English and by the Western devaluation of the Japanese approach and their journal *Primates* (this volume and in press). She characterizes the Japanese perception as being that they must "negotiate airtime." In the context of our present discussion, it seems fair to say that at least some Japanese primatologists perceive themselves as marginalized by Western scientists. However, it is clear that the ideas/approaches of Japanese primatologists do not arise from their marginalized position—rather that their ideas and approaches, arising from their own cultural history, are marginalized by the international science that is dominated by North American and European nations.

If the apparent similarity between Japanese primatological approaches and "feminine" science are not explicable by shared marginality, how might they be explained? First of all, the purported similarities are not likely to be accurate descriptions for either "group." As Wylie noted in our e-mail exchanges, the idea of a "feminine" science has been extensively criticized as essentialist and oversimplified by feminist analysts of science, and Asquith has argued in earlier publications (1991) that different schools of Japanese primatology have internally variable histories, approaches, and ideas. Secondly, we may be inappropriately conceptualizing approaches to science within an oversimplified framework of dichotomies—objective versus subjective, holistic versus reductionist, fact versus value, internal versus external, etc. Although traditional Western science certainly developed and continues to be framed within such dichotomies (e.g., mind versus body, reason versus emotion), this is no doubt an oversimplification, and sharply limits our options for approaches to science. Within this framework, one must be either connected or separate, contextual or independent, and so forth. Thus, Japanese approaches and "feminine" approaches may appear similar only in their contrast to mainstream, traditional Western science.

Harding (1986) has evaluated the supposed similarities between a "feminine" worldview and the worldview of other cultural traditions such as Africans, Native Americans, and the Chinese (e.g., connectedness rather than individuation, cooperation with nature rather than domination, and a high evaluation of personal knowledge and personal relations) and concludes that this curious coincidence is the result of patriarchal conceptual schemes, such that "femininity" and "Africans" are concepts that are socially created to emphasize "otherness." Analogously, Fee (1991) has evaluated why the feminist critique of science sounds similar to other radical epistemologies (Marxism, Third World perspectives) and concludes that they sound the same not because the critiquing groups are the same, but because the science being critiqued is the same. In other words, all radical critiques of science attack different aspects of the same beast—masculine, bourgeois, European science—and in differentiating themselves (feminist rather than masculinist, Marxist rather than bourgeois, developing nation rather than Euro-American) from this beast, they all fall into "the opposite" set of traits. All of these analyses seem to suggest that the purported similarities are the result of gross oversimplifications, and that categories of "others" do not hold similar ideas because of their marginalized standpoints—rather that their ideas are marginalized because of who they are.

In her chapter for this book, Emília Yamamoto suggested a different way that women in Euro-American science and scientists from developing nations may share the effects of marginality, not in terms of similar approaches to science, but in terms of a similar position in science. She argued that the role of gender has not been much of an issue in Brazilian science precisely because all Brazilian scientists, men and women alike, feel themselves to be in such a powerless position vis-à-vis European and North American scientists. In our discussions, Yamamoto drew parallels between Euro-American women scientists and Brazilian scientists in their struggle for recognition—the struggle to have one's articles published in good international journals, and then to have one's ideas recognized and cited by the international research community. Unlike the case for Japanese primatology, which originated and took root independently of Euro-American primate studies, it seems possible that the development of Brazilian primatology may have been shaped at least in part by its marginal position and late arrival on the international scientific stage.

During our e-mail discussions, Zuleyma Tang-Martinez also drew attention to the many underrepresented groups within North American science—not only women, but also people of color, ethnic minorities, immigrants, disabled people, members of impoverished socioeconomic classes, lesbians and gays, to name a few. Zuleyma also asked that we consider what it means that some Euro-American scientists (mostly male, white, and middle class) have gendered these groups of "others" similarly to their gender stereotypes of women (i.e., emotional, irrational, subjective, passive, submissive, less intelligent)? And she asked us to imagine what would be the impact on science if any or all of these underrepresented groups were encouraged and given unlimited access to science and scientific careers? How would science change if these groups were no longer underrepresented?

Alison Wylie added that the consensus from many studies of science is that science is better if the community doing the science is more inclusive—and not just inclusive in the sense of tolerating difference, but also in the sense of constructive engagement of a diversity of standpoints. In order for formerly marginalized groups to have an impact, they have to be not just tolerated, but constructively engaged (Longino 1990). Then Wylie went on to ask: "Given the range of factors that have been identified as making a difference to primatology . . . what sorts of inclusiveness might really make a difference now and in the future? What kinds of institutional/disciplinary mechanisms might foster really effective, engaged inclusiveness?" This resembles, albeit in a modified form, the question raised by Naomi Quinn—What can we do to change the reward structures of science such that previously marginalized and underrepresented groups will be encouraged to participate fully?

Picking up the threads from the issues raised by Latour, Wylie, Quinn, Tang-Martinez and others, Brian Noble wove them together into this comprehensive and action-oriented question: "Did women make an interesting difference in our ideas about primates, do women continue to change the discipline of primatology, and what does this indicate practically about the differences we should be working to make in science now?" I would add: And what does our case study of primatology indicate about the interaction of scientific practice and ideals with gender identity, gender role, and gender symbolism? Perhaps we are finally beginning to arrive at questions that can be widely agreed upon by those with an interest in science and gender.

Where Do We Go from Here?

Although some contributors felt that we had spent too much time on gender issues (or that any time spent on gender issues is too much time!), and despite the fact that we all became periodically impatient with the very real difficulties of talking across disciplines, our discussions helped to flesh out a more complete picture of the nature of gender, and of science, and of their interactions. And, as we had predicted in designing our Wenner-Gren workshop, it was clear that everyone had an important piece to add to the puzzle—primatologists, ethologists, animal behaviorists, psychologists, cultural anthropologists, media specialists, science-analysts, the list goes on. Not surprisingly, the feminist scholars had devoted the most time and effort to understanding the role(s) of gender in various scientific fields. We learned much from the research that has been conducted by our contributors on how gender has operated in the fields of primate studies (Donna Haraway), archaeology (Alison Wylie), cultural anthropology (Naomi Quinn and Sydel Silverman), and developmental biology (Evelyn Fox Keller).

Evelyn Fox Keller noted in her chapter for this volume that three important distinctions—between sex and gender, between women and gender, and between the socializing and symbolic work of gender—have helped to organize much of the work that has been done in gender and science. And she suggested that these three distinctions provide a logical basis for a tripartite subdivision of the research on gender and science: (1) women in science; (2) the science of gender; and (3) gender in science. I have concluded from numerous conversations with primatological colleagues that what many of them want to know is whether gender is relevant to their work as scientists, and if so, how? In other words, what difference does gender make to science or science to gender? Taking my cue from these concerns of my colleagues, from the suggestions of the contributors, and from Evelyn's three-way structure, I suggest that it would be productive to investigate the following set of questions:

Women and Men in Primatology. What difference does it make if few, many, or no women/men are present and practicing in primatology and related disciplines? If women are spread evenly across the specialties or concentrated in a few areas? If individual women, or individual men, have acted as influential role models for younger primatologists?

The Primatology of Gender. What difference does it make if Western cultural understandings of masculinity and femininity have both influenced and been influenced by primatological constructions of sex differences and of gender? If the work of particular women or men primatologists has changed our ideas about sex and gender? If the feminist critique of science has changed our ideas about sex and gender?

Gender in Primatology. What difference does it make if the practices and institutional aspects of primatology are gendered masculine or feminine? If power in primatology is gendered? If, in the past two decades, primatology itself has come to be gendered feminine by outside observers?

Further, I would ask, What can we generalize from this case study of one science, and how can we apply our findings to making science more inclusive in all senses of that term? The participants in our Wenner-Gren workshop and subsequent e-mail exchanges may differ radically in their perspectives on gender issues, but they do find common ground in their respect for science, and in their desire to comprehend how science functions in society and how to make it better. In my view, one way to improve science is to better understand its multifaceted interactions with gender identities, roles, and symbolic systems.

Acknowledgements

My research is funded by an ongoing operating grant from the Natural Sciences and Engineering Research Council of Canada (NSERCC). I thank Naomi Quinn, Shirley Strum, Alison Wylie, and Sandra Zohar for many helpful suggestions that improved this chapter, and all the participants in our Teresopolis workshop for their stimulating contributions to our discussions of gender and science.

NOTES TO CHAPTER TWENTY-THREE

1. For more complete histories and analyses of the term "gender," see Haraway 1991b; Hawkesworth 1997.

2. As a primatologist, this reminds me somewhat of the argument that dominance rank does not adhere in individuals, rather dominance rank refers to the relationship between individuals. Thus, it is argued that an individual does not walk around all its life as the "#3 monkey," s/he is only coded as the #3 monkey in a given social context. However, the difference is that whereas we have not shown that those individuals ranked #3 in a hierarchy will behave in a consistent manner, we have developed the cultural expectation that all human individuals coded "female" or "male" will behave in a consistent manner that we call "masculine" or "feminine."

SECTION 7

Conclusions and Implications

24 Implications: Future Encounters of the Primate Kind

Shirley C. Strum and Linda M. Fedigan

Introduction

Instead of the usual "conclusion," we would like to end the book with a consideration of the "future" as seen through the eyes of the workshop participants. Our chapters "Science Encounters" and "Gender Encounters" have already explored the implications of the workshop for the central questions of our project. But much more emerged in Brazil and during the e-mail exchanges. Although not part of the original agenda, these thoughts and concerns are relevant. They situate and connect the more academic issues to the bigger context of science and society. Rather than try to summarize what was said, we use e-mails to let individuals speak for themselves in their characteristic lively fashion.

The section is organized around six topics. The order of presentation is not a statement of priorities or importance. In most cases there are no definitive answers or solutions but the conversations are illuminating. We begin with *"The Media and Science,"* which carries forward what remains an unresolved debate. This is followed by *"Gender and Science on the Periphery."* The two topics are placed together because they seem to share many similar themes. *"Science Wars"* is next. The tension between different views of science and the work that was necessary to be able to talk to each other has already been introduced. Here we see what participants felt had been accomplished that could provide insights for others. During the science exchanges we came to consider the *"Value of Primate Studies."* Scientists and science analysts differed about how to judge their worth but agreed that the study of primates was valuable. Everyone was less optimistic about *"The Future of Primates and Primate Studies,"* both of which are intimately connected to the growing biodiversity crisis and global politics.

Worries about the future quietly punctuated our more academic discussions. It was as if connecting science rather than isolating it unleashed a Pandora's box of suppressed anxieties not just about the future of primates and primate studies but about the future of other disciplines, of the scientific enterprise in all its dimensions, of society, and of the natural world. Among the wealth of opinions about how to proceed, we did agree on one thing—the future would need *"New Teams."* The multidisciplinary, cross-cultural, and historical process that was "Teresopolis" exposed the shortcomings of thinking and acting in isolation.

The e-mails below reflect the nature of discussions on these various subjects. As with other e-mails, they have been edited to highlight key points.

The Media and Science

Throughout the workshop, science analysts offered a new way to think about the media and primates, about science and popular culture. For those actually involved with the media and trying to educate the public about primates, it was unclear exactly how to proceed. Understanding the process is only a first step. Even if we discard as inaccurate the "diffusionist" metaphor in which the media is required to faithfully "represent" scientific results to a passive public [as discussed in the e-mails exchanges in section 5], what should be put in its place? How do we make media and science a joint effort in which, if no one has control, at least everyone has "responsibility." During our e-mail exchanges we had the opportunity to see how scientific results get transformed in press releases in the popular media: "Chimps Cheat, Study Shows. UCSD Team Says Females Routinely Sneak Off for Sex . . ." read the headlines in the local newspaper. That afternoon, CBS, NBC, NPR, ITV, BBC had live or telephone interviews with the scientist who did the work. This particular episode of media and science upset many of us, but it offered a concrete example of what happens and made us think about what could be changed. Certainly, as the world gets smaller and space and resources for both nonhuman and human primates decline, there are new challenges, new opportunities, and new responsibilities for both scientists and the media.

To: <teresopolis@majordomo.srv.ualberta.ca>
From: "Shirley Strum" <sstrum@weber.ucsd.edu>
. . . understanding the process (which I don't really yet despite my extensive familiarity with it) is only a first step for me. I want to know how to get the scientific and cultural meanings closer together, running more in tandem. When the negotiations transform "chimps in unexpected and not so unexpected ways," I don't feel like I want to sit back and say "isn't popular culture interesting, isn't popular culture powerful . . ." Is it silly of me to want to make the connection "more appropriate," to have scientific images help to make better cultural images (and vice versa)? . . . I still don't see what it is that I can, should, am supposed to do . . .

Recently, in another forum, Evelyn Fox Keller discussed how the goals of the history of science have changed from "making science better" to understanding the process/practice for its own sake. Fine, but as a scientist I want to take that knowledge and make the process better. With some aspects of the practice of science, I think I know how. I certainly don't when it comes to science, the media, and popular culture. So tell me, if I want to be an interventionist scientist, what steps should I take based on this new understanding of the process of popular culture and the media.
PS. By the end of the workshop, Dick Byrne retracted the comment that he made that his colleagues get him wrong more than the media!!!

To: <teresopolis@majordomo.srv.ualberta.ca>
From: "Linda Fedigan" <Linda.Fedigan@ualberta.ca>
I would like to comment on Gregg's suggestion . . . that it is inappropriate for us to assume that the press is just there to interpret scientific findings to the public and that we have to realize that popular culture will produce its own meanings ("chimps in popular culture are different animals from those within primate studies.") I can accept that there are different understandings of chimps and I can accept that the media is not just a mouthpiece for scientists. But what gives me a knot in my stomach when I interact with the press is that the public, for the most part, also assumes that what is reported in the press is the scientist's view of the world. So if the popular press wants to represent female chimpanzees as adulterous, and if the public develops a view of chimps as "cheating" on their partners, that doesn't bother me (just as the anthropomorphic representation of monkeys in *Curious George* children's books doesn't bother me). But IF the press explicitly or implicitly presents the view that x, the scientist who did the work, thinks of chimps as adulterous and studies "cheating" rather than paternity patterns in chimps, that does bother me. In other words, I think that the press has some responsibility to try and present the scientist's point of view when they are reporting the results of a scientific study. They are not just representing chimps, they are representing the scientist. An ongoing study at my university has found that the general reader's accuracy in understanding the science items reported in newspapers is largely

determined by how the information is presented—their conclusion is that the better educated the reporter is in science, the more accurate will be the public's understanding of what is reported. I may be falling into the diffusionist model that Gregg has warned us against, but I agree with Shirley that my goal is to have scientific images and cultural images running more in tandem and I want to know how to better get my views across.

To: <teresopolis@majordomo.srv.ualberta.ca>
From: "Thelma Rowell" <thelma@ingleton.demon.co.uk>
. . . the whole [chimp mating] story challenges the idea of some great joint effort between media and science to me. Is it because of all the social sciences converging in primate studies? . . .

To: <teresopolis@majordomo.srv.ualberta.ca>
From: "Gregg Mitman" <gmitman@ou.edu>
. . . I don't think I have any forthright solutions to the concerns about wanting to make the scientific and the cultural run more in tandem, except that by knowing what are the dominant narratives (e.g., about chimps) in popular culture, and thinking ahead what possible spins a reporter might take on a story to make it sell, one could be more savvy about the presentation of one's results, by leaking the angles or suggesting possible leads one wants to get out. . . . But many of you have had much more experience, both positive and negative, with the media than I. Perhaps it would be useful to hear about stories that people were satisfied with and why they worked . . .

To: <teresopolis@majordomo.srv.ualberta.ca>
From: "Brian Noble" <brian.noble@ utoronto.ca>
I think that Gregg's and Bruno's pushing for a completely revised approach is the sort of move that will really make a difference, and Donna Haraway's work is a striking case in point. Each intervener is bound to affect things differently.

As soon as I began to consider the production and circulation of primatological and dinosaur knowledges, I began to see that no one in particular had "control," only the possibility of intervening, and rather, that there were what one might call "systemic" or "corporatizing" powers at play. I agree . . . that close analyses, network studies like Bruno has done, discourse analyses, deconstructions, all help to contour the processes and relationalities. This is about trading, the very sort of trading that the Teresopolis meetings have begun to facilitate. And since many of these studies do deal with power differentials in the multiple mediations taking place, they also can be read for guidance on how complexly situated interventions take place. Bruno's multihorizon picture in the meetings [see chapter 17] offered something that clearly avoided the rather inadequate idea of the one-way translation of pure nature into mixed-up culture via science and media etc., rather suggesting a nexus of agencies, a redistribution of "nature" and "culture" at every stage, and multiple resources working to produce current "setups."

I also think that Alison Wylie's addition of positioned "subjects'" (see chapter 12) (e.g.,

Shirley wanting to be an interventionist) is important as a resource in the setup, as is the consideration that the setup is being altered continually through changing historical conditions. The surprising thing is that once you begin to look into such matters, you find that scientific truth, while remarkably powerful and predictive within the terms of particular less-articulated setups, is also ephemeral, shaky, and a very limited resource in much larger networks producing and very forcefully conditioning "what counts as nature."

I don't know who has time to study primates and also do this, so perhaps the idea of trading, exchanging, and good listening remains the most efficient means. Alison Wylie's positioned subject gets all the more crucial, though, as everyone is going to bring different "agendas" to their interventions, spinning things in ever-alternate manners . . .

So, on the "cheater" [chimps], who will intervene in the contradictions, how and with what stories?

[from another e-mail] . . . It seems to me that the last ten to fifteen years has seen the heating up of primatology in political terms (I mean in how primatology exchanges with the polity, the people). There's more wide-ranging and controversial media play, conservation and endangered species issues always on the rise, gender and critical analyses intensifying, debates and practices around primates as disease vectors or as surrogates for biomedical testing, postcolonial conditions of conducting fieldwork, etc. It's a great big political swirl . . .

At the same time, more and more primatologists are actively (rather than passively) engaging the media, the politics, the publics, the policies in designing and conducting their studies. Others are finding they have no choice since it seems to be found everywhere, in grant applications, in fieldwork, in media presentations, in the news. And I don't mean politics as in bureaucratic hurdles or office politicking, but rather committing to something because you care. This can be caring about the animals, the discipline, who gets represented in the discipline, or even how things might be relevant to other people, the public(s).

As more scientists do take such active responsibility, the otherwise "pure science" or passive questioning gets more and more backgrounded. It gets increasingly difficult (and many would also suggest, irresponsible) to ignore the media, the public, politics, changing environments, etc.

. . . if this in any way characterizes what indeed people feel is going on . . . today, what do other people think is at stake? for primates, for primatology, for science. . . .

Gender and Science on the Periphery

Difference and inclusiveness were recurrent themes in Brazil and in the e-mails. There are many ways to have "difference." While we might not agree on whether "a difference makes a difference" in science or whether it makes an "interesting difference," or even the advantages of other visions of scientific enterprise, we almost reached a consensus that science should become more inclusive. But how? Who and what should be included and why have they been excluded up to now? Science on the periphery provided a new "take" on gender and Euro-American science. As well, issues of gender and science in North America appeared remarkably similar to those of "science on the periphery" experienced by those in developing countries. Unique and separate disputes merged into a vital struggle over "power" in science (see chapter 23, "Science Encounters"). Obviously the transfer of scientific technology is not enough in the development of global science. Attitudes have to change as well. The study of primates is seen as exemplary in extending scientific inclusiveness. Participants argued that it should continue to push the "envelope" of possibilities. Thus an interest in the role that women may have played in changing scientific ideas about primate society led us to broader considerations: the importance of standpoint and the variety of standpoints that have, can, and possibly should make a difference in science.

To: <teresopolis@majordomo.srv.ualberta.ca>
From: "Alison Wylie" <awylie@julian.uwo.ca>

. . . If there is any point of consensus in the studies I was referring to, it is that science itself is better if the community doing the science is more inclusive, and not just in the sense of tolerating difference. What's needed is constructive engagement of a diversity of standpoints. I should say, too, that the striking thing about both the workplace studies and the content analyses in archeology is how quickly these have led beyond gender . . .

. . . Given the range of factors that have been identified as making a difference to primatology (in its development and currently), what sorts of inclusiveness might really make a difference now or in the future? And what kinds of institutional/disciplinary mechanisms might foster really effective, engaged inclusiveness?

To: <teresopolis@majordomo.srv.ualberta.ca>
From: "Brian Noble" <brian.noble@utoronto.ca>

. . . if it isn't just who participates but the very stuff they participate with that sets up the conditions of exclusion, shouldn't we be taking the forms themselves into account in a very active way, as much as we do the participants and their differences, keeping the linkages and de-linkages between the two clear all the while?

. . . Isn't primatology, as such, a vanguard locale for reconfiguring science in relation to the gendering of scientific practices and science as culture?

To: <teresopolis@majordomo.srv.ualberta.ca>
From: "Zuleyma Tang-Martinez" <szthalp@umslvma.umsl.edu>

. . . The question for me is whether we should be encouraging the notion that men and women do science differently because of their biological sex (the essentialist approach critiqued by Alison Wylie), and I hope it has been clear that my answer to that is a definite "no." Whether we should encourage thinking that there is a male versus a female way of doing science (whether because of biological sex or gender differences)—I do not believe this is a good approach either. And lastly, whether we need to recognize that gender differences can lead to different approaches to science—I believe that the answer is yes, but not necessarily that one approach is better or superior to the other. As I have said before, to me it is the diversity of approaches and what each individual (based on all aspects of who they are, not only sex or gender) brings into science that gives a broader perspective and richness to our scientific endeavors.

. . . What kind of science do we want in the future?? One in which sex and gender (and class and race, etc., etc.) play a role, or one in which they do not? And would the latter even be possible?

To: <teresopolis@majordomo.srv.ualberta.ca>
From: "Brian Noble" <brian.noble@utoronto.ca>
. . . I would take "inclusiveness" to mean not only the inclusion of women or other marginalized people, but also the inclusion of those nonhuman features such as marginalized forms, practices, schemas, stories, spaces, etc. (and marginalized animals for ethologists, like "sheep"!) which will complement the inclusion of women or other marginalized folks in the continual shifting. Taking both people and nonpeople into account might help to get at what Bruno was referring to . . . as "the critical edge." Or to put it another way, as other sorts of people are included, wouldn't it be most interesting if science also opened up to nonhuman stuff like different forms of practice, storytelling, etc. which they bring along with them and which are around us all the time anyway?

To: <tereopolis@majordomo.srv.ualberta.ca>
From: "Zuleyma Tang-Martinez" <szthalp@umslvma.umsl.edu>
. . . Another related point that has already been alluded to by others: many feminists find Japanese science or American Indian approaches to nature very attractive because of the perceived emphasis on connectedness, empathy, holism, etc., all characteristics that we have gendered "female." But does this really mean that such approaches are more "feminine"? I think it does only if we set up the gendering prescribed by Euro-American cultures as the "norm" by which other societies and endeavors carried out in other cultures are measured and interpreted. Such an approach not only ignores the great complexity of other approaches to science and nature . . . but [is] also somewhat chauvinistic in its insistence that we interpret and judge other cultures and approaches through our culturally bound lens of gender. I think it might also be interesting to ask ourselves why it is that in some countries with a long history of machismo, this does not carry over to science, at least in terms of the systematic exclusion of women as women (although there may be exclusion based on social class which disproportionately disadvantages women).

[from the next day's e-mail] . . . At one point, Maria Emília [Yamamoto] pointed out that Euro/American science takes on what is essentially a gendered stance (still from an Euro/

American perspective) to science and scientists in the developing countries. The science and scientists in the developing countries are devalued and considered inferior, their methods are often treated as being less objective (i.e., less scientific), and their findings and observations are largely ignored (e.g., it is very difficult for Latin American scientists to have their findings published in U.S. journals). Interestingly, a Venezuelan behavioral ecologist who did his Ph.D. in the U.K. and is now back in Venezuela, told me that while he was using the U.K. address on his manuscripts he had no problems publishing his research in U.S. journals, but as soon as he began using his Venezuelan address (even though he claims the research, etc., were the same) he began having great difficulties in getting anything published. He finally gave up in disgust and now sends his research to Latin American journals. I might add that he is at the best Venezuelan university and is considered one of the two top behavioral ecologists in Venezuela . . . there is the assumption that Euro-American science needs to "rescue" science in the developing world. Thus, Euro-American science assumes the "male" role (objective, superior, dominant, etc.) while Third World science is assigned the "female" role (inferior, less objective, dependent on "the North," and the "maiden in distress needing to be rescued by the knight in shining armor"). Note also the underlying themes of hierarchy and control in all this. Third World science is viewed as a flawed and defective version of Euro-American science just like some men have viewed women as flawed and defective versions of men.

To: <teresopolis@majordomo.srv.ualberta.ca>
From: "Brian Noble" <brian.noble@utoronto.ca>
A point that captivated me most in the meetings was the differential interest mounted by American (Washburn, anthropological), British (Maddingly, zoological), gendered (heterogeneous), Japanese (Kyoto school, "primatographic"), and Brazilian (postcolonial, conservation ecology) positions. The interest of the latter of these two was quite outweighed by the interest of the others. I heartily agree that primatologists should feel good about . . . the tremendous number of connections . . . gained, but the differential interest earned across these positions is also something worthy of addressing. . . . Women in primatology have continued to win such redress, and the discipline is, of course, still thriving . . . why not keep extending it to other interests?

Science Wars

We began with suspicion and lack of understanding, with different styles and different ideas about what science is. We unwittingly recreated the "science wars" in our small, isolated world. But we also explored the possibility of overcoming these differences, forging a new community—one populated by scientists and analysts who can listen to each other and carry on a productive discussion. And we began to understand why it was necessary to build new understandings, acting cooperatively rather than antagonistically. Certainly some of the fight over science is just "silly" and prevents us from dealing with the more pressing issues in science and in today's world.

To: <teresopolis@majordomo.srv.ualberta.ca>
From: "Alison Jolly" <ajolly@arachne.Princeton.edu>

. . . we might all want to look at the *Nature* of May 22, volume 387, issue 6631 [at] a long briefing on the Science Wars, pp. 331–336, and a report on Norton Wise being turned down for the chair at the Institute for Advanced Study, p. 325.

My initial reaction is that the "Briefing" article is much less interesting than Teresopolis, because it is so empty of content: it is almost "Tis-Tisn't"—as in science is or isn't true/searching for truth/making progress. Most of the case studies so far have been in physics, and are not explained. Our stuff with the passion fruit mousse of primatology at least makes more intuitive sense of what is worth arguing about.

To: <teresopolis@majordomo.srv.ualberta.ca>
From: "Gregg Mitman" <gmitman@ou.edu>

. . . [the] issue of science as a social construction . . . seemed to paralyze so many of our discussions in Teresopolis. The fact that there are no pristine field sites for primatologists anymore, and that the very alteration of the landscape by humans has changed primates, humans studying them, and the relationships between them, is a nice example where both human and nonhuman actors are historical agents in the co-production of knowledge, made real, by the material interaction between them, which of course is dynamic. It moves us to an understanding of science as process. . . . The . . . point raised by Alison [Jolly; see e-mail in "The Future of Primates and Primate Studies," below] is that primatologists will then likely make inferences about how current habitat conditions will lead one to make larger theoretical claims about natural selection for primate behavior patterns taking place in extreme situations, without recognizing the very historically contingent nature upon which one is making these universal claims. In many respects, I think there is much common ground between primatologists abandoning the notion of studying primates in pristine nature and grappling with the issue of human-altered landscapes and talk in science studies about production and the co-construction of nature. And it seems to me that we are all pretty inept at breaking out of the nature/culture dichotomy, which I take to be at the heart of these discussions, but a move we absolutely must make (if we could only figure out how) in terms of survival—both of human and nonhuman species—in the twenty-first century.

To: <teresopolis@majordomo.srv.ualberta.ca>
From: "Brian Noble" <brian.noble@utoronto.ca>

. . . I wonder whether we are now seeing the rise of a) a community of science + science studies calibration and cooperation, and b) the emergence of hybrid critically engaged scientists/science studiers. Thank goodness for primatology. There is lots and lots going on at the intersections other than mock battles. Noting there is a contest in the emergent spaces is one thing . . . taking up arms is quite another.

As you can see, I am promoting the position that the war is over, or it should be, if indeed it was ever such a thing.

To: <teresopolis@majordomo.srv.ualberta.ca>
From: "Gregg Mitman" <gmitman@ou.edu>
. . . It is the practice of community—defined not as unity through sameness, but as unity in difference—that enables a collective conversation to take place that integrates yet respects historical situatedness. It is through interaction with the natural environment, in the form of living, working and playing, that communities are made. We place too little value on these forms of experiential knowledge within the academic community: historians know little, from the perspective of bodily knowledge, about the real work that gets done by a biologist in the field. Similarly, biologists often fail to appreciate the real work of hours spent by historians in cramped and dusty archives, pouring over illegible script, looking for that important gem. Perhaps this is why, as we write . . . amidst the brushfires ignited by the supposed science wars, the relationship among academics resembles more the tower of Babel than true community. If we really seek to shatter the science/humanities and nature/culture dichotomies that paralyze action, we need to spend a lot more time living, working, and playing in each other's fields . . .

To: <teresopolis@majordomo.srv.ualberta.ca>
From: "Shirley Strum" <sstrum@weber.ucsd.edu>
. . . The science wars, while a great concern of mine, appear somewhat trivial when put in the context of the real wars going on. . . . Moving science into the developing world in a new way, not just transfer of technology, but science-in-action with poor, uneducated people actually involved in a variety of partnerships, really changes the context and the frame. It makes me rethink the nature of science in terms of its entire process (from research to application or rejection) just as seeing science in the context of science studies revolutionized my ideas about science.

Does putting the "science wars" into an international, North/South perspective change the arguments or the sensitivities and emotions?

The Value of Primate Studies

Although we never specifically asked about the "value of primate studies," the debate about science and about the media elicited the topic. Scientists disagreed about the need to justify (and how to justify) studying primates. Despite the angst of some primate watchers, no one felt primate studies should be abandoned. Perhaps the most ringing praise came from the science analysts who pointed out the multitude of benefits that have accrued to science and to society from this work, regardless of the original intentions or scientific justifications of the research.

To: <teresopolis@majordomo.srv.ualberta.ca>
From: "Thelma Rowell" <thelma@ingleton.demon.co.uk>
. . . When applying for NIH funding, one used to have to fill in a bit about how one's research was going to help the health of US citizens. One will do almost anything for money, and if congressmen will only give it on those conditions, I can justify my request in those terms . . .

[from another e-mail] . . . [I] gave up applying for primate research grants; found I could do it pretty well on tax refunds and that saved a lot of effort and let me ponder my own questions in peace. Not popular in the department, that. I wonder if grant-funded research is a bubble? Have watched it expand from almost nothing, then, like a bubble, get stretched thinner and thinner—what will happen when it bursts?

To: <teresopolis@majordomo.srv.ualberta.ca>
From: "Charis Thompson Cussins" <cmc34@cornell.edu>
On the subject raised by Thelma of what is necessary to tell congress to get them to pay for one's scientific research, perhaps researchers should remember whose money congress is disposing of, and take some responsibility for assuring accountability to those constituents. Constituents of science (i.e., all of us who live in and by its truths, in the main) do not need research to serve us (not all research needs to be a cure for AIDS) but we do need enough transparency to assure that our money is being reasonably spent—that we are not paying for too many atomic bombs, Nazi science, Cyril Burts or superconducting supercolliders. Good science and workable democracies flourish together, it is often said, and both need at least a certain amount of transparency and accountability.

To: <teresopolis@majordomo.srv.ualberta.ca>
From: "Shirley Strum" <sstrum@weber.ucsd.edu>
Just back from a week of watching baboons to read the interesting exchange. . . . I've thought a bit about some of the issues and wonder whether who pays for the work isn't a major mover. Thelma gave up trying to get funding for her Kakamega work and perhaps part of her perspective comes from the freedom to follow her nose and her personal interests that she got by paying for her work herself. I have traveled the same route in many recent years.

I stopped to think how much money has been spent on my "modest" long-term research project over twenty-five years. . . . Can I justify it? I'm not certain. I have justified it in the past in the usual academic, scientific and even conservation terms. But as the . . . social and scientific ethos changes, what am I supposed to have accomplished and to whom should I be responsible? In primatology there are diverse constituencies. Are there more than in other fields? Is it enough to say that we know more about baboons, that I teach better classes, that the public is more knowledgeable about them through my efforts (no matter how unhappy I might be with media representation) . . . that Kenya is better because of the training, education, employment, conservation options that baboon research has offered, that the world is better because I explored key conservation issues for primates and helped test new approaches?

I love being in the "seemingly" isolated world of baboons when I watch them, but the connections are so many and so powerful now, as compared to twenty-five years ago it seems to me, that I only have "brief" journeys into that isolated world—little fantastic fantasies. When I'm back in the real world of which the baboon world is a part, the questions change and I am forced to think about whether "watching baboons is good for something." And the answer seems to be that what is at stake changes in relationship to each of these constituencies . . . and even within them in the last twenty-five years (maybe perhaps in the last twenty-five minutes!!). So, if we pay for the research ourselves, do we exit the (responsibility/justification) loop and return to the kind of science of the early wealthy naturalists?

To: <teresopolis@majordomo.srv.ualberta.ca>
From: "Thelma Rowell" <thelma@ingleton.demon.co.uk>
I had a professor of paleontology at Cambridge, an expert on the synapsids/early mammals, who, he said, was frequently asked what was the usefulness of his work. He used to reply that at least he was sure that he did no harm, and felt that was a rarely valid claim.

Your point is good, Shirley, but perhaps we shouldn't take ourselves too seriously. As a good capitalist, if someone will give you the money, then it is worth it to someone. If we are, au fond, in the entertainment industry, we are bargain basement indeed, compared with your standard sports or pop star! and at least as good value.

[from another e-mail] . . . and a lot cheaper. I don't think science should be expected to be good for something.

To: <teresopolis@majordomo.srv.ualberta.ca>
From: "Brian Noble" <brian.noble@utoronto.ca>
I've been trying hard for a couple of days to come to terms with what's been going on in the discussion of the last few days re: media, funding, public, and the "good" of science, etc. I'd like to offer some thoughts . . .

Clearly there's the matter of convergent zoological and anthropological lineages, but I'm wondering if there isn't more yet animating even that contrast. (Thelma, you've very bravely allowed your own views to be the lightning rod for the current exchanges, and I hope you will indulge things a little more . . . thank you for this).

. . . The position you've espoused Thelma: studying primates as animals, pursuing scientific problems because they interest you in particular, conducting science because it should be seen as intrinsically valuable, not because it should be "good for something," not because it should tell us about being human or human issues . . . that position seems increasingly difficult to sustain in the increasing torrent of public (i.e., political) involvements . . .

all the apparatuses which used to bolster a sort of "pure science" of primatology are losing ground in the swirl. Lots of folk in contemporary science studies are seriously addressing this as well . . .

To: <teresopolis@majordomo.srv.ualberta.ca>
From: "Bruno Latour" <latour@paris.ensmp.fr>
Sorry friends, I am not sure I understand this nostalgia, self despondency, and hesitation on the worth of your work. You are terrific and without you the primates would not be part of our lives, of our sociology, biology, comparative ethology, ethics, conservation, neuropsychology. They would be confined to pests to be destroyed by natives, to trophies to be hung on the walls of their villas by hunters, and to constrained lab animals just a bit more expensive than guinea pigs. Is this nothing? What worth do you want in addition? Interest my dear fellows, is not a given, it is made, and scientists, you in particular have been very good in rendering these animals interesting to a much wider constituency, turning all your sites into a conglomerate of different disciplines that would never have dreamed to come together if not for the field sites (who would have connected social theory, endocrinology, and ecology without them?).

Now of course this does not solve the question of who should fund whom and what is the return on investment, but what is amazing to outsiders like me, is, on the contrary, the incredible amount of interest you have been able to mount, grant application after grant application over fifty years for your animals—I guess this interest would not have been elaborated so strongly if you had paid for all of it only with your own money . . .

To: <teresopolis@majordomo.srv.ualberta.ca>
From: "Brian Noble" <brian.noble@utoronto.ca>
I agree with Bruno on the terrific work of primatologists. I also take note, Bruno, of the way in which you use the word "interest," as an "added value." I tried to use the word "politics" in a related way, to try and get at what primatologists and publics care about. . . . I think "interest" may be more useful for the moment since it rolls together curiosity with value and participation. That gloss changes the idea of "public interest in primates," just as it changes the gloss of Thelma's or Shirley's "interest" which is gained by use of personal or unofficial moneys. . . . Thelma's choice to go it alone in relation to funding, or Jane Goodall's noting of humanish ways of chimps supported by the National Geographic complex, or Sarah Hrdy's infanticide and stalking scenarios, or Robert Sussman's concentration on race, are interests born of connections. In all this, then, theory, method, gender (and media as well)—the organizers' themes—are interests that have been gained, added in along the way.

The Future of Primates and Primate Studies

The future of primates and primate studies are interconnected, perhaps mutually constructed as some of the participants suggest below. Although we focused on how and why ideas about primate society have changed, there was a persistent overriding concern about a factor that was not on our list: conservation. The natural world has altered dramatically in this century. How these worldly changes have affected the animals, their behavior (and our ideas about their society), is unknown. Questions about what is "natural," as well as how we define "nature," simmered just below the surface of workshop discussions but were never directly addressed. We could not, and did not, ignore "conservation" however; the current biodiversity crisis threatens the future of both primates and primate watchers, who now live in a world of habitat loss and fragmentation. Everywhere, recently diminished small populations of wild animals are vulnerable to the normal causes of extinction. Increasingly they are now also threatened by human aggression, indirectly because of civil wars and directly through the growing conflict between nonhuman and human primates over declining and limited resources. Science is changing as a result of this new global reality. There is less concern about academic controversies, like questions about the origins of different types of social organization, and more concern about population and habitat viability. Attitudes about science are also shifting. Conservation is what young primatologists want to do. As some participants suggest, there are new opportunities in these changing circumstances. But there is also a new sense of responsibility. Primatologists increasingly use their knowledge in order to help save endangered species. New techniques rely on the best of science applied to the thorniest of conservation problems, whether it is the obstacles to captive propagation, the design of successful translocation and reintroduction experiments, or predicting the future vulnerability of a population using sophisticated computer models. Everyone at Teresopolis felt we had a role to play in the conservation of the natural world through science and personal commitment. It was not clear, however, exactly what it was that "we" could or should do.

To: <teresopolis@majordomo.srv.ualberta.ca>

From: "Thelma Rowell" <thelma@ingleton.demon.co.uk>

. . . I think you underestimate the effect of people. A tracker I worked with in Cameroon said authoritatively that there is no forest without paths—human paths. For Africa I think he is right, and there is no modern habitat whose present characteristics are not largely due to human activity, just as in Europe. That being so, and adding the fact that the human population of East Africa has approximately quadrupled since independence thirty-odd years ago, it would be ecologically astonishing if there had not been ramifications throughout the continent, whether settled by people or not.

Now, it is also quite probable that the population of people 150 years ago was more similar to that of today, and there has been a crash at colonization preceding the present expansion—and then there was the rinderpest epidemic, the result of that colonization, but having its own immense effects in "unoccupied" areas. All other animals and plants must have been reacting to these changes, so to take the situation in 1960, or 1930, as a previously unchanging baseline is unlikely to be realistic.

Beyond all that, however, there is the effect of research activity, at Gombe and Mahale in particular. I would be surprised if the effect of provisioning lasted only as long as the food supply. . . . I do wonder if my behavior would have remained pristine if I was followed everywhere by a chimpanzee with a notebook for years on end.

A most powerful experience was going out with Shirley's baboons and realizing after a while that they had arranged a place for me in their movement.

To: teresopolis@majordomo.srv.ualberta.edu
From: "Alison Jolly" <ajolly@arachne.Princeton.edu>
I feel strongly about this—the theme I took for a President's Peroration at the Wisconsin IPS meetings was "Behavior in the Bottleneck." The gist of it was that we will more and more be studying nonhuman primates as affected by human primates. Willy-nilly, we will make a virtue of necessity. We will (and have) linked conservation biology with small group genetics and viability. But also on the behavioral side we will study migration between isolated forests, the effects of sparse or highly crowded populations, edge effects. And then claim that probably a lot of natural selection has always taken place in extreme circumstances: during the drought, or after the fire or cyclone that wipes out some troops' ranges and then in a while promotes new leaf growth. This is all part of the general ecological shift from believing in stability to being impressed by patchiness, change, catastrophe, at different scales. Looking at primate social behavior in the face of human-induced change will soon be respectable.

I had started on this theme before going to Teresopolis, but it obviously owes a lot to your paper, Karen, and to visiting your muriqui. But I have hardly ever seen a primate field site that wasn't massively impacted by humans . . .

(from the next day's e-mail) . . . I was thinking about the Gombe population of chimps which is a population of what I understand to be about 150 chimpanzees, in an isolated forest now surrounded by agriculture. Also about last years' epidemic of respiratory disease which killed nine (?) chimps of the northern community and which was transmitted apparently from the surrounding villagers, many of whom also died though with less international attention.

It could be argued that the vulnerability of the chimp population influences demography and genetics, rather than behavior. . . . I do not think that the components of chimp, or muriqui, or ringtailed lemur behavior change because they are in small reserves. However, even a shift in the frequencies of, say, aggressive behavior related to crowding in a small reserve, may be significant to the animals themselves. . . .

Shirley had asked in an earlier e-mail why it was that the Japanese scientists did not seem to be as involved in the growing conservation movement as other scientists despite working in the midst of vanishing habitats while studying their animals. Hiroyuki Takasaki responded as follows:

To: <teresopolis@majordomo.srv.ualberta.ca>
From: "Hiroyuki Takasaki" <takasaki@big.ous.ac.jp>

A brief response to Shirley's comment on the little concern/awareness about conservation issues among Japanese primatologists. From the very start, after establishment in 1985, the Primate Society of Japan is one of the most concerned in conservation activities among academic societies in Japan. By Japanese standards, it is rather radical. The president does not hesitate to write letters to mayors and governors to stop "pest control," i.e., removal of crop-raiding monkeys. The annual meeting always has a workshop on primate conservation, in particular of the Japanese macaque, which has become a crop raider in many places where the majority of farmhands are now aged people. These days, papers on conservation occupy a considerable number of pages in every issue of *Primate Research,* the Society's journal. The Society has a specialized committee and circulates newsletters on conservation . . .

Outside Japan, the Japanese primatologists' conservation activities are limited. However, that does not mean they are unaware. Following the Mahale case, Sugiyama has succeeded in getting money for conservation of the Bossou chimpanzees from JICA (Japanese aid). Also Izawa did the same in Colombia in order to create a new reserve. These were news in Japan, but did not hit the media overseas, I believe. News media circulate news which their subscribers want to hear or read as news. Nobody cares that the Japanese government pays money for conservation.

As funds are very limited in private sectors in Japan because of the current tax system, it is most realistic to try to involve the Japanese government in conservation efforts at present. However, as basic human needs have priorities in this kind of aid, it is not an easy task to convince bureaucrats to care about the needs of nonhuman primates.

Fundraising and media campaigns, however, do not help much in many primate habitats. Where does all the money go? Just see what happened in the case of Zaire under Mobutu's regime. Although the Wamba area became a reserve after Kano and his colleagues' repeated requests to the government, the only effective means of conservation has been the continuation of research (on bonobos) there. Any conservation approach which greatly affects the local residents' "subsistence" will never work. Even in the presence of researchers they will poach if they feel they have the right inherited from their ancestors to do so. Being a butterfly hunter, I can understand how they feel. This means fighting with the local government and people is a waste of effort. Only patience will work.

Any political trouble means the discontinuation of research. Japan's main resource is its people. The Japanese government will not fund nationals who make trouble overseas. It is dependent on the importation of most of its resources. Its food self-sufficiency is less than two-thirds. Without importation, one-third of the population will starve. The domestic Japanese nationals' survival depends on the steady import of goods from overseas, which are changed into export goods to buy food. It has to be that way until the self-sufficiency imbalance gets corrected. . . . Thus no primatologist will be funded by the government if he/she criticizes any foreign government which has essential mining or natural resources. This is usually an unspoken rule, but understood by all Japanese primatologists who work overseas.

Becoming a persona non grata means discontinuation of research, which will never help conservation. Therefore, once getting involved in a particular primate habitat, continuation of research there by all means is the best and most practical thing a field primatologist can do with limited manpower and funds for conservation. This is an implicit consensus among all Japanese field primatologists, I believe. This explains part of their tendency to adhere to particular habitats; site sweepers are rare among them . . .

To: terespolis@majordomo.srv.ualberta.ca
From: "Alison Jolly" <ajolly@arachne.Princeton.edu>
Should we at some point include "force" as an influence on primatology. I have just received a message from Drs. Takeshi Furuichi and Chie Hasimoto sent to Sue Savage-Rumbaugh . . . which makes me upset. It seems that the civil war in [what was] Zaire has disrupted everything including all the missions. The missionaries are a genial host to all the bonobo researchers. The mission supervised not only a cure of souls, but the only hospital and functioning schools in the region. Without their help, the bonobo researchers cannot get to their field site or function adequately there. Roads and airfields have been destroyed and no one knows how much else. The bonobo research has been put on hold, indefinitely, but they hope to return as soon as possible to pay their workers and to let people know that they are still intending to continue their research and conservation activities.

New Teams

Innovative collaborations creating new teams will be a necessity in the future. Our experience at the workshop indicates that this is not an easy goal. It requires hard work, sustained effort, basic trust, and good will. Even then, the road will be bumpy. These teams are needed in all the areas we considered at Teresopolis. Collaborations in science are needed, ranging from the most basic tasks, doing science (with new questions, new methods . . .) and understanding science (situated in its complexity), to the more serious challenge of promoting and safeguarding science. Equally critical will be collaborations in conservation, which projects require diverse sets of expertise and even more far-ranging partnerships. The ability of scientists to make their work and the conservation of their subjects meaningful to society obliges us to find new types of media involvement (and new levels of media and science responsibility). It also challenges us to break down the barriers between mainstream science and science on the periphery and possibly even develop new notions of what science is and how it can be done most effectively.

We began by wanting to understand how and why ideas about primate society have changed. We ended by realizing that to find the answer to what we thought was a relatively straightforward set of questions meant we had to consider topics as far ranging as the nature of science, the effect of the biodiversity crisis, the issue of marginalization by gender and by culture, the antagonistic relationship between different groups of scientists and between scientists and science analysts. And much more. Fortunately, we no longer have to do this enormous task alone.

References

Abbott, D. H. 1984. Behavioral and physiological suppression of fertility in subordinate marmoset monkeys. *Amer. J. Primatol.* 6: 169–186.

Abbott, D. H., and J. P. Hearn. 1978. Physical, hormonal and behavioural aspects of sexual development in the marmoset monkey (*Callithrix jacchus*). *J. Endocrinol.* 117: 329–339.

Abbott, D. H., J. Barret, and L. M. George. 1993. Comparative aspects of the social suppression of reproduction in female marmosets and tamarins. In *Marmosets and tamarins: Systematics, behaviour and ecology,* ed. A. B. Rylands. Oxford: Oxford University Press, pp. 152–163.

Abu-Lughod, L. 1993a. Can there be a feminist ethnography? *Critique of Anthrop.* 13: 7–27.

———. 1993b. *Writing women's worlds: Bedouin stories.* Berkeley: University of California Press.

Acitelli, L. K., and A. M. Young. 1996. Gender and thought in relationships. In *Knowledge structures in close relationships: A social psychological approach,* ed. G. J. O. Fletcher and J. Fitness. Mahwah: L. Erlbaum Associates, pp. 147–168.

Adams, E. R., and G. W. Burnett. 1991. Scientific vocabulary divergence among female primatologists working in East Africa. *Social Studies of Science* 21: 547–560.

Adang, O. M. J., J. A. B. Wensing, and J.A.R.A.M. van Hooff. 1987. The Arnhem Zoo chimpanzee colony: Development and management aspects. *International Zoo Yearbook* 26: 236–248.

Adler, N. T. 1969. Effects of the male's copulatory behavior on successful pregnancy of the female rat. *Journal of Comparative and Physiological Psychology* 69: 613–622.

Agre, P. 1995. From high tech to human tech: Empowerment, measurement, and social studies of computing. *Computer Supported Cooperative Work* 3: 167–195.

Aguirre, A. C. 1971. *O mono Brachyteles arachnoides (E. Geoffroy):*

Situação atual da espécie no Brasil. Rio de Janeiro: Academia Brasileira de Ciências.

Alencar, A. I. 1995. Relações de dominância e fertilidade de fêmeas cativas de *Callithrix jacchus.* M.Sc. diss. Universidade Federal do Rio Grande do Norte, Brazil.

Alencar, A. I., M. E. Yamamoto, M. S. Oliveira, F. A. Lopes, M. B. Sousa, and N. G. Silva. 1995. Behavior and progesterone levels in *Callithrix jacchus* females. *Braz. J. Med. Biol. Res.* 28: 591–595.

Alexander, R. D. 1986a. Biology and law. *Ethology and Sociobiology* 7: 19–25.

———. 1986b. Ostracisms and indirect reciprocity: The reproductive significance of humor. *Ethology and Sociobiology* 7: 105–122.

———. 1987. *The Biology of moral systems.* New York: Aldine.

Allen, C., and M. Bekoff. 1998. *Species of minds.* Cambridge, MA: MIT Press.

Allen, G. E. 1994. Modern biological determinism: The violence initiative, the human genome project, and the new eugenics. *Ybk. Soc. Sci.*

Almquist, J. O., and E. B. Hale. 1956. An approach to the measurement of sexual behavior and semen production of dairy bulls. *Proc. Int. Congr. Anim. Reprod.* 3: 50–59.

Alonso, C. S., and S. Porfírio. 1993. Um caso de poliginia de *Callithrix kuhli* com criação simultânea de filhotes. *Biotemas* 6: 89–99.

Altmann, J. 1974. Observational study of behavior: Sampling methods. *Behavior* 49: 227–265.

———. 1980. *Baboon mothers and infants.* Cambridge, MA: Harvard University Press.

Altmann, J., G. Hausfater, and S. A. Altmann. 1988. Determinants of reproductive success in savannah baboons. In *Reproductive success,* ed. T. H. Clutton-Brock. Chicago: University of Chicago Press, pp. 403–418.

Altmann, S. A. 1962. A field study of the sociobiology of rhesus monkeys, *Macaca mulatta. Annals of the New York Academy of Sciences* 102: 338–435.

———. 1965. Sociobiology of rhesus monkeys. II. Stochastics of social communications. *Journal of Theoretical Biology* 8: 490–522.

———, ed. 1967. *Social communication among primates.* Chicago: University of Chicago Press.

———. 1974. Baboons, space, time, and energy. *American Zoologist* 14: 221–248.

———. 1979a. Demographic constraints on behavior and social organization. In *Primate ecology and human origins,* ed. E. O. Smith and I. S. Bernstein. New York: Garland Press, pp. 47–63.

———. 1979b. Baboon progressions, order or chaos? A study of one-dimensional group geometry. *Anim. Behav.* 27: 46–80.

Altmann, S. A., and J. Altmann. 1970. *Baboon ecology: African field research.* Chicago: University of Chicago Press.

Andersson, M. 1994. *Sexual selection.* Princeton, NJ: Princeton University Press.

Andrew, R. 1962. The situations that evoke vocalizations in primates. *Ann. N.Y. Acad. Sci.* 102: 296–315.

Angier, N. 1997. In the society of female chimps, subtle signs of vital status. *New York Times,* August 12, B11.

Apted, M. [Director]. 1988. [Motion Picture]. Gorillas in the mist: The adventure of Dian Fossey. Los Angeles: Universal Pictures.

Araújo, A. S. 1996. Influence des facteurs écologiques, comportamentaux et démographiques sur la dispersion de *Callithrix jacchus.* Ph.D. diss., Université Paris-Nord, France.

Ardrey, R. 1966. *The territorial imperative.* New York: Athenaeum.

Arnold, A. P. 1996. Genetically triggered sexual differentiation of brain and behavior. *Hormones and Behavior* 30: 495–505.

Arnold, S., and M. Wade. 1984. On the measurement of natural selection: Theory. *Evolution* 38: 709–719.

Ascher, R. 1960. Archaeology and the public image. *American Antiquity* 25 (3): 401–403.

Ashburner, M. 1993. Epilogue. In *The development of Drosophila melanogaster,* ed. M. Bate and A. M. Arias. Cold Spring: Harbor Laboratory Press.

Asquith, P. J. 1981. *Some aspects on anthropomorphism in the terminology and philosophy underlying Western and Japanese studies of the social behaviour of non-human primates.* D. Phil. diss., Oxford University.

———. 1984. Directions in primatology. *Shiso* 717: 36–51. (J)

———. 1986a. Anthropomorphism and the Japanese and Western traditions in primatology. East and west. In *Primate ontogeny, cognition, and behavior: Developments in field and laboratory research,* ed. J. Else and P. Lee. New York: Academic Press, pp. 61–71.

———. 1986b. Imanishi's impact in Japan. *Nature* 323: 675–676.

———. 1991. Primate research groups in Japan: Orientations and East-West differences. In *The Monkeys of Arashiyama. Thirty-five Years of Research in Japan and the West,* ed. L. M. Fedigan and P. J. Asquith. New York: SUNY Press, pp. 81–98.

———. 1994. The intellectual history of field studies in primatology. In *Strength and diversity: A reader in Physical Anthropology,* ed. A. Herring and L. Chan. Toronto: Scholar's Press, pp. 49–75.

———. 1995. Of monkeys and men: Cultural views in Japan and the West. In *Ape, Man, Apeman: Changing Views, 1600–2000,* ed. R. Corbey and B. Theunissen. Leiden: Royal Netherlands Academy of Arts and Sciences, pp. 308–325.

———. 1996. Japanese science and western hegemonies: Primatology and the limits set to questions. In *Naked science: Anthropological inquiry into boundaries, power, and knowledge,* ed. L. Nadler. New York: Routledge, pp. 239–256.

———. 1999. The "world system" of anthropology and "professional others." In *Anthropological theory in North America,* ed. E. L. Cerroni-Long. Westport, CT: Greenwood Publ., pp. 31–49.

Aureli, F. 1992. Post-conflict behavior among wild long-tailed macaques (*Macaca fascicularis*). *Behavioral Ecology and Sociobiology* 31: 329–337.

———. 1997. Post-conflict anxiety in nonhuman primates: The mediating role of emotion in conflict resolution. *Aggressive Behavior* 23: 315–328.

Aureli, F., and C. P. van Schaik. 1991a. Post-conflict behaviour in long-tailed macaques (*Macaca fascicularis*): I. The social events. *Ethology* 89: 89–100.

———. 1991b. Post-conflict behaviour in long-tailed macaques (*Macaca fascicularis*): II. Coping with uncertainty. *Ethology* 89: 101–114.

Aureli, F., M. Das, and H. C. Veenema. 1997. Differential kinship effect on reconciliation in three species of macaques (*Macaca fascicularis, M. fuscata,* and *M. sylvanus*). *J. Comp. Psychol.* 111 (1): 91–99.

Aureli, F., C. P. van Schaik, and J.A.R.A.M. van Hooff. 1989. Functional aspects of reconciliation among captive long-tailed macaques (*Macaca fascicularis*). *Am. J. Primatol.* 19: 39–51.

Aureli, F., R. Cozzolino, C. Cordischi, and S. Scucchi. 1992. Kin-oriented redirection among Japanese macaques: An expression of a revenge system? *Anim. Behav.* 44: 283–291.

Aureli, F., M. Das, D. Verleur, and J.A.R.A.M. van Hooff. 1994. Post-conflict social interactions among barbary macaques (*Macaca sylvanus*). *Int. J. Primatol.* 15: 471–485.

Aureli, F., H. C. Veenema, J. C. van Panthaleon van Eck, and J.A.R.A.M. van Hooff. 1993. Reconciliation, consolation, and redirection in Japanese macaques (*Macaca fuscata*). *Behaviour* 124: 1–21.

Austin, C. R. 1957. Fate of spermatozoa in the uterus of the mouse and rat. *J. Endocrinol.* 14: 335–342.

Axelrod, R., and W. D. Hamilton. 1981. The evolution of cooperation. *Science* 211: 1390–1396.

Babcock, B. A., and Parezo, N.J. 1988. Daughters of the desert: Women anthropologists and the Native American Southwest 1880–1980. Albuquerque: University of New Mexico Press.

Bachmann, C., and H. Kummer. 1980. Male assessment of female choice in hamadryas baboons. *Behav. Ecol. Sociobiol.* 6: 315–321.

Backman, C. W. 1985. Interpersonal congruency theory revisited: A revision and extension. *Journal of Social and Personal Relationships* 2: 489–505.

Bacus, E. A., et. al., eds. 1993. *A gendered past: A critical bibliography of gender in archaeology.* Ann Arbor, MI: University of Michigan Museum of Anthropology Report Series.

Baerends, G. P. 1941. Fortpflanzungsverhalten und Orientierung der Grabwespe *Ammophila campestris* Jur. *Tijdschrift voor Entomologie* 84: 68–275.

———. 1958. Comparative methods and the concept of homology in the study of behaviour. *Archives Néerlandaises de Zoologie* 13: 401–417.

Baerends, G. P., and J. M. Baerends-van Roon. 1950. An introduction to the study of the ethology of cichlid fishes. *Behaviour, Suppl.* 1: 1–243.

Baier, A. 1985. *Postures of the mind: Essays on mind and morals.* Minneapolis: University of Minnesota Press.

Baker, A. J., J. M. Dietz, and D. G. Kleiman. 1993. Behavioural evidence for monopolization of paternity in multi-male groups of lion tamarins. *Anim. Behav.* 46: 1091–1103.

Baker, R. R., and M. A. Bellis. 1989. Number of sperm in human ejaculates varies in accordance with sperm competition theory. *Anim. Behav.* 37: 867–869.

Balda, R. P., I. M. Pepperberg, and A. C. Kamil. 1998. *Animal cognition in nature: The convergence of psychology and biology in laboratory and field.* San Diego: Academic Press.

Ball, J. 1937. A test for measuring the sexual excitability in the female rat. *Comparative Psychology Monographs* 14: 1–37.

Baltzer, F. 1967. *Theodor Boveri.* Translated by Dorthea Rudnick. Berkeley: University of California Press.

Bandura, A. 1973. *Aggression: A social learning analysis.* Englewood Cliffs, NJ: Prentice-Hall.

Barash, D. P. 1986. *The hare and the tortoise: Culture, biology, and human nature.* New York: Viking.

Barfield, R. J., and L. A. Geyer. 1972. Sexual behavior: Ultrasonic postejaculatory song of the male rat. *Science* 176: 1349.

Barnes, B. 1971. Making out in industrial research. *Science Studies* 1: 157–175.

———. 1985. *About science.* Oxford: Blackwell.

Barnes, B., and D. Edge, eds. 1982. *Science in context: Readings in the sociology of science.* Milton Keynes: Open University Press.

Barnes, B., and S. Shapin, eds. 1979. *Natural order: Historical studies in scientific culture.* London: Sage.

Barrat, C. L. R., A. E. Bolton, and I. D. Cooke. 1990. Functional significance of white blood cells in the male and female reproductive tract. *Hum. Reprod.* 5: 639–648.

Bartholomew, G. A. Jr., and J. B. Birdsell. 1953. Ecology and the protohominids. *Am. Anthropol.* 55: 481–498.

Bartlett, T. Q., R. W. Sussman, and J. M. Cheverud. 1993. Infant killing in primates: A review of observed cases with specific reference to the sexual selection hypothesis. *Am. Anthropol.* 95: 958–990.

Barton, R., A. Whiten, S. C. Strum, T. W. Byrne, and A. J. Simpson. 1992. Habitat use and resource availability in baboons. *Anim. Behav.* 43: 831–844.

Bass, K. 1995. *The monkey in the mirror.* Bristol: BBC.

Bateman, A. J. 1948. Intrasexual selection in *Drosophila. Heredity* 2: 349–368.

Bauer, P., and J. Mandler. 1989. One thing follows another: Effects of temporal structure on 1- and 2-year-olds' recall of events. *Developmental Psychology* 25: 197–206.

Beach, F. A. 1937. The neural basis of innate behavior. I. Effects of cortical lesions upon the maternal behavior patterns in the rat. *Journal of Comparative Psychology* 24: 393–434.

———. 1938. Techniques useful in studying the sex behavior of the rat. *Journal of Comparative Psychology* 26: 355–359.

———. 1939. The neural basis of innate behavior. II. Comparison of learning ability and instinctive behavior in the rat. *Journal of Comparative Psychology,* 28: 225–262.

———. 1940. Effects of cortical lesions upon the copulatory behavior of male rats. *Journal of Comparative Psychology* 29: 193–245.

———. 1942. Effects of testosterone proprionate upon the copulatory behavior of sexually inexperienced male rats. *Journal of Comparative Psychology* 33: 227–247.

———. 1943. Effects of injury to the cerebral cortex upon the display of masculine and feminine mating behavior by female rats. *Journal of Comparative Psychology* 35: 169–198.

———. 1944. Effects of injury to the cerebral cortex upon sexually receptive behavior in the female rat. *Psychosomatic Medicine* 6: 40–54.

———. 1946. Mating behavior in male rats castrated at various ages and injected with androgen. *Journal of Experimental Zoology* 101: 91–141.

———. 1947. Review of physiological and psychological studies of sexual behavior in mammals. *Physiological Reviews* 27: 240–307.

———. 1950. The Snark was a Boojum. *The American Psychologist* 5: 115–124.

———. 1961. Sex differences in the physiological basis of mating behavior in animals. In *The Physiology of Emotions.* Illinois: Charles Thomas.

———. 1974. Human sexuality and evolution. In *Reproductive behavior,* ed. W. Montagna and W. A. Sadler. Conference on Reproductive Behavior, Oregon Regional Primate Research Center, 1973. New York: Plenum Press.

———. 1976. Sexual attractivity, proceptivity, and receptivity in female mammals. *Hormones and Behavior* 7: 105–138.

———. 1981. Historical origins of modern research on hormones and behavior. *Hormones and Behavior* 15: 325–376.

Beach, F. A., and L. Jordan. 1956a. Sexual exhaustion and recovery in the male rat. *Quarterly Journal of Experimental Biology* 8: 121–133.

———. 1956b. Effects of sexual reinforcement upon the performance of male rats in a straight runway. *Journal of Comparative and Physiological Psychology* 49: 105–110.

Beach, F. A., and B. J. LeBoeuf. 1967. Coital behavior in dogs. I. Preferential mating in the bitch. *Animal Behavior* 15: 546–558.

Beardsley, T. 1991. Smart Genes. *Scientific American* (August): 87–95.

Bedford, J. M. 1983. Significance of the need for sperm capacitation before fertilization in eutherian mammals. *Biol. Reprod.* 28: 108–120.

Behar, R. 1995. Introduction: Out of exile. In *Women writing culture,* ed. R. Behar and D. A. Gordon. Berkeley: University of California Press, pp. 1–29.

Behar, R., and D. A. Gordon, eds. 1995. *Women writing culture.* Berkeley: University of California Press.

Belenkey, M. F., B. M. Clinchy, N. R. Goldgerger, J. M. Tarule, eds. 1986. *Women's ways of knowing. The development of self, voice and mind.* New York: Basic Books.

Bell, D. 1984. *Daughters of the dreaming.* Melbourne: McPhee Gribble/George Allen, and Unwin.

Bellis, M. A., R. Baker, and M. J. G. Gage. 1990. Variation in rat ejaculates consistent with the kamikaze-sperm hypothesis. *J. Mammal.* 71: 479–480.

Bem, S. L. 1981. Gender schema theory: A cognitive account of sex-typing. *Psychological Review* 88: 369–371.

Bercken, J. H. L. van den, and A. R. Cools. 1980a. Information-statistical analysis of social interaction and communication: An analysis of variance approach. *Anim. Behav.* 28: 127–188.

———. 1980b. Information-statistical analysis of factors determining ongoing behaviour and social interaction in Java-monkeys (*Macaca fascicularis*). *Anim. Behav.* 28: 188–200.

Bercovitch, F. B. 1991. Social stratification, social strategies, and reproductive success in primates. *Ethol. Sociobiol.* 12: 315–333.

Berg, M. 1997. *Rationalizing medical work.* Cambridge, MA: MIT Press.

Bermant, G. 1961. Response latencies of female rats during sexual intercourse. *Science* 133: 1771–1773.

Bermant, G., S. E. Glickman, and J. M. Davidson. 1968. Copulatory activity of male rats: Effects of limbic lesions. *Journal of Comparative and Physiological Psychology* 65: 118–125.

Bernstein, I. S. 1981. Dominance: The baby and the bathwater. *Beh. Brain Sci.* 4: 419–457.

Betzig, L., M. Borgerhoof Mulder, and P. Turke, eds. 1988. *Human reproductive behaviour: A Darwinian perspective.* Cambridge, UK, and New York: Cambridge University Press.

Bhaskar, R. 1978. *A realist theory of science.* Sussex: Harvester Press.

Biagioli, M. 1990. Galileo the emblem maker. *ISIS* 81: 230–258.

Bielby, W. 1991. Sex and careers: Is science a special case? In *The outer circle: Women in the scientific community,* ed. H. Zuckerman, J. R. Cole, and J. T. Bruer. New York: W. W. Norton.

Bierens de Haan, J. A. 1929. *Animal Psychology for Biologists.* London: London University Press.

———. 1931. Werkzeuggebrauch und Werkzeugherstellung bei einem niederen Affen (*Cebus hypoleucus* HUMB.). *Zeitschrift für vergleienden Physiologie* 13.

———. 1935. Zahlbegriff und Handlungsrhythmus bei einem Affen. *Zoologische Jahrbücher. Abt. allgemeine Zoologie und Physiologie* 54.

———. 1936. Notion de nombre et faculté de compter chez les animaux. *Journal de Psychologie* 33.

———. 1937a. Über den Begriff des Instinktes in der Tierpsychologie. *Folia Biotheoretica* 2, 1–16.

———. 1937b. *Labyrinth und Umweg. Ein Kapitel aus der Tierpsychologie.* Leiden: Brill.

———. 1937c. Versuche über das Erfassen räumlicher Beziehungen in waagerechter und senkrechter Richtung bei einigen niederen Affen. *Zeitschrift für Tierspychologie* 1.

———. 1940. *Die tierischen Instinkte und ihre Umbau durch Erfahrung,* Leiden: Brill.

———. 1945. *Instinct en Intelligentie bij Dieren.* Gorinchem: Noorduijn.

———. 1948. Animal psychology and the science of animal behaviour. *Behaviour* 1: 71–80.

Bijker, W., T. Hughes, and T. Pinch. 1987. *The social construction of technological systems: New directions in the sociology and history of technology.* Cambridge, MA: MIT Press.

Bingham, H. C. 1932. Gorillas in a native habitat. *Carn. Inst. Pub.* 426: 1–66.

Biology and Gender Study Group. 1989. The importance of feminist critique for contemporary cell biology. In *Feminism and Science,* ed. Nancy Tuana. Bloomington: Indiana University Press, pp. 172–187.

Birch, C. 1988. The Postmodern challenge to biology. In *The reenchantment of science: Postmodern proposals,* ed. D. R. Griffin. Albany: State University of New York Press, pp. 69–78.

Birkhead, T. R., and F. Fletcher. 1995. Depletion determines sperm number in male zebra finches. *Anim. Behav.* 49: 451–456.

Birkhead, T. R., and A. P. Moller. 1992. *Sperm competition in birds: Evolutionary causes and consequences.* San Francisco: Academic Press.

———. 1993. Female control of paternity. *Trends Ecol. Evol.* 8: 100–104.

Birkhead, T. R., H. D. M. Moore, and J. M. Bedford. 1997. Sex, science, and sensationalism. *Trends in Ecology and Evolution* 12: 121–122.

Birkhead, T. R., A. P. Moller, and W. J. Sutherland. 1993. Why do females make it so difficult for males to fertilize their eggs? *J. Theor. Biol.* 161: 51–60.

Bishop, A. 1962. Control of the hand in lower primates. *Ann. N.Y. Acad. Sci.*102: 316–337.

Bjorkqvist, K., and P. Niemela, eds. 1992. *Of mice and women: Aspects of female aggression.* San Diego: Academic Press.

Bjornson, R., and M. Waldman. 1983. *A feeling for the organism: The life and work of Barbara McClintock.* San Francisco: Freeman.

———. 1989. The gender/science system: Or, is sex to gender as nature is to science? In *Feminism and science,* ed. N. Tuana. Bloomington, and Indianapolis: Indiana University Press. 33–44.

Bloor, D. 1991. *Knowledge and social imagery,* 2d ed. Chicago: University of Chicago Press.

Blum, A. S. 1993. *Picturing nature: American nineteenth-century zoological illustration.* Princeton: Princeton University Press.

Blurton Jones, N. G. 1967. An ethological study of some aspects of social behavior of children in nursery school. In *Primate ethology,* ed. D. Morriss. Chicago: Aldine Pub. Co., pp. 437–463.

Boaz, N. T. 1982. American research on australopithecines and early *Homo,* 1925–1980. In *A history of American physical anthropology, 1930–1980,* ed. F. Spencer. New York: Academic Press, pp. 239–260.

Boekhorst, I. J. A. te, and P. Hogeweg. 1994a. Effects of tree size on travelband formation in orang-utans: Data analysis suggested by a model study. In *Artificial life IV,* ed. R. A. Brooks and P. Maes. Cambridge, MA: MIT Press, pp. 119–129.

———. 1994b. Self-structuring in artificial "Chimps" offers new hypotheses for male grouping in chimpanzees. *Behaviour* 130: 229–253.

Boekhorst, I. J. A. te, C. L. Schürmann, and J. Sugardjito. 1990. Residential status and seasonal movements of wild orang-utans in the Gunung Leuser Reserve (Sumatera, Indonesia). *Anim. Behav.* 39: 1098–1109.

Bontius, J. 1658. *Historiae naturalis et medicae Indiae orientalis. Libri sex, Pars V: Historia Animalum,* ed. W. Piso. Amsterdam: Lodewijk and Daniel Elsevier.

Booth. 1962. Some observations on behavior of Cercopithecus monkeys. *Ann. N.Y. Acad. Sci.* 102: 477–487.

Bordo, S. 1990. Feminism, postmodernism and gender-skepticism. In *Feminism/Postmodernism,* ed. L. Nicholson. New York: Routledge, pp. 133–176.

Boring, E. G. 1950. *A history of experimental psychology,* 2d ed. New York: Appleton-Century-Crofts.

Bourlière, F. 1962. The need for a new conservation policy for wild primates. *Ann. N.Y. Acad. Sci.* 102: 185–189.

Bowlby, J. 1944. Forty-four juvenile thieves: Their characters and home life. *International Journal of Psycho-analysis* 25: 19–52, 107–127.

Box, H. O. 1975a. Quantitative studies of behavior within captive groups of marmoset monkeys (*Callithrix jacchus*). *Primates* 16: 155–174.

———. 1975b. A social developmental study of young monkeys (*Callithrix jacchus*) within a captive family group. *Primates* 16: 419–435.

———. 1977. Social interactions in family groups of captive marmosets (*Callithrix jacchus*). In *The biology and conservation of the Callitrichidae,* ed. D. G. Kleiman. Washington, DC: Smithsonian Institution Press, pp. 239–249.

Boysen, S., and E. Capaldi. 1993. *The emergence of numerical competetnce: Animal and human models.* Hillsdale, NJ: Lawrence Erlbaum.

Bradbury, J. W., and S. L. Vehrencamp. 1977. Social organization and foraging in emballonurid bats. *Beh. Ecol. Sociobiol.* 2: 1–17.

Bradbury, J. W., R. M. Gibson, and I. M. Tsai. 1986. Hot-spots and the evolution of leks. *Anim. Behav.* 34: 1694–1709.

Brain, C. 1992. Deaths in a desert baboon troop. *Int. J. Primatol.* 13: 593–599.

Bramblett, C. A. 1994. *Patterns of primate behavior.* Prospect Heights, IL: Waveland Press.

Brody, J. E. 1996. Gombe chimps archived on video and CD-ROM. *New York Times,* February 20, B5, 10.

Brosius, J. P. 1996. Analysis and interventions: Anthropological engagements with environmentalism. Unpublished conference ms. Presented at 1996 American Anthropological Association Meetings, San Francisco.

Brosius, J. P., A. Tsing, and C. Zerner. 1996. Representing communities: Histories and politics of community-based resource management. Document prepared for conference of the same name, held June 1–3, 1997.

Brown, D. E. 1991. *Human universals.* Philadelphia: Temple University Press.

Brown, J. 1975. Iroquois women: An ethnohistorical note. In *Toward an anthropology of women,* ed. R. R. Reiter. New York: Monthly Review Press, pp. 235–251.

Brownlee, S. 1987. These are real swinging primates. *Discover* April 8: 66–77.

Bshary, R., and R. Noë. 1997a. Anti-predation behaviour of red colobus monkeys in the presence of chimpanzees. *Behav. Ecol. Sociobiol.* 41: 321–333.

———. 1997b. Red colobus and diana monkeys provide mutual protection against predators. *Anim. Behav.* 54: 1461–1474.

Buettner-Janusch, J., ed. 1962. The relatives of man: Modern studies on the relation of the evolution of nonhuman primates to human evolution. *Ann. New York Acad. Sci.* 102: 108–514.

Bulgar, J., and W. J. Hamilton III. 1988. Inbreeding and reproductive success in a natural chacma baboon, *Papio cynocephalus ursinus,* population. *Anim. Behav.* 36: 574–578.

Burkhardt, R. W. 1997. The founders of ethology and the problem of animal subjective experience. In *Animal Consciousness and Animal Ethics,* ed. M. Dol, S. Kasanmoentalib, S. Lijmbach, E. Rivas, and R. van den Bos. Assen: van Gorcum, pp. 1–16.

Burton, F. D. 1994. In the footsteps of Anaximander: Qualitative research in primatology. In *Strength in diversity: A reader in physical anthropology,* ed. A. Herring and L. Chan. Toronto: Scholar's Press, pp. 77–102.

Butler, J. 1990. *Gender trouble: Feminism and the subversion of identity.* London: Routledge.

———. 1993. *Bodies that matter: On the discursive limits of "sex."* New York and London: Routledge.

Buytendijk, F. J. J. 1920. *Psychologie der Dieren.* Haarlem: Bohn (also 1932). French ed.: *Traité de Psychologie des Animaux.* Paris, 1928.

———. 1953. *Traité de psychologie des animaux.* Paris.

Bygott, J. D. 1979. Agonistic behavior, dominance, and social structure in wild chimpanzees of the Gombe National Park. In *The great apes,* ed. D. Hamburg and E. McCown. Menlo Park, CA: Benjamin/Cummings, pp. 405–428.

Byrne, D. 1971. *The attachment paradigm.* New York: Academic Press.

Byrne, D., D. Nelson, and K. Reeves. 1966. Effects of consensual validation and invalidation on attraction as a function of verifiability. *Journal of Experimental Social Psychology* 2: 98–107.

Byrne, R. W. 1990. Tactical deception in primates: The 1990 database. *Primate Report* 1990: 27.

———. 1994. The evolution of intelligence. In *Behavior and evolution,* ed. P. J. B. Slater and T. R. Halliday. Cambridge, UK: Cambridge University Press, pp. 223–265.

———. 1995. *The thinking ape. Evolutionary origins of intelligence.* Oxford: Oxford University Press.

———. 1997. What's the use of anecdotes? Distinguishing psychological mechanisms in primate tactical deception. In *Anthropomorphism, anecdotes, and animals,* ed. R. W. Mitchell, N. S. Thompson, and H. Lyn Miles. New York: SUNY: 134–150.

Byrne, R. W., and J. M. E. Byrne. 1991. Hand preferences in the skilled gathering tasks of mountain gorillas (*Gorilla g. beringei*). *Cortex* 27: 521–546.

———. 1993. Complex leaf-gathering skills of mountain gorillas (*Gorilla g. beringei*): Variability and standardization. *Am. J. Primatol.* 31: 241–261.

Byrne, R. W., and A. Russon. 1998. Learning by imitation: A hierarchical approach. *Behavioral and Brain Sciences* 21: 667–721.

Byrne, R. W., and M. Tomasello. 1995. Do rats ape? *Animal Behavior* 50: 1417–1420.

Byrne, R. W., and A. Whiten. 1990. Tactical deception in primates: The 1990 database. *Primate Report* 27: 1–101.

———, eds. 1988. *Machiavellian intelligence. Social expertise and the evolution of intellect in monkeys, apes, and humans.* Oxford: Clarendon Press.

Callon, M. 1986. Some elements of a sociology of translation: Domestication of the scallops and fishermen of St. Breuc Bay. In *Power, action, and belief: A new sociology of knowledge?* ed. J. Law. London: Routledge and Kegan Paul, pp. 196–233.

Camper, P. 1782. *Natuurkundige Verhandelingen over den Orang Outang en eenige andere Aap-soorten.* Amsterdam: Meijer and Warnars.

———. 1790. *Redevoeringen over de Wijze om de onderscheidene Hartstogten op onze Wezens te Verbeelden.* Utrecht: B. Wild.

Canguilhem, G. 1991. *The normal and the pathological.* New York: Zone Books.

Carpenter, C. R. 1934. A field study of the behavior and social relation of the howling monkeys (*Alouatta palliata*). *Comp. Psychol. Monogr.* 10: 1–168.

———. 1935. Behavior of red spider monkeys in Panama. *J. Mamm.* 16: 171–180.

———. 1940. A field study in Siam of the behavior and social relations of the gibbon (*Hylobates lar*). *Comp. Psychol. Monogr.* 16: 1–212.

———. 1964a. A field study of the behavior and social relations of howling monkeys (1934). In *Naturalistic behavior of nonhuman primates,* ed. C. R. Carpenter. University Park: Pennsylvania State University Press, pp. 3–92.

———. 1964b. *Naturalistic behavior of nonhuman primates.* University Park: Pennsylvania State University Press.

———. 1965. The howlers of Barro Colorado Island. In *Primate behavior,* ed. I. De-Vore. New York: Holt, Rinehart, and Winston, pp. 250–291.

Carson, R. L. 1950. *The sea around us.* Reprint, 1989. New York: Oxford University Press.

Carter, C. S., D. M. Witt, B. Kolb, and I. Q. Whishaw. 1982. Neonatal decortication and adult female sexual behavior. *Physiology and Behavior* 29: 763–766.

Cartmill, M. 1990. Human uniqueness and theoretical content in paleontology. Int. J. Primatol. 11: 173–192.

Cartwright, L. 1992. Women, x-rays, and the public culture of prophylactic imaging. *Camera Obscura* 29: 19–56.

———. 1995. *Screening the body: Tracing medicine's visual culture.* Minneapolis: University of Minnesota Press.

Cartwright, N. 1983. *How the laws of physics lie.* Oxford: Clarendon.

Cavalieri, P., and P. Singer. 1993. *The great ape project.* New York: St. Martin's Press.

Chalmers, N. R. 1986. Group composition, ecology and daily activities of free-living mangabeys in Uganda. *Folia primatol.* 8: 247–262.

Chan, L. K. 1992. Problems with socioecological explanations of primate social diversity. In *Social processes and mental abilities in non-human primates,* ed. F. D. Burton. New York: Edwin Mellon Press, pp. 1–30.

———. 1993. A phylogenetic interpretation of reproductive parameters and mating patterns in Macaca. *Am. J. Primatol.* 30: 303–304.

Chapais, B. 1991. Matrilineal dominance in Japanese macaques: The contribution of an experimental approach. In *The monkeys of Arashiyama,* ed. L. M. Fedigan and P. Asquith. Albany, NY: SUNY Press, pp. 251–273.

Chapman, C. 1989. Ecological constraints on group size in three species of neotropical primates. *Folia Primatol.* 73: 1–9.

Chapman, C., L. M. Fedigan, L. Fedigan, and L. J. Chapman. 1989. Post-weaning resource competition and sex ratios in spider monkeys. *Oikos* 54: 315–319.

Cheney, D, and R. Seyfarth. 1990. *How monkeys see the world.* Chicago: University of Chicago Press.

Chevalier-Skolnikoff, S. 1977. A Piagetian model for describing and comparing socialization in monkey, ape and human infants. In *Primate bio-social development: Biological, social and ecological determinants,* ed. S. Chevalier-Skolinikoff and F. E. Poirier. New York: Garlans, pp. 159–187.

Chivers, D. J. 1974. *The siamang in Malaya. A field study of a primate in a tropical rain forest. Contributions to Primatology #4.* Basel: S. Karger.

Chodorow, N. 1974. Family structure and feminine personality. In *Woman, culture, and society,* ed. M. Z. Rosaldo and L. Lamphere. Stanford: Stanford University Press, pp. 43–66.

———. 1989. *Feminism and psychoanalytic theory.* New Haven: Yale University Press.

Cirne, M. F. A., M. E. Yamamoto, A. I. Alencar, M. F. Sousa, and A. Araújo. 1995. Aspectos da dominância em primatas: Diferenças inter e intra-sexo, estilos de subordinação e sua relação com estados fisiológicos e sistemas de acasalamento. *Anais de Etologia* 13: 178–191.

Claassen, C. 1992. Bibliography of archaeology and gender: Papers delivered at archaeology conferences 1964–1992. *Annotated Bibliographies for Anthropologists* 1 (2).

———, ed. 1994. *Women in archaeology.* Philadelphia: University of Pennsylvania Press.

Clarke, A. 1991. Social worlds/arenas theory of organizational theory. In *Social organization and social process: Essays in honour of Anselm Strauss,* ed. D. R. Maines. New York: de Gryter, pp. 119–148.

Clarke, A., and J. Fujimura, eds. 1993. *The right tools for the job: At work in twentieth-century life sciences.* Princeton: Princeton University Press.

Clarke, A., and T. Montini. 1993. The many faces of RU486. *Science, technology, and human values* 18 (1): 42–78.

Clarke, M. R., and K. E. Glander. 1984. Female reproductive success in a group of free-ranging howling monkeys (*Alouatta palliata*) in Costa Rica. In *Female primates,* Ed. M. F. Small. New York: Alan R. Liss, pp. 111–126.

Clifford, J. 1997. *Routes: Travel and translation in the late twentieth century.* Cambridge, MA: Harvard University Press.

Clifford, J., and G. Marcus, eds. 1986. *Writing culture: The poetics and politics of ethnography.* Berkeley: University of California Press.

Clutton-Brock, T. H. 1974. Primate social organization and ecology. *Nature* 250: 539–542.

Clutton-Brock, T. H., and P. H. Harvey. 1976. Evolutionary rules and primate societies. In *Growing points in ethology,* ed. P. P. G. Bateson and R. A. Hinde. Cambridge, UK: Cambridge University Press, pp. 195–237.

———. 1977. Primate ecology and social organization. *J. Zool.* 183: 1–39.

———, eds. 1988. *Reproductive success: Studies of individual variation in contrasting breeding systems.* Chicago: University of Chicago Press.

Clynes, M. E., and N. S. Kline. 1960. Cyborgs and space. *Astronautics* (Sept. 26–27): 75–76.

Code, L. 1991. *What can she know? Feminist theory and the construction of knowledge.* Ithaca: Cornell University Press.

Cohen, J. 1984. Immunological aspects of sperm selection and transport. In *Immunological aspects of reproduction in mammals,* ed. D. B. Crighton. London: Butterworths, pp. 77–89.

Cold Spring Harbor Symposia on Quantitative Biology. 1951. Volume 15. New York: The Biological Laboratory, Long Island Biological Association.

Collias, N., and C. W. Southwick. 1952. A field study of population density and social organization in howling monkeys. *Proceedings of the American Philosophical Society* 96: 143–156.

Collier, J. F. 1974. Women in politics. In *Woman, culture, and society,* ed. M. Z. Rosaldo and L. Lamphere. Stanford: Stanford University Press, pp. 89–96.

———. 1988. *Marriage and inequality in classless society.* Stanford: Stanford University Press.

Collier, J. F., and M. Z. Rosaldo. 1981. Politics and gender in simple societies. In *Sexual meanings: The cultural construction of gender and sexuality,* ed. S. B. Ortner and H. Whitehead. Cambridge, UK: Cambridge University Press, pp. 275–329.

Collier, J., and S. Yanagisako, eds. 1987. *Gender and kinship: Essays toward a unified analysis.* Stanford: Stanford University Press.

Collins, H. 1985. *Changing order: Replication and induction in scientific practice.* London: Sage.

Collins, P. H. 1991a. *Black feminist thought: Knowledge, consciousness, and the politics of empowerment.* New York: Routledge, Chapman, Hall.

———. 1991b. Learning from the outsider within: The sociological significance of black feminist thought. In *Beyond methodology: Feminist scholarship as lived research,* ed. M. M. Fonow and J. A. Cook. Bloomington, IN: Indiana University Press, pp. 35–59.

Conkey, M. W., and J. D. Spector. 1984. Archaeology and the study of gender. In *Advances in archaeological method and theory,* vol. 7, ed. M. B. Schiffer. New York: Academic Press, pp. 1–38.

Conte, F. A., M. M. Grumbach, Y. Ito, C. R. Fisher, and E. R. Simpson. 1994. A syndrome of female pseudohermaphrodism, hypergonadotropic hypogonadism, and multicystic ovaries associated with missense mutations in the gene encoding aromatase (P450arom). *Journal of Clinical Endocrinology and Metabolism* 78: 1287–1292.

Coontz, S., and P. Henderson. 1986. Property forms, political power, and female labour in the origins of class and state societies. In *Women's work, men's property:*

The origins of gender and class, ed. S. Coontz and P. Henderson. London: Verso, pp. 108–155.

Cooter, R., and S. Pumfrey. 1994. Separate spheres and public spaces: Reflections on the history of science popularization and science in popular culture. *History of Science* 32: 237–267.

Cords, M. 1988. Resolution of aggressive conflicts by immature long-tailed macaques (*Macaca fascicularis*). *Anim. Behav.* 36: 1124–1135.

Cords, M., and F. Aureli. 1993. Patterns of reconciliation among juvenile long-tailed macaques. In *Juvenile primates, life history, development and behavior,* ed. M. E. Pereira, and L. A. Fairbanks. New York, Oxford University Press, pp. 271–284.

Coss, R. G., and R. O. Goldthwaite. 1995. The persistence of old designs for perception. In *Perspectives in ethology,* vol. 11, ed. N. S. Thompson. New York, Plenum Press, pp. 83–148.

Couffer, J. 1963. *Songs of wild laughter.* London: Constable and Co.

Cowlishaw, G., and R. I. M. Dunbar. 1991. Dominance rank and mating success in primates. *Anim. Behav.* 41: 1045–1056.

Cowlishaw, G., and S. M. O'Connell. 1996. Male-male competition, paternity certainty and copulation calls in female baboons. *Anim. Behav.* 51: 235–238.

Cox, C. R., and B. J. LeBoeuf. 1977. Female incitation of male competition: A mechanism of sexual selection. *Am. Nat.* 111: 317–335.

Craig, W. 1918. Apetites and aversions as constituents of instincts. *Biological Bulletin* 34: 91–107.

Crary, Jonathan. 1994. *Techniques of the observer: On vision and modernity in the nineteenth century.* Cambridge, MA: MIT Press.

Crisler, L. 1958. *Arctic wild.* New York: Harper and Bros.

Crockett, C. M. 1985. Population studies of red howler monkeys (*Alouatta seniculus*). *National Geographic Research* 1: 264–273.

Crockett, C. M., and J. F. Eisenberg. 1987. Howlers: Variations in group size and demography. In *Primate societies,* ed. B. Smuts, D. L. Cheney, R. M. Seyfarth, R. W. Wrangham, and T. T. Struhsaker. Chicago: University of Chicago Press, pp. 54–68.

Crockett, C. M., and R. Sekulic. 1984. Infanticide in red howler monkeys (*Alouatta seniculus*). In *Infanticide: Comparative and evolutionary perspectives,* ed. G. Hausfater and S. B. Hrdy. Hawthorne, NY: Aldine, pp. 173–191.

Crook, J. H. 1966. Gelada baboon herd structure and movement. *Symp Zool Soc. London* 18: 237–258.

———. 1970a. Social organization and the environment, aspects of contemporary social ethology. *Anim. Behav.* 18: 197–209.

——— 1970b. The socio-ecology of primates. In *Social behaviour in birds and mammals,* ed. J. H. Crook. London: Academic Press, pp. 103–166.

——— 1989. Introduction: Socioecological paradigms, evolution, and history: Perspectives for the 1990s. In *Comparative socioecology,* ed. V. Standen and R. A. Foley. Boston: Blackwell Scientific Publications, pp. 1–36.

Crook, J. H., and S. Gartlan. 1966. Evolution of primate societies. *Nature* 210: 1200–1203.

Cussins, A. 1992. Content, embodiment, and objectivity: The theory of cognitive trails. *MIND* 101: 651–688.

Cussins, Thompson C. In press. Elephants, biodiversity and complexity: Amboseli National Park, Kenya. In *Complexities in science, technology, and medicine,* ed. A-M. Mol and J. Law. Duke University Press.

Custance, D., A. Whiten, and K. A. Bard. 1995. Can young chimpanzees imitate arbitrary actions? Hayes and Hayes (1952) revisited. *Behavior* 132: 837–859.

Daly, M., and M. Wilson. 1983. *Sex, evolution and behavior.* Boston: Willard Grant.

———. 1988. *Homicide.* New York: Aldine de Gruyter.

Dart, R. 1925. *Australopithecus africanus:* The man-ape of South Africa. *Nature* 115: 195–199.

Dart, R., and A. Keith. 1925. On the Taungs skull: An exchange of letters. *Nature* 116: 462.

Darwin, C. 1859. *On the origin of species by means of natural selection.* London: Murray.

———. 1871. *The descent of man and selection in relation to sex.* London: Murray.

———. 1874. *The descent of man and selection in relation to sex.* Chicago: The Henneberry Company.

———. 1890. *Het Uitdrukken der Gemoedsaandoeningen bij den Mensch en de Dieren.* Arnhem-Nijmegen: Cohen (Dutch ed. of *The Expression of the Emotions in Man and Animals.* London: Murray, 1872).

Das, M., Zs. Penke, and J.A.R.A.M. van Hooff. 1997. Affiliation between aggressors and third parties following conflicts in long-tailed macaques (*Macaca fascicularis*). *Int. J. Primatol.* 18: 157–179.

———. 1998. Postconflict affiliation and stress-related behavior of long-tailed macaque aggressors (*Macaca fascicularis*). *International Journal of Primatolology* 19: 53–72.

Dasser, V. 1988. Mapping social concepts in monkeys. In *Machiavellian intelligence,* ed. R. Byrne and A. Whiten. Oxford: Clarendon Press, pp. 85–93.

Datta, S. B. 1986. The role of alliances in the acquisition of rank. In *Primate ontogeny, cognition, and behavior: Developments in field and laboratory research,* ed. J. Else and P. Lee. New York: Academic Press, pp. 219–225.

Davies, E. M., and P. D. Boersma. 1984. Why lionesses copulate with more than one male. *Am. Nat.* 123: 594–611.

Davies, N. B. 1978. Territorial defence in the speckled wood butterfly (*Pararge aegeria*): The resident always wins. *Anim Behav.* 26: 138–147.

Dawkins, R. 1976a. *The selfish gene.* Oxford: Oxford University Press.

———. 1976b. Hierarchical organization: A candidate principle for ethology. In *Growing points in ethology,* ed. P. P. G. Bateson and R. A. Hinde. Cambridge, UK: Cambridge University Press, pp. 7–54.

Dawkins, R., and J. R. Krebs. 1981. Animal signals: Information or manipulation? In *Behavioural Ecology: An evolutionary approach,* ed. J. R. Krebs and N. B. Davies. Sunderland, MA: Sinauer Associates, pp. 282–315.

Dawson, B. V., and B. M. Foss. 1965. Observational learning in budgerigars. *Anim. Behav.* 13: 470–474.

Dawson, C., and A. S. Woodward. 1913. On the discovery of a Paleolithic human skull and mandible in a flint-bearing gravel overlying the Wealden (Hastings Beds) at Piltdown, Fletching, Sussex. *Quart. J. Geo. Soc.* 69: 117–144.

De Beauvoir, S. 1952. *The second sex.* New York: Knopf.

de Waal, F. B. M. 1977a. Agonistisch gedrag binnen een groep Java-apen. In *Agressief Gedrag, Oorzaken en Functies,* ed. P. R. Wiepkema, and J.A.R.A.M. van Hooff. Utrecht, Bohn, Scheltema, and Holkema, pp. 165–182.

———. 1977b. The organization of agonistic relations within two captive groups of Java-monkeys (*Macaca fascicularis*). *Zeitschrift für Tierpsychologie,* 44: 225–282.

———. 1978. Exploitative and familiarity-dependent support strategies in a colony of semi-free living chimpanzees. *Behaviour* 66: 268–312.

———. 1982. *Chimpanzee politics: Power and sex among apes.* London: Unwin Paperbacks.

———. 1984. Sex differences in the formation of coalitions among chimpanzees. *Ethology and Sociobiology* 5: 239–255.

———. 1986a. The brutal elimination of a rival among captive male chimpanzees. *Ethol. and Sociobiol.* 7: 251–273.

———. 1986b. The integration of dominance and social bonding in primates. *Quarterly Review of Biology* 61: 459–479.

———. 1987a. Coalitions in monkeys and apes. In *Coalition Formation,* ed. H. A. M. Wilke. Elsevier-North Holland, Amsterdam, pp. 1–28.

———. 1987b. Dynamics of social relationships. In *Primate societies,* ed. B. B. Smuts, D. L. Cheney, R. M. Seyfarth, R. W. Wrangham, and T. T. Struhsaker. Chicago: University of Chicago Press, pp. 421–429.

———. 1987c. Tension regulation and nonreproductive functions of sex among captive bonobos (*Pan paniscus*). *National Geographic Research* 3: 318–335.

———. 1988. The communicative repertoire of captive bonobos (*Pan paniscus*), compared to that of chimpanzees. *Behaviour* 106: 183–251.

———. 1989a. *Peacemaking among primates.* Cambridge, MA: Harvard University Press.

———. 1989b. Food sharing and reciprocal obligations among chimpanzees. *Journal of Human Evolution* 18: 433–549.

———. 1989c. The chimpanzee's sense of social regularity and its relation to the human sense of justice. *American Behav. Scientist* 34: 335–349.

———. 1991. Complementary methods and convergent evidence in the study of primate social cognition. *Behaviour* 118: 297–320.

———. 1992. Coalitions as part of reciprocal relations in the Arnhem chimpanzee colony. In *Coalitions and alliances in humans and other animals,* ed. A. H. Harcourt and F. B. M. de Waal. New York: Oxford University Press, pp. 233–257.

———. 1993. Reconciliation among primates: A review of empirical evidence and theoretical issues. In *Primate social conflict,* ed. W. A. Mason and S. P. Mendoza. Albany, NY: SUNY Press, pp. 111–144.

———. 1996a. *Good natured: The origins of right and wrong in humans and other animals.* Cambridge, MA: Harvard University Press.

———. 1996b. Conflict as negotiation. In *Great ape societies,* ed. W. C. McGrew, T. Nishida, and L. F. Marchant. New York: Cambridge University Press, pp. 159–172.

de Waal, F. B. M., and F. Aureli. 1997. Conflict resolution and distress alleviation in monkeys and apes. In "The integrative neurobiology of affiliation," ed. C. S. Carter. *Annals of the New York Academy of Sciences* 807: 317–328.

de Waal, F. B. M., and J. Hoekstra. 1980. Contexts and predictability of aggression in chimpanzees. *Anim. Behav.* 28: 929–937.

de Waal, F. B. M., and J.A.R.A.M. van Hooff. 1981. Side-directed communication and agonistic interactions in chimpanzees. *Behaviour* 77: 164–198.

de Waal, F. B. M., and L. Luttrell. 1988. Mechanisms of social reciprocity in three primate species: Symmetrical relationship characteristics or cognition? *Ethology and Sociobiology* 9: 101–118.

de Waal, F. B. M., and R. M. Ren. 1988. Comparison of the reconciliation behavior of stumptail and rhesus macaques. *Ethology* 78: 129–142.

de Waal, F. B. M., and A. van Roosmalen. 1979. Reconciliation and consolation among chimpanzee. *Behavioral Ecology and Sociobiology* 5: 55–66.

de Waal, F. B. M., and D. Yoshihara. 1983. Reconciliation and redirected affection in rhesus monkeys. *Behaviour* 85: 224–241.

Dear, P. 1985. Totius in Verba: Rhetoric and authority in the early royal society. *ISIS* 76: 145–161.

Degler, C. N. 1991. *In search of human nature.* New York: Oxford University Press.

Dennis, M. A. 1994. "Our first line of defense": Two university laboratories in the post-war American state. *ISIS* 85: 427–455.

DeRousseau, C. J., ed. 1990. *Primate life history and evolution. Monographs in Primatology,* vol. 14. New York: Wiley-Liss.

deRuiter, J. R., E. J. Wickings, W. Scheffrahn, N. Menard, M. Bruford, and M. Inoue. 1993. Symposium: Paternity, male social rank, and sexual behavior. *Primates* 34: 469–555.

Descartes, R. 1637. *Discourse on method.* Reprint, 1993. Indianapolis: Hackett Publishing.

Despret, V. 1996. *Naissance d'une théorie éthologique.* Paris: Les Empêcheurs de penser en rond.

DeVore, I., ed. 1965. *Primate behavior: Field studies of monkeys and apes.* New York: Holt, Rinehart and Winston.

DeVore, I., and K. R. L. Hall. 1965. Baboon ecology. In *Primate behaviour. Field studies of monkeys and apes,* ed. I. DeVore. New York: Holt, Rinehart, and Winston, pp. 20–52.

DeVore, I., and S. L. Washburn. 1960. *Baboon behavior.* Berkeley: University of California Press.

———. 1963. Baboon ecology and human evolution. In *African ecology and human evolution,* ed. F. C. Howell and F. Bourliere. Chicago: Aldine, pp. 335–367.

Dewsbury, D. A. 1978. The comparative method in studies of reproductive behavior. In *Sex and behavior: Status and prospectus,* ed. T. E. McGill, D. A. Dewsbury, and B. D. Sachs. New York: Plenum Press, pp. 83–112.

———. 1981. On the function of the multiple-intromission, multiple-ejaculation copulatory patterns of rodents. *Bull. Psychon. Soc.* 18: 221–223.

———. 1982. Ejaculate cost and male choice. *Am. Nat.* 119: 601–610.

———. 1984. Sperm competition in muroid rodents. In *Sperm competition and the evolution of animal mating systems,* ed. R. L. Smith. Orlando, FL: Academic Press, pp. 546–571.

Di Fiore, A., and D. Rendall. 1994. Evolution of social organization: A reappraisal for primates by using phylogenetic methods. *Proceedings of he National Academy of Science* 91: 9941–9945.

di Leonardo, M. 1991. Introduction: Gender, culture, and political economy: Feminist anthropology in historical perspective. In *Gender at the crossroads of knowledge: Feminist anthropology in the postmodern era,* ed. M. di Leonardo. Berkeley: University of California Press, pp. 1–48.

Dickinson, A. 1980. *Contemporary animal learning theory.* Cambridge, UK: Cambridge University Press.

Dickson, D. 1997. Champions or challengers of the cause of science. *Nature* 387: 333–334.

Dietz, J. M., and A. J. Baker. 1993. Polygyny and female reproductive success in golden lion tamarins, *Leontopithecus rosalia. Anim. Behav.* 46: 1067–1078.

Digby, L. J. (1995). Infant care, infanticide, and female reproductive strategies in polygynous groups of common marmosets (*Callithrix jacchus*). *Behav. Ecol. Sociobiol.* 37: 51–61.

Digby, L. J., and S. F. Ferrari. 1994. Multiple breeding females in free-ranging groups of *Callithrix jacchus. Int. J. Primatol.* 15: 389–397.

Dijkgraaf, S. 1943. Over een merkwaardige functie van den gehoorzin bij vleermuizen. *Verslagen Nederlandsche Akademie van Wetenschappen, Afd. Natuurkunde* 52, 622–627.

———. 1946. Die Sinneswelt der Fledermäuse. *Experientia* 2: 438–448.

———. 1949. Spallanzani und die Fledermäuse. *Experientia* 5: 90–92.

———. 1952. Über die Schallwahrnehmung bei Meeresfischen. *Zeitschrift für vergleichende Physiologie* 34: 104–122.

———. 1957. Sinnesphysiologische Beobachtungen an Fledermäusen. *Acta Physiologica Pharmacologica Neerlandica* 6: 675–684.

———. 1960. Spallanzani's unpublished experiments on the sensory basis of object perception in bats. *Isis* 51: 9–20.

———. 1963. The functioning and significance of the lateral-line organs. *Biological Review* 38: 51–105.

Dijkgraaf, S., and A. J. Kalmijn. 1963. Untersuchungen über die Funktion der Lorenzinischen Ampullen an Haifischen. *Zeitschrift für vergleichende Physiologie* 47: 316–338.

Disney, W. 1953. What I've learned from the animals. *American Magazine* 155: 23, 106–109.

Dixson, A. F. 1977. Observations on the evolution of the genitalia and copulatory behaviour in male primates. *J. Zool. Lond.* 213: 423–443.

Dobzhansky, T. 1944. On species and races of living and fossil man. *Am. J. Phys. Anthropol.* 2: 251–265.

Dolhinow, P., ed. 1972. *Primate patterns*. New York: Holt, Rinehart, and Winston.

Dollard, J. C., and N. E. Miller. 1950. *Personality and Psychotherapy.* New York: McGraw-Hill.

Doty, R. L. 1974. A cry for the liberation of the female rodent: Courtship and copulation in rodentia. *Psychological Review* 81: 159–172.

Douglas, M. 1966. *Purity and danger.* New York: Praeger.

———. 1970. *Natural symbols*. Harmondsworth: Penguin.

Douglas, W. O. 1960. *My Wilderness: The Pacific West.* Garden City, New York: Doubleday, and Co.

Draper, P. 1975. !Kung women: Contrasts in sexual egalitarianism in foraging and sedentary contexts. In *Toward an anthropology of women,* ed. R. R. Reiter. New York: Monthly Review Press, pp. 77–109.

Drea, C. M., J. Hawk, and S. E. Glickman. 1996. Aggression decreases as play increases in infant spotted hyenas (*Crocuta crocuta*): Preparation for joining the clan. *Anim. Behav.* 51: 1323–1336.

Drea, C., A. Neves, V. Lopez, and S. E. Glickman. 1996. Cooperation in captive spotted hyenas. Paper presented at the 1996 meetings of the Animal Behavior Society.

Drea, C. M., M. L. Weldele, N. G. Forger, E. M. Coscia, L. G. Frank, P. Licht, and S. E. Glickman. 1998a. Androgens and masculinization of the genitalia in the spotted hyaena (*Crocuta crocuta*). 2. Effects of prenatal anti-androgens. *Journal of Reproduction and Fertility* 113: 118–128.

Drea, C. M., E. M. Coscia, and S. E. Glickman. 1998b. Hyenas. In *Encyclopedia of reproduction,* ed. E. Knobil, J. Neill, and P. Licht. San Diego: Academic Press.

Dubois, E. 1898. The brain-cast of *Pithecanthropus erectus. Proc. Intl. Cong. Zool.*

Dunbar, R. I. M. 1984a. *Reproductive decisions: An economic analysis of gelada baboon social strategies.* Princeton: Princeton University Press.

———. 1984b. Use of infants by male gelada in agonistic contexts: Agonistic buffering, progeny protection or soliciting support? *Primates* 25: 28–35.

———. 1988. *Primate social systems.* Ithaca: Cornell University Press.

———. 1989. Social systems as optimal strategy sets: The costs and benefits of sociality. In *Comparative socioecology,* ed. V. Standen and R. A. Foley. Oxford: Blackwell, pp. 131–149.

———. 1990. The apes as we want to see them. *New York Times Book Review* (1990), Jan. 10.

Dupré, J. 1993. *The disorder of things: Metaphysical foundation of the disunity of science.* Cambridge, MA: Harvard University Press.

Eaton, R. L. 1978. Why some felids copulate so much: A model for the evolution of copulation frequency. *Carnivore* 1: 42–51.

Eberhard, W. G. 1996. *Females control: Sexual selection by cryptic female choice.* Princeton, NJ: Princeton University Press.

Edwards, J., S. Franklin, E. Hirsch, F. Price, and M. Strathern. 1993. *Technologies of procreation: Kinship in the age of assisted conception.* Manchester and New York: Manchester University Press.

Eisenberg, J. F., N. A. Muckenhirn, and R. Rudran. 1972. The relation between ecology and social structure in primates. *Science* 176: 863–874.

Ellefson, J. O. 1974. A natural history of white-handed gibbons in the Malayan Peninsula. In *Gibbons and siamang,* vol. 3, ed. D. M. Rumbaugh. Basel: S. Karger, pp. 2–136.

Elton, C. 1927. *Animal ecology.* London: Sidgwick and Jackson.

Emlen, S. T. 1995. An evolutionary theory of the family. *Proc. Nat. Acad. Sci.* 92: 8092–8099.

Engelstad, E. 1991. Images of power and contradiction: Feminist theory and post-processual archaeology. *Antiquity* 65: 502–514.

Enomoto, T. 1978. On social preference in sexual behavior of Japanese monkeys (*Macaca fuscata*). *J. Human Evol.* 7: 283–293.

Epple, G. 1973. The role of pheromone in the social communication of marmoset monkeys (Callithricidae). *J. Reprod. Fert.* Suppl. 19: 447–454.

———. 1975a. Parental behavior in *Saguinus fuscicollis ssp.* (Callithricidae). *Folia Primatol.* 24: 221–238.

———. 1975b. The behavior of marmoset monkeys (Callitrichidae). In *Primate Behavior,* vol. 4, ed. L. A. Rosenblum. New York: Academic Press, pp. 195–239.

Epple, G., and Y. Katz. 1984. Social influences on estrogen excretion and ovarian cyclicity in saddle-back tamarins (*Saguinus fuscicollis*). *Amer. J. Primatol.* 6: 215–227.

Epstein, S. 1995. The construction of lay expertise: AIDS activism and the forging of credibility in the reform of clinical trials. *Science, Technology, and Human Values* 20: 408–437.

———. 1996. *Impure science: AIDS, activism, and the politics of knowledge.* Berkeley: University of California Press.

Etienne, M., and E. Leacock, eds. 1980. *Women and colonization: Anthropological perspectives.* New York: Praeger Publishers.

Evans, S. 1983. The pair-bond of the common marmoset, *Callithrix jacchus jacchus:* An experimental investigation. *Anim. Behav.* 31: 651–658.

Evans-Pritchard, E. E. 1965. The position of women in primitive societies and in our own. In *The position of women in primitive societies and other essays in social anthropology,* ed. E. E. Evans-Pritchard. New York: Free Press (Macmillan), pp. 37–58.

Ezrahi, Y. 1990. *The descent of Icarus: Science and the transformation of contemporary democracy.* Cambridge, MA: Harvard University Press.

Faithorn, E. 1975. The concept of pollution among the Kafe of the Papua New Guinea Highlands. In *Toward an anthropology of women,* ed. R. R. Reiter. New York: Monthly Review Press, pp. 127–140.

Fausto-Sterling, A. 1985. *Myths of gender: Biological theories about women and men* New York, NY: Basic Books.

———. 1995. Gender, race, and nation: The comparative anatomy of "Hottentot" women in Europe, 1815–1817. In *Deviant bodies: Critical perspectives ion difference in science and popular culture,* ed. J. Terry and J. Urla. Bloomington: Indiana University Press, pp. 19–48.

Favell, J. 1973. *The Power of Positive Reinforcement: A Handbook of Behavior Modification.* Springfield, Ill.: Charles Thomas.

Fedigan, L. M. 1982. *Primate paradigms: Sex roles and social bonds.* Montreal: Eden Press.

———. 1983. Dominance and reproductive success in primates. *Yearbook of Physical Anthropology* 26: 91–129.

———. 1986a. Demographic trends in the *Alouatta palliata* and *Cebus capucinus* populations in Santa Rosa Park, Costa Rica. In *Primate ecology and conservation,* ed. J. Else and P. Lee. Cambridge, UK: Cambridge University Press, pp. 285–293.

———. 1986b. The changing role of women in models of human evolution. *Ann. Rev. Anthropol.* 15: 25–66.

———. 1992. *Primate paradigms: Sex roles and social bonds.* Chicago: University of Chicago Press.

———. 1994. Science and the successful female: Why there are so many women primatologists. *American anthropologist* 96: 10–20.

———. 1997. Is primatology a feminist science? In *Women in human evolution,* ed. L. Hager. New York: Routledge, pp. 56–75.

Fedigan, L. M., and P. J. Asquith, eds. 1991. *The monkeys of Arashiyama: 35 years of research in Japan and the West.* Albany, NY: SUNY Press.

Fedigan, L. M., and J. Baxter. 1984. Sex differences and social organziation in free-ranging spider monkeys (*Ateles geoffroyi*) at Tikal, Guatemala. *Primates* 25: 279–294.

Fedigan, L., and S. Strum. 1997. Changing images of primate societies: The role of theory, method, and gender. *Current Anthropology* 38: 677–681.

Fedigan, L. M., L. Fedigan, C. Chapman, and K. Glander. 1988. Spider monkey home ranges: A comparison of radio telemetry and direct observation. *Am. J. Primatol.* 16: 19–29.

Fedigan, L. M., L. Fedigan, H. Gouzoules, S. Gouzoules, and N. Koyama. 1986. Lifetime reproductive success in female Japanese macaques. *Folia Primatol.* 47: 143–157.

Fee, E. 1986. Critiques of modern science: The relationship of feminism to other radical epistemologies, In *Feminist approaches to science,* ed. R. Bleier. New York: Pergamon Press, pp. 42–56.

Feenberg, A., and A. Hannay, eds. 1995. *Technology and the politics of knowledge.* Bloomington: Indiana University Press.

Ferrari, S. F., and V. H. Diego. 1989. A re-evaluation of the social organization of the Callitrichidae, with reference to the ecological differences between genera. *Folia Primatol.* 52: 132–147.

———. 1992. Long-term changes in a wild marmoset group. *Folia Primatol.* 58: 215–218.

Finch, C. 1973. *The art of Walt Disney.* New York: Harry N. Abrams.

Findlen, P. 1991. The economy of scientific exchange in early modern Italy. In

Patronage and institutions: Science, technology, and medicine at the European court, 1500–1750, ed. B. T. Moran. Rochester, NY: Boydell Press, pp. 5–24.

Fischer, D. H. 1970. *Historian's fallacies: Towards a logic of historical thought.* New York: Harper and Row.

Fisher, R. A. 1930. *The genetical theory of natural selection.* Oxford: Clarendon Press.

Fiske, J. 1989. *Reading the popular.* New York: Routledge.

Fleck, L. 1935. *Genesis and development of a scientific fact.* Chicago: University of Chicago Press.

Foley, R. 1986. Anthropology and behavioural ecology. *Anthropology Today* 2: 13–15.

Fonseca, G. A. B. 1985. The vanishing Brazilian Atlantic forest. *Biological Conservation* 34: 17–34.

Ford, C. S., and F. A. Beach. 1951. *Patterns of sexual behavior.* New York: Harper.

Forster, D., and S. C. Strum. 1994. Sleeping near the enemy: Patterns of sexual competition in baboons. *Proceedings of the International Congress of Primatology. Current Primatology,* vol. 2, pp. 19–24.

Fossey, D. 1972. Vocalizations of the mountain gorilla. *Anim. Behav.* 20: 36–53.

———. 1979. Development of the mountain gorilla (*Gorilla gorilla beringei*): The first 36 months. In *The great apes,* ed. D. A. Hamburg and E. R. McCown. Menlo Park, CA: Benjamin Cummings Publishing Co., pp. 139–192.

———. 1983. *Gorillas in the mist.* Boston: Houghton and Mifflin.

Foucault, M. 1979. *Discipline and punish,* trans. A. Sheridan. New York: Vintage Books.

———. 1994. *The birth of the clinic: An archaeology of medical perception.* New York: Vintage.

Fouts, R. S. 1973. Acquisition and testing of gestural signs in four young chimpanzees. *Science* 180: 978–980.

Fox, R. 1977. *Kinship and marriage,* trans. from the English (1967) into the Japanese by K. Kawanaka. Tokyo: Shisakusha. (J)

Frank, L. G. 1986a. Social organization of the spotted hyaena (*Crocuta crocuta*). I. Demography. *Anim. Behav.* 34: 1500–1509.

———. 1986b. Social organization of the spotted hyaena (*Crocuta crocuta*) II. Dominance and reproduction. *Anim. Behav.* 34: 1510–1527.

Frank, L. G., S. E. Glickman, and C. J. Zabel. 1989. Ontogeny of female dominance in the spotted hyaena. *Symposia of the Zoological Society of London* 61: 127–146.

Frank, L. G., S. E. Glickman, and I. Powch. 1990. Sexual dimorphism in the spotted hyaena. *Journal of Zoology London* 221: 308–313.

Frank, L. G., S. E. Glickman, and P. Licht. 1991. Fatal sibling aggression, precocial development, and androgens in neonatal spotted hyenas. *Science* 252: 702–704.

Franklin, S. 1995. Science as culture, cultures of science. *Annual Review of Anthropology* 24: 163–184.

———. 1997. *Embodied progress: A cultural account of Assisted conception.* London: Routledge.

Franklin, S., and H. Ragoné, eds. 1998. *Reproducing reproduction.* Philadelphia: University of Pennsylvania Press.

French, J. A., and J. A. Stribley. 1987. Synchronization of ovarian cycles within and between social groups in golden lion tamarins (*Leontopithecus rosalia*). *Amer. J. Primatol.* 2: 473–478.

Freund, M. 1963. Effect of frequency of emission on semen output and an estimate of daily sperm production in man. *J. Reprod. Fert.* 6: 269–286.

Fuchs, H. 1995. Psicologia animal no Brasil: O fundador e a fundação. *Psicologia USP* 6: 15–42.

Furuichi, T. 1988. *In the forest of Bilya.* Tokyo: Tokyo-kagaku-dojin. (J)

Gaddum-Rosse, P. 1981. Some observations on sperm transport through the utero-tubal junction of the rat. *Am. J. Anat.* 160: 333–341.

Gage, M. J. G. 1991. Risk of sperm competition directly affects ejaculate size in the Mediterranean fruit fly. *Anim. Behav.* 42: 1036–1037.

Gailey, C. 1987. *Kinship to kingship: Gender hierarchy and state formation in the Tongan Islands.* Austin: University of Texas Press.

Galdikas, B. M. F. 1979. Orangutan adaptation at Tanjung Puting Reserve: Mating and ecology. In *The great apes,* ed. D. A. Hamburg and E. R. McCown. Menlo Park, CA: Benjamin Cummings Publishing Co., pp. 195–233.

Galef, B. G. 1988. Imitation in animals: History, definitions, and interpretation of data from the psychological laboratory. In *Social learning: Psychological and biological perspectives,* ed. T. Zentall and B. G. Galef Jr. Hillsdale, NJ: Lawrence Erlbaum, pp. 3–28.

Galef, B. G., L. A. Manzig, and R. M. Field. 1986. Imitation learning in budgerigars: Dawson and Foss (1965) revisited. *Behavioral Processes* 13: 191–202.

Galison, P. 1997. *Image and logic: The material culture of modern physics.* Chicago: University of Chicago Press.

Galison, P., and D. Stump, eds. 1995. *The disunity of science.* Stanford: Stanford University Press.

Gallup, G. G. 1970. Chimpanzees: Self-recognition. *Science* 167: 341–343.

Gallup, G. G., D. J. Povinelli, S. D. Suarez, J. R. Anderson, L. Lethmate, and E. W. Menzel. 1995. Further reflections on self-recognition in primates. *Anim. Behav.* 50 (6): 1525–1532.

Gallup, G. G., M. K. McClure, S. D. Hill, and R. A. Bundy. 1970. Capacity for self-recognition in differentially reared chimpanzees. *Psychological Record* 21: 69–74.

Garber, P. A. 1987. Foraging strategies among living primates. *Annual Review of Anthropology* 16: 339–364.

———. 1994. Phylogenetic approach to the study of tamarin and marmoset social systems. *Am. J. Primatol.* 34: 199–219.

Garber, P. A., L. Moya, and C. Malaga. 1984. A preliminary field study of the moustached tamarin monkey (*Saguinus mystax*) in northeastern Peru: Questions concerned with the evolution of a communal breeding system. *Folia Primatol.* 42: 17–32.

Garcia, J., and R. A. Koelling. 1966. Relation of cue to consequence in avoidance learning. *Psychonomic Science* 4: 123–124.

Garcia, J., F. R. Ervin, and R. A. Koelling. 1966. Learning with prolonged delay of reinforcement. *Psychonomic Science* 5: 121–122.

Gardner, H. 1985. *The mind's new science.* New York: Basic Books.

Gardner, R. A., and B. T. Gardner. 1969. Teaching sign language to a chimpanzee. *Science* 165: 664–672.

Garfinkel, H. 1967. *Studies in ethnomethodology.* Englewood Cliffs, NJ: Prentice-Hall.

Garfinkel, H., M. Lynch, and E. Livingstone. 1981. The work of discovering science constructed with material from the optically discovered pulsar. *Philosophy of the Social Sciences* 11: 131–158.

Garner, R. 1896. *Gorillas and chimpanzees.* London: Osgood, McIlvaine, and Co.

Gartlan, J. S. 1968. Structure and function in primate society. *Folia Primatol.* 8: 89–120.

———. 1974. Adaptive aspects of social structure in *Erythrocebus patas. Symposium of the 5th Congress of the International Primatological Society* (1974), pp. 161–171.

———. 1986. The integration of dominance and social bonding in primates. *Quarterly Review of Biology* 61: 459–479.

Gartlan, J. S., and C. K. Brain. 1968. Ecology and social variability in *Cercopithecus aethiops* and *C. mitis.* In *Primates: Studies in adaptation and variability,* ed. P. Jay. New York: Holt Rinehart and Winston, pp. 253–292.

Gautier-Hion, A. 1988. Polyspecific associations among forest guenons: Ecological, behavioural, and evolutionary aspects. In *A primate radiation,* ed. A. Gautier-Hion, F. Bourliere, J. P. Gautier, and J. Kingdon. Cambridge, UK: Cambridge University Press, pp. 452–476.

Gautier-Hion, A., F. Bourliere, J. P. Gautier, and J. Kingdon, eds. 1988. *A primate radiation.* Cambridge, UK: Cambridge University Press.

Gavan, J. A., ed. 1955. *The nonhuman primates and human evolution.* Detroit: Wayne State University.

Gerall A. A., J. L. Dunlap, and S. E. Hendricks. 1973. Effect of ovarian secretions on female behavioral potentiality in the rat. *Journal of Comparative and Physiological Psychology* 82: 449–465.

Gero, J. M. 1983. Gender bias in archaeology: A cross-cultural perspective. In *The Socio-Politics of Archaeology,* ed. J. M. Gero, D. M. Lacy, and M. L. Blakey. Research Report Number 23. Amherst, MA: Department of Anthropology, University of Massachusetts, pp. 51–58.

———. 1985. Socio-politics and the woman-at-home ideology. *American Antiquity* 50 (2): 342–350.

Gero, J. M., and M. W. Conkey, eds. 1991. *Engendering archaeology: Women and prehistory.* Oxford: Basil Blackwell Press.

Gibbs, W. W. 1995. Lost science in the Third World. *Scientific American* 273.: 92–99.

Gieryn, T. 1983. Boundary work and the demarcation of science from non-science: Strains and interests in professional ideologies of scientists. *American Sociological Review* 48: 781–795.

Gieryn, T., and A. Fiegert. 1990. Ingredients for the theory of science in society: O-rings, ice water, c-clamp, Richard Feynman, and the press. In *Theories of science in society,* ed. S. Cozzens and T. Gieryn. Bloomington: Indiana University Press, pp. 67–97.

Gilbert, S. 1997. Bodies of knowledge. In *Changing life,* ed. P. Taylor, S. Halfon, and P. Edwards. Minneapolis: University of Minnesota Press, pp. 36–52.

Gilchrist, R. 1992. Review of *Experiencing the past: The character of archaeology,* by Michael Shanks. *Archaeological Reviews from Cambridge* 11.1: 188–191.

———. 1993. *Gender and material culture: The archaeology of religious women.* London: Routledge.

Gilligan, C. 1982. *In a different voice.* Cambridge, MA: Harvard University Press.

Gilmore, H. A. 1981. From Radcliffe-Brown to sociobiology: Some aspects of the rise of primatology within physical anthropology. *Am. J. Phys. Anthropol.* 56: 387–392.

Ginsberg, J. R., and D. I. Rubenstein. 1990. Sperm competition and variation in zebra mating behavior. *Behav. Ecol. Sociobiol.* 26: 427–434.

Ginsburg, F., and A. L. Tsing, eds. 1990. *Uncertain terms: Negotiating gender in American culture.* Boston: Beacon Press.

Glander, K. E. 1978. Howling monkey feeding behaviour and plant secondary compounds: A study of strategies. In *The ecology of arboreal folivores,* ed. G. C. Montgomery. Washington, DC: Smithsonian Institute Press, pp. 561–574.

———. 1982. The impact of plant secondary compounds on primate feeding behavior. *Yearbook of Physical Anthropology* 25: 1–18.

———. 1992. Dispersal patterns in Costa Rican mantled howling monkeys. *Int. J. Primatol.* 13: 415–436.

Glander, K. E., L. M. Fedigan, L. Fedigan, and C. Chapman. 1991. Field methods for capture and measurement of three monkey species in Costa Rica. *Folia Primatol.* 57: 70–82.

Glickman, S. E., L. G. Frank, S. Pavgi, and P. Licht. 1992. Hormonal correlates of "masculinization" in female spotted hyaenas (*Crocuta crocuta*): I. Infancy through sexual maturity. *Journal of Reproduction and Fertility* 95: 451–462.

Glickman, S. E., E. M. Coscia, L. G. Frank, M. L. Weldele, and C. M. Drea. 1998. Androgens and masculinization of the genitalia in the spotted hyaena (*Crocuta crocuta*). III. Effects of juvenile gonadectomy. *Journal of Reproduction and Fertility* 113: 129–135.

Glickman, S. E., L. G. Frank, J. M. Davidson, E. R. Smith, and P. K. Siiteri. 1987. Androstenedione may organize or activate sex-reversed traits in female spotted hyenas. *Proceedings of the National Academy of Sciences* 84: 3444–3447.

Goldfoot, D. A., and D. A. Neff. 1985. On measuring behavioral sex differences in social contexts. In *Handbook of behavioral neurobiology,* ed. N. Adler, D. Pfaff, and R. W. Goy. 7: 767–783.

Goldizen, A. W. 1987. Facultative polyandry and the role of infant carrying in wild saddle-back tamarins (*Saguinus fuscicollis*). *Behav. Ecol. Sociobiol.* 20: 99–109.

Goldizen, A. W., J. Mendelson, M. van Vlaadirgen, and J. Terborgh. 1996. Saddle-back tamarin (*Saguinus fuscicollis*) reproductive strategies: Evidence from a thirteen-year study of a marked population. *Amer. J. Primatol.* 38: 57–83.

Gomendio, M., and E. R. S. Roldan. 1993. Mechanisms of sperm competition: Linking physiology and behavioral ecology. *Trends Ecol. Evol.* 8: 95–100.

Goodall, J. 1962. Nest-building behavior in the free-ranging chimpanzee. *Ann. N.Y. Acad. Sci.* 102: 455–467.

———. 1963a. My life among the wild chimpanzees. *National Geographic* (August), 124: 272–308.

———. 1963b. Feeding behaviour of wild chimpanzees, a preliminary report. *Symp. Zool. Soc. Lond.* 10: 39–47.

———. 1965a. New discoveries among Africa's chimpanzees. *National Geographic* (December), 127: 802–831.

———. 1965b. Chimpanzees of the Gombe Stream Reserve. In *Primate behavior,* ed. I. DeVore. New York: Holt Rinehart and Winston, pp. 425–447.

———. 1967a. Mother-offspring relationships in chimpanzees. In *Primate ethology,* ed. D. Morris. London: Weidenfeld and Nicolson, pp. 287–346.

———. 1967b. *My friends the wild chimpanzees.* Washington, DC: National Geographic.

———. 1968. The behaviour of free-living chimpanzees of the Gombe Stream Reserve. *Anim. Behav. Monographs* 1: 161–311.

———. 1977. Infant-killing and cannabilism in free-living chimpanzees. *Folia Primatol.* 28: 259–282.

———. 1986. *The chimpanzees of Gombe: Patterns of behavior.* Cambridge, MA: Belknap Press of Harvard University Press.

———. 1990. *Through a window: Thirty years with the chimpanzees of Gombe.* London: Weidenfeld and Nicolson.

Gooding, D., T. Pinch, and S. Shaffer, eds. 1989. *The uses of experiment.* Cambridge, UK: Cambridge University Press.

Goodman, N. 1955. *Fact, fiction, and forecast.* Indianapolis: Bobbs-Merrill.

Goodman, S. M. 1994. The enigma of antipredator behavior in lemurs: Evidence of a large extinct eagle on Madagascar. *Int. J. Primatol.* 15: 129–134.

Goody, J. 1982. Alternative paths to knowledge in oral and written cultures. In *Spoken and written language,* ed. D. Tannen. Cambridge, UK: Cambridge University Press.

Gordon, D. A. 1995. Conclusion: Culture writing women: Inscribing feminist anthropology. In *Women writing culture,* ed. R. Behar and D. A. Gordon. Berkeley: University of California Press, pp. 429–441.

Goss-Custard, J. D., R. I. M. Dunbar, F. Pelham, and G. Aldrich-Blake. 1972. Survival, mating, and rearing strategies in the evolution of primate social structure. *Folia Primatol.* 17: 1–19.

Götz, W., H. Kummer, and W. Angst. 1978. Schutz der Paarbildung durch Rivalenhemmung bei Mantelpavianen (Gehege- und Freilandexperimente). Commentary text to Film D 1168, Göttingen, Institut für den Wissenschaftlichen Film, *Publ. Wiss. Film, Sekt. Biol.* 11 (8): 1–22.

Gouzoules, S., H. Gouzoules, and P. Marler. 1984. Rhesus monkey (*Macaca mulatta*) screams: Representational signaling in the recruitment of agonistic aid. *Anim. Behav.* 32: 182–193.

Gowaty, P. A. 1994. Architects of sperm competition. *Trends Ecol. Evol.* 9: 160–162.

Gowaty, P. A., and W. C. Bridges. 1991. Behavioral, demographic, and environmental correlates of extra-pair fertilizations in eastern bluebirds, *Sialia sialis. Behav. Ecol.* 2: 339–350.

Gray, J. P. 1985. *Primate sociobiology.* New Haven, CT: HRAF Press.

Green, S. 1975. Dialects in Japanese monkeys: Vocal learning and cultural transmission of locale-specific vocal behavior? *Zeitschrift fur Tierpschologie* 38: 304–314.

Greenspan, R. J. 1995. Understanding the genetic construction of behavior. *Sci. Am.* April: 72–78.

Gregory, W. K. 1949. The bearing of the *Australopithecinae* upon the problem of man's place in nature. *Am. J. Phys. Anthropol.* 7: 485–512.

Griffin, D. R. 1958. *Listening in the dark.* New Haven: Yale University Press.

———. 1976. *The question of animal awareness: Evolutionary continuity of mental experience.* New York: Rockefeller University Press.

———. 1984. *Animal thinking.* Cambridge, MA: Harvard University Press.

———. 1992. *Animal minds.* Chicago: University of Chicago Press.

Gross, P. R., and N. Levitt. 1994. *Higher superstition: The academic left and its quarrels with science.* Baltimore and London: Johns Hopkins University Press.

Grossberg, L., C. Nelson, and P. Treichler, eds. 1992. *Cultural studies.* New York: Routledge.

Gwynne, D. T. 1991. Sexual competition among females: What causes courtship reversal? *Trends Ecol. Evol.* 6: 118–121.

Hacking, I. 1983. *Representing and intervening: Introductory topics in the philosophy of natural science.* Cambridge, UK: Cambridge University Press.

———. 1985. Styles of scientific reasoning. In *Post-Analytic Philosophy,* ed. J. Rajchman and C. West. New York: Columbia University Press, pp. 145–163.

———. 1992. The self-vindication of the laboratory sciences. In *Science as practice and culture,* ed. A. Pickering. Chicago: University of Chicago Press, pp. 29–64.

Haecker, V. 1926. Phanogenetisch gerichtete bestrebungen in Amerika (phenogenetic directed efforts in America). *Z. Induht. Abst. Vererb.* 41: 232–238.

Hager, L. 1997a. Sex and gender in paleoanthropology. In *Women in Human Evolution,* ed. L. Hager. London: Routledge.

Hager, L., ed. 1997b. *Women in human evolution.* New York: Routledge.

Hall, K. R. L. 1962a. Sexual, agonistic, and derived social behaviour patterns of the wild chacma baboon, *P. ursinus. Proc. Zool. Soc. Lond.* 139: 283–327.

———. 1962b. Numerical data, maintenance activities and locomotion of the wild chacma baboon, *P. ursinus. Proc. Zool. Soc. Lond.* 139: 181–220.

———. 1963. Variations in the ecology of the chacma baboon, *Papio ursinus. Symp. Zool. Soc. Lond.* 10: 1–28.

———. 1966. Distribution and adaptation of baboons. *Symp. Zool. Soc. Lond.* 17: 49–73.

———. 1968. Behaviour and ecology of the wild patas monkeys, *Erythrocebus patas,* in Uganda. In *Primates: Studies in adaptation and variability,* ed. P. Jay. New York: Holt, Rinehart, and Winston, pp. 32–119.

Hall, K. R. L., and I. DeVore. 1965. Baboon social behaviour. In *Primate behaviour,* ed. I. DeVore. New York: Holt, Rinehart, and Winston, pp. 53–110.

Halliday, T. R. 1976. The libidinous newt: An analysis of variation in the sexual behavior of the male smooth newt. *Triturus vulgaris. Anim. Behav.* 24: 398–414.

Hamilton, W. D. 1964. The genetical evolution of social behavior I and II. *Journal of Theoretical Biology* 7: 1–52.

Hanen, M., and Kelley, J. 1992. Gender and archaeological knowledge. In *Metaarchaeology: Reflections by archaeologists and philosophers,* ed. Lester Embree. Boston: Kluwer, pp. 195–227.

Haraway, D. 1983. Signs of dominance: From physiology to a cybernetics of primate society, C. R. Carpenter 1930–70. *Studies in the History of Biology* 6: 129–219.

———. 1989. *Primate visions: Gender, race, and nature in the world of modern science.* New York: Routledge.

———. 1991a. *Simians, cyborgs, and women: The reinvention of nature.* London and New York: Routledge.

———. 1991b. "Gender" for a Marxist dictionary: The sexual politics of a word. In *Simians, cyborgs and w. The reinvention of nature.* New York: Routledge, pp. 127–148.

———. 1997. Haraway, D. 1997. *Modes_witness@second_Millennium. Femaleman_meets_oncomouse.* London: Routledge.

Harcourt, A. H. 1979a. Social relationships between adult male and female mountain gorillas in the wild. *Anim. Behav.* 27: 325–342.

———. 1979b. The social relations and group structure of wild mountain gorillas. In *The great apes,* ed. D. A. Hamburg and E. R. McCown. Menlo Park, CA: Benjamin Cummings, pp. 187–192.

———. 1988. Alliances in contests and social intelligence. In *Machiavellian intelligence,* ed. R. Byrne and A. Whiten. Oxford: Clarendon, pp. 132–152.

———. 1989. Social influences on competitive ability: Alliances and their consequences. In *Comparative socioecology,* ed. V. Standen and R. A. Foley. Oxford: Blackwell, pp. 223–242.

Harcourt, A. H., and F. B. M. DeWaal. 1992. *Coalitions and alliances in humans and other animals.* Oxford: Oxford University Press.

Harding, S., ed. 1983. Why has the sex/gender system become visible only now? In *Discovering reality: Feminist perspectives on epistemology, metaphysics, methodology, and philosophy of science,* ed. S. Harding and M. B. Hintikka. Boston: D. Reidel, pp. 311–324.

———. 1986. *The science question in feminism.* Ithaca, NY: Cornell University Press.

———. 1991. *Whose science? Whose knowledge? Thinking from women's lives.* Buckingham: Open University Press.

———. 1993. *The Racial Economy of Science: Toward a Democratic Future.* Bloomington: Indiana Press.

———. 1997a. Multicultural and global feminist philosophies of science: Resources and challenges. In *Feminism, Science and the Philosophy of Science,* ed. L. H. Nelson and J. Nelson. Dordrecht: Kluver Publications, pp. 263–287.

———. 1997b. Women's standpoints on nature: What makes them possible? *Osiris* 12: 186–200.

Harrison, M. J. S. 1983. Territorial behaviour in the green monkey, *Cercoptihecus sabaeus:* Seasonal defense of local food supplies. *Beh. Ecol. Sociobiol.* 12: 85–94.

Hartouni, V. 1991. Containing women: Reproductive discourse in the 1980s. In

Technoculture, ed. C. Penley and A. Ross. Minneapolis: University of Minnesota Press, pp. 27–56.

Hartsock, N. 1983. The feminist standpoint: Developing the ground for a specifically feminist historical materialism. In *Discovering reality: Feminist perspectives on epistemology, metaphysics, methodology, and philosophy of science,* ed. S. Harding and M. B. Hintikka. Boston: D. Reidel, pp. 238–310.

———. 1987. Rethinking modernism: Minority vs. majority theories. *Cultural Critique* 7: 187–206. (Special Issue on the Nature and Context of Minority Discourse II.)

Harvey, D. 1989. *The condition of postmodernity.* Oxford: Basil Blackwell.

Hasegawa, T. 1990. Sex differences in ranging patterns. In *The Chimpanzees of the Mahale Mountains: Sexual and life history strategies,* ed. T. Nishida. Tokyo: Univerisity of Tokyo Press, pp. 99–114.

Hauser, M. D. 1990. Do chimpanzee copulatory calls incite male-male competition? *Anim. Behav.* 39: 596–597.

Hausfater, G. 1975. *Dominance and reproduction in baboons* (*Papio cynocephalus*). Basel: S. Karger.

Hawkesworth, M. 1997. Confounding gender. *Signs* 22: 649–685.

Hayaki, H. 1990. *The human within the chimpanzee.* Tokyo: Shokabo. (J)

Hayes, H. T. P. 1986. The dark romance of Dian Fossey. *Life* 9 (11) (November): 64–70.

Hayes, K. J., and C. Hayes. 1951. The intellectual development of a home-reared chimpanzee. *Proceedings of the American Philosophical Society* 95: 105–109.

Hayes, K. J., and C. Hayes. 1952. Imitation in a home-raised chimpanzee. *Journal of Comparative and Physiological Psychology* 45: 450–459.

Hays, S. P. 1987. *Environmental politics in the United States, 1955–1985.* Cambridge, UK: Cambridge University Press.

Hays-Gilpin, K. 1996. Archaeology of gender, feminist archaeology, women in prehistory, women in archaeology. *Diotima's archaeology bibliography.* Http://www.uky.edu/ArtsSciences/Classics/biblio/genderarch.html.

Hays-Gilpin, K., and D. S. Whitely, eds. 1998. *Reader in gender archaeology.* New York: Routledge.

Hazama, N. 1965. Wild Japanese monkeys at Mt. Hiei. In *Monkeys: Sociological studies,* ed. S. Kawamura and J. Itani. Tokyo: Chuokoronsha, pp. 375–401. (J).

Heckman, S. 1997. Truth and method: Feminist standpoint theory revisited. *Signs* 22.2: 341–365. (With commentaries by N. C. M. Hartsock, P. H. Collins, S. Harding, D. E. Smith, and a reply by Heckman: pp. 366–402.)

Hediger, H. 1942. *Wildtiere in Gefangenschaft. Ein Grundriß der Tiergartenbiologie.* Basel: Benno Schwabe (English edition, 1950, *Animals in Captivity.* London, Butterworths).

———. 1954. *Skizzen zu einert Tierpsychologie im Zoo und im Zirkus.* Zürich: Gutenberg.

Heijden, P. G. M. van der, H. de Vries, and J.A.R.A.M. van Hooff. 1990. Correspondence analysis of transition matrices, with particular reference to missing entries and asymmetries. *Anim. Behav.* 40: 49–64.

Hemelrijk, C. K. 1990a. Models of and tests for reciprocity, unidirectionality and other social interaction patterns at group level. *Anim. Behav.* 39: 1013–1029.

———. 1990b. A matrix partial correlation test used in investigations of reciprocity and other social interaction patterns at group level. *J. Theor. Biol.* 143: 405–420.

———. 1991. Letter to the editor. Interchange of "altruistic" acts as an epiphenomenon. *J. Theor. Biol.* 153: 137–139.

———. 1996. Reciprocation in apes: From complex cognition to self-structuring. In *Great ape societies,* ed. W. C. McGrew, T. Nishida, and L. F. Marchant. New York: Cambridge University Press, pp. 185–195.

———. 1997. Cooperation without genes, games or cognition. In *4th Conference on Artificial Life, ed.* P. Husbands, and I. Harvey. Cambridge, MA: MIT Press, pp. 511–520.

Hemelrijk, C. K., and A. Ek. 1991. Reciprocity and interchange of grooming and "support" in captive chimpanzees. *Anim. Behaviour.* 41: 923–935.

Hemelrijk, C .K., G. J. van Laere, and J.A.R.A.M. van Hooff. 1992. Sexual exchange relationships in captive chimpanzees? *Behavioral Ecology and Sociobiology* 30, 269–275.

Hemelrijk, C. K., T. J. M. Klomberg, J. J. Nooitgedagt, and J.A.R.A.M. van Hooff. 1991. Side-directed behaviour and recruitment of support in captive chimpanzees. *Behaviour* 118: 89–102.

Henzi, S. P. 1996. Copulation calls and paternity in chacma baboons. *Anim. Behav.* 51: 233–234.

Hernstein-Smith, B. 1997. The microdynamics of incommensurability: Philosophy of science meets science studies. In *Mathematics, science, and post-classical theory,* ed. B. Herstein-Smith and A. Plotnisky. Durham, NC: Duke University Press, pp. 243–266.

Herschberger, R. 1948. *Adam's rib.* New York: Harper and Row.

Hess, D. 1997. *Science studies: An advanced introduction.* New York: New York University Press.

Heyes, C. M. 1993. Imitation, culture and cognition. *Anim. Behav.* 46: 999–1010.

———. 1995. Imitation and flattery: A reply to Byrne and Tomasello. *Anim. Behav.* 50: 1421–1424.

Heyes, C. M., G. R. Dawson, and T. Nokes. 1992. Imitation in rats: Initial responding and transfer evidence. *Quarterly Journal of Experimental Psychology* 45B: 229–240.

Hildebolt, C. F., J. E. Phillips-Conroy, and C. J. Jolly. 1993. Alveolar bone loss in wild baboons. *American Journal of Physical Anthropology* 29: 61–72.

Hilgartner, S. 1990. The dominant view of popularization: Conceptual problems, political uses. *Social Studies of Science* 20: 519–539.

Hill, D. A., and J.A.R.A.M. van Hooff. 1994. Affiliative relationships between males in groups of nonhuman primates: A summary. *Behaviour* 130 (3–4): 143–151.

Hill, D. A., and N. Okayasu. 1996. Determinants of dominance among female macaques: Nepotism, demography and danger. In *Evolution and ecology of macaque societies,* ed. J. E. Fa and D. G. Lindburg. Cambrdige, UK: Cambridge University Press, pp. 459–472.

Hill, R. A., and R. I. M. Dunbar. 1998. An evaluation of the roles of predation rate and predation risk as selective pressures on primate grouping behaviour. *Behaviour* 135: 411–430.

Hinde, R. A. 1978. Dominance and role—two concepts with dual meanings. *Journal of Social and Biological Structures* 1: 27–38.

———. 1983. *Primate social relationships: An integrated approach.* Oxford: Blackwell Scientific Publications.

———. 1991. A biologist looks at anthropology. *Man* 26: 583–608.

———. 1996. Primatology had multiple sources. Paper prepared for Wenner-Gren Foundation for Anthropological Research, Symposium no. 120, Changing Images of Primate Societies: The Role of Theory, Method, and Gender. Teresopolis, Brazil.

———. 1998a. Humans and human habitats. In *Mind, brain, and the environment,* ed. B. Cartledge. Oxford: Oxford University Press, pp. 6–27.

———. 1998b. Mind and artefact: A dialectical perspective. In *Cognition and material culture: The archaeology of symbolic storage,* ed. C. Renfrew and C. Scarre. Cambridge: McDonald Institute, pp. 175–180.

Hinde, R. A., and J. Fisher. 1951. Further observations on the opening of milk bottles by birds. *British Birds* 34: 393–396.

Hinde, R. A., and L. McGinnis. 1977. Some factors influencing the effects of temporary mother-infant separation—some experiments with rhesus monkeys. *Psychological Medicine* 7: 197–222.

Hinde, R. A., and J. Stevenson-Hinde. 1963. *Constraints on learning.* London: Academic Press.

Hinde, R. A., T. Rowell, and Y. Spencer-Booth. 1964. Behaviour of socially-living rhesus monkeys in their first six months. *Proceedings of the Zoological Society of London* 143: 609–649.

Hiraiwa-Hasegawa, M. 1983. *Maternal care of the Japanese macaque.* Tokyo: Kaimeisha. (J)

———. 1990. Maternal investment before weaning. In *The chimpanzees of the Mahale Mountains,* ed. T. Nishida. Tokyo: University of Tokyo Press, pp. 257–266.

———. 1992. Sociobiology and Japanese primatology: A case of a struggle for survival in a conformist society; or a case of the founder effect. Paper given at Princeton University, unpubl. ms.

Hobson, R. P. 1996. On not understanding minds. In "What young chimpanzees know about seeing," *Monographs of the Society for Research in Child Development,* ed. Povinelli, D. J., and T. J. Eddy, 61 (3, ser. 247): 153–160.

Hodder, I., ed. 1982. *Symbolic and structural archaeology.* Cambridge, UK: Cambridge University Press.

Hofer, T. 1968. Anthropologists and native ethnographers in Central European villages: Comparative notes on the professional personality of two disciplines. *Curr. Anthropol.* 9 (4): 311–315.

Hogeweg, P. 1988. Mirror beyond mirror, puddles of life. In *Proceedings of Artificial Life,* ed.G. C. Langton. Redwood City: Addison-Wesley, pp. 297–316.

Hogeweg, P., and B. Hesper. 1979. Heterarchical, selfstructuring simulation systems:

Concepts and applications in biology. In *Methodologies in systems modelling and simulation.* Amsterdam: North-Holland, pp. 221–231.

Holenweg, A. K., R. Noë, and M. Schabel. 1996. Waser's gas model applied to associations between red colobus and diana monkeys in the Taï National Park, Ivory Coast. *Folia primatol.* 67: 125–136.

Holloway, R. L. 1974. *Primate aggression, territoriality, and xenophobia. A comparative perspective.* New York: Academic Press.

Holmes, J. G., and J. K. Rempel. 1989. Trust in close relationships. In *Close relationships,* ed. C. Henrick. Newbury Park: Sage, pp. 187–220.

Höner, O., L. Leumann, and R. Noë. 1997. Polyspecific associations of red colobus (*Colobus badius*) and diana monkey (*Cercopithecus diana*) groups in the Taï National Park, Ivory Coast. *Primates* 38: 281–291.

Honigmann, J. J. 1963. *Understanding culture.* New York: Harper and Row.

Hooff, J.A.R.A.M. van. 1962. Facial expressions in higher primates. *Symp. Zool.Soc. London* 8: 97–125.

———. 1967a. The facial displays of catarrhine monkeys and apes. In *Primate ethology,* ed. D. Morris. London: Weidenfeld and Nicolson, pp. 7–68.

———. 1967b. The care and management of captive chimpanzees with special emphasis on the ecological aspects. *Aeromed. Res. Lab. TR,* 67–15. Alamogordo: Aeromed. Res. Lab.

———. 1970. A component analysis of the structure of the social behaviour of a semi-captive chimpanzee group. *Experientia* 26: 549–550.

———. 1972. A comparative approach to the phylogeny of laughter and smiling. In *Non-verbal communication,* ed. R. A. Hinde. London: Cambridge University Press, pp. 209–241.

———. 1973a. A structural analysis of the social behaviour of a semi-captive group of chimpanzees. In *Social communication and movement, ed.* M. von Cranach, and I. Vine. London: Academic Press, pp. 75–162.

———. 1973b. The Arnhem Zoo chimpanzee consortium; an attempt to create an ecologically and socially acceptable habitat. *International Zoo Yearbook* 13: 195–204.

———. 1976a. The comparison of facial expressions in man and higher primates. In *Methods of inference from animal to human behaviour,* ed. M. von Cranach. The Hague/Chicago: Mouton-Aldine, pp. 165–196.

———. 1976b. A comparative approach to the phylogeny of laughter and smiling. In *play, its role in development and evolution,* ed. J. S. Brumer, A. Jolly, and K. Silva. Harmondsworth: Penguin Books, pp. 130–139.

———. 1989. Laughter and humour, and the "duo-in-uno" of nature and culture. In *The nature of culture,* ed. W. A. Koch. Bochum: Brockmeyer, pp. 120–149.

———. 1990. Social behaviour and its influences on reproduction and growth in primates. In *Psychoneuroendocrinology of growth and development,* ed. A. K. Slob, and M. J. Baum. Rotterdam: Medicon, pp. 187–198.

———. 1990. Intergroup competition and conflict in animals and man. In *Sociobiology and Conflict,* ed. V. Falger, and H. van der Dennen. London: Chapman and Hall, pp. 23–54.

———. 1990. Organisation sociale chez les simiens: Stratégies adaptatives écologiques et sociales. In *Primates, recherches actuelles,* ed. J. J. Roeder and J. R. Anderson. Paris: Hasson, pp. 182–194.

———. 1992. Competing for progeny: The socioecology of primate mating systems. In *Sex matters,* ed. W. Bezemer, P. Cohen-Kettenis, K. Slob, and N. van Son-Schoones. Amsterdam: Elsevier Science Publ., pp. 93–96.

———. 1994. Understanding chimpanzee understanding. In *Chimpanzee cultures,* ed. R. W. Wrangham, W. C. McGrew, F. B. M. de Waal, and P. C. Heltne. Cambridge MA: Harvard University Press. pp. 267–284.

———. 1996. The orangutan, a social outsider: A socio-ecological test case. In *The neglected ape,* ed. R. D. Nadler, B. F. M. Galdikas, L. K. Sheeran and N. Rosen. New York: Plenum Press, pp. 153–163.

———. 1997. The socio-ecology of sex ratio variation in primates: Evolutionary deduction and empirical evidence. *Applied Animal Behavior Science* 51 (3–4): 293–307.

Hooff, J.A.R.A.M. van, and F. Aureli. 1994. Social homeostasis and the regulation of emotion. In *Emotions: Essays on emotion theory,* ed. S. H. M. van Goozen, N. E. van de Poll, and J. A. Sergeant. Hillsdale, NJ: Lawrence Erlbaum, pp. 197–218.

Hooff, J.A.R.A.M. van, and C. P. van Schaik. 1992. Cooperation in competition: The ecology of primate bonds. In *Coalitions and alliances in humans and other animals,* ed. S. Harcourt, and F. B. M. de Waal. Oxford: Oxford University Press, pp. 357–391.

———. 1994. Male bonds: Afilliative relationships among nonhuman primate males. *Behaviour* 130: 143–151.

Hooks, B. 1990. *Yearning: Race, gender, and cultural politics.* Boston: South End Press.

Hooton, E. 1937. *Apes, men, and morons.* New York: G. P. Putnam's Sons.

———. 1942. *Man's poor relations.* New York: Doubleday.

———. 1947. *Up from the ape.* New York: Macmillan.

———. 1955. The importance of primate studies in anthropology. In *The nonhuman primates and human evolution,* ed. J. A. Gavan. Detroit: Wayne State University Press, pp. 1–10.

Howell, F. C., and F. Bourliere, eds. 1963. *African ecology and human evolution.* Chicago: Aldine.

Howells, W. W., ed. 1962. *Ideas on human evolution: Selected essays 1949–1961.* Cambridge, MA: Harvard University Press.

Hrdy, S. B. 1976. Care and exploitation of nonhuman primate infants by conspecifics other than the mother. In *Advances in the study of behavior,* vol. 6, ed. D. S. Lehrman, R. A. Hinde, and E. Shaw. New York: Academic Press, pp. 101–158.

———. 1977. *The langurs of Abu: Female and male strategies of reproduction.* Cambridge, MA: Harvard University Press.

———. 1979. Infanticide among animals: A review, classification, and examination of the implications for reproductive strategies for females. *Ethol. Sociobiol.* 1: 13–40.

———. 1981. *The Woman That Never Evolved.* Cambridge, MA: Harvard University Press.

———. 1984a. Assumptions and evidence regarding the sexual selection hypothesis: A reply to Boggess. In *Infanticide: Comparative and evolutionary perspectives,* ed. G. Hausfater and S. B. Hrdy. New York: Aldine, pp. 315–319.

———. 1984b. Introduction, In *Female primates. Studies by women primatologists,* ed. M. F. Small. New York: Alan R. Liss, pp. 103–109.

———. 1986. Empathy, polyandry, and the myth of the coy female. In *Feminist approaches to science,* ed. R. Bleier. New York: Pergamon Press, pp. 119–146.

———. 1988. Raising Darwin's consciousness: Females and evolutionary theory. In *The evolution of sex,* ed. R. Bellig and G. Stevens. San Fransisco: Harper and Row, pp. 161–169.

———. 1996. Raising Darwin's consciousness: Female sexuality and the prehominid origins of patriarchy. Paper prepared for Wenner-Gren Foundation for Anthropological Research, Symposium no. 120, Changing Images of Primate Societies: The Role of Theory, Method, and Gender. Teresopolis, Brazil.

Hrdy, S. B., and G. Hausfater. 1984. Comparative and evolutionary perspective on infanticide: Introduction and overview. In *Infanticide: Comparative and evolutionary perspectives,* ed. G. Hausfater and S. B. Hrdy. New York: Aldine, pp. xiii–xxxv.

Hrdy, S. B., and D. B. Hrdy. 1976. Hierarcical relations among female Hanuman langurs. *Science* 193: 913–915.

Hrdy, S. B., and P. L. Whitten. 1987. Patterning of sexual activity. In *Primate societies,* ed. B. B. Smuts, D. L. Cheney, R. M. Seyfarth, R. W. Wrangham, and T. T. Struthsaker. Chicago: University of Chicago Press, pp. 370–384.

Hrdy, S. B., and G. C. Williams. 1983. Behavioral biology and the double standard. In *Social behavior of female vertebrates,* ed. S. K. Wasser. New York: Academic Press, pp. 3–17.

Hrdy, S. B., C. Janson, and C. van Schaik. 1995. Infanticide: Let's not throw out the baby with the bath water. *Evol. Anthropol.* 3: 151–154.

Hubbard, R. 1989. Science, facts, and feminism. In *Feminism and science,* ed. N. Tuana. Bloomington: Indiana University Press, pp. 119–131.

Huber, M. H. R., F. H. Bronson, and C. Desjardins. 1980. Sexual activity of aged male mice: Correlation with level of arousal, physical endurance, pathological status, and ejaculatory capacity. *Biol. Reprod.* 23: 305–316.

Huck, U. W., R. D. Lisk, J. C. Allison, and C. G. Van Dongen. 1986a. Determinants of mating success in the golden hamster (*Mesocricetus auratus*): Social dominance and mating tactics under seminatural conditions. *Anim. Behav.* 34: 971–989.

Huck, U. W., R. D. Lisk, E. J. Parents, and D. E. Principato. 1986b. Determinants of mating success in the golden hamster (*Mesocricetus auratus*): III. Female acceptance of multiple mating partners. *J. Comp. Psychol.* 100: 128–136.

Huffman, M. A. 1991a. History of the Arashiyama Japanese macaques in Kyoto, Japan. In *The monkeys of Arashiyama,* ed. L. M. Fedigan and P. J. Asquith. Albany: State University of New York Press, pp. 21–53.

———. 1991b. Mate selection and partner preferences in female Japanese macaques. In *The monkeys of Arashiyama,* ed. L. M. Fedigan and P. J. Asquith. Albany: State University of New York Press, pp. 101–122.

———. 1993. An investigation of the use of medicinal plants by wild chimpanzees: Current status and future prospects. *Primate Research* 9: 179–187. (J)

Hughes, T. 1983. *Networks of power: Electrification in Western society, 1880–1930.* Baltimore: Johns Hopkins University Press.

Humphrey, N. 1976. The social function of intellect. In *Growing points in ethology,* ed. P. P. G. Bateson and R. A. Hinde. Cambridge, UK: Cambridge University Press, pp. 303–317.

Hunte, W., and J. A. Horrocks. 1987. Kin and non-kin interventions in the aggressive disputes of vervet monkeys. *Behavioural Ecology and Sociobiology* 20: 257–263.

Hunter, F. M., M. Petrie, M. Otronen, T. Birkhead, and A. P. Moller. 1993. Why do females copulate repeatedly with one male? *Trends Ecol. Evol.* 8: 21–26.

Huxley, J. S. 1942. *Evolution, the modern synthesis.* London: Allen and Unwin.

———. 1973. *Memories II.* New York: Harper and Row.

Huxley, T. H. 1959. *Man's place in nature.* Ann Arbor: University of Michigan.

Idani, G. 1990. Relations between unit-groups of bonobos at Wamba, Zaire: Encounters and temporary fusions. *Afr. Study Monogr.* 11: 153–186.

Imanishi, K. 1941a. *Seibutsu no Sekai.* In *The world of living things.* Tokyo: Kobunsha. (J)

———. 1941b. *The world of living things.* Tokyo: Kobunsha. (J)

———. 1949a. *Seibutsu Shakai no Ronri.* In *The logic of living societies.* Tokyo: Mainichi Shinbunsha. (J)

———. 1949b. *The logic of living societies.* Tokyo: Rikusuisha (J)

———. 1951. *Prehuman societies.* Tokyo: Iwanami-shoten. (J)

———. 1952. Evolution of humanity. In *Man,* ed. K. Imanishi. Tokyo: Mainichi-shinbunsha, pp. 36–94.

———. 1954. Nomadism, an ecological interpretation. *Silver jubilee volume of the Zinbun-Kagaku-Kenkyusyo, Kyoto Univ.,* pp. 466–479.

———. 1955. The horses of Toimisaki. In *Nihon dobutsuki,* ed. K. Imanishi, vol. 1. Tokyo: Kobunsha, pp. 9–140. (J)

———. 1957. *Reichôrui kenkyû group no tachiba* (The standpoint of the Primates Research Group). *Shizen* 12 (2): 1–9.

———. 1960a. *The gorilla.* Tokyo: Bungeishunjusha. (J)

———. 1960b. Social organization of sub-human primates in their natural habitat. *Curr. Anthrop.* 1: 399–407.

———. 1965. The origin of human family: A primatological approach. In *Monkeys: Sociological studies,* ed. S. Kawamura and J. Itani. Tokyo: Chuokoronsha, pp. 3–69. (J).

———. 1966. The purpose and method of our research in Africa. *Kyoto Univ. Afr. Studies* 1: 1–10.

———. 1970. Field studies on primate societies. In *Profiles of Japanese science and scientists,* ed. H. Yukawa. Tokyo: Kodansha, pp. 29–42.

———. 1984. A proposal for *Shizengaku:* The conclusion to my study of evolutionary theory. *J. Social Biol. Struct. 7:* 357–368.

———. 1974–1975, 1993. *The complete work of Imanishi Kinji,* Vols. 1–14. Tokyo: Kodansha. (J)

Imanishi, K., and S. A. Altmann, eds. 1965. *Japanese monkeys:* A collection of translations. Chicago: S. A. Altmann.

Irigaray, L. 1974 (trans. 1985). Women, science's unknown. In *Speculum of the other woman,* ed. L. Irigaray. Ithaca: Cornell University Press, pp. 13–25.

Irwin, A., and B. Wynne, eds. 1996. *Misunderstanding science? The public reconstruction of science and technology.* Cambridge, UK: Cambridge University Press.

Isbell, L. A. 1991. Contest and scramble competition: Patterns of female aggression and ranging behavior among primates. *Behavioural Ecology* 2: 143–155.

Itani, J. 1954. The monkeys of Takasakiyama. In *Nihon dobutsuki,* vol. 2, ed. K. Imanishi. Tokyo: Kobunsha. (J)

———. 1957. Personality in Japanese monkeys. *Iden* 11 (1): 29–33. (J)

———. 1961. *The forest of gorillas and the pygmies.* Tokyo: Iwanami-shoten. (J)

———. 1972. *Primate social structure.* Tokyo: Kyoritsu-shuppan. (J)

———. 1975. Twenty years with Mount Takasaki monkeys. In *Primate utilization and conservation,* ed. G. Bermount and D. G. Lindburg. New York: John Wiley and Sons, pp. 101–125.

———. 1977a. *In search of chimpanzees in the wilderness.* Tokyo: Heibonsha. (J)

———. 1977b. Evolution of primate social structure. *J. Human Evol.* 6: 235–243.

———, ed. 1977c. *The chimpanzee.* Tokyo: Kodansha. (J)

———. 1980. Social structure of African great apes. *J. Reprod. Fert.,* Suppl. 28: 33–41.

———. 1983. Sociological studies of Japanese monkeys. In *Recent progress of natural sciences in Japan,* vol. 8, *Anthropology.* Tokyo: The Fourth Division, Science Council of Japan, pp. 89–94.

———. 1985a. The evolution of primate social structure. *Man* (n.s.) 20: 539–611.

———. 1985b. Japanese primatology in retrospect and in prospect. *Primate Research,* 1: 5–14.

———. 1993. On the shore of Lake Tanganyika. *Primate Research* 9: 215–224.

Itani, J., and S. Akira. 1967. The social unit of chimpanzees. *Primates* 8: 355–382.

Itani, J., and A. Nishimura. 1973. The study of infrahuman culture in Japan. In *Symposia of the Fourth International Congress of Primatology,* vol. 1, *Precultural primate behavior,* ed. E. W. Menzel. Basel: Karger, pp. 26–55.

Itani, J., and A. Suzuki. 1967. The social unit of chimpanzees. *Primates* 8: 355–381.

Itani, J., and K. Tokuda. 1958. The Japanese monkeys of Koshima Island: Their sexual behavior. In *Nihon Dobutsuki,* vol. 3, ed. K. Imanishi. Tokyo: Kobunsha. pp. 1–233. (J)

Izawa, K. 1977. The chimpanzees in the Kasakati Basin—II: Unit groups and their home ranges. In *The chimpanzee,* ed. J. Itani. Tokyo: Kodansha, pp. 187–248.

———. 1979. *Animals in the country of water and forest.* Tokyo: Dobutsusha.

———. 1991. Another evolutionary path: The diverse societies of Neotropical monkeys. In *Cultural Historiography of Monkeys,* ed. T. Nishida, K. Izawa, and T. Kano. Tokyo: Heibonsha, pp. 57–83. (J)

Jackson, S., and D. A. Dewsbury. 1979. Recovery from sexual satiety in male rats. *Anim. Learn. Behav.* 7: 119–124.

James, W. T. 1890. *Principles of psychology,* vol. 1. New York: Henry Holt.

———. 1907. *Pragmatism. A new name for some old ways of thinking followed by the meaning of truth.* Cambridge, MA: Harvard University Press.

Janik, V. M., and P. J. B. Slater. 1997. Vocal learning in mammals. *Advances in the Study of Behavior* 26: 59–99.

Janson, C. H. 1984. Female choice and mating system in the brown capuchin monkey, *Cebus apella. Zeitschrift für Tierpsychologie* 65: 177–200.

———. 1992. Evolutionary ecology of primate social structure. In *Evolutionary ecology and human behavior,* ed. E. A. Smith and B. Winterhalder. New York: Aldine de Gruber, pp. 95–130.

———. 1998. Testing the predation hypothesis for vertebrate sociality: Prospects and pitfalls. *Behaviour* 135: 389–410.

Jasanoff, S. 1990. *The fifth branch: Science advisors as policymakers.* Cambridge, MA: Harvard University Press.

———, ed. 1994. *Learning from disaster: Risk management after Bhopal.* Philadelphia: University of Pennsylvania Press.

———. 1995. *Science at the Bar.* Cambridge, MA: Harvard University Press.

———. 1996a. Beyond epistemology: Relativism and engagement in the politics of science. *Social Studies of Science* 26: 396–418.

———. 1996b. Is science socially constructed—and can it still inform public policy? *Science and Engineering Ethics* 2: 263–276.

———. 1996c. Science and norms in global environmental regimes. In *Earthly goods: Environmental change and social justice,* ed. F. O. Hampson and J. Reppy. Ithaca: Cornell University Press, pp. 173–197.

Jasanoff, S., G. Markle, J. Peterson, and T. Pinch, eds. 1995. *Handbook of science and technology studies.* London: Sage.

Jay, P. C. 1962. Aspects of maternal behavior among langurs. *Ann. N.Y. Acad. Sci.* 102: 468–476.

———. 1965. The common langurs of north India. In *Primate behavior,* ed. I. DeVore. New York: Holt, Rinehart, and Winston, pp. 197–249.

———, ed. 1968. *Primates: Studies in adaptation and variability.* New York: Holt, Rinehart, and Winston.

Jenks, S. M. M. Weldele, L. G. Frank, and S. E. Glickman. 1995. Acquisition of matrilineal rank in captive spotted hyenas (*Crocuta crocuta*): Emergence of a natural social system in peer-reared animals and their offspring. *Anim. Behav.* 50: 893–904.

JICA. 1980. *Mahale: Study for the proposed Mahale Mountains National Park, final report.* Tokyo: The Japan International Cooperation Agency.

Jolly, A. 1964a. Choice of cue in prosimian learning. *Anim. Behav.* 12: 571–577.

———. 1964b. Prosimians' manipulation of simple object problems. *Anim. Behav.* 12: 56–57.

———. 1966a. *Lemur behavior.* Chicago: University of Chicago Press.

———. 1966b. Lemur social behavior and primate intelligence. *Science* 153: 501–506.

———. 1967. Breeding synchrony in wild *Lemur catta.* In *Social communication among primates,* ed. S. A. Altmann. Chicago: University of Chicago Press, pp. 3–14.

———. 1984. The puzzle of female feeding priority. In *Female primates,* ed. M. F. Small. New York: Alan R. Liss.

Jolly, A., P. Oberle, and R. Albignac. 1984. *Madagascar.* New York: Pergamon Press.

Jones, C. B. 1980. Seasonal parturition, mortality, and dispersal in the mantled howler monkey, *Alouatta palliata. Brenesia* 17: 1–10.

Jordanova, L. J. 1989. Medical images of the female body. In *Sexual visions: Images of gender in science and medicine between the eighteenth and twentieth centuries,* ed. L. J. Jordanova. Madison, WI: University of Wisconsin Press, pp. 134–159.

Judge, P. J. 1991. Dyadic and triadic reconciliation in pigtail macaques. *Am. J. Primatol.* 23: 225–237.

Jullien, F. 1995. *The propensity of things. Toward a history of efficacy in China.* Cambridge, MA: Zone Books.

Kanagawa, H., E. S. E. Hafez, M. Nawar, and S. Jaszczak. 1972. Patterns of sexual behavior and anatomy of copulatory organs in macaques. *Z. Tierpsychol.* 31: 449–460.

Kano, T. 1972. Distribution and adaptation of the chimpanzee in the open country on the eastern shore of Lake Tanganyika. *Kyoto Univ. Afr. Studies* 7: 37–129.

———. 1983. An ecological study of the pygmy chimpanzees (*Pan paniscus*) of Yalosodi, Republic of Zaire. *Int. J. Primatol.* 4 (1): 1–32.

———. 1986. *The last ape: Pygmy chimpanzee behavior and ecology.* Tokyo: Dobutsuha. (J)

———. 1987. *The fire of Elya: Wonderful folktales in an African forest.* Tokyo: Dobutsusha. (J)

———. 1997. *One man's tales of the forest.* Tokyo: University of Tokyo Press. (J)

Kappeler, P. M. 1993. Sexual selection and lemur social systems. In *Lemur social systems and their ecological basis,* ed. P. M. Kappeler and J. U. Ganzhorn. New York: Plenum, pp. 223–240.

Kappeler, P. M., and C. P. van Schaik. 1992. Methodological and evolutionary aspects of reconciliation research among primates. *Ethology* 92: 51–69.

Karmiloff-Smith, A. 1993. *Beyond modularity: A developmental perspective on cognitive science.* Massachusetts: Bradford/MIT.

Kawabe, S. 1964. *Monkey babies.* Tokyo: Chuokoron-sha. (J)

Kawai, M. 1955. The domestic rabbit. In *Nihon dobutsuki,* ed. K. Imanishi, vol. 1. Tokyo: Kobunsha, pp. 141–283. (J)

———. 1961. *The gorilla expedition.* Tokyo: Kobunsha. (J)

———. 1964, revised 1969. *Ecology and society of Japanese monkeys.* Tokyo: Kawadeshobo.

———. 1979a. *Forest produced the primates.* Tokyo: Heibonsha. (J)

———, ed. 1979b. *Ecological and sociological studies of gelada baboons.* Contribution to *Primatology,* vol. 16. Basel: Karger.

———, ed. 1990. *Prehuman sociology—probing into African primate societies.* Tokyo: Kyoikusha. (J)

———. 1992. *The descent of man.* Tokyo: Kodansha. (J)

Kawamura, S. 1957. The deer of Nara Park. In *Nihon dobutsuki,* ed. K. Imanishi, vol. 4. Tokyo: Kobunsha. (J)

Kawanaka, K. 1981. Infanticide and cannibalism in chimpanzees, with sepcial reference to the newly observed case in the Mahale Mountains. *Afr. Study Monogr.* 1: 69–99.

———. 1984. Association, ranging, and social unit in chimpanzees of the Mahale Mountains, Tanzania. *Int. J. Primatol.,* 5: 411–434.

Kawanaka, K., and T. Nishida. 1975. Recent advances in the study of inter-unit-group relationships and social structure of wild chimpanzees of the Mahale Mountains. In *Proc. Symp. 5th Congr. Int. Primat. Soc.,* ed. S. Kondo, M. Kawai, and Y. Ehara. Tokyo: Japan Science Press, pp. 173–186.

Kazdin, A. E. 1978. *History of behavior modification: Experimental foundations of contemporary research.* Baltimore: University Park Press.

Keddy, A. C. 1986. Female mate choice in vervet monkeys (*Cerocopithecus aethips sabeus*). *Am. J. Primatol.* 10: 125–134.

Keith, A. 1912. Modern problems relating to the antiquity of man, *Report of the British Association.*

———. 1931. *New discoveries: The antiquity of man.* London: Williams and Norgate.

———. 1949. *A new theory of human evolution.* New York: Philosophical Library.

Keith, A., G. E. Smith, A. S. Woodward, and W. J. H. Duckworth. 1925. The fossil anthropoid from Taungs. *Nature* 115: 234–236.

Keller, E. F. 1983. *A Feeling for the Organism. The Life and Work of Barbara McClintock.* New York: W. H. Freeman and Co.

———. 1985. *Relections on gender and science.* New Haven: Yale University Press.

———. 1989a. The gender/science system: Or, is sex to gender as nature is to science? In *Feminism and Science,* ed. N. Tuana. Bloomington, IN: Indiana University Press, pp. 33–44.

———. 1989b. The wo/man scientist: Issues of sex and gender in the pursuit of science. In *Rethinking patterns of knowledge,* vol. 6, *Papers in comparative studies,* ed. R. Bjornson and M. Waldman. Columbus: The Ohio State University. 109–118.

———. 1992a. *Secrets of life, secrets of death: Essays on language, gender and science.* New York: Routledge.

———. 1992b. Introduction—gender and science: An update. In *Secrets of life, secrets of death: Essays on language, gender, and science,* ed. London: Routledge, pp. 1–36.

———. 1995. *Refiguring life: Metaphors of twentieth-century biology.* New York: Columbia University Press.

———. 1996. *Drosophila* embryos as transitional objects: The work of Donald Poulson and Christiane Nüsslein-Volhard. *Hist. Studies in the Physical and Biological Sciences* 26 (2): 313–346.

———. 1997. Developmental biology as feminist cause. *Women, Gender, and Science, Osiris* 12: 16–28.

Keller, L., and H. K. Reeve. 1995. Why do females mate with multiple males? The sexually selected sperm hypothesis. *Adv. Study Behav.* 24: 291–315.

Kelly, R. 1993. *Constructing inequality: The fabrication of a hierarchy of virtue among the Etoro.* Ann Arbor: University of Michigan Press.

Kennedy, J. S. 1992. *The new Anthropomorphism.* Cambridge, UK: Cambridge University Press.

Kessler, S. 1996. The medical construction of gender: Case management of intersexed infants. In *Gender and scientific authority,* ed. B. Laslett, S. G. Kohlstedt, H. Longino, and E. Hammonds. Chicago: University of Chicago Press, pp. 340–363.

Kevles, D. 1998. *The Baltimore case: A trial of politics, science and character.* New York: Norton.

Kinzey, W. G. 1982. Distribution of primates and forest refuges. In *Biological diversification in the tropics,* ed. G. T. Prance. New York: Columbia University Press, pp. 455–482.

———, ed. 1987. *The evolution of human behavior: Primate models.* Albany: SUNY Press.

Kinzey, W., and E. P. Cunningham. 1994. Variability in Platyrrhine social organization. *Am. J. Primatol.* 34: 185–198.

Kitahara-Frisch. 1991. Culture and primatology: East and West. In *The monkeys of Arashiyama,* ed. L. M. Fedigan and P. J. Asquith. Albany: State University of New York Press, pp. 74–80.

Kitamura, K. 1977. Persistent spatial proximity among individual Japanese monkeys. *Kikan-jinruigaku* 8: 3–39. (J)

———. 1983. Pygmy chimpanzee association patterns in ranging. *Primates* 24: 1–12.

———. 1990. Interactional synchrony: A fundamental condition for communication. *Senri Ethnol. Stuides* 29: 123–140.

Kleiman, D. G. 1977a. Monogamy in mammals. *Q. Rev. Biol.* 52: 39–69.

———. 1977b. Characteristics of reproduction and sociosexual interactions in pairs of lion tamarins (*Leontopithecus rosalia*) during the reproductive cycle. In *The Biology and Conservation of the Callitrichidae,* ed. D. G. Kleiman. Washington: Smithsonian Institution Press, pp. 181–190.

Knorr-Cetina, K. 1992. The couch, the cathedral, and the laboratory in science. In *Science as practice and culture,* ed. A. Pickering. Chicago: University of Chicago Press, pp. 113–138.

Kohlberg, L. 1981. *The philosophy of moral development.* San Francisco: Harper and Row.

Köhler, W. 1921. *Intelligenzprüfungen an Menschenaffen.* Berlin: Springer.

———. 1922. Zur Psychologie des Schimpansen. *Psychologische Forschung* 1: 2–46.

———. 1927. The mentality of apes. London: Routledge and Kegan Paul.

Kohts, N. 1937. La conduite du petit chimpanzé et de l'enfant de l'homme. *Journal de Psychologie Normale et Pathologiques* 34: 494–531.

Komarovsky, M. 1988. The new feminist scholarship: Some precursors and polemics. *Journal of Marriage and the Family* 50: 585–593.

Koprowski, J. L. 1993. Alternative reproductive tactics in male eastern gray squirrels: "Making the best of a bad job." *Behav. Ecol.* 4: 165–171.

Kortlandt, A. 1940. Wechselwirkung zwischen Instinkten. *Archives Néerlandaises de Zoologie* 4: 442–520.

———. 1954. Cosmologie der dieren. *Vakblad voor Biologen* 34: 1–14.

———. 1959. An attempt at clarifying some controversial notions in animal psychology and ethology. *Archives Néerlandaises de Zoologie* 13: 196–229.

———. 1961. Cimpansees, geen mensen, maar ook geen dieren—maar wat dan wel? *De Syllabus* 25: 167–169; 171–173; 175–178.

———. 1967a. Handgebrauch bei freilebenden Schimpansen. In *Handgebrauch und Verständigung bei Affen und Frühmenschen,* ed. B. Rensch. Bern/Stuttgart: Hans Huber, pp. 59–102.

———. 1967b. Experimentation with chimpanzees in the wild. In *Neue Ergebnisse der Primatologie/Progress in Primatology,* ed. D. Starck, R. Schneider, and H.-J Kuhn. Stuttgart: Gustav Fischer, pp. 208–224.

———. 1972. *New perspectives on ape and human evolution.* Amsterdam: Stichting voor Psychobiologie.

———. 1980. How might early hominids have defended themselves against large predators and food competitors. *Journal of Human Evolution* 9: 79–112.

Koyama, N. 1967. On dominance rank and kinship of a wild Japanese monkey troop in Arashiyama. *Primates* 8: 189–216.

———. 1970. Changes in dominance rank and division of a wild Japanese monkey troop in Arashiyama. *Primates* 11: 335–390.

———. 1991. Lemur societies in Madagascar: With special reference to the ring-tailed lemur. In *Cultural historiography of monkeys,* ed. T. Nishida, K. Izawa, and T. Kano. Tokyo: Heibonsha, pp. 33–56. (J)

Kracauer, S. 1960. *Theory of film: The redemptions of physical reality.* New York: Oxford University Press.

Krebs, J. R., and N. B. Davies. 1981. *An introduction to behavioural ecology.* Sunderland, MA: Sinauer Associates.

Krebs, J. R., and N. B. Davies. 1993. *An introduction to behavioral ecology.* Oxford: Blackwell.

Kroeber, A. 1928. Sub-human culture beginnings. *Quart. Rev. Biol.* 3: 325–342.

Kruuk, H. 1972. *The spotted hyena.* Chicago: The University of Chicago Press.

Kuhl, P. K., and A. N. Meltzoff. 1982. The bimodal perception of speech in infancy. *Science* 218: 1138–1141.

———. 1984. The intermodal representation of speech in infants. *Infant Behavior and Development* 7: 361–381.

Kuhn, T. 1970. *The structure of scientific revolutions.* Chicago: University of Chicago Press.

———. 1977. *The essential tension: Selected studies in scientific tradition and change.* Chicago: University of Chicago Press.

Kummer, H. 1967. Tripartite relations in hamadryas baboons. In *Social communication among primates,* ed. S. A. Altmann, reprinted in *Machiavellian intelligence,* 1991, ed. R. Byrne and A. Whiten. Oxford: Clarendon, pp. 113–121.

———. 1968. Social organization of hamadryas baboons. *Bibliotheca Primatologica* 6: 1–189.

———. 1971a. *Primate societies: Group techniques of ecological adaptations.* Chicago: Aldine.

———. 1971b. Immediate causes of primate social systems. In *Proceedings of the Third International Congress of Primatology, Zurich, 1970.* Basel: S. Karger.

———. 1973. Dominance versus possession. An experiment on hamadryas ba-

boons. In *Proceedings of the Fourth International Congress of Primatology,* vol. 1. Basel: S. Karger, pp. 226–231.
———. 1995. In quest of the sacred baboon: A scientist's journey. Princeton: Princeton University Press.
Kummer, H., and F. Kurt. 1963. Social units of a free-living population of hamadryas baboons. *Folia primatol.* 1: 4–19.
Kummer, H., V. Dasser, and P. Hoyningen Huenen. 1990. Exploring primate social cognition: Some critical remarks. *Behaviour* 112: 84–98.
Kummer, H., W. Goetz, and W. Angst. 1970. Cross-species modifications of social behavior in baboons. In *Old World monkeys. Evolution, systematics, and behaviour,* ed. J. R. Napier and P. R. Napier. London: Academic Press, pp. 351–364.
———. 1972. Anpassung eines Anubis-Weibchens an das Haremsssystem der Mantelpaviane (Freilandexperimente). Commentary text to Film D 1095, Göttingen, Institut für den Wissenschaftlichen Film, *Publ. Wiss. Film, Sekt. Biol.* 6 (1): 1–9.
———. 1974. Triadic differentiation: An inhibitory process protecting pair bonds in baboons. *Behaviour* 49: 62–87.
Kuroda, S. 1982. *The pygmy chimpanzee: The unknown ape.* Tokyo: Chikumashobo. (J)
Lack, D. 1954. *The natural regulation of animal numbers.* Oxford: Clarendon.
———. 1966. *Population studies of birds.* Oxford: Oxford University Press.
Lancaster, J., and R. B. Lee. 1965. The annual reproductive cycle in monkeys and apes. In *Primate behavior,* ed. I. DeVore. New York: Holt, Rinehart, and Winston, pp. 486–513.
Lancaster, R.N, and M. di Leonardo, eds. 1997.*The gender sexuality reader: Culture, history, political economy.* New York: Routledge.
Lande, R., and S. Arnold. 1983. The measurement of selection on correlated characters. *Evolution* 37: 1210–1226.
Lanier, D. L., D. Q. Estep, and D. A. Dewsbury. 1975. Copulatory behaviors of golden hamsters: Effects on pregnancy. *Physiol. Behav.* 15: 209–212.
Lashley, K. S. 1951. The problem of serial order in behavior. In *Cerebral mechanisms in behavior: The Hixon Symposium,* ed. L. A. Jeffress. New York: Wiley, pp. 112–136.
Latour, B. 1987. *Science in action.* Cambridge, MA: Harvard University Press.
———. 1988. *The pasteurization of France,* trans. A. Sheridan and J. Law. Cambridge, MA: Harvard University Press.
———. 1993. *We have never been modern.* Cambridge, MA: Harvard University Press.
———. 1994. Les objets ont-ils une histoire? Rencontre de Pasteur et de Whitehead dans un bain d'acide lactique. In *L'effet Whitehead,* ed. I Stengers. Paris: J. Vrin, pp. 197–217.
———. 1995a. Joliot history and physics mixed together. In *History of scientific thought,* ed. M. Serres. 611–635. London: Blackwell.
———. 1995b. The "Pédofil" of Boa Vista: A photo-philosophical montage. *Common Knowledge* 4 (1): 144–187.
———. 1996a. Do scientific objects have a history? Pasteur and Whitehead in a bath of lactic acid. *Common Knowledge* 5 (1): 76–91.

———. 1996b. *Petite réflexion sur le culte moderne des dieux Faitiches.* Paris: Les Empêcheurs de penser en rond.

———. 1996c. *Aramis, or the love of technology.* Harvard: Harvard University Press.

———. 1997. Socrates' and Callicles' settlement or the invention of the impossible body politic. *Configurations* (spring) 2: 189–240.

———. 1998. From the world of science to the world of research? *Science 280:* 208–209.

———. 1999. *Pandora's hope. Essays on the reality of science studies.* Cambridge, MA: Harvard University Press.

Latour, B., and S. C. Strum. 1986. Human social origins: Please tell us another story. *Journal of Social and Biological Structures* 9: 169–187.

Latour, B., and S. Woolgar. 1986. *Laboratory life: The construction of scientific facts.* Princeton: Princeton University Press.

Law, J., ed. 1986. *Power, action, and belief: A new sociology of knowledge?* London: Routledge and Kegan Paul.

———, ed. 1991. *A sociology of monsters: Essays on power, technology, and domination.* London and New York: Routledge.

Lawes, M. J., and S. P. Henzi. 1995. Inter-group encounters in blue monkeys: How territorial must a territorial species be? *Anim. Behav.* 49: 240–243.

Lawick-Goodall, Baroness J. van. 1967. *My friends the wild chimpanzees.* Washington, DC: National Geographic Society.

———. 1971. *In the shadow of man.* Boston: Houghton Mifflin Co.

Layne, L., ed. 1998. Special issue: Anthropological approaches in science and technology studies. *Science, Technology, and Human Values* 23.

Le Gros Clark, W. E. 1947. Anatomy of the *Australopithecinae. J. Anat.* 81: 300–333.

———. 1948. Observations on the anatomy of the fossil *Australopithecinae. Ybk. Phys. Anthropol.* 3: 143–177.

Leacock, E. 1972. Introduction. In *The origin of the family, private property and the state,* ed. Friedrich Engels. New York: International Publishers, pp. 7–67.

———. 1978. Women's status in egalitarian societies: Implications for social evolution. *Curr. Anthrop.* 19: 247–275.

———. 1986. Women, Power and Authority. In *Visibility and power: Essays on women in society and development,* ed. L. Duby, E. Leacock, and S. Ardener. Dehli: Oxford University Press, pp. 107–135.

Leahey, T. H. 1992. *A history of psychology: Main currents in psychological thought,* 3d ed. Englewood Cliffs, NJ: Prentice-Hall.

LeBihan, J. 1992. Gorilla girls and chimpanzee mothers: Sexual and cultural identity in the primatologist's field. *Journal of Commonwealth Literature* 27 (1): 139–148.

Lee, R. B., and I. DeVore, eds. 1968. *Man the hunter.* Chicago: Aldine.

Lehrman, D. S. 1953. A critique of Konrad Lorenz's theory of instinctive behavior. *Quarterly Review of Biology* 28: 337–363.

———. 1974. Can psychiatrists use ethology? In *Ethology and psychiatry,* ed. N. F. White. Toronto: University of Toronto Press, pp. 187–196.

Leis, N. B. 1974. Women in groups: Ijaw women's associations. In *Woman, culture, and society,* ed. M. Z. Rosaldo and L. Lamphere. Stanford: Stanford University Press, pp. 223–242.

Lemos de Sá, R. M. 1991. População de *Brachyteles arachnoides* (Primates, Cebidae) da Fazenda Esmerald, Rio Casca, Minas Gerais. In *A primatologia no Brasil-3,* ed. A. B. Rylands and A. T. Bernardes. Belo Horizonte, Brasil: Fundação Biodiversitas, pp. 235–238.

Lemos de Sá, R. M., and K. E. Glander. 1993. Capture techniques and morphometrics for the woolly spider monkey, or muriqui (*Brachyteles arachnoides,* E. Geoffroy 1806). *Am. J. Primatol.* 29: 145–153.

Leone, M. P. 1982. Some opinions about recovering mind. *American Antiquity* 47: 742–760.

Leone, M. P., P. B. Potter, and P. A. Shakel. 1987. Toward a critical archaeology. *Current Anthropology* 28.3: 283–302.

Levins, R. 1966. The strategy of model building in population biology. *American Scientist* 54.4: 421–431.

Lévy-Bruhl, L. 1985. *How natives think.* Princeton: Princeton University Press.

Lewenstein, B., ed. 1992. *When science meets the public.* Washington: AAAS.

Licht, P., L. G. Frank, S. Pavgi, T. Yalcinkaya, P. K. Siiteri, and S. E. Glickman. 1992. Hormonal correlates of "masculinization" in female spotted hyaenas (*Crocuta crocuta*). II. Maternal and fetal steroids. *Journal of Reproduction and Fertility* 95: 463–474.

Licht, P., T. Hayes, Pei-San Tsai, G. R. Cunha, S. Hayward, M. Martin, R. Jaffe, M. Golbus, H. S. Kim, and S. E. Glickman. 1998. Androgens and masculinization of the genitalia in the spotted hyaena (*Crocuta crocuta*). 1. Fetal urogenital anatomy and placental metabolism. *Journal of Reproduction and Fertility* 113: 106–117.

Limoges, C. 1994. Milne-Edwards, Darwin, Durkheim and the division of labor. In *The natural and the social sciences,* ed. I. B. Cohen. Dordrecht: Kluwer, pp. 317–343.

Linden, E. 1974. *Apes, men and language.* Harmondsworth: Penguin Books.

Livingstone, F. B. 1973. Did the australopithecines sing? *Curr. Anthropol.* 14: 25–26.

Lloyd, E. 1996. Objectivity and the double standard for feminist epistemologies. *Synthese* 104: 351–381.

Long, J. A., and H. M. Evans. 1922. The oestrous cycle in the rat and its associated phenomena. *Mem. University of California* 6: 1–137.

Longino, H. E. 1989. *Science as social knowledge: Values and objectivity in scientific inquiry.* Princeton: Princeton University Press.

———. 1994. In search of feminist epistemology. *The Monist* 77.4: 472–485.

Lorenz, K. 1937. Über den Begriff der Instinkthandlung. *Folia Biotheoretica* 2: 17–50.

———. 1941. Vergleichende Bewegungsstudien an Anatiden. *Zeitschrift für Ornithologie* 89: 194–294.

———. 1942. Induktive und teleologische Psychologie. *Die Naturwissenschaften* 30: 133–143.

———. 1950. The comparative method in studying innate behavior patterns. *Symp. Soc. Exp. Biol.* 4: 221–268.

———. 1951. Ausdrucksbewegungen höherer Tiere. *Die Naturwissenschaften* 38: 113–116.

———. 1953. Die Entwicklung der vergleichenden Verhaltensforschung in den letzten 12 Jahren. *Zoologischer Anzeiger, Suppl.* 16: 36–58.

———. 1959. Gestaltwahrnehmung als Quelle wissenschaftlicher Erkenntnisse. Reprinted in *Über tierisches und menschliches Verhalten,* vol. 2, ed. K. Lorenz, 1965. München: Piper, pp. 255–300.

———. 1963. Haben Tiere ein subjektives Erleben? Reprinted in *Über tierisches und menschliches Verhalten,* vol. 2, ed. K. Lorenz, 1965 München: Piper, pp. 359–374.

———. 1991. *On life and living.* Translated by Richard D. Bosley. New York: St. Martin's Press.

Lorenz, K., and N. Tinbergen. 1938. Taxis und Instinkthandlung in der Eirollbewegung der Graugans. *Zeitschrift für Tierpsychologie* 2: 1–29.

Lovejoy, C. O. 1981. The origin of man. *Science* 211: 341–348.

Lutz, A., and J. L. Collins. 1993. *Reading National Geographic.* Chicago: University of Chicago Press.

Lutz, C. 1995. The gender of theory. In *Women writing culture,* ed. R. Behar and D. A. Gordon. Berkeley: University of California Press, pp. 249–266.

Lynch, M. 1985a. Discipline and the material form of images: An analysis of scientific visibility. *Social Studies of Science* 15: 37–66.

———. 1985b. *Art and artefact in laboratory science: A study of shop work and shop talk in a research laboratory.* London: Routledge and Kegan Paul.

Lynch, M., and S. Woolgar, eds. 1990. *Representation in scientific practice.* Cambridge, MA: MIT Press.

MacCormack, C., and M. Strathern, eds. 1980. *Nature, culture, and gender.* Cambridge, UK: Cambridge University Press.

Macilwain, C. 1997a. "Science wars" blamed for loss of post. *Nature* 387: 325.

———. 1997b. Campuses ring to a stormy clash over truth and reason. *Nature* 387: 331–333.

MacKenzie, D. 1990. *Inventing accuracy: A historical sociology of nuclear missile guidance.* Cambridge, MA: MIT Press.

Mannheim, K. 1936. *Ideology and Utopia: An introduction to the sociology of knowledge,* trans. L. Wirth and E. Shils. London: Kegan Paul, Trench, Trubner, and Co.

Manson, J. H. 1992. Measuring female mate choice in Cayo Santiago rhesus macaques. *Anim. Behav.* 44: 405–416.

———. 1997. Primate consortships: A critical review. *Current Anthropology* 38: 366.

Marais, E. 1968. *The soul of the ape.* New York: Atheneum.

Marcus, G. 1995. Ethnography in/of the world system: The emergence of multi-sited ethnography. *Annual review of anthropology.* 24: 95–117.

Martin, E. 1992. The end of the body? *American Ethnologist* 19 (1): 121–140.

———. 1994. *Flexible bodies.* Boston: Beacon Press.

———. 1996. The egg and the sperm: How science has constructed a romance based on stereotypical male-female roles. In *Gender and scientific authority,* ed. B. Laslett, S. G. Kohlstedt, H. Longino, and E. Hammonds. Chicago: University of Chicago Press, pp. 323–339.

Martin, P., and P. Bateson. 1986. *Measuring behavior: An introductory guide.* Cambridge, UK: Cambridge University Press.

Martinez-Alier, V. 1974. Marriage, class, and colour in nineteenth-century Cuba: A study of racial attitudes and sexual values in a slave society. Cambridge, UK: Cambridge University Press.

Maruhashi, T., and H. Takasaki. 1996. Socio-ecological dynamics of Japanese macaque troop ranging. In *Evolution and ecology of macaque societies,* ed. J. E. Fa and D. G. Lindburg. Cambridge, UK: Cambridge University Press, pp. 146–159.

Maruhashi, T., J. Yamagiwa, and T. Furuichi. 1986. *Wild Japanese monkeys of Yakushima.* Tokyo: Tokai University Press. (J)

Masataka, N. 1983. Psycholinguistic analysis of alarm calls of Japanese monkeys (*Macaca fuscata fuscata*). *Am. J. of Primatol.* 5 (2): 111–126.

———. 1988. The response of red-chested moustached tamarins to long calls from their natal and alien populations. *Anim. Behav.* 36 (1): 55–61.

Mascia-Lees, F. E., P. Sharpe, and C. B. Cohen. 1989. The postmodernist turn in anthropology: Cautions from a feminist perspective. *Signs* 15: 7–33.

———. 1991. Reply to Kirby. *Signs* 16: 401–408.

Mason, W. A. 1990. Premises, promises, and problems of primatology. *Am. J. Primatol.* 22: 123–138.

Mason, W. A., and S. P. Mendoza, eds. 1993. *Primate social conflict.* Albany, NY: SUNY Press.

Masood, E. 1997. Gunfire echoes in debates on public understanding. *Nature* 387: 335.

Masui, K. 1976. Records on the troop size and composition of the Japanese monkey (*Macaca fuscata*): An examination of the speculations concerning the problems of troop size and population. *Physiol. Ecol., Jpn.* 17: 185–194. (J)

Masui, K., Y. Sugiyama, A. Nishimura, H. Ohsawa. 1975. The life table of Japanese monkeys at Takasakiyama: A preliminary report. In *Contemporary primatology,* ed. S. Kondo, M. Kawai, and Y. Ehara. Basel: Karger, pp. 401–406.

Matsuzawa, T. 1985. The use of numbers by a chimpanzee. *Nature* 315: 57.

———. 1994. Field experiments on use of stone tools by chimpanzees in the wild. In *Chimpanzee cultures,* ed. R. W. Wrangham, W. C. McGrew, F. B. M. de Waal, and P. C. Heltne. Cambridge MA: Harvard University Press, pp. 351–370.

———. 1995. *Chimpanzees are human: Ai and her fellows in Africa.* Tokyo: Iwanamishoten. (J)

———. 1996. Chimpanzee intelligence in nature and in captivity. In *Great ape societies,* ed. W. C. McGrew, T. Nishida, and L. F. Marchant. New York: Cambridge University Press, pp.196–212.

Matsuzawa, T., T. Hasegawa, S. Gotoh, and K. Wada. 1983. One-trial long-lasting food aversion learning in wild Japanese monkeys (*Macaca fuscata*). *Behavioral and Neural Biology* 39: 155–159.

Maynard-Smith, J. 1978. Evolution and the theory of games. In *Readings in sociobiol-*

ogy, ed. T. H. Clutton-Brock and P. H. Harvey. San Fransisco: W. H. Freeman and Co., pp. 258–270.

Mayo, D. G., and R. D. Hollander, eds. 1991. *Acceptable evidence: Science and values in risk management.* Oxford: Oxford University Press.

Mayr, E. 1942. *Systematics and the origin of species.* New York: Columbia University Press.

McClintock, M. A. 1984. Group mating in the domestic rat as a context for sexual selection: Consequences for the analysis of sexual behavior and neuroendocrine responses. *Adv. Study Behav.* 14: 1–50.

———. 1987. A functional approach to behavioral endocrinology in rodents. In *Psychobiology of reproductive behavior: An evolutionary perspective,* ed. D. Crews. Englewood Cliffs, NJ: Prentice-Hall. pp. 176–203.

McClintock, M. K., and N. T. Adler. 1978. The role of the female during copulation in wild and domestic Norway rats (*Rattus norvegicus*). *Behaviour* 67: 67–96.

McClintock, M. K., J. J. Anisko, and N. T. Adler. 1982. Group mating among Norway rats. II. The social dynamics of copulation: Competition, cooperation, and mate choice. *Anim. Behav.* 30: 410–425.

McDougall, W. 1923. *An outline of psychology.* London: Methuen.

McEvoy, J. P. 1955. McEvoy in Disneyland. *Reader's Digest* 66: 19–26.

McGinnis, P. R. 1979. Sexual behavior in free-ranging chimpanzees: Consort relationships. In *The great apes,* ed. D. A. Hamburg and E. R. McCown. Menlo Park, CA: Benjamin Cummings, pp. 429–439.

McGraw W. S., and R. Noë. 1995. The Tai forest monkey project. *African Primates* 1: 17–19.

McGrew, W. C. 1972. *An ethological study of children's behavior.* New York: Academic Press.

McKay, F. E. 1971. Behavioral aspects of population dynamics in unisexual-bisexual *Poeciliopsis* (Pisces: Poeciliidae). *Ecology* 52: 778–790.

McShane, J. 1980. *Learning to talk.* Cambridge, UK: Cambridge University Press.

Mead, M. 1949. *Male and female.* New York: William Morris.

Mealey, L., R. K. Young, and L. L. Betzig. 1985. Commentary on: Despotism and differential reproduction by L. L. Betzig. *Ethology and Sociobiology* 6: 75–6.

Melnick, D. J., and K. K. Kidd. 1983. The genetic consequences of social groups fission in a wild population of rhesus monkeys (*Macaca mulatta*). *Behavioural Ecology and Sociobiology* 12: 229–236.

Melnick, D. J., K. K. Kidd, and M. C. Pearl. 1987. Cercopithecus in multimale groups: Genetic diversity and population structure. In *Primate societies,* ed. B. B. Smuts, D. L. Cheney, R. M. Seyfarth, R. W. Wrangham, and T. T. Struhsaker. Chicago: University of Chicago Press, pp. 121–134.

Melosh, B. 1993. Introduction. In *Gender and American history since 1890,* ed. B. Melosh. London: Routledge, pp. 1–28.

Menezes, E. D. B. 1992. Professores estrangeiros no Brasil: Uma perspectiva histórica. *Ciência Hoje* 14: 39–46.

Merton, R. K. 1942. Science and technology in a democratic order. *Journal of Legal and Political Sociology* 1: 15–26.

Merton, R. K. 1970. *Science, technology, and society in seventeenth century England.* New York: Harper and Row.

———. 1973a. Science and the social order. In *The sociology of science: Theoretical and empirical investigations,* ed. N. W. Storer. Chicago: University of Chicago Press, pp. 254–266.

———. 1973b. The normative structure of science. In *The sociology of science: Theoretical and empirical investigations,* ed. N. W. Storer. Chicago: University of Chicago Press, pp. 267–278.

Miczek, K. A., W. Tornatzky, and J. Vivian. 1991. Ethology and neuropharmacology: Rodent ultrasounds. In *Advances in pharmacological sciences.* Basel: Birkhauser Verlag.

Miles, H. L. 1986. Cognitive development in a signing orangutan. *Primate Report* 14: 179–180.

Millar, R. 1972. *The Piltdown men.* New York: Ballantine.

Miller, B. D. 1993. The anthropology of sex and gender hierarchies. In *Sex and gender hierarchies,* ed. B. D. Miller. Cambridge, UK: Cambridge University Press, pp. 3–31.

Miller, P. 1995. Crusading for chimps and humans . . . Jane Goodall. *National Geographic Magazine* (December) 188: 102–129.

Millet, K. 1969. *Sexual politics.* London: Rupert Hart-Davis.

Milton, K. 1981. Distribution patterns of tropical plant foods as an evolutionary stimulus to primate mental development. *American Anthropologist* 83: 534–548.

———. 1988. Foraging behaviour and the evolution of primate intelligence. In *Machiavellian intelligence,* ed. R. Byrne and A. Whiten. Oxford: Clarendon, pp. 285–305.

Mineka, S.1987. A primate model of phobic fears. In *Theoretical foundations of behavior therapy,* ed. H. Eysenck and I. Martin. New York: Plenum, pp. 81–111.

Mitani, M. 1996. *The Ndoki Forest: The last virgin forest in Africa.* Tokyo: Dobutsusha. (J)

Mitchell, C. J., C. M. Heyes, M. R. Gardner, and G. R. Dawson. 1999. Limitations of a bidirectional control procedure for the investigation of imitation in rats: Odour cues on the manipulandum. *Quarterly Journal of Experimental Psychology* 52B: 193–202.

Mitchell, R. W., and N. S. Thompson. 1986. *Deception: Perspectives on human and nonhuman deceit.* Albany, NY: SUNY Press.

Mitman, G. 1993. Cinematic nature: Hollywood technology, popular culture, and the American Museum of Natural History. *Isis* 84: 637–661.

———. 1996. When nature is the zoo: Vision and power in the art and science of natural history. *Osiris* 11: 117–143.

———. 1999. *Reel nature: America's romance with wildlife on film.* Cambridge, MA: Harvard University Press.

Mittermeier, R. A., A. B. Rylands, and A. F. Coimbra-Filho. 1988. Systematics: Species and subspecies—an update. In *Ecology and Behavior of Neotropical Primates,* vol. 2, ed. R. A. Mittermeier, A. B. Rylands, A. F. Coimbra-Filho, and G. A. B. Fonseca. Washington: World Wildlife Fund, pp. 13–75.

Miyadi, D. 1964. Social life of Japanese monkeys. *Science 143* (Feb.21): 783–786.

Modleski, T. 1993 (1991). Cinema and the dark continent: Race and gender in popular film. In *American feminist thought at century's end: A reader,* ed. L. S. Kaufman. Cambridge, MA: Blackwell. 73–92.

Moghaddam, F. M. 1998. *Social psychology: Exploring universals across cultures.* New York: W. H. Freeman and Co.

Moller, A. P. 1994. *Sexual selection and the barn swallow.* Oxford: Oxford University Press.

Money, J., and A. A. Ehrhardt. 1972. *Man and woman, boy and girl.* Baltimore, MD: Johns Hopkins University Press.

Montgomerie, R., and R. Thornhill. 1989. Fertility advertisement in birds: A means of inciting male-male competition? *Ethology* 81: 209–220.

Montgomery, S. 1991. *Walking with the great apes: Jane Goodall, Dian Fossey, Biruté Galdikas.* Boston: Houghton Mifflin.

Moore, A. 1990. The evolution of sexual dimorphism by sexual selection: The separate effects of intrasexual selection and intersexual selection. *Evolution* 44: 315–331.

Moore, H. L. 1994. *A passion for difference: Essays in anthropology and gender.* Bloomington: Indiana University Press.

Moore, J. 1982. Coalitions in langur all-male bands. *Int. J. Primatol.* 3: 314.

———. 1984. Female transfer in primates. *Int. J. Primatol.* 5: 537–589.

———. 1992. Dispersal, nepotism, and primate social behavior. *Int. J. Primatol.* 13: 361–378.

Morell, V. 1993. Called "trimates," three bold women shaped their field. *Science* 260 (16 April 1993): 420–425.

Morgan, C. L. 1894. *Introduction to animal psychology.* London: Scott.

Mori, A. 1992. *Forest storytellers in Cameroon.* Tokyo: Heibonsha. (J)

Mori, U. 1974. The inter-individual relationships observed in social play of the young Japanese monkeys of the natural troop in Koshima islet. *J. Anthrop. Soc. Nippon* 82: 303–318. (J)

Mori, U., and H. Kudo. 1986. *Social development and relationships of Japanese macaque females.* Tokyo: Tokai University Press. (J)

Morin, P. A., J. J. Moore, R. Chakraborty, L. Jin, J. Goodall, and D. S. Woodruff. 1994. Kin selection, social structure, gene flow, and the evolution of chimpanzees. *Science* 265: 1193–1201.

Morris, D. 1967a. *The naked ape.* London: Cape.

———, ed. 1967b. *Primate ethology.* London: Weidenfeld and Nicolson.

———. 1970. *Patterns of Reproductive Behaviour.* Collected Papers. London: Jonathan Cape.

Morris, R. 1995. All made up: Performance theory and the new anthropology of sex and gender. *Annual review of anthropology* 24: 567–592.

Morse, M. 1995. *Women changing science: Voices from a field in transition.* New York: Plenum Press.

Moser, S. 1996. Science, stratigraphy and the deep sequence: Excavation vs. regional survey and the question of gendered practice in archaeology. *Antiquity* 70: 813–823.

———. 1998. The cultural dimensions of archaeological practice: The role of fieldwork and its gendered associations. Paper presented at the School of American Research Seminar, "Doing Archaeology as a Feminist," Santa Fe, NM.

Muir, J. 1954. Camera in the wilderness. *True: The Man's Magazine* (May) 43: 104–109.

Mukerji, C. 1989. *A fragile power: Scientists and the state.* Princeton: Princeton University Press.

Mukhopadhyay, C. C., and P. Higgins. 1988. Anthropological studies of women's status revisited, 1977–1987. *Ann. Rev. Anthropol.* 17: 461–495.

Mulkay, M. 1976. Norms and ideology in science. *Social Science Information* 15: 637–656.

———. 1979. *Science and sociology of knowledge.* London: Allen and Unwin.

Muller, H. J. 1929. The gene as the basis of life. Paper presented at symposium entitled The Genetic International Congress of Plant Sciences, Section of Genetics, August 19, 1926, Ithaca, NY. Published in *Proceedings of the International Congress of Plant Science* 1 (1929): 897–921.

Muller, V. 1977. The formation of the state and the oppression of women: A case study in England and Wales. *Rev. Radical Pol. Economics* 93: 7–21.

———. 1985. Origins of class and gender stratification in northwest Europe. *Dialectical Anthrop.* 10: 1–2, 93–106.

Murdock, G. P. 1967. *Ethnographic atlas.* Pittsburgh: University of Pittsburgh Press.

Murie, A. 1944. *The wolves of Mt. McKinley.* Washington, DC: Government Printing Office.

Murie, M. E. 1962. *Two in the far north.* New York: Alfred A. Knopf.

Murie, M., and O. Murie. 1966. *Wapiti wilderness.* New York: Alfred A. Knopf.

Murie, O. 1953. Wild country as a national asset. *The Living Wilderness* 45: 1–29.

———. 1973. *Journeys to the far north.* Palo Alto: Wilderness Society and American West Publishing Co.

Murphy, R. C. 1952. Water birds. *Natural History* (September): 330.

Murray, S. L., and J. G. Holmes. 1996. The construction of relationship realities. In *Knowledge structures and interaction in close relationships,* ed. G. J. O. Fletcher and J. Fitness. Hillsdale, NJ: Lawrence Erlbaum, pp. 91–120.

Nakagawa, N. 1994. *The food table of monkeys: An introduction to feeding ecology.* Tokyo: Heibonsha. (J)

Nakamichi, M., H. Fujii, and T. Koyama. 1983. Development of a congenitally malformed Japanese monkey in a free-ranging group during the first four years of life. *Am. J. Primatol.* 5 (3): 205–210.

Napier, J. R. 1960. Studies of the hands of living primates. *Proc. Zool. Soc. Lond.* 134: 647–657.

———. 1961. Prehensility and opposability in the hands of primates. *Symp. Zool. Soc. Lond.* 5: 115–132.

Napier, J. R., and N. A. Barnicot, eds. 1963. *The primates: Zoological Society of London,* no. 10. London: Zoological Society of London.

Narayan, U. 1988. Working together across difference: Some considerations on emotions and political practice. *Hypatia* 3.2: 31–47.

Nash, R. 1977. The exporting and importing of nature: Nature-appreciation as a commodity, 1850–1980. *Perspectives in American History* XII: 517–560.

———. 1982. *Wilderness and the American mind,* 3d ed. New Haven: Yale University Press.

Needham, J. (1990) 1953. Thoughts on the social relations of science and technology in China. In *A selection from the writings of Joseph Needham,* ed. M. Davis. Jefferson, NC: MacFarland, pp. 360–365.

Neely, B. 1993. *Blanche on the lam.* New York: Viking Penguin.

Neisser, U. 1976. *Cognition and reality.* San Francisco: W. H. Freeman.

Nelson, M. C., S. M. Nelson, and A. Wylie, eds. 1994. *Equity issues for women in archeology.* Archeological Papers of the American Anthropological Association, Number 5. Washington DC: American Anthropological Association.

Nelson, S. M. 1997. *Gender in archaeology: Analyzing power and prestige.* Walnut Creek, CA: AltaMira Press.

Neville, M. K., K. E. Glander, F. Braza, and A. B. Rylands. 1987. The howling monkeys, genus *Alouatta.* In *Ecology and behaviour of neotropical primates,* ed. B. A. B. de Fonesca, A. B. Rylands, R. A. Mittermeier, and A. F. Coimbra-Filho. Rio de Janeiro: Academie Brasil Ciencias, pp. 349–453.

Newell, A., J. C. Shaw, and H. A. Simon. 1958. Elements of a theory of human problem solving. *Psychological Review* 65: 151–166.

Newton, P. N. 1988. The variable social organization of Hanuman langurs (*Presbytis entellus*): Infanticide and the monopolization of females. *Int. J. Primatol.* 9: 59–77.

Nijhout, H. F. 1990. Metaphors and the role of genes in development. *BioEssays* 129: 441–446.

Nishida, T. 1968. The social group of wild chimpanzees in the Mahale Mountains. *Primates* 9: 167–224.

———. 1973a. The ant-gathering behaviour by the use of tools among wild chimpanzees of the Mahale Mountains. *J. Human Evol. 2:* 357–370.

———. 1973b. *The children of the mountain spirits.* Tokyo: Chikumashobo. (J)

———. 1976. The bark eating habits in primates, with special reference to their status in the diet of wild chimpanzees. *Folia Primatol.* 25 (4): 277–287.

———. 1979. The social structure of chimpanzees of the Mahale Mountains. In *The great apes,* ed. D. A. Hamburg and E. R. McCown. Menlo Park, CA: Benjamin Cummings, pp. 73–122.

———. 1981. *The world of wild chimpanzees.* Tokyo: Chuokoronsha. (J)

———. 1987. Local traditions and cultural transmission. In *Primate societies,* ed. B. B. Smuts, D. L. Cheney, R. M. Seyfarth, R. W. Wrangham, and T. T. Struhsaker. Chicago: University of Chicago Press, pp. 462–474.

———. 1990a. A quarter century of research in the Mahale Mountains: An overview. In *The chimpanzees of the Mahale Mountains: Sexual and life history strategies,* ed. T. Nishida. Tokyo: University of Tokyo Press: 3–35.

———, ed. 1990b. *The chimpanzees of the Mahale Mountains: Sexual and life history strategies.* Tokyo: University of Tokyo Press.

———. 1994. *Thirty-four stories of wild chimpanzees.* Tokyo: Kinokuniya-shoten. (J)

Nishida, T., and K. Kawanaka. 1972. Inter-unit-group relationships among wild chimpanzees of the Mahale Mountains. *Kyoto Univ. Afr. Studies* 7: 131–169.

Nishida, T., K. Izawa, and T. Kano, eds. 1991. *Cultural historiography of monkeys.* Tokyo: Heibonsha. (J)

Nishimura, A. 1991. A patrilineal nonhuman society in South America: From the social interactions of woolly monkeys. In *Cultural historiography of monkeys,* ed. T. Nishida, K. Izawa, and T. Kano. Tokyo: Heibonsha, pp. 85–107. (J)

Nissen, H. W. 1931. A field study of the chimpanzee. *Comp. Psych. Mono.* 8: 1–122.

Noble, B. E. 1993. Learning and unlearning separation: From Jane Goodall to epistemic possibility. Presented at the symposium Theorizing Women's Intercultural Narratives, University of Alberta, Edmonton, Canada.

———. 1996. Leaky visions of gender, nature, and apes: The persistence of Fossey's mist. Contributed to 1996 Wenner-Gren symposium, Changing Images of Primate Societies: The Role of Theory, Method, and Gender. Teresopolis, Brazil.

———. In press. *Dinosaurs, modernity, and the lost world: The public politics of monstrous fascination.* Ann Arbor: University of Michigan Press.

Noë, R. 1986 Lasting alliances among adult male savannah-baboons. In *Primate ontogeny, cognition and social behaviour,* ed. J. G. Else and P. C. Lee, Cambridge, UK: Cambridge University Press, pp. 381–392.

———. 1990. A veto game played by baboons: A challenge to the use of the Prisoners' Dilemma as a paradigm for reciprocity and coopeartion. *Anim. Behav.* 39: 78–90.

———. 1992. Alliance formation among male baboons: Shopping for profitable partners, In *Coalitions and alliances in humans and other animals,* ed. A. H. Harcourt and F. B. M. de Waal. Oxford Scientific Publ., pp. 285–323.

Noë, R., and R. Bshary. 1997. The formation of red colobus—diana monkey associations under predation pressure from chimpanzees. *Proc. R. Soc.* B. 264: 253–259.

Noë, R., and P. Hammerstein. 1994. Biological markets: Supply and demand determine the effect of partner choice in cooperation, mutualism and mating. *Behav. Ecol.Sociobiol,* in press.

Noë, R., and A. A. Sluijter. 1995. Which adult male savanna baboons form coalitions? *Int. J. Primatol.* 16 (1): 77–105.

Noë, R., C. P. van Schaik, and J.A.R.A.M. van Hooff. 1991. The market effect, an explanation for pay-off asymmetries among collaborating animals. *Ethology* 87: 97–118.

Noller, P. 1987. Non-verbal communication in marriage. In *Intimate relationships,* ed. D. Perlman and S. Duck. Beverley Hills: Sage, pp. 149–175.

Noordwijk, M. A. van, and C. P. van Schaik. 1987. Competition among female long-tailed macaques, *Macaca fascicularis. Anim. Behav.* 35: 577–589.

———. 1988. Male careers in Sumatran long-tailed macaques. *Behaviour* 107: 24–43.

Norikoshi, K., and N. Koyama. 1975. Group shifting and social organization among Japanese monkeys. In *Proc. Symp. 5th Congr. Int. Primat. Soc.,* ed. S. Kondo, M. Kawai, and Y. Ehara. Tokyo: Japan Science Press, pp. 43–61.

Northrop, F. S. C. 1965. *The logic of the sciences and humanities.* Cleveland, OH: Meridian.

Norton, G., R. J. Rhine, G. W. Wynn, and R. D. Wynn. 1987. Baboon diet: A five year study of stability and variability in the plant feeding and habitat of yellow baboons (*Papio cynocephalus*) of Mikumi National Park, Tanzania. *Folia Primatol.* 48: 78–120.

Norwood, V. 1993. *Made From this earth: American women and nature.* Chapel Hill: University of North Carolina Press.

Noske, B. 1997 (1989). *Beyond boundaries: Humans and animals,* rev. ed. Montreal and New York: Black Rose Books.

Nowak, M., and K. Sigmund. 1993. A strategy of win-stay, lose-shift that outperforms tit-for-tat in the prisoner's dilemma. *Journal of Theoretical Biology* 168: 219–226.

Oakley, A. 1972. *Sex, gender and society.* New York: Harper and Row.

———. 1997. A brief history of gender. In *Who's afraid of feminism? Seeing through the backlash,*ed. A. Oakley and J. Mitchell. New York: The New Press, pp. 29–55.

Oakey A., and J. Mitchell. 1997. *Who's afraid of feminism? Seeing through the backlash.* New York: The New Press.

Oakley, K. P. 1961. *Man the tool-maker.* Chicago: University of Chicago Press.

Oates, J. F., T. Swain, and J. Zantovska. 1977. Secondary compounds and food selection by colobus monkeys. *Biochemical Systematics and Ecology* 5: 317–321.

O'Connell, S. M., and G. Cowlishaw. 1994. Infanticide avoidance, sperm competition and mate choice: The function of copulation calls in female baboons. *Anim. Behav.* 48: 687–694.

Ogburn, M. 1991. The life cycle of *National Geographic Magazine.* From the proceedings of the national convention of the Association for Education in Journalism and Mass Communication, August, 1991.

Oreskes, N. 1997. Objectivity or heroism? On the invisibility of women in science. *Science in the Field, Osiris* 11: 87–113.

Orians, G. H. 1969. On the evolution of mating systems in birds and mammals. *Am. Nat.* 103: 589–603.

Ortner, S. B. 1974. Is female to male as nature is to culture? In *Woman, culture, and society,* ed. M. Z. Rosaldo and L. Lamphere. Stanford: Stanford University Press, pp. 67–88.

Packer, C. 1977. Reciprocal altruism in *Papio anubis. Nature* 265: 441–443.

Packer, C., D. A Collins, A. Sindimwo, and J. Goodall. 1995. Reproductive constraints on aggressive competition in female baboons. *Nature* 373: 60–63.

Palombit, R. 1994. Dynamic pair bonds in hylobatids: Implications regarding monogamous social systems. *Behaviour* 128: 65–101.

Parker, S. T. 1977. Piaget's sensorimotor period in an infant macaque: A model for comparing unstereotyped behavior and intelligence in human and nonhuman primates. In *Primate bio-social development: Biological, social and ecological determinants,* ed. S. Chevalier-Skolnikoff and F. E. Poirier. New York: Garland, pp. 43–112.

Parker, S. T., and K. R. Gibson. 1977. Object manipulation, tool use, and sensorimotor intelligence as feeding adaptations in early hominids. *Journal of Human Evolution* 6: 623–641.

———. 1979. A developmental model for the evolution of language and intelligence in early hominids. *The Behavioral and Brain Sciences* 2: 367–408.

———, eds. 1990. *"Language" and intelligence in monkeys and apes.* Cambridge, UK: Cambridge University Press.

Pasternak, B. 1976. *Introduction to kinship and social organization.* Englewood Cliffs, NJ: Prentice-Hall.

Pasternak, B., C. R. Ember, and M. Ember. 1997. *Sex, gender, and kinship: A cross-cultural perspective.* Upper Saddle River, NJ: Prentice-Hall.

Patterson, F., and E. Linden. 1981. *The education of Koko.* New York: Holt, Rinehart, and Linden.

Patterson, T. C. 1995. *Toward a social history of archaeology.* Philadelphia: Harcourt Brace.

Pearson, G. 1996.Of sex and gender. *Science* 274: 328–9 (letter to the editor).

Pedersen, J., S. E. Glickman, L. G. Frank, and F. A. Beach. 1990. Sex differences in play of immature spotted hyenas. *Hormones and Behavior* 24: 403–420.

Pels, P., and L. Nencel. 1991. Introduction: Critique and the deconstruction of anthropological authority. In *Constructing knowledge: Authority and critique in social sciences,* ed. L. Nencel and P. Pels. London: Sage Publications, pp. 1–21.

Pereira, M. E. 1995. Development and social dominance among group-living primates. *Amer. J. Primatol.* 37: 143–176.

Pereira M. E., and M. L. Weiss. 1991. Female mate choice, male migration, and the threat of infanticide in ringtailed lemurs. *Beh. Bio. Sociobiol* 18: 141–152.

Peterson, D., and J. Goodall. 1993. *Visions of caliban: On chimpanzees and people.* Boston: Houghton Mifflin.

Petter, J. J. 1962. Recherches sur l'écologie et l'éthologie des lémuriens malgaches. *Mém Mus. Natl. Hist. Nat.* (Paris) 27: 1–146.

Petter-Rousseaux, A. 1962. Recherches sur la biologie de la réproduction des primates inférieurs. *Mammalia* A: 3794: 1–87.

Phillips, P., and S. Arnold. 1989. Visualizing multivariate selection. *Evolution* 43: 1209–1222.

Piaget, J. 1951. *Play, dreams, and imitation in childhood.* New York: Norton.

Pickering, A. 1990. Knowledge, practice and mere construction. *Social Studies of Science* 20.4: 682–729.

———, ed. 1992. *Science as practice and culture.* Chicago: University of Chicago Press.

———. 1995. *The mangle of practice.* Chicago: University of Chicago Press.

Pinch, T. 1985. Towards an analysis of scientific observation: The externality and evidential significance of observation reports in physics. *Social Studies of Science* 15: 167–187.

Pinch, T., and W. Bijker. 1984. The social construction of facts and artifacts, or how the sociology of science and the sociology of technology might benefit each other. *Social Studies of Science* 14: 399–441.

Pitnick, S., and T. L. Karr. 1996. Sperm caucus. *Trends Ecol. Evol.* 11: 148–151.

Popp, J. L. 1978. Male baboons and evolutionary principles. Unpublished Ph.D. thesis, Harvard University.

Portielje, A. F. J. 1927. Zur Ethologie bezw. Psychologie von *Phalacrocrax carbo subcormorans* BREHM. *Ardea* 16: 107–123.

———. 1938. *Dieren zien en leren kennen.* Amsterdam: Nederlandse Keurboekerij.

Povinelli, D. J. 1993. Reconstructing the evolution of mind. *American Psychologist* 48: 493–509.

Povinelli, D. J., and T. J. Eddy. 1996. What young chimpanzees know about seeing. *Monographs of the Society for Research in Child Development* 61: (3, ser. 247).

Povinelli, D. J., K. Nelson, and S. Boysen. 1990. Inferences about guessing and knowing by chimpanzees (*Pan troglodytes*). *Journal of Comparative Psychology* 104: 203–210.

Povinelli, D. J., A. Rulf, and D. Bierschwale. 1994. Absence of knowledge attribution and self-recognition in young chimpanzees (*Pan troglodytes*). *Journal of Comparative Psychology* 108: 74–80.

Pratt, M. L. 1992. *Imperial eyes: Travel writing and transculturation.* London and New York: Routledge.

Premack, D.1971. Language in chimpanzee? *Science* 172: 808–822.

———. 1988. "Does the chimpanzee have a theory of mind?" Revisited. In *Machiavellian intelligence,* ed. R. Byrne and A. Whiten. Oxford: Clarendon, pp. 160–179.

Premack, D., and G. Woodruff. 1978. Does the chimpanzee have a theory of minf? *Behavioral and Brain Sciences* 1: 515–526.

Preucel, R. W., ed. 1991. *Processual and postprocessual archaeologies: Multiple ways of knowing the past.* Carbondale IL: Center for Archaeological Investigations, Southern Illinois University.

Preuschoft, S. 1992. "Laughter" and "smile" in Barbary macaques (*Macaca sylvanus*). *Ethology* 91 (3): 220–236.

Preuschoft, S., and J.A.R.A.M. van Hooff. 1997a. Homologizing primate facial displays: A critical review of methods. *Folia Primatol.* 65: 121–137.

———. 1997b. The social function of "smile" and "laughter": Variations across primate species and societies. In *Nonverbal communication: Where nature meets culture,* ed. U. Segerstråle, and P. Molnár, Mahwah, NJ: Lawrence Erlbaum, pp. 171–191.

Preuschoft, S., and H. Preuschoft. 1995. Primate nonverbal communication: Our communicatory heritage. In *Origins of semiosis,* ed. W. Nöth. Berlijn: Mouton de Gruyter, pp. 61–100.

Prins, B. 1995. The ethics of hybrid subjects: Feminist constructivism according to Donna Haraway. *Science, Technology, and Human Values* 20: 352–367.

Provost, M., T. G. Wynn, T. P. Huber, and W. C. McGrew. 1993. Seasonal and sex differences in Gombe chimpanzee ranging. Paper presented at the 63rd Annual Meeting of the American Association of Physical Anthropologists, Denver.

Pusey, A. 1979. Intercommunity transfer of chimpanzees in Gombe National Park. In *The great apes,* ed. D. A. Hamburg and E. R. McCown. Menlo Park, CA: Benjamin Cummings, pp. 465–480.

Pusey, A. E., and C. Packer. 1987. Dispersal and philopatry. In *Primate societies,* ed. B. B. Smuts, D. L. Cheney, R. M. Seyfarth, R. W. Wrangham, and T. T. Struhsaker. Chicago: University of Chicago Press, pp. 250–266.

Pusey, A., J. Williams, and J. Goodall. 1997. The influence of dominance rank on the reproductive success of female chimpanzees. *Science* (August 8): 828–831.

Pyke, G. H., H. R. Pulliam, and E. L. Charnov. 1977. Optimal foraging: A selective review of theories and tests. *Quarterly Review of Biology* 52: 137–154.

Queller, D. C. 1987. The evolution of leks through female choice. *Anim. Behav.* 35: 1424–1432.

Quine, W. V. O. 1953. Two dogmas of empiricism. In *From a logical point of view,* ed. W. V. O. Quine. Cambridge, UK: Cambridge University Press, pp. 20–46.

———. 1969. *Ontological relativity and other essays.* New York: Columbia University Press.

Quinn, N. 1977. Anthropological studies on women's status. *Ann. Rev. Anthropol.* 6: 181–225.

Ransom, T. W. 1979. *The Beach Troop of Gombe.* Lewisburg: Bucknell University Press.

Ransom, T. W., and B. S. Ransom. 1971. Adult male infant relations among baboons, *Papio anubis. Folia Primatol.* 16: 179–195.

Ransom, T. W., and T. Rowell. 1972. Early social development of feral baboons. In *Primate Socialization,* ed. F. E. Poirier. New York: Random House.

Rapp, R. 1995. Heredity, or: Revising the facts of life. In *Naturalizing power: Essays in feminist cultural analysis,* ed. S. Yanagisako and C. Delaney. New York and London: Routledge, pp. 69–86.

Reader, J. 1988. *Missing links.* London: Penguin.

Reichard, U. 1995. Extra-pair copulations in a monogamous gibbon, *Hylobates lar. Ethology* 100: 99–112.

Reiter, R. R., ed. 1975. *Toward an anthropology of women.* New York: Monthly Review Press.

Rendall, D., and A. DiFiore. 1996. The road less traveled: Phylogenetic perspectives in primatology. *Evolutionary Anthropology* 4: 43–52.

Reynolds, V. 1967. *The apes. The gorilla, chimpanzee, orangutan, and gibbon, their history and their world.* New York: Harper and Row.

Reynolds, V., and F. Reynolds. 1965. Chimpanzees in the Budongo Forest. In *Primate Behavior,* ed. I. DeVore. New York: Holt, Rinehart, and Winston.

Rhine, R. J. 1975. The order of movement of yellow baboons (*Papio cynocephalus*). *Folia Primatol.* 23: 72–104.

Rhine, R. J., and B. J. Westlund. 1981. Adult male positioning in baboon progressions: Order and chaos revisited. *Folia Primatol.* 35: 77–116.

Ribnick, R. 1982. A short history of primate field studies: Old World monkeys and apes. In *A history of physical anthropology, 1930–1980,* ed. F. Spencer. New York: Academic Press, pp. 49–73.

Richard, A. F. 1978. *Behavioral variation: Case study of a Malagasy lemur.* Lewisburg: Bucknell University Press.

———. 1981. Changing assumptions in primate ecology. *American Anthropologist* 83: 517–533.

———. 1985. *Primates in nature.* New York: Freeman.

Richard, A. F., P. Rakotomanga, and M Schwartz. 1991. Demography of *Propithecus verreauxi* at Beza Mahafaly, Madagascar: Sex ratio, survival, and fertility 1984–88. *American Journal of Physical Anthropology* 84: 307–322.

Rijksen, H. D. 1978. A field study of Sumatran orang-utans (*Pongo pygmaeus abelii,* LESSON 1827): Ecology, behaviour and conservation. *Meded. Landbouwhogesch. Wageningen* 78: 1–420.

Ristau, C. A. 1991. *Cognitive Ethology.* Hillsdale, NJ: Lawrence Erlbaum.

Ristau, C. A., and D. Robbins. 1982. Language in great apes: A critical review. In *Advances in the study of behavior,* vol. 12, ed. J. S. Rosenblatt, R. A. Hinde, C. Beer, and M. C. Busnel. New York: Plenum Press, pp. 141–255.

Roda, S. A. 1989. Ocorrência de duas fêmeas reprodutivas em grupos selvagens de *Callithrix jacchus* (Primates, Callitrichidae). Paper presented at the meeting of the Brazilian Zoological Society, Universidade Federal da Paraíba, João Pessoa, Brazil.

Rodman, P. 1988. Resources and group sizes in primates. In *The ecology of social behavior,* ed. C. N. Slobodchikoff. New York: Academic Press.

Roëll, D. R. 1996. *De wereld van het instinct: Nico Tinbergen en het ontstaan van de ethologie in Nederland (1920–1950).* Rotterdam: Erasmus Publishing.

Roitblat, H. L. 1987. *Introduction to comparative cognition.* New York: W. H. Freeman.

Roldan, E. R. S., M. Gomendio, and A. D. Vitullo. 1992. The evolution of eutherian spermatozoa and underlying selective forces: Female selection and sperm competition. *Biol. Rev.* 67: 551–593.

Roosmalen, M. G. M. van. 1980. *Habitat preferences, diet, feeding strategy, and social organization of the black spider monkey (Ateles p. paniscus Linnæus 1758) in Surinam.* Ph.D. thesis, Agricultural University Wageningen.

Rorty, R. 1979. *Philosophy and the mirror of nature.* Princeton: Princeton University Press.

Rosaldo, M. Z. 1974. Woman, culture, and society: A theoretical overview. In *Woman, culture, and society,* ed. M. Z. Rosaldo and L. Lamphere. Stanford: Stanford University Press, pp. 17–42.

———. 1980. The use and abuse of anthropology: Reflections on feminism and cross-cultural understanding. *Signs* 5: 389–417.

Rosaldo, M. Z., and L. Lamphere, eds. 1974. *Woman, culture, and society.* Stanford: Stanford University Press.

Rosenberger, A. L. 1979. Phylogeny, evolution, and classification of New World monkeys (Platyrrhini, Primates). Unpublished Ph.D. thesis, City University of New York.

Rosenberger, A. L., and K. B. Strier. 1989. Adaptive radiation of the ateline primates. *Journal of Human Evolution* 18: 717–750.

Rosenqvist, G., and A. Berglund. 1992. Is female behaviour a neglected topic? *Trend Ecol. Evol.* 7: 174–176.

Ross, C. 1991. Life history patterns of New World monkeys. *Int. J. Primatol.* 12: 481–502.

Rosser, S. 1990. *Female-friendly science: Applying women's studies methods and theories to attract students.* New York: Pergamon Press.

———. 1997. Possible implications of feminist theories for the study of evolution, In *Feminism and evolutionary biology,* ed. P. A. Gowaty. New York: Chapman and Hall, pp. 21–41.

Rossiter, M. W. 1995. *Women scientists in America: Before affirmative action 1940–1972.* Baltimore, MD: Johns Hopkins University Press.

———. 1997. Which science, which women? *Women, gender, and science, Osiris* 12: 169–185.

Rothe, H., and A. Koenig. 1991. Variability of social organization in captive common marmosets (*Callithrix jacchus*). *Folia Primatol.* 57: 28–33.

Rothenberg, P. S. 1998. *Race, class and gender in the U.S.* New York: St. Martin's Press.

Rowan, A. 1951. Husband and wife camera team. *American Cinematographer:* 263, 277.

Rowell, T. E. 1966. Forest living baboons in Uganda. *Journal of Zoology* 149: 344–364.

———. 1967. Female reproductive cycles and the behavior of baboons and rhesus macaques. In *Social communication among primates,* ed. S. A. Altmann. Chicago: University of Chicago Press, pp. 15–32.

———. 1972. *The social behavior of monkeys.* Middlesex, England: Penguin Books.

———. 1974. The concept of dominance. *Behavioural Biology* 11: 131–154.

———. 1995. Choosy or promiscuous—it depends on the time scale. In *Current primatology,* vol. 2, *Social development, learning and behaviour,* ed. J. J. Roeder, B. Theirry, J. R. Anderson, and Herrenschmidt. Strasbourg: University Louis Pasteur, pp. 11–18.

———. 1999. The myth of peculiar primates. In *Mammalian social learning: Comparative and ecological perspectives. Symp. Zool. Soc. Lond.* 73. Cambridge University Press.

Rowell, T. E., and R. A. Hinde. 1963. Responses of rhesus monkeys to mildly stressful situations. *Anim. Behav.* 11: 235–243.

Rowell, T. E., and C. A. Rowell. 1993. The social organization of feral *Ovis aries* ram groups in the pre-rut period. *Ethology* 95: 213–232.

Rubin, G. 1975. The traffic in women. In *Toward an anthropology of women,* ed. R. R. Reiter. New York: Monthly Review Press, pp. 157–210.

Rudwick, M. 1985. *The great Devonian controversy: The shaping of scientific knowledge among gentlemanly specialists.* Chicago: University of Chicago Press.

Ruiter, J. R. de, and E. Geffen. 1998. Relatedness of matrilines, dispersing males and social groups in long-tailed macaques (*Macaca fascicularis*). *Proc. R. Soc. Lond.* B 265: 79–87.

Ruiter, J. R. de, J.A.R.A.M. van Hooff, and W. Scheffrahn. 1994. Social and genetic aspects of paternity in wild long-tailed macaques (*Macaca fascicularis*). *Behaviour* 129 (3–4): 203–225.

Ruiter, J. R. de, W. Scheffrahn, G.J.J.M. Trommelen, A. G. Uitterlinden, R. D. Martin, and J.A.R.A.M. van Hooff. 1991. Male social rank and reproductive success in wild long-tailed macaques. In *Paternity in primates: Genetic tests and theories,* ed. R. D. Martin, A. F. Dixson, E. J. Wickings, pp. 175–191.

Rumbaugh, D. 1977. *Language learning by a chimpanzee.* New York: Academic Press.

Ruse, M.. 1986. *Taking Darwin seriously: A naturalistic approach to philosophy.* Oxford: Blackwell.

Russell, P. J. 1996. *Genetics.* New York: Harper Collins.

Russell, R. J. 1993. *The lemur's legacy: The evolution of power, sex, and love.* New York: G. P. Putnam's Sons.

Russet, C. E. 1989. *Sexual science: The Victorian construction of womanhood.* Cambridge, MA: Harvard University Press.

Russon, A. E. 1996. Imitation in everyday use: Matching and rehearsal in the spontaneous imitation of rehabilitant orangutans (*Pongo pygmaeus*). In *Reaching into thought: The minds of the great apes,* ed. A. E. Russon, K. A. Bard, and S. T. Parker. Cambridge, UK: Cambridge University Press, pp. 152–176.

Russon, A. E., and B. M. F. Galdikas. 1993. Imitation in ex-captive orangutans (*Pongo pygmaeus*). *Journal of Comparative Psychology* 107: 147–161.

Rylands, A. B. 1986. Infant carrying in a wild marmoset group, *Callithrix humeralifer:* Evidence for a polyandrous mating system. In *A Primatologia no Brasil,* vol. 2, ed. M. T. de Mello. Brasília: Soc. Bras. de Primatologia, pp. 131–144.

———. 1989. Evolução do sistema de acasalamento em Callitrichidae. In *Etologia: de Animais e de Homens,* ed. C. Ades. São Paulo: Edicon/Edusp, pp. 87–108.

———. 1996. Habitat and the evolution of social and reproductive behavior in Callitrichidae. *Amer. J. Primatol.* 38: 5–18.

Rylands, A. B., R. A. Mittermeier, and E. R. Luna. 1995. A species list for the New World primates (Platyrrhini): Distribution by country, endemism, and conservation status according to the Mace-Land system. *Neotropical Primates* 3 (supplement): 113–160.

Sacks, K. 1974. Engels revisited: Women, the organization of production, and private property. In *Woman, culture, and society,* ed. M. Z. Rosaldo and L. Lamphere. Stanford: Stanford University Press, pp. 207–222. (Reprinted in *Toward an anthropology of women,* ed. R. R. Reiter, pp. 211–234.)

———. 1979. *Sisters and wives: The past and future of sexual equality.* Urbana: University of Illinois Press.

Sade, D. 1967. Determinants of dominance in a group of free-ranging rhesus monkeys. In *Social communication among primates,* ed. S. A. Altmann. Chicago: University of Chicago Press, pp. 99–114.

———. 1972. A longitudinal study of social relations of rhesus monkeys. In *Functional and evolutionary biology of primates,* ed. R. H. Tuttle. Chicago: Aldine-Atherton.

Sakura, O., T. Sawaguchi, H. Kudo, and S. Yoshikubo. 1986. Declining support for Imanishi. *Nature* 323: 586.

Sanday, P. R. 1974. Female status in the public domain. In *Woman, culture, and society,* ed. M. Z. Rosaldo and L. Lamphere. Stanford: Stanford University Press, pp. 189–206.

Sander, K. 1986. The role of genes in ontogenesis. In *History of embryology,* ed. T. J.

Horder, J. A. Witkowski, and C. C. Wylie. Cambridge, UK: Cambridge University Press, pp. 363–395.

Sangren, P. S. 1988. Rhetoric and the authority of ethnography. *Curr. Anthrop.* 29: 405–24.

Sapolsky, R. 1989. Hypercortisolism among socially-subordinate wild baboons originates at the CNS level. *Archives of General Psychiatry* 46: 1047–1051.

———. 1990. Adrenocortical function, social rank, and personality among wild baboons. *Biological Psychiatry* 28: 862–879.

———. 1993. The physiology of dominance in stable versus unstable social hierarchies. In *Primate social conflict,* ed. W. A. Mason and S. P. Mendoza. New York: State University of New York Press, pp. 171–204.

Savage-Rumbaugh, E. S. 1986. *Ape language: From conditional response to symbol.* New York: Columbia University Press.

Savage-Rumbaugh, S., and R. Lewin. 1994. Kanzi: The ape at the brink of the human mind. New York: John Wiley.

Savage-Rumbaugh, E. S., J. Murphy, R. A. Sevcik, K. E. Brakke, S. L. Williams, and D. A. Rumbaugh. 1993. Language comprehension in ape and child. *Monographs of the Society for Research in Child Development* 58: (3–4, ser. 233).

Scarr, S. 1993. Biological and cultural diversity: The legacy of Darwin for development. *Child Development.* 64: 1333–1353.

Schaik, C. P. van. 1983. Why are diurnal primates living in groups? *Behaviour* 87: 120–144.

———. 1989. The ecology of social relationships amongst female primates. In *Comparative socioecology, the behavioural ecology of humans and other mammals,* ed. V. Standen and G. R. A. Foley. Oxford: Blackwell, pp. 195–218.

———. 1996. Social evolution in primates: The role of ecological factors and male behaviour. *Proceedings of the British Academy* 88: 9–31.

Schaik, C. P. van, and R. I. M. Dunbar. 1990. The evolution of monogamy in large primates: A new hypothesis and some crucial tests. *Behaviour* 115: 30–62.

Schaik, C. P. van, and J.A.R.A.M. van Hooff. 1983. On the ultimate causes of primate social systems. *Behaviour* 85: 91–117.

———. 1996. Toward an understanding of the orangutan's social system. In *Great ape societies,* ed. W. C. McGrew, T. Nishida, and L. F. Marchant. New York: Cambridge University Press, pp. 3–16.

Schaik, C. P. van, and M. Hörsterman. 1994. Predation risk and the number of adult males in a primate group: A comparative test. *Behavioral Ecology and Sociobiology* 24: 265–276.

Schaik, C. P. van, and P. M. Kappeler. 1993. Life history, activity period and lemur social systems. In *Lemur social systems and their ecological basis,* ed. P. M. Kappeler and J. U. Ganzhorn. New York: Plenum, pp. 241–260.

———. 1997. Infanticide risk and the evolution of male-female association in primates. *Proceedings of the Royal Society, London* B 264: 1687–1694.

Schaik, C. P. van, and M. A. van Noordwijk. 1986. The hidden costs of sociality: Intra-group variation in feeding strategies in Sumatran long-tailed macaques (*Macaca fascicularis*). *Behaviour* 99: 296–315.

———. 1988. Scramble and contest among female long-tailed macaques in a Sumatran rain forest. *Behaviour* 105: 77–98.

———. 1989. The special role of male *Cebus* monkeys in predation avoidance, and its effect on group composition. *Behavioral Ecology and Sociobiology* 24: 265–276.

Schaik, C. P. van, and J.A.G.M. de Visser. 1990. Fragile sons or harassed daughters? Sex differences in mortality among juvenile primates. *Folia Primatol.* 55: 10–23.

Schaik, C. P. van, W. J. Netto, A. J. J. van Amerongen, and H. Westland. 1989. Social rank and sex ratio of captive long-tailed macaque females (*Macaca fascicularis*). *Amer. J. Primatol.* 19: 147–161.

Schaik, C. P. van, M. A. van Noordwijk, T. van Bragt, and M. A. Blankenstein. 1991. A pilot study of the social correlates of levels of urinary cortisol, prolactin and testosterone in wild long-tailed macaques. *Primates* 32: 345–356.

Schaik, C. P. van, M. A. van Noordwijk, B. Warsono, and E. Sutriono. 1983. Party size and early detection of predators in Sumatran forest primates. *Primates* 24: 211–221.

Schaller, G. B. 1963. *The mountain gorilla: Ecology and behavior.* Chicago: University of Chicago Press.

———. 1964. *The year of the gorilla.* Chicago: University of Chicago Press.

———. 1965a. The behavior of the mountain gorilla. In *Primate behavior,* ed. I. DeVore. New York: Holt, Rinehart, and Winston, pp. 324–367.

———. 1965b. Behavioral comparisons of the apes. In *Primate behavior,* ed. I. DeVore. New York: Holt, Rinehart, and Winston, pp. 474–481.

———. 1972. The behavior of the mountain gorilla. In *Primate patterns,* ed. P. Dolhinow. New York: Holt, Rinehart, and Winston, pp. 85–124.

Schatten, G., and H. Schatten. 1983. The energetic egg. *The Sciences* 23: 28–34.

Scheffrahn W., J. R. de Ruiter, and J.A.R.A.M. van Hooff. 1996. Genetic relatedness within and between populations of *Macaca fascicularis* on Sumatra and off-shore islands. In *Evolution and ecology of Macaque societies,* ed. J. E. Fa, and D. G. Lindburg. Cambridge, UK: Cambridge University Press, pp. 20–42.

Scheffrahn, W., W. W. Socha, J. R. de Ruiter, and J.A.R.A.M. van Hooff. 1987. Blood genetic markers in Sumatran *Macaca fascicularis* populations. *Genetica* 73: 179–180.

Schenkel, R. 1947. Ausdrucksstudien an Wölfen. *Behaviour* 1: 81–129.

———. 1956. Zur Deutung der Phasianidenbalz. *Der ornithologische Beobachter* 53: 182–201.

———. 1958. Zur Deutung der Balzleistungen einiger Phasianiden und Tetraoniden. *Der ornithologische Beobachter* 55: 65–95.

Schenkel, R., and L. Schenkel-Hulliger. 1967. On the sociology of free-ranging *colobus* (*Colobus guereza caudatus*) Thomas 1885. In *Neue ergebnisse der primatologie,* ed. D. Stark, R. Schneider and H.-J. Kuhn. Stuttgart: Fischer, pp. 185–194.

Schiebinger, L. 1993. *Nature's body: Gender in the making of modern science.* Boston: Beacon Press.

———. 1999. *Has feminism changed science? If so how, if not why not?*

Schluter, D. 1988. Estimating the form of natural selection on a quantitative trait. *Evolution* 45: 849–861.

Schubert, G., and R. D. Masters, eds. 1991. *Primate politics.* Carbondale: Southern Illinois Press.

Schultz, R. J. 1967. Gynogenesis and triploidy in the viviparous fish *Poeciliopsis. Science* 157: 1564–1567.

Schürmann, C. L. 1981. Mating behaviour of wild orang-utans in Sumatra. In *Primate behavior and sociobiology,* ed. A. B. Chiarelli and R. S. Corruccini. Berlin: Springer, pp. 130–135.

———. 1982. Mating behaviour of wild oranguntans. In *The Orangutan: Its biology and conservation,* ed. L. de Boer. Den Haag: Junk, pp. 271–286.

Schürmann, C. L., and J.A.R.A.M. van Hooff. 1986. Reproductive strategies of the orang-utan: New data and a reconsideration of existing socio-sexual models. *International Journal of Primatology* 7: 265–287.

Schutz, A. 1967. *The phenomenology of the social world.* Evanston: Northwestern University Press.

Schutz, A., and T. Luckmann. 1973. *The structures of the life-world,* vol. 1. Evanston: Northwestern University Press.

Schwagmeyer, P. L., and S. J. Wootner. 1986. Scramble competition polygyny in thirteen-lined ground squirrels: The relative contributions of overt conflict and competitive mate searching. *Behav. Ecol. Sociobiol.* 9: 359–364.

Science Magazine. 1993. Anthropology: Nature-culture battleground (Jobs in social science II, V. Morell) 261: 1798–1802.

Segal, L. 1987. *Is the future female? Troubled thoughts on contemporary feminism.* New York: Peter Bedrick Books.

Seifert, D., ed. 1991. *Gender in historical archaeology,* special issue of *Historical Archaeology* 25.4.

Seligman, M. E. P., and J. L. Hager. 1972. *Biological boundaries of learning.* New York: Appleton-Century-Crofts.

Sent, S., ed. 1978. *Morality as a biological phenomenon.* Berlin: Dahlem Konferenzen.

Seton, E. T. 1917. *Wild animals I have known.* New York: Charles Scribner's Sons.

Seyfarth, R. M. 1978. Social relationships among adult male and female baboons. II. Behavior throughout the female reproductive cycle. *Behaviour* 64: 227–247.

Seyfarth, R. M., and D. Cheney. 1984. Grooming, alliances, and reciprocal altruism in vervet monkeys. *Nature* 308: 541–543.

———. 1988. Do monkeys understand their relations? In *Machiavellian intelligence,* ed. R. Byrne and A. Whiten. Oxford: Clarendon Press.

Shanks, M. 1992. *Experiencing the past: The character of archaeology.* London: Routledge, Chapman, Hall.

Shapin, S. 1982. History of science and its sociological reconstructions. *History of Science* 20: 157–211.

———. 1984. Pump and circumstance: Robert Boyle's literary technology. *Social Studies of Science* 14: 481–520.

———. 1992. Discipline and bounding: The history and sociology of science as seen through the externalism/internalism debate. *History of Science* 30: 333–369.

———. 1994. *A social history of truth: Civility and science in seventeenth-century England.* Chicago: University of Chicago Press.

———. 1999. Scientific antlers. *London Review of Books* 21 (5): 27–28.

Shapin, S., and S. Schaffer. 1985. *Leviathan and the air-pump: Hobbes, Boyle and the experimental life.* Princeton NJ: Princeton University Press.

Shapiro, D. Y., A. Marconato, and T. Yoshikawa. 1994. Sperm economy in a coral reef fish, *Thalassoma bifasciatum. Ecology* 75: 1334–1344.

Sheffield, F. D., J. J. Wulff, and R. Backer. 1951. Reward value of copulation without sex drive reduction. *Journal of Comparative and Physiological Psychology* 44: 3–8.

Sherry, D. F., and B. G. Galef. 1984. Cultural transmission without imitation: Milk bottle opening by birds. *Anim. Behav.* 32: 937–938.

Shields, O. 1967. Hilltopping. *J. Res. Lepidopt.* 6: 69–178.

Shiva, V. 1993. *Monocultures of the mind.* London: Zed Books.

Shively, C. 1985. The evolution of dominance hierarchies in nonhuman primate society. In *Power, dominance, and nonverbal behavior,* ed. S. Ellyson and I. Davido. Berlin: Springer-Verlag, pp. 67–87.

Shozu, M., K. Akasofu, T. Harada, and Y. Kubota. 1991. A new cause of female pseudohermaphroditism: Placental aromatase deficiency. *Journal of Clinical Endocrinology and Metabolism* 72: 560–566.

Sibatani, A. 1983. Kinji Imanishi and species identity. *Rivista di Biologia* 76 (1): 25–42.

Silk, J. B. 1982. Altruism among female *Macacca radiata:* Explanations and analysis of patterns of grooming and coalition formation. *Behaviour* 79: 162–187.

Silverberg, J., and J. P. Gray. 1992. *Aggression and peacefulness in humans and other primates.* Oxford: Oxford University Press.

Silverblatt, I. 1978. Andean women in Inca society. *Fem. Stud.* 4: 37–61.

———. 1987. *Moon, sun, and witches: Gender ideologies and class in Inca and colonial Peru.* Princeton: Princeton University Press.

———. 1988. Women in states. *Ann. Rev. Anthrop.* 17: 427–460.

Simpson, G. G. 1949. *The meaning of evolution.* New Haven, CT: Yale University Press.

Simpson, M. J. A. 1973. The social grooming of male chimpanzees. In *Comparative ecology and behavior of primates,* ed. R. Michael and J. Crook. London: Academic Press, pp. 411–505.

Sims, S. R. 1979. Aspects of mating frequency and reproductive maturity in *Papilio zelicaon. Am. Midl. Nat.* 102: 36–50.

Sippi, D. 1989. Aping Africa: The mist of immaculate miscegenation. *CineAction!* 18, Fall 1989.

Skinner, B. F. 1938. *The Behavior of Organisms.* New York: Appleton-Century-Crofts.

———. 1959. A case history in scientific method. In *Psychology: A study of a science,* vol. 2, ed. S. Koch. New York: McGraw-Hill.

Smale, L., K. E. Holekamp, M. Weldele, L. G. Frank, and S. E. Glickman. 1995. Competition and cooperation between littermates in the spotted hyaena (*Crocuta crocuta*). *Anim. Behav.* 50: 671–682.

Small, M. F. 1984. *Female primates: Studies by women primatologists.* New York: Alan R. Liss.

———. 1989. Female choice in nonhuman primates. *Yearbook of Physical Anthropology* 32: 103–127.

———. 1990. Promiscuity in barbary macaques (*Macaca sylvanus*). *Am. J. Primatol.* 20: 267–282.

———. 1993. *Female choices: Sexual behavior of female primates.* Ithaca, NY: Cornell University Press.

———. 1995. Making a monkey of human nature. *New Scientist* (10 June 1995): 30–33.

Smith, B. C. 1996. *On the origins of objects.* Cambridge, MA: MIT Press.

Smith, D. E. 1974. Women's perspective as a radical critique of sociology. *Sociological Inquiry* 44: 7–13.

———. 1987.*The everyday world as problematic: A feminist sociology.* Toronto: University of Toronto Press.

Smith, G. E. 1912. Presidential Address, Anthropological Section, *Report of the British Association.*

———. 1913. The Piltdown skull and braincast. *Nature* 92: 267.

———. 1917. Fourth note on the Piltdown gravel, with evidence of a second skull of *Eanthropus dawsoni. Quart. J. Geol. Soc.* 73: 1–10.

Smith, S. M. 1988. Extra-pair copulations in black-capped chickadees: The role of the female. *Behaviour* 197: 15–23.

Smuts, B. B. 1983a. Dynamics of social relationships between adult male and female olive baboons: Selective advantages. In *Primate social relationships,* ed. R. A. Hinde. Oxford: Blackwell, pp. 112–115.

———. 1983b. Special relationships between adult male and female baboons. In *Primate social relationships,* ed. R. A. Hinde. Oxford: Blackwell, pp. 262–266.

———. 1985. *Sex and friendship in baboons.* New York: Aldine.

———. 1987a. Gender, aggression, and influence. In *Primate societies,* ed. B. B. Smuts, D. L. Cheney, R. M. Seyfarth, R. W. Wrangham, and T. T. Struhsaker. Chicago: University of Chicago Press, pp. 306–317.

———. 1987b. Sexual competition and mate choice. In *Primate societies,* ed. B. B. Smuts, D. L. Cheney, R. M. Seyfarth, R. W. Wrangham, and T. T. Struhsaker. Chicago: University of Chicago Press, pp. 385–399.

———. 1992. Male aggression against women: An evolutionary perspective. *Human Nature* 3: 1–44.

Smuts, B. B., and R. W. Smuts. 1993. Male aggression and sexual coercion of females in nonhuman primates and other mammals: Evidence and theoretical implications. *Advances in the Study of Behavior* 22: 1 –63.

Smuts, B. B., D. L. Cheney, R. M. Seyfarth, R. W. Wrangham, and T. T. Struhsaker, eds. 1987. *Primate societies.* Chicago: University of Chicago Press.

Smuts, G. L., J. Hanks, and I. J. Whyte. 1978. Reproduction and social organization of lions from the Kruger National Park. *Carnivore* 1: 17–28.

Snowdon, C. T. 1990a. Language capacities of nonhuman primates. *Yearbook of Physical Anthropology* 33: 215–243.

———. 1990b. Mechanisms maintaining monogamy in monkeys. In *Contemporary Issues in Comparative Psychology,* ed. D. A. Dewsbury. Sunderland: Sinauer, pp. 225–251.

Sommer V. 1994. Infanticide among the langurs of Jodhpur: Testing the sexual selection hypothesis with a long-term record. In *Infanticide and parental care,* ed. S. Parmigiani and F. vom Saal. London: Harwood Academic Publishers, pp. 155–198.

Sonneborn, T. M. 1941. Sexuality in unicellular organisms. In *Protozoa in biological research,*ed. G. N. Calkins and F. M. Summers. Chicago: University of Chicago Press, pp. 666–709.

Sonnert, G. 1995a. *Gender Differences in Science Careers. The Project Access Study.* New Brunswick, NJ: Rutgers University Press.

——— 1995b. *Who Succeeds in Science? The Gender Dimension.* New Brunswick, NJ: Rutgers University Press.

Southwick, C. H., ed. 1963. *Primate social behavior.* New York: Van Nostrand Reinhold.

Southwick, C. H., and R. B. Smith. 1986. The growth of primate field studies. In *Comparative primate biology,* vol. 2A, *Behavior, conservation, and ecology,* ed. G. Mitchell and J. Erwin. New York: Alan R. Liss, pp. 73–91.

Spector, J. D. 1994. *What this awl means.* Minneapolis: Minnesota Historical Society.

Spence, J. T., and R. L Helmreich. 1978. *Masculinity and femininity.* Austin, Texas: University of Texas Press.

Spence, K. W. 1937. Experimental studies of learning and higher mental processes in infra-human primates. *Psychological Bulletin* 34: 806–850.

Spencer, F. 1990. *Piltdown: A scientific forgery.* London: Oxford University Press.

Sperling, S. 1991. Baboons with briefcases vs. langurs in lipstick. Feminism and functionalism in primate studies. In *Gender and the crossroads of knowledge: Feminist Anthropology in the postmodern era,* ed. M di Leonardo. Berkeley: University of California Press, pp. 204–234.

Spiro, M. 1997. Review of *After the fact: Two countries, four decades, one anthropologist,* by Clifford Geertz (Cambridge, MA: Harvard University Press, 1995.) *Society* (March/April).

Sprague, D. S. 1989. *Male intertroop movement during the mating season among the Japanese macaques of Yakushima Island, Japan.* Ph.D. diss., Yale University.

———. 1993. Applying GIS and remote sensing to wildlife management: Assessing habitat quality for the Japanese monkey (*Macaca fuscatta*). *Research Reports,* vol. 3, *Division of Changing Earth and Agroenvironment, National Institute of Agroenvironmental Sciences,* pp. 89–100.

Standen, V., and R. A. Foley, eds. 1989. *Comparative socioecology: The behavioural ecology of humans and other mammals.* Oxford: Blackwell.

Stanford, C. B. 1998. Predation and male bonds in primate societies. *Behaviour* 135: 513–533.

Star, S. L. 1990. Layered space, formal representation, and long distance control: The politics of information. *Fendamenta Scientiae* 10: 125–155.

———. 1991. Power, technologies and the phenomenology of conventions. In *A sociology of monsters,* ed. J. Law. London: Routledge, pp. 25–56.

Star, S. L., and J. Griesemer. 1989. Institutional ecology, translations, and boundary objects: Amateurs and professionals in Berkeley's Museum of Vertebrate Zoology, 1907–1939. *Social Studies of Science* 19: 387–420.

Stathern, M., and S. Franklin. 1993. "Kinship and the new genetic technologies: An assessment of existing anthropological research." A report compiled for the Commission of the European Communities, Medical Research Division (Human Genome Analysis Programme).

Steenbeek, R. 1996. What a maleless group can tell us about the constraints on female transfer in Thomas's langurs (*Presbytis thomasi*). *Folia Primatol.* 67 (4): 169–181.

Steenbeek, R., and P. Assink. 1998. Individual differences in long-distance calls of male wild Thomas langurs (*Presbytis thomasi*). *Folia Primatol.* 69: 77–80.

Steklis, H. 1993. Primate socioecology from the bottom up. In *Milestones in human evolution,* ed. A. J. Almquist and A. Manyak. Prospect Heights, Ill: Waveland Press.

Stellar, E. 1960. The marmoset as a laboratory animal: Maintenance, general observations of behavior and simple learning. *J. Comp. Physiol. Psychol.* 53: 1–10.

Stengers, I. 1996. *Cosmopolitiques,* vol. 1, *La guerre des sciences.* Paris: La Découverte: Les Empêcheurs de penser en rond.

Stepan, N. L. 1996. Race and gender: The role of analogy in science. In *Feminism and Science,* ed. E. F. Keller and H. E. Longino. Oxford: Oxford University Press, pp. 121–136.

Sterck, E. H. M. 1997. Determinants of female dispersal in Thomas langurs. *Am. J. Primatol.* 42: 179–198.

———. 1998. Female dispersal, social organization, and infanticide in langurs: Are they linked to human disturbance? *Amer. J. Primatology* 44 (4): 235–254.

Sterck, E. H. M., and R. Steenbeek. 1997. Female dominance relationships and food competition in the sympatric Thomas langur and long-tailed macaque. *Behaviour* 134: 749–774.

Sterck, E. H. M., D. P. Watts, and C. P. van Schaik. 1997. The evolution of female social relationships in nonhuman primates. *Behav. Ecol. Sociobiol.* 41: 291–309.

Stevenson-Hinde, J., and M. Zunz. 1978. Subjective assessment of individual rhesus monkeys. *Primates* 19: 473–482. See also *Primates* 21: 498–509.

Stolcke, V. 1981. Women's labours: The naturalization of social inequality and women's subordination. In *Of marriage and the market: Women's subordination in international perspective,* ed. K. Young, C. Wolkowitz, and R. McCullagh. London: CSE Books, pp. 30–48.

Stoller, R. 1968. *Sex and gender.* New York: Science House.

Stone, C. P. 1922. Congenital sexual behavior of young male albino rats. *Journal of Comparative Psychology* 2: 95–153.

———. 1927. The retention of copulatory ability in male rats following castration. *Journal of Comparative Psychology* 7: 369—387.

Strathern, M. 1987. An awkward relationship: The case of feminism and anthropology. *Signs* 12: 276–292 (special issue on Reconstructing the Academy).

———. 1988. *The gender of the gift.* Berkeley: University of California Press.

———. 1992. *Reproducing the future: Anthropology, kinship, and the new reproductive technologies.* Manchester: Manchester University Press.

———. 1994. The new modernities. Paper for the Conference of the European Society for Oceanists, Basel, December.

———, ed. 1995. *Shifting contexts: Transformations in anthropological knowledge.* London: Routledge.

Strauss, A. 1991. *Creating sociological awareness: Collective images and symbolic representations.* New Brunswick, N. J.: Transaction.

Strauss, C., and N. Quinn. 1997. *A cognitive theory of cultural meaning.* Cambridge, UK: Cambridge University Press.

Strier, K. B. 1990. New World primates, new frontiers: Insights from the wooly spider monkey, or muriqui (*Brachyteles arachnoides*). *Int. J. Primatol.* 11: 7–19.

———. 1992a. *Faces in the forest: The endangered muriqui monkeys of Brazil.* New York: Oxford University Press.

———. 1992b. Atelinae adaptations: Behavioral strategies and ecological constraints. *American Journal of Physical Anthropology* 88: 515–524.

———. 1994a. Myth of the typical primate. *Yrb. Phys. Anthrop* 37: 233–272.

———. 1994b. Brotherhoods among atelins: Kinship, affiliation, and competition. *Behaviour* 130: 151–167.

———. 1996a. Viability analysis of an isolated population of muriqui monkeys (*Brachyteles arachnoides*): Implications for primate conservation and demography. *Primate Conservation* 14–15 (1993–1994): 43–52.

———. 1996b. Male reproductive strategies in New World primates. *Human Nature* 7: 105–123.

———. 1997. Behavioral ecology and conservation biology of primates and other animals. *Advances in the Study of Behavior* 26: 101–158.

———. 1999. Why is female kin bonding so rare: Comparative sociality of New World primates. In *Comparative primate socioecology,* ed. P. C. Lee. Cambridge, UK: Cambridge University Press, pp. 300–319.

Strier, K. B., and G. A. B. Fonseca. 1996/1997. The endangered muriquis of Brazil's Atlantic forest. *Primate Conservation* 17: 131–137.

Strier, K. B., and T. E. Ziegler. 1994. Insights into ovarian function in wild muriqui monkeys (*Brachyteles arachnoides*). *Am. J. Primatol.* 32: 31–40.

Strier, K. B., and T. E. Ziegler. 1997. Behavioral and endocrine characteristics of reproductive cycles in wild muriqui monkeys, *Brachyteles arachnoides. Am. J. Primatol.* 42: 299–310.

Strier, K. B., F. D.C. Mendes, J. Rímoli, and A. O. Rímoli. 1993. Demography and social structure in one group of muriquis (*Brachyteles arachnoides*). *Int. J. Primatol.* 14: 513–526.

Struhsaker, T. T. 1967. Auditory communication among vervet monkeys (*Cercopithecus aethiops*). In *Social communication among primates,* ed. S. A. Altmann. Chicago: University of Chicago Press, pp. 281–324.

———. 1969. Correlates of ecology and social organization among African cercopithecines. *Folia Primatol.* 9: 123–134.

Strum, S. C. 1975a. Life with the Pumphouse Gang: New insights into baboon behavior. *National Geographic Magazine* 147: 672–691.

———. 1975b. Primate predation: Interim report on the development of a tradition in a troop of olive baboons. *Science* 187: 755–757.

———. 1982. Agonistic dominance in male baboons: An alternate view. *Int. J. Primatol.* 3: 175–202.

———. 1983a. Use of females by male olive baboons (*Papio anubis*). *Am. J. Primatol.* 5: 93–109.

———. 1983b. Why males use infants. In *Primate paternalism,* ed. D. M. Taub. New York: Van Nostrand, pp. 146–185.

———. 1987. *Almost human: A journey into the world of baboons.* New York: Random House.

———. 1989. Longitudinal data on patterns of consorting among male olive baboons. *American Journal of Physical Anthropology* 78: 310.

———. 1994a. Reconciling aggression and social manipulation as means of competition, part 1, Life history perspective. *Int. J. Primatol.* 91: 387–389.

———. 1994b. Lessons learned. In *Natural connections: Perspectives in community-based conservation,* ed. D. Western, R. M. Wright, and S. Strum. Washington, DC: Island Press. 512–523.

———. 1996. Commentary on Changing Images of Primate Societies. Wenner-Gren Symposium no. 120, Teresopolis, Brazil. Circulated August 1996.

Strum, S. C., and L. M. Fedigan. 1996. Theory, method, and gender: What changed our views of primate society? Paper for the Wenner-Gren Symposium no. 120, Changing Images of Primate Societies. The Role of Theory, Method, and Gender. Teresopolis, Brazil.

Strum, S. C., D. Lindburg, and D. Hamburg, eds. 1998. *The new physical anthropology: Science, humanism and critical reflection.* Englewood Cliffs, NJ: Prentice-Hall.

Sturtevant, A. H. 1932. The use of mosaics in the study of the developmental effects of genes. *Proceedings of the Sixth International Congress of Genetics,* p. 304.

Suchman, L. 1994. Working relations of technology production and use. *Computer Supported Cooperative Work* 2: 21–39.

Sugardjito, J., and J.A.R.A.M. van Hooff. 1986. Age-sex class differences in positional behaviour of Sumatran orang-utans (*Pongo pygmaeus abelii*) in the Gunung Leuser National Park, Indonesia. *Folia primatol.* 47: 14–25.

Sugardjito, J., I. J. A. te Boekhorst, and J.A.R.A.M. van Hooff. 1987. Ecological constraints on the grouping of wild orang-utans (*Pongo pygmaeus*) in the Gunung Leuser National Park, Indonesia. *Int. J. Primatol.* 8: 17–41.

Sugawara, K. 1980. The structure of social encounters among solitary males of Japanese monkeys. *Kikan-jinruigaku,* 11: 3–70. (J)

———. 1991. The economics of social life among the Central Kalahari San (G//snakhwe and G/wikhwe) in the sedentary community at !Koi!Kom. *Senri Ethnol. Studies* 30: 91–116.

Sugiyama, Y. 1965a. On the social change of Hanuman langurs (*Presbytis entellus*) in their natural condition. *Primates* 6: 381–418.

———. 1965b. Short history of the ecological and sociological studies on nonhuman primates in Japan. *Primates* 6: 457–460.

———. 1967. Social organization of Hanuman langurs. In *Social communication among primates,* ed. S. A. Altmann. Chicago: University of Chicago Press, pp. 221–236.

———. 1978. *The people and chimpanzees of the Bossou Village.* Tokyo: Kinokuniyashoten. (J)

———. 1980. *Eco-ethology of infanticide.* Tokyo: Hokuto-shuppan. (J)

———. 1981. *Stuides on behavior and society of wild chimpanzees: A path to the Evolution of Man.* Tokyo: Kodansha. (J)

———. 1985. On the start and background of The Primate Society of Japan. *Primate Research* 1: 39–44. (J)

Sussman, R. W. 1977. Feeding behavior of *Lemur catta* and *Lemur fulvus.* In *Primate feeding behavior,* ed. T. H. Clutton-Brock. London: Academic Press, pp. 1–36.

———, ed. 1979. *Primate ecology: Problem-oriented field studies.* New York: Wiley.

———. 1991. Demography and social organization of free-ranging *Lemur catta* in the Beza Mahafaly Reserve, Madagascar. *American Journal of Physical Anthropology* 84: 43–58.

———. 1992. Male life history and intergroup mobility among ringtailed lemurs (*Lemur catta*). *Int. J. Primatol.* 13: 395–354.

Sussman, R. W., and P. A. Garber. 1987. A new interpretation of the social organization and mating system of the *Callitrichidae. Int. J. Primatol.* 8: 73–93.

Sussman, R. W., and W. G. Kinzey. 1984. The ecological role of Callitrichidae: A review. *Amer. J. Phys. Anthropol.* 64: 419–449.

Sussman, R. W., J. M. Cheverud, and T. Q. Bartlett. 1995. Infant killing as an evolutionary strategy: Reality or myth? *Evol. Anthropol.* 3: 149–150.

Sussman, R. W., G. M. Green, and L. Sussman. 1994. Satellite imagery, human ecology, anthropology, and deforestation in Madagascar. *Human Ecology* 22: 333–354.

Suzuki, A. 1971. Carnivority and cannibalism observed among forest-living chimpanzees. *J. Anthrop. Soc. Nippon* 79: 30–48.

———. 1977. *The way to omnivory: Ecology of wild chimpanzees.* Tokyo: Tamagawa University Press. (J)

———. 1985. The social structure of orang-utans. *Kagaku* 55 (5): 308–314. (J)

———. 1992. *Chimpanzees watching the sunset.* Tokyo: Maruzen.

Symington, M. M. 1988. Demography, ranging patterns, and activity budgets of black spider monkeys (*Ateles paniscus chamek*) in the Manu National Park, Peru. *Am. J. Primatol.* 15: 45–67.

Takahata, M., H. Komatsu, and M. Hisada. 1984. Positional orientation determined by the behavioural context in *Procambarus clarkii* Girard. *Behaviour* 88: 240–265.

Takahata, Y. 1985. *The life and observation of Japanese monkeys.* Tokyo: Nyu-saiensusha. (J)

Takasaki, H. 1981. Troop size, habitat quality, and home range area in Japanese macaques. *Behav. Ecol. Sociobiol.* 9: 277–281.

———. 1984. A report from the Imanishian seminar meetings. *Seibutsu-kagaku* 36 (2): 94–98. (J)

———. 1996. Traditions of the Kyoto school of field primatology in Japan. Paper for the Wenner-Gren conference, Changing Images of Primate Societies.

Takasaki, H., and Takenaka, O. 1991. Paternity testing in chimpanzees with DNA amplification from hairs and buccal cells in wadges: A preliminary note. In *Primatology today,* ed. A. Ehara, T. Kimura, O. Takenaka, and M. Iwamoto. Amsterdam: Elsevier, pp. 612–616.

Tambiah, S. 1990. *Magic, science, religion and the scope of rationality.* Cambridge, UK: Cambridge University Press.

Tanaka, I. 1995. Matrilineal distribution of louse-egg-handling techniques during grooming in free-ranging Japanese macaques. *Am. J. Phys. Anthropol.* 98: 197–201.

Tanaka, J., and M. Kakeya, eds. 1991. *Natural historiography of man.* Tokyo: Heibonsha. (J)

Tang-Martinez, Z. 1997. The curious courtship of sociobiology and feminism: A case of irreconcilable differences. In *The biological basis of human behavior,* ed. R. W. Sussman. Needham Heights, MA: Simon and Schuster, pp. 336–353.

Tattersall, I. 1995. *The fossil trail.* New York: Oxford University Press.

Tax, S., ed. 1960. *The evolution of man.* Chicago: University of Chicago Press.

Teleki, G. 1981. C. Raymond Carpenter, 1905–1975. *Am. J. Phys. Anthropol.* 56: 383–386.

Terborgh, J. 1983. *Five New World primates. A study in comparative ecology.* Princeton: Princeton University Press.

Terborgh, J., and A. W. Goldizen. 1985. On the mating system of the cooperatively breeding saddle-back tamarin (*Saguinus fuscicollis*). *Behav. Ecol. Sociobiol.* 16: 293–299.

Terborgh, J., and C. H. Janson. 1986. The socioecology of primate groups. *Annual Review of Ecology and Systematics* 17: 11–135.

Terrall, M. 1995. Gendered spaces, gendered audiences: Inside and outside the Paris Academy of Sciences. *Configurations* 3: 207–232.

Thorpe, W. H. 1956/1963. *Learning and instinct in animals.* London: Methuen.

———. 1961. *Bird-song.* Cambridge, UK: Cambridge University Press.

Tiefer, L. 1978. The context and consequences of contemporary sex research: A feminist perspective. In *Sex and behavior: Status and prospectus,* ed. McGill, W., Dewsbury, D., and Sachs, B. New York: Plenum Press. Pp. 363–385.

———. 1988. In honor of him. *Hormones and Behavior* 22: 440–442.

Tilley, C., ed. 1993. *Interpretive archaeology.* Oxford: Berg Publishers.

Timmermans, P. J. A., E. L. Röder, and P. Hunting. 1986. The effect of absence of the mother on the acquisition of phobic behaviour in in cynomolgus monkeys (*Macaca fascicularis*). *Behav. Res. Ther.* 24: 67–72.

Timmermans, P. J. A., J. D. Vochteloo, and J. M. H. Vossen. 1997. Mobility of surrogate mothers and persistent neophobia in cynomolgus monkeys (*Macaca fascicularis*). *Primates* 38: 139–148.

Tinbergen, N. 1940. Die Übersprungbewegung. *Zeitschrift für Tierpsychologie* 4: 1–10.

———. 1942. An objectivistic study of the innate behaviour of animals. *Bibiotheca biotheoretica* 1: 39–98.

———. 1950. The hierarchical organization of nervous mechanisms underlying instinctive behaviour. *Symposia of the Society for Experimental Biology* 4: 304–312.

———. 1951. *The Study of Instinct.* Oxford: Oxford University Press.

———. 1952. Derived activities: Their causation, biological significance, origin and emancipation during evolution. *Quarterly Review of Biology* 27: 1–32.

———. 1959. Comparative studies of the behaviour of gulls (*Laridae*): A progress report. *Behaviour* 15: 1–70.

———. 1963. On aims and methods of ethology. *Zeitschrift für Tierpsychologie* 20: 410–433.

Titchener, W. B. 1898. The postulates of a structural psychology. *Philosophical Review* 7: 449–465.

Tokuda, K. 1962. A primatological expedition to the Amazon. *Shizen* 17 (12): 80–87.

Tolman, E. C. 1932. *Purposive Behavior in Animals and Men.* New York: Century.

Tomasello, M. 1990. Cultural transmission in the tool use and communicatory signaling of chimpanzees? In *"Language" and intelligence in monkeys and apes,* ed. S. T. Parker and K. R. Gibson. Cambridge, UK: Cambridge University Press, pp. 274–311.

———. 1996. Chimpanzee social cognition. In "What young chimpanzees know about seeing," *Monographs of the Society for Research in Child Development, ed.* Povinelli, D. J., and T. J., 61 (3, ser. 247): 161–173.

Tomasello, M., M. Davis-Dasilva, L. Camak, and K. Bard. 1987. Observational learning of tool-use by young chimpanzees. *Human Evolution* 2: 175–183.

Tooby, J., and L. Cosmides. 1992. The psychological foundations of culture. In *The adapted mind: Evolutionary psychology and the generation of culture,* ed. J. H. Barkow, L. Cosmides, and J. Tooby. New York: Oxford University Press, pp. 19–136.

Tooby, J., and I. DeVore. 1987. The reconstruction of hominid behavioral evolution through strategic modeling. In *The evolution of human behavior: Primate models,* ed. W. G. Kinzey. Albany: SUNY Press, pp. 183–237.

Torgovnick, M. 1996. A passion for the primitive: Dian Fossey among the animals. *The Yale Review.* 84 (1996).

Traweek, S. 1988. *Beamtimes and lifetimes: The world of high energy physicists.* Cambridge, MA: Harvard University Press.

———. 1993. An introduction to cultural, gender, and social studies of science and technologies. *Culture, Medicine, and Psychiatry* 17: 3–25.

Trivers, R. L. 1971. The evolution of reciprocal altruism. *Quarterly Review of Biology* 46: 35–57.

———. 1972. Parental investment and sexual selection. In *Sexual selection and the descent of Man, 1871–1971,* ed. B. Campbell. Chicago: Aldine, pp. 136–179.

———. 1974. Parent-offspring conflict. *Am. Zool.* 14: 249–264.

Tsumori, A. 1967. Newly acquired behavior and social interactions of Japanese monkeys. In *Social communication among primates,* ed. S. A. Altmann. Chicago: University of Chicago Press, pp. 207–220.

Tulp, N. 1641. Een Indiaansche Satyr. In *Geneeskundige Waarnemingen 3,* Amsterdam, ed. Lodewijk and Daniel Elsevier, pp. 370–379.

Turke, P. W., and L. L. Betzig. 1985. Those who can do: Wealth, status, and reproductive success on Ifaluk. *Ethology and Sociobiology* 6: 79–87.

Turner, T. 1981. Blood protein variation in a population of Ethiopian vervet monkeys. *American Journal of Physical Anthropology* 55: 225–232.

Tutin, C. E. G. 1979. Mating patterns and reproductive strategies in a community of wild chimpanzees. *Behavioral Ecology and Sociobiology* 6: 29–38.

Tuttle, R. H. 1975. *Socioecology and psychology of primates.* Paris: Mouton Publishers.

Tylor, E. B. 1871. *Primitive culture.* London: John Murray.

Uehara, S. 1981. The social unit of wild chimpanzees: A reconsideration based on the diachronic data accumulated at Kasoje in the Mahale Mountains, Tanzania. *J. Afr. Studies* 20: 15–32. (J)

Uexküll, J. von. 1921. *Umwelt und Innenwelt der Tiere.* Berlin: Springer.

Utami, S. S., and J.A.R.A.M. van Hooff. 1997. Meat-eating by adult female Sumatran orangutans (*Pongo pygmaeus abelii*). *Am. J. Primatol.* 43: 159–165.

Utami, S. S., and T. Mitra Setia. 1996. Behavioral changes in wild male and female Sumatran orangutans (*Pongo pygmaeus abelii*) during and following a resident male take-over. In *The neglected ape,* ed. R. D. Nadler, B. F. M. Galdikas, L. K. Sheeran, and N. Rosen. New York: Plenum Press, pp. 183–190.

Utami, S. S., S. A. Wich, E. H. M. Sterck, and J.A.R.A.M. van Hooff. 1997. Food competition between wild orangutans in large fig trees. *Int. J. Primatol.* 18: 909–927.

Van Voorhies, W. A. 1992. Production of sperm reduces nematode lifespan. *Science* 360: 456–458.

Veenema, H. C., M. Das, and F. Aureli. 1994. Methodological improvements for the study of reconciliation. *Behavioural Processes* 31: 29–37.

Veenema, H. C., B. M. Spruijt, W. H. Gispen, and J.A.R.A.M. van Hooff. 1997. Ageing, dominance history and social behaviour in Java-monkeys (*Macaca fascicularis*). *Neurobiololy of Aging* 18: 509–515.

Visalberghi, E., and D. M. Fragaszy. 1990. Do monkeys ape? In *"Language" and intelligence in monkeys and apes,* ed. S. T. Parker and K. R. Gibson. Cambridge, UK: Cambridge University Press, pp. 247–273.

Vochteloo, J. D., P. J. A. Timmermans, J. A. Duijghuisen, and J. H. M. Vossen. 1993. Effects of reducing the mother's radius of action on the development of mother-infant relationships in long-tailed macaques. *Anim. Behav.* 45: 603–612.

———. 1995. The range of action of the mother and the avoidance of big novel objects in long-tailed macaques. *Int. J. Primatol.* 16: 277–293.

———. 1996. The development of range of action in infant cynomolgus monkeys (*Macaca fascicularis*) reared by restrained mothers. *Primates* 37: 167–173.

Vosmaer, A. 1768. *Beschrijving van ene zeldzaame Amerikaansche aap-soort, genaamd Quatta.* Amsterdam: Meijer.

———. 1770. *Beschrijving van ene zeldzaame, nog niet beschreven aap-soort, genaamd de fluiter.* Amsterdam: Meijer.

———. 1778. *Beschrijving van de zoo zeldzaame als zonderlinghe aap-soort, genaamd Orang Outang, van het eiland Borneo.* Amsterdam: Meijer and Warnars.

Vygotsky, L. 1978. *Mind in society: The development of higher psychological processes.* Cambridge, MA: Harvard University Press.

Wachter, B., M. Schabel, and R. Noë. 1977. Diet overlap and polyspecific associations of red colobus and Diana monkeys in the Taï National Park, Ivory Coast. *Ethology* 103: 514–526.

Wada, K. 1964. Some observertaions on the life of Japanese monkeys in the snowy district in Japan. *Physiol. Ecol., Jpn.* 12: 151–174. (J)

Walker, P. L., and D.C. Cook. 1998. Gender and sex: Vive la difference. *American Journal of Physical Anthropology* 106: 255–259.

Wall, D. Z. 1994. *The archaeology of gender: Separating the spheres in urban America.* New York: Plenum Press.

Wallen, K. 1996. Nature needs nurture: The interaction of hormonal and social influences on the development of behavioral sex differences in rhesus monkeys. *Hormones and Behavior* 30: 364–378.

Walt Disney Studios. 1952. *The story of Walt Disney's true-life adventure series.* Burbank, CA: Walt Disney Productions.

Walters, J. R., and R. M. Seyfarth. 1987. Conflict and cooperation. In *Primate societies,* ed. B. B. Smuts, D. L. Cheney, R. M. Seyfarth, R. W. Wrangham, and T. T. Struhsaker. Chicago: University of Chicago Press, pp. 306–317.

Warden, C. J. 1931. *Animal Motivation.* New York: Columbia University Press.

Warner, L. H. 1931. A study of sex behavior in the white rat by means of the obstruction method. In *Animal Motivation,* ed. C. J. Warden. New York: Columbia University Press. Pp. 119–178.

Warner, R. R. 1987. Female choice of sites versus mates in a coral reef fish, *Thalassoma bifasciatum. Anim. Behav.* 35: 1470–1478.

Warren, A. 1997a. The dance of the sifaka. Bristol: Partridge Films.

———. 1997b. A lemur's tale. Bristol: Partridge Films.

Waser, P., and R. H. Wiley. 1980. Mechanisms and evolution of spacing in animals. In *Handbook of behavioral neurobiology,* vol. 3, ed. P. Marler and J. G. Vandenbergh. New York: Plenum Press, pp. 159–233.

Washburn, M. F. 1908. *The animal mind: A text-book of comparative psychology.* New York: The Macmillan Company. A second edition was published in 1917, a third in 1926, and a fourth in 1936.

Washburn, S. L. 1951. The new physical anthropology. *Transactions of the New York Academy of Science, Series II* 13: 298–304.

———, ed. 1961. *Social life of early man.* Chicago: Aldine.

———. 1962. The analysis of primate evolution with particular reference to the origin of man. Cold Spring Harbor Symposia on Quantitative Biology (1951). In *Ideas on evolution: Selected essays, 1941–61,* ed. W. W. Howells. Cambridge, MA: Harvard University Press, pp. 154–171.

———, ed. 1963. *Classification and human evolution.* Chicago: Aldine.

Washburn, S. L., and I. DeVore. 1961. Social behavior of baboon and early man. In *The social life of early man,* ed. S. L. Washburn. New York: Wenner-Gren Foundation for Anthropological Research, pp. 91–105.

Washburn, S. L., and S. C. Lancaster. 1968. The evolution of hunting. In *Man the hunter,* ed. R. B. Lee and I. DeVore. Chicago: Aldine, pp. 293–303.

Wasser, S. K., ed. 1983a. *Social behavior of female vertebrates.* New York: Academic Press.

———. 1983b. Reproductive competition and cooperation among female yellow baboons. In *Social behavior of female vertebrates,* ed. S. K. Wasser. New York: Academic Press.

Wasser, L. M., and S. K. Wasser. 1995. Environmental variation and development among free-ranging yellow baboons (*Papio cynocephalus*). *Am. J. Primatol.* 35: 15–30.

Wasser, S. K., S. L. Monfort, and D. E. Wildt. 1991. Rapid extraction of fecal steroids for measuring reproductive cyclicity and early pregnancy in free-ranging yellow baboons (*Papio cynocephalus cynocephalus*). *Journal of Reproductive Fertility* 92: 415–423.

Wasser, S. K., L. Risler, and R. A. Steiner. 1988. Excreted steroids in primate feces over the menstrual cycle and pregnancy. *Biology of Reproduction* 39: 862–872.

Watanabe, K. 1993. Bibliography: Field research on Japanese monkeys, 1975–1992, with some remarks on the history of researches. *Primate Research* 9: 33–60. (J)

Watanabe, K., A. Mori, and M. Kawai. 1992. Characteristic features of the reproduction of Koshima monkeys, *Macaca fuscata fuscata:* A summary of thirty-four years of observation. *Primates* 33: 1–32.

Watson, J. B. 1913. Psychology as the behaviorist views it. *Psychological Review* 20: 158–177.

Watson-Verran, H., and D. Turnbull. 1995. Science and other indigenous knowledge systems. In *Handbook of science and technology studies,* ed. S. Jasanoff, G. Markle, J. Peterson, and T. Pinch. London: Sage, pp. 115–139.

Watts, D. P. 1990. Ecology of gorillas and its relation to female transfer in mountain gorillas. *Int. J. Primatol.* 11: 21–45.

Weber, M. 1948. Science as a vocation. In *From Max Weber: Essays in sociology,* ed. H. H. Gerth and C. W. Mills. New York: Oxford University Press, pp. 129–156.

Weidenreich, F. 1946. *Apes, giants, and man.* Chicago: University of Chicago Press.

Weiner, J. S., F. P. Oakley, and W. E. Le Gros Clark. 1953. The solution of the Piltdown problem. *Bull. Brit. Mu. Nat. Hist.* 2: 141–146.

Western, J. D., and S. C. Strum. 1983. Sex, kinship, and the evolution of social manipulation. *Ethology and Sociobiology* 4: 19–28.

Western, D., R. M. Wright, and S. C. Strum, eds. 1994. *Natural connections. Perspectives in community-based conservation.* Washington, DC: Island Press.

Westman, R. S. 1990. Proof, poetics, and patronage: Copernicus' preface to *De revolutionibus.* In *Reappraisals of the scientific revolution,* ed. D. Lindberg and R. Westman. Cambridge, UK: Cambridge University Press, pp. 167–206.

White, R. 1995. *The organic machine: The remaking of the Columbia River.* New York: Hill and Wang.

Whitehead, A. N. 1929. *Process and reality: An essay in cosmology.* New York: Free Press. Reprinted 1978 (New York: Free Press).

Whiten, A., ed. 1993. *Natural theories of mind: Evolution, development, and stimulation of everyday mindreading.* Cambridge, MA: Blackwell Publishers.

Whiten, A., and R. W. Byrne. 1986. The St. Andrews catalogue of tactical deception in primates. *St. Andrews Psychological Reports,* no. 10.

———. 1988. The Machiavellian intelligence hypotheses: Editorial. In *Machiavellian intelligence: Social expertise and the evolution of intellect in monkeys, apes, and humans,* ed. R. W. Byrne and A. Whiten. Oxford: Clarendon, pp. 1–9.

Whiten, A., D. M. Custance, J.-C. Gomez, P. Teixidor, and K. A. Bard. 1996. Imitative learning of artificial fruit processing in children (*Homo sapiens*) and chimpanzees (*Pan troglodytes*). *Journal of Comparative Psychology* 110: 3–14.

Whyte, M. K. 1978. *The status of women in preindustrial societies.* Princeton: Princeton University Press.

Wiepkema, P. R. 1961. An ethological analysis of the reproductive behaviour of the bitterling (*Rhodeus amarus* BLOCH). *Archives Néerlandaises de Zoologie* 14, 103–199.

———. 1977. Agressief gedrag als regelsysteem. In *Agressief Gedrag, Oorzaken en Functies,* ed. P. R. Wiepkema and J.A.R.A.M. van Hooff. Utrecht: Bohn, Scheltema, and Holkema, pp. 69–78.

Wiley, H. R., and J. Poston. 1996. Indirect mate choice, competition for mates, and coevolution of the sexes. *Evolution* 50: 1371–1381.

Williams, G. C. 1975. *Sex and evolution.* Princeton, NJ: Princeton University Press.

———. 1966. *Adaptation and natural selection.* Princeton: Princeton University Press.

Williams, L. 1967. *The dancing chimpanzees: A study in the origins of primitive music.* New York: Norton.

Willis, R. G., ed. 1994 (1989). *Signifying animals: Human meaning in the natural world.* London: Routledge.

Wilson, E. O. 1975. *Sociobiology: The new synthesis.* Cambridge, MA: Harvard University Press.

———. 1992. *The diversity of life.* Cambridge, MA: Harvard University Press.

Wimsatt, W. C. 1987. False models as means to truer theories. In *Neutral models in biology,* ed. M. H. Nitecki and A. Hoffman. Oxford: Oxford University Press, pp. 23–55.

Wittenberger, J. F. 1981. *Animal social behavior.* Boston: Duxbury Press.

Wolf, M. 1992. *A thrice-told tale: Feminism, postmodernism, and ethnographic responsibility.* Stanford: Stanford University Press.

Wolpert, L. 1992. *The unnatural nature of science.* London: Faber and Faber.

Wood-Jones, F. 1941. *The hand.* London: Balliere, Tindall and Cox.

Woodward, A. S. 1913. Missing links among extinct animals. *Report of the British Association.*

World Resources Institute, et al. 1993. *Biodiversity prospecting* (a contribution to the WRI/IUCN/UNEP global biodiversity strategy, May 1993).

Wrangham, R. W. 1979a. On the evolution of ape social systems. *Soc. Sci. Infor.* 18 (3): 335–369.

———. 1979b. Sex differences in chimpanzee dispersion. In *The great apes,* ed. D. Hamburg and E. McCown. Menlo Park, CA: Benjamin/Cummings, pp. 481–489.

———. 1980. An ecological model of female-bonded primate groups. *Behaviour* 75: 262–300.

———. 1987. Evolution of social structure. In *Primate societies,* ed. B. B. Smuts, D. L. Cheney, R. M. Seyfarth, R. W. Wrangham, and T. T. Struhsaker. Chicago: University of Chicago Press, pp. 282–296.

———. 1997. Subtle, secret female chimpanzees. *Science* 277 (8 August): 774–775.

Wrangham, R. W., and D. Peterson. 1996. *Demonic males.* Boston: Houghton Mifflin.

Wrangham, R. W., and B. Smuts. 1980. Sex differences in the behavioral ecology of chimpanzees in Gombe National Park. *Journal of Reproduction and Fertility* 28, suppl.: 13–31.

Wrangham, R. W., W. C. McGrew, F. B. M. de Waal, and P. C. Heltne. 1994. *Chimpanzee Cultures.* Cambridge, MA: Harvard University Press.

Wright, R. 1994. *The moral animal: Evolutionary psychology and everyday life.* New York: Vintage.

Wright, R. P., ed. 1995. *Gender and archaeology.* Philadelphia: University of Pennsylvania Press.

Wylie, A. 1991a. Gender theory and the archaeological record: Why is there no archaeology of gender? In *Engendering archaeology: Women and prehistory,* ed. J. M. Gero and M. W. Conkey. Oxford: Basil Blackwell, pp. 31–54.

———. 1991b. Feminist critiques and archaeological challenges. In *The archaeology of gender: Proceedings of the 22nd Annual Chacmool Conference,* ed. D. Walde and N. Willows. Calgary: The Archaeological Association of the University of Calgary, pp. 17–23.

———. 1994. The trouble with numbers: Workplace climate issues in archaeology. In *Equity issues for women in archaeology,* ed. M. C. Nelson, S. M. Nelson, and A. Wylie. Washington DC: Archeological Papers of the American Anthropological Association, no. 5, pp. 66–71.

———. 1995. The contexts of activism on "climate" issues. In *Breaking anonymity: The chilly climate for women faculty,* ed. The Chilly Collective. Waterloo, ON: Wilfrid Laurier University Press, pp. 29–60.

———. 1996. The constitution of archaeological evidence: Gender politics and science. In *The disunity of science: Boundaries, contexts, and power,* ed. P. Galison and D. J. Stump. Stanford, CA: Stanford University Press, pp. 311–343.

———. 1997a. The engendering of archaeology: Refiguring feminist science studies. *Osiris* 12: 80–99.

———. 1997b. Good science, bad science, or science as usual? Feminist critiques of science. In *Women in human evolution,* ed. L. Hager. London: Routledge, pp. 29–55.

Yalcinkaya, T. M., P. K. Siiteri, J-L. Vigne, P. Licht, S. Pavgi, L. G. Frank, and S. E. Glickman. 1993. A mechanism for virilization of female spotted hyenas *in utero*. *Science* 260: 1929–1931.

Yamada, M. 1966. Five natural troops of Japanese monkeys on Shodoshima Island: I. Distribution and social organization. *Primates* 7: 315–362.

Yamagiwa, J. 1984. *The gorilla: Silverback in forest*. Tokyo: Heibonsha. (J)

———. 1993. *Between gorilla and man*. Tokyo: Kodansha. (J)

———. 1994. *Origin of human family*. Tokyo: University of Tokyo Press. (J)

Yamagiwa, J., and D. A. Hill. 1998. Intraspecific variations in social organization of Japanese macaques: Past and present scope of field studies in natural habitats. *Primates* 39 (3): 257–273.

Yamamoto, M. E., and A. Araújo. 1991. Organização social dos calitriquídeos: Integração de dados de campo e de cativeiro. *Biotemas* 4: 37–52.

Yamamoto, M. E., M. F. Arruda, M. B. C. Sousa, and A. I. Alencar. 1996. Mating systems and reproductive strategies in *Callithrix jacchus* females. Paper presented at the 18th Congress of the International Primatological Society, Madison, WI.

Yanagisako, S., and C. Delaney, eds. 1995. *Naturalizing power: Essays in feminist cultural analysis*. London: Routledge.

Yerkes, R. M. 1916. The mental life of monkeys and apes: A study of ideational behavior. *Behavior Monographs* 3.

———. 1927. The mind of a gorilla. Parts I and II. *Genetic Psychology Monographs* 2: 1–193, 375–551.

———. 1928. The mind of a gorilla. Part III. *Comparative Psychology Monographs* 5: 1–92.

———. 1939. Social dominance and sexual status in the chimpanzee. *Quart. Rev. Biol.* 14: 115–136.

———. 1943. *Chimpanzees: A laboratory colony*. New Haven: Yale University Press.

Yerkes, R. M., and A. W. Yerkes. 1929. *The great apes: A study of anthropoid life*. New Haven: Yale University Press.

York, A. D., and T. E. Rowell. 1988. Reconciliation following aggression in patas monkeys, *Erythrocebus patas*. *Anim. Behav.* 36: 502–509.

Young, R. 1986. *Darwin's metaphor*. Cambridge, UK: Cambridge University Press.

Zabel, C. J., S. E. Glickman, L. G. Frank, K. B. Woodmansee, and G. Keppel. 1992. Coalition formation in a colony of prepubertal spotted hyenas. In *Coalitions and alliances in humans and other animals,* ed. A. H. Harcourt, and F. B. M. De Waal. Oxford: Oxford University Press, pp. 113–136.

Ziegler, T. E., A. Savage, G. Scheffler, and C. T. Snowdon. 1987. The endocrinology of puberty and reproductive functioning in female cotton-top tamarins (*Saguinus oedipus*) under varying social conditions. *Biol. Reprod.* 37: 618–627.

Zihlman, A. 1978. Women in human evolution. Part II, Subsistence and social organization among early hominids. *Signs. Journal of Women in Culture and Society* 4: 4–20.

———. 1987. Sex, sexes and sexism in human evolution. *Yearbook of Physical Anthropology* 30: 11–19.

———. 1997. The paleolithic glass ceiling: Women in human evolution. In *Women in human evolution,* ed. L. Hager. London: Routledge, pp. 91–113.

Zilsel, E. 1942. The sociological roots of modern science. *American Journal of Sociology* 47: 245–279.

Zuberbühler, K., R. Noë, and R. M. Seyfarth. 1997. Diana monkey long-distance calls: Messages for conspecifics and predators. *Anim. Behav.* 53: 589–604.

Zuckerman, S. 1932. *The social life of monkeys and apes.* London: Routledge and Kegan Paul.

———. 1933. *Functional affinities of man, monkeys, and apes.* London: Kegan Paul.

———. 1963. Concluding remarks of Chairman. *Symp. Zool. Soc. Lond.* 10: 119–123.

Contributors

Anuska Irene Alencar
Departamento de Psicologia
Universidade Potiguar
Natal, RN
Brazil

Pamela Asquith
Department of Anthropology
University of Alberta
Edmonton, Alberta
Canada T6G 2H4

Richard Byrne
Department of Psychology
University of St. Andrews
St. Andrews, Fife
Scotland KY16 9JU

Charis Thompson Cussins
Department of Sociology and Women's Studies Program
University of Illinois at Urbana-Champagne
Urbana, Illinois
USA 61801

Linda Marie Fedigan
Department of Anthropology
University of Alberta
Edmonton, Alberta
Canada T6G 2H4

Stephen Glickman
Department of Psychology
University of California
Berkeley, California
USA 94720

Donna Haraway
History of Consciousness Department
University of California
Santa Cruz, California
USA 95064

Robert A. Hinde
Sub-Department of Animal Behaviour
Madingley and St. John's College
Cambridge
England CB2 1TP

Alison Jolly
Department of Ecology and Evolutionary Biology
Princeton University
Princeton, New Jersey
USA 08544

Evelyn Fox Keller
Program in Science, Technology, and Society
Massachusetts Institute of Technology
Cambridge, Massachusetts
USA 02139

Bruno Latour
Centre de Sociologie de L'Innovation
École des Mines
Paris
France

Gregg Mitman
Department of History of Science
University of Oklahoma
Norman, Oklahoma
USA 70319

Brian Noble
Department of Anthropology and Sociology/Museum of Anthropology
University of British Columbia
Vancouver, British Columbia
Canada

Naomi Quinn
Department of Cultural Anthropology
Duke University Box 90091
Durham, North Carolina
USA 27708-0091

Thelma Rowell
Department of Integrative Biology
University of California at Berkeley
Berkeley, California
USA 94720

Karen Strier
Department of Anthropology
University of Wisconsin-Madison
Madison, Wisconsin
USA 53706

Shirley C. Strum
Department of Anthropology
University of California, San Diego
La Jolla, California
USA 92093-0532

Robert Sussman
Department of Anthropology
Washington University
St. Louis, Missouri
USA 63130

Hiroyuki Takasaki
Laboratory of Anthropology
Okayama University of Science
Okayama 700-0005
Japan

Zuleyma Tang-Martinez
Department of Biology
University of Missouri–St. Louis
St. Louis, Missouri
USA 63121

Jan A.R.A.M. van Hooff
Universiteit Utrecht
Utrecht
Pb 80-086, 3508 TB
The Netherlands

Alison Wylie
Department of Philosophy
Washington University
St. Louis, Missouri
USA

Maria Emília Yamamoto
Departamento de Psicologia
Universidade Potiguar
Natal, RN
Brazil

Index